# Wireless Communications
## TDMA versus CDMA

# Wireless Communications

## TDMA versus CDMA

by

Savo G. Glisic
University of Oulu

and

Pentti A. Leppänen
University of Oulu

KLUWER ACADEMIC PUBLISHERS
BOSTON / DORDRECHT / LONDON

A C.I.P. Catalogue record for this book is available from the Library of Congress

ISBN 0-7923-8005-3

Published by Kluwer Academic Publishers,
P.O. Box 17, 3300 AA Dordrecht, The Netherlands.

Sold and distributed in the U.S.A. and Canada
by Kluwer Academic Publishers,
101 Philip Drive, Norwell, MA 02061, U.S.A.

In all other countries, sold and distributed
by Kluwer Academic Publishers Group,
P.O. Box 322, 3300 AH Dordrecht, The Netherlands.

*Printed on acid-free paper*



Printed in the Netherlands

# CONTENTS

# PREFACE

This book is a collection of individual contributions from a number of authors all around the world. The material covers some of the major issues related to modern wireless communications.

The intention of the editors was to gather the most prominent specialist for wireless communications and present the major, up to date, issues in this field. The material is grouped into four chapters: Communication theory, covering coding and modulation, Wireless communications, Antenna & Propagation and Advanced Systems & Technology.

This is the time when considerable research, business and standardization efforts are focused on defining the third generation of mobile communication systems. Within this book we are trying to present the atmosphere in all these circles and point out the major dilemmas they are suppose to resolve.

At this moment it seems that both Time Division Multiple Access (TDMA) and Code Division Multiple Access (CDMA) still remain viable candidates for application in future systems. We believe that material presented in this book provides enough information for proper understanding of the arguments in favor of one or another multiple access technique. One should be aware that the final decision about which of the two technique should be employed depends not only on technical arguments but also on the amount of new investments needed and compatibility with previous systems and their infrastructures.

Another intention of the book is to point out a number of applications of wireless communications. Contributions like: Wideband Alternatives for Wireless Data Communications, Wireless ATM Technology, Wireless Data Services, Wireless in Local Loops and Wireless Communication Technology in Intelligent Transport Systems provide a comprehensive insight into a number of emerging possibilities for new applications of wireless systems.

In a number of contributions covering RF IC design and VLSI signal processing for wireless communications a review of the state of the art of the technology in this field is presented.

The editors of the book would like to thank to all authors for their contributions. We believe this book will serve as a reference in wireless communications for some time until the fast progress in this field makes it out of date.

Special thanks to Mrs. Pirjo Kumpumäki, Roman Pichna, Jari Sillanpää and all people of Telecommunications Laboratory and the Centre for Wireless Communications at the University of Oulu who helped them in handling word processing for this book.

Editors

# INTRODUCTION
# TDMA versus CDMA

Savo G. Glisic
Pentti A. Leppänen

## 1 BACKGROUND

In Time Division Multiple Access (TDMA), within a given time frame a particular user is allowed to transmit within a given time slot. This technique is used in most of the second-generation digital mobile communication systems. In Europe the system is known as GSM, in USA as DAMPS and in Japan as MPT. In Code Division Multiple Access (CDMA) every user is using a distinct code so that it can occupy the same frequency bandwidth at the same time with other users and still can be separated on the basis of low correlation between the codes. These systems like IS-95 in USA are also developed and standardized within the second generation of the mobile communication systems. CDMA systems within a cellular network can provide higher capacity [1-28] and for this reason they become more and more attractive. In the process of defining the third generation of mobile communication systems TDMA and CDMA still remain the major rival multiple access techniques. This applies for both, land and satellite communications.

Satellite projects based on CDMA are Globalstar, Odissey, Constellation, Ellipsio and Archimedes. Globalstar is planed for global roaming as an extension to terrestrial cellular networks. The services will be routed and billed through existing carriers. User services charges of more than 20 million USD per month are expected from 7 million users. The system will have 48 satellites in 8 LEO orbits with 8 beams per satellite. The overall project budget is 1.5 billion USD. The main companies involved in the project are: Loral Corporation, Qualcomm Inc., Alcatel NV, Alenia Spazio, Aerospatiale and Deutshe Aerospace.

Odissey is intended for voice, data and paging to hand held terminals on the market including North America, Pacific Rim countries and Europe, followed in a later stage by North Africa and South America. The expected market size is 2.3

million subscribers and user charges 0.60 USD per minute. The system will have 12 satellites in 3 MEO orbits with 19-37 beams per satellite. The overall budget is 0.8 billion USD. The main companies involved in the project are: TRW Space & Technology Group and Mitra Marconi Space.

Constellation is intended for wireless services to customers worldwide where no viable alternatives exists. Market size should be about 900,000 users by year 2000. The expected user charge is 1-5 USD per minute (voice). The system will have 48 satellites in 4 LEO orbits with 7 beams per satellite. The overall project budget is 0.5 billion USD. The main companies involved in the project are: Constellation Corp. (CTA/DSI Inc., Pacific Communications Sciences, International MicroSpace Inc.).

Ellipsio is intended as a complement to existing cellular systems for the market including USA and the Northern Hemisphere. The system will have 24 satellites in HEO orbit with 4 beams per satellite. The overall project budget is 0.28 billion USD. The main companies involved in the project are: Ellipsat Corp.( Mobile Communications Holdings Inc., Matra Group and Fairchild).

Archimedes is intended for the market on the Northern Hemisphere (North America, Europe and Japan). The system will have 6 satellites in one (6 or 16 hours) orbit. The main institution involved in the project is the European Space Agency.

The most known satellite systems based on TDMA technology are: Iridium, Calling and Inmarsat.

Iridium is intended for voice, data and paging to hand held terminals. World wide coverage is planed for the market size of 1.8 million subscribers after 5 years of service. The expected user charges 3 USD per minute. The system will have 66 satellites in 6 planes in LEO orbit with 48 beams per satellite. The initial budget of 3.37 billion USD has been revised and considerably increased. The main companies involved in the project are: Motorola, Japanese group, Saudi Arabian group, Lockheed Missile Space, Raytheon and others.

Calling is intended for rural telephony for developing countries and remote parts of industrialized nations (hand held terminals). Expected market size is 20 million subscribers. The system will have 840 satellites in 21 planes in LEO orbits with 64 beams per satellite. The overall budget is 6.5 billion USD and the main companies involved in the project are: Calling Communications Corporation (AT&T and McCaw Cellular Communications Inc.).

Inmarsat is intended for hand held terminals for business travelers and professionals on the move. Expected market size is 2.5 million users by year 2000. User charges under 1 USD per minute. The system will have 12-15 satellites in MEO orbit or 4 satellites in GSO with 61/109 beams per satellite. The budget of the project is 1-1.5 billion USD.

Comparison between CDMA and TDMA is very similar in the field of land mobile communications. So far, the submitted proposals to ETSI, for Universal Mobile Telephone System (UMTS) Terrestrial Radio Access (UTRA), can be grouped into the following main categories

- wideband CDMA,
- TDMA,
- hybrid CDMA/TDMA,
- OFDM and
- ODMA (Opportunity Driven Multiple Access).

PCS1900) standard can already provide bit rates of up to 115.2 kbit/s using GPRS and HSCSD. Furthermore, the enhanced GSM carrier can provide bit rates up to 256 kbit/s within the 200 kHz carrier. For IS-95 the higher bit rates beyond 9.6 kbit/s are currently standardized. The first step is Phase 1 providing bit rates up to 64 kbit/s. Phase 2 will go up to 144 kbit/s. Phase 3, will be wideband CDMA with a carrier spacing of 5 MHz. Several companies have made wideband CDMA proposals including Nokia, Lucent, Motorola, Nortel and Qualcomm. IS-136 is currently supporting data rates of up to 14.4 kbit/s. Further evolution is expected up to 57.8 kbit/s. However, the evolution path towards FPLMTS/IMT-2000 is unclear. The US PCS standard IS-656, wideband CDMA with a bandwidth of 5 MHz, has been used as a basis for the Core B proposal in Japan.

Bearing in mind this persistent rivalry between CDMA and TDMA in this book we represent a number of contributions discussing different issues in modern wireless communications by treating these two techniques in parallel.

## 2 AN OVERVIEW OF THE BOOK CONTENT

### 2.1 Communication Theory

The increasing interest for digital mobile-radio transmission systems has led to an extensive study of coding and modulation (C/M) for fading channels. For these channels the paradigms developed for the Gaussian channel may not be valid anymore, and a fresh look at the design philosophies is needed. Specifically, in the past system designer choices were driven by their knowledge of the behavior of modulation and coding over the Gaussian channel. Consequently, they tried to apply to radio channels solutions that were far from optimum on channels where non-linearities, Doppler shifts, fading, shadowing, and interference from other users make the channel far from Gaussian.

A considerable body of work has been reversing this perspective, and it is now being widely recognized that C/M solutions for the fading channel should be selected by taking into account the basic features of the channel. One early example of this is the design of modulation schemes: starting from standard phase-shift keying which has an excellent performance over the Gaussian channel. Variations of its basic scheme were derived to cope with non-linear effects, tight bandwidth constraints, etc., resulting into effective modulations like GMSK. Another, more recent, example is the study of "fading codes," i.e., code-modulation schemes that are specifically optimized for a Rayleigh channel, and hence do not attempt to maximize the Euclidean distance between error events, but rather their Hamming distance.

As a result, the channel model turns out to have a considerable impact on the choice of the preferred solution for C/M schemes. For example, the fading model may depend on the antenna system being used. To be more specific, assume that intelligent antennas are available. Then the system may end up having radiation patterns that are so directive that a propagation model close to LOS will be a good one. This can be possibly coupled with an "on-off channel," that is, a channel which for a percentage of time is a good (viz, Gaussian) channel, and for the remaining time is so bad that reliable transmission is not possible. In this situation, the standard solution of transmitting signals with the same energy all the time is often a bad choice.

In addition to UMTS activities, GSM (TDMA based system) continues evolution towards the third generation. The enhanced GSM system, currently under a feasibility study in the ETSI, will offer wide area coverage up to 144 kbit/s and even higher bit rates for local coverage. This will facilitate the introduction of a subset of UMTS services already into the second generation systems. Of course, a new revolutionary wideband access will provide the full set of UMTS services with increased spectrum efficiency and flexibility.

In Japan the situation is very similar. The Future Public Land Mobile Telephone System (FPLMTS) Study Committee in ARIB (Association of Radio Industries and Business) was established in April 1993 to coordinate the Japanese research and development activities for the FPLMTS system. In October 1994, the committee established the Radio Transmission Special Group for radio transmission technical studies and the production of a draft specification for FPLMTS. The Special Group consists of two Ad Hoc groups, CDMA and TDMA.

Originally, there were 13 different CDMA proposals. In the beginning of 1995, they were merged into three CDMA/FDD (Core A, B and C) and into one TDD proposal. The Core B system is based on the wideband CDMA standardized for US Personal Communications System (PCS) known as IS-665. In the end of 1996, the four schemes were further combined into a one single proposal where the main parameters are from the Core A. Respective companies have contributed three of the original wideband CDMA proposals from Japan (2 FDD and 1 TDD concept) also to ETSI.

Both laboratory tests and field trials have been conducted for the Japanese wideband CDMA proposals during 1995 and 1996. E.g. Core A has tested speech and 384 kbit/s video transmission in field trials. 2 Mbit/s transmission has also been performed in laboratory conditions using a 20 MHz bandwidth. In the second phase of Core A trials, currently on-going, an interference canceller is also tested. The Core B proposal has only been validated for speech service in the framework of PCS standardization in the USA. Core C has been tested for data rates up to 32 kbit/s. Also, the CDMA/TDD proposal was tested in laboratory conditions.

For TDMA, there were seven proposals for the pedestrian environment, six for the office environment and another six for the vehicular environment. From these proposals the group compiled a single carrier TDMA system, Multimode and Multimedia TDMA (MTDMA). A prototype equalizer for the MTDMA system with a carrier bit rate of 1.536 Mbit/s and user bit rate of 512 kbit/s has been build and tested. Since one of the TDMA proposals was based on multicarrier OFDM technology Band Division Multiple Access (BDMA) proposal it was decided to continue studies of this technology. BDMA has also been proposed to the ETSI.

The main conclusion of the FPLMTS Radio Transmission Technology Special Group is that detailed studies on CDMA shall be started, i.e., wideband CDMA is the main technology choice of Japan for FPLMTS/IMT-2000. Further studies will be made for MTDMA high speed transmission in office and pedestrian environments.

The focus of US research activities has been on PCS. Recently, the third generation systems have also gained more attention. Since there exists three main second generation technologies, namely GSM (TDMA), IS-95 (CDMA) and IS-136, most likely we will see several paths towards FPLMTS/IMT-2000. The FPLMTS frequency band is already partially used for PCS systems and thus the evolution of PCS systems within that frequency band is important. GSM1900 (also known as

If the channel model is uncertain, or not stable enough in time to design a C/M scheme closely matched to it, then the best proposition may be a solution that provides suboptimum performance on a wide variety of channel models. The use of antenna diversity with maximal-ratio combining provides good performance on a wide variety of fading environments. Another such solution is offered by bit-interleaved coded modulation (BICM). These issues are discussed by Ezio Biglieri et al within the contribution " Coding and Modulation for The Fading Channel".

In the same line is a contribution by Branka Vucetic entitled "Iterative Decoding Algorithms". In the paper she presents iterative algorithms which can be used for the decoding of concatenated codes. The decoding operation is based on either a maximum a posteriori (MAP) algorithm or a Viterbi algorithm generating a weighted soft estimate of the input sequence. The iterative algorithm performs the information exchange between the two component decoders. The performance gain of the MAP algorithm over the Viterbi algorithm at low SNR leads to a slight performance advantage. However the MAP algorithm is computationally much more complex than the Viterbi algorithm. The operations in the MAP algorithm are multiplications and exponentiations while in the Viterbi algorithm they are simple add, compare and select operations. As an example these algorithms are applied to decoding of turbo codes and their performance is compared on a Gaussian channel.

Andreas Polydoros and Keith M. Chugg discuss the principle of Per-Survivor Processing (PSP) that provides a general framework for the approximation of likelihood-based data detection (search) algorithms whenever the presence of unknown quantities prevents the precise use of the classical Viterbi Algorithm (VA). The PSP concept is based on the realization that decision-directed estimation of the relevant unknown parameters should be embedded into the structure of the search algorithm itself, thus joining together the two main functions of the receiver. It is argued that PSP not only offers superior and robust performance versus the classical segregated-task architectures in many challenging environments, but that it is also the natural way of approaching the decoding problem from first principles, thus making it the appropriate vehicle for integrated-task receiver design.

Akbar M. Sayeed and Behnaam Aazhang discuss time-frequency presentations (TFP) for processing time-varying signals and systems. They are two-dimensional (2D) signal representations jointly parameterized in terms of both time and frequency. As such, TFRs provide a natural framework for representing the time-varying mobile wireless channel and the optimal signaling and receiver structures for communications over such channels.

In this paper, the authors present a time-frequency perspective of wireless communications over time-varying multipath channels. The discussion is primarily in the context of CDMA systems because of their well-known ability to combat multipath fading. Starting with a time-frequency description of the mobile wireless channel, they arrive at a canonical finite-dimensional time-frequency representation of the channel that will serve as the backbone of our treatment.

The canonical time-frequency-based channel representation shows that spread-spectrum signaling over time-varying multipath channels possesses additional degrees of freedom that are not exploited by existing communications systems. Essentially, CDMA systems possess a large time-bandwidth product (TBP) that can be exploited to provide diversity against channel fading. The state-of-the-art RAKE receiver that achieves multipath diversity exploits only the large bandwidth but not the large TBP of CDMA systems. The time-frequency channel representation identifies Doppler as another dimension for diversity, and facilitates the exploitation of joint multipath-Doppler diversity by fully utilizing the available TBP in CDMA systems. Thus, a time-frequency approach to communication over multipath channels has the potential of delivering substantial gains over conventional techniques in virtually all aspects of system performance.

The fundamental time-frequency channel representation can be exploited in a variety of aspects of communication system design and analysis ranging from new signaling and receiver structures to multiuser detection, timing-acquisition and interference-suppression to information-theoretic issues related to multipath fading channels. The power of the time-frequency paradigm is illustrated by focusing on two main themes: 1) novel signaling and receiver structures, and 2) a new approach to multiuser timing acquisition and interference suppression that fully incorporates the multipath channel.

In order to successfully demodulate a digitally modulated signal, the receiver must first synchronize its oscillators and clocks to the received signal. If direct-sequence spread-spectrum modulation is used, then it is also necessary to synchronize to the code sequence that was used to produce the spectral spreading. Methods used to achieve coarse synchronization to the code sequence, also called acquisition of the code sequence, are discussed by Dilip Sarwate in his paper entitled “Acquisition of Direct-Sequence Spread-Spectrum Signals”. Serial search methods require comparatively little hardware but can take a long time to achieve acquisition. On the other hand, parallel methods can acquire the code sequence very quickly, but can be somewhat expensive to implement. A survey of some of the basic ideas is presented and some improvements on known methods are suggested.

## 2.2 Wireless Communications

We are emerging to a new and exciting era for wireless LANs (WLAN). After a decade of self realization for the WLAN industry, today markets for WLAN and inter-LAN bridges are finding their way in the Health Care, Manufacturing, Finance, and Education. The IEEE 802.11 standard for WLANs, operating in the unlicensed ISM bands, is emerging as a mature standard presenting a well defined technology that is adopted by the manufacturers and accepted by the users. Chipsets are developed for the 802.11 that makes it available for further software creativity to develop new applications and expand the market. ETSI's RES-10 has defined another alternative technology, HIPERLAN I, that is more focused toward the ad-hoc networking applications and supports higher data rates up to 23Mbps. An ad-hoc network provides an environment to set up a temporary wireless network among a group of users equipped with WLANs. In the last year, research around wireless ATM soared like an epidemic engaging numerous companies in examining the suitability of yet another alternative technology for the implementation of wireless LANs. The continual demand of the wireless LAN industry for more unlicensed bands in useful spectrums, initiated by WINForum, resulted in the release of 20 MHz of unlicensed band around 2GHz for asynchronous and isochronous applications and the U-NII (formerly SUPERNet) won 300 MHz of unlicensed bands in 5 GHz earlier this year. On the other hand, the pan-European third generation cellular service UMTS is considering connectionless packet switched networks as class D bearer services and under its research arm ACTS (Advanced Communications Technology and Services) the MEDIAN, WAND, SAMBA, and AWACS projects are addressing WLAN services. The third generation systems will incorporate WLAN standards into the universal mobile services. These issues are discussed in the contribution by Kaveh Pahlavan et al entitled "Evolving Wireless LAN Industry Products and Standards". This paper provides an updated overview of all the activities around the WLAN industry.

One major issue in cellular radio system is the problem of handoffs. A mobile leaving one cell and entering a neighboring cell must transfer its call to the new cell. Traditional handoff schemes are threshold policies. The signal strengths from the current and candidate base stations are measured, and the difference between these two measurements is compared with a fixed threshold to determine the handoff action. Because of the statistical fluctuations in signal strength, the threshold must be large enough to reduce the number of bouncing between two cells. However, increasing the threshold may result in higher probability of forced termination because the signal from the current base station may have been too weak before the threshold criterion is met.

A handoff between two cells will change the number of active users in each cell, and the traffic conditions in the cells will also affect the handoff decision (for example, handoff can not be made if the new cell has no capacity left). Furthermore, the call quality depends on the other user interference in a CDMA network. Therefore, the handoff and channel assignment problems should be considered together to obtain an optimal policy, which will take into account the switching cost, the blocking rate and the call quality in a CDMA cellular system. This formulation and the resulting optimization of joint handoff and code assignment policies based on Markov decision processes are the contribution of the paper entitled "Optimal Joint Handoff and Code Assignment in CDMA Cellular Networks" by Evaggelos Geraniotis and Yu-Wen Chang.

A distinct advantage of CDMA cellular networks is the soft handoff feature. The mobile remains connected to both base stations during the transition period of switching from one base station to another. It has been shown that CDMA soft handoff can provide larger cell coverage than hard handoff, but no optimization or performance evaluation of policies for hard or soft handoff was carried out.

In the above paper, the authors compare the blocking rates of the traffic of two cells and the number of handoffs for both the optimal hard and soft handoff schemes used in conjunction with CDMA code assignment and compare them to each other and to traditional fixed threshold/direct-admission policies.

Broadband wireless network technologies such as wireless ATM are motivated by the increasing importance of portable computing/telecommunications applications in both business and consumer markets. The rapid penetration of cellular phones and laptop PC's during the previous decade is proof that users place a significant value on portability as a key feature which enables tighter integration of such technologies with their daily lives. In the last few years, first generation multimedia capabilities (such as CD-ROM drives and MPEG decoders) have become available on portable PC's, reflecting the increasingly mainstream role of multimedia in computer applications. As multimedia features continue their inevitable migration to portable devices such a laptop PC's and PDA's, wireless extensions to broadband networks will be required to support user requirements. Such broadband wireless services could first start in the private local area network scenario, gradually moving to microcellular public PCS systems if the technology proves feasible for general consumer use. The basic idea is to provide a seamless wired + wireless networking environment with uniform protocols and applications across both mobile and fixed devices.

Fundamental network architecture and radio technology innovations are required to realize the above vision for a seamless wired + wireless broadband wireless network. At the architectural level, wireless systems need to migrate from the current model of single-application networks with custom protocol stacks towards generic integrated service networks with "plug-in" software and hardware components to support radio access and mobility. In terms of radio technology, it will be necessary to re-examine basic assumptions about frequency band, cell size, modem complexity, etc. to achieve service speeds of the order of 10-100 Mbps at reasonable reliability and spectrum utilization levels. On the network side, mobility which has traditionally been treated as an external function, must be recognized as a first-class feature to be considered during all phases of protocol and hardware design. Integrated mobility support within standard network protocols such as IP or ATM will facilitate a move towards the use of generic network API's and common applications on both mobile and fixed terminals. Finally, mobile multimedia services will require a networking framework which supports real-time streams with quality-of-service (QoS) in both wired and wireless segments of the system.

Wireless ATM, is a specific broadband wireless solution which substantially meets the architectural and performance goals outlined above. Work on wireless ATM has been motivated by the increasing acceptance of ATM switching technology as a basis for broadband networks which support integrated services with QoS control. The 53 byte ATM cell turns out to be quite reasonable for use as the basic transport unit over high bit-rate radio channels, taking into account both error control and medium access requirements. ATM signaling protocol (e.g. Q.2931) for connection establishment and QoS control also provide a suitable basis for mobility extensions such as handoff and location management. Early research results in this area indicated that it is indeed possible to use standard ATM protocols to support seamless wired + wireless networking. This should be done via incorporation of new wireless specific protocol sublayers (e.g. medium access control and data link control) into the ATM data plane, together with a limited number of mobility extensions to existing ATM control protocol layers. Subsequent proof-of-concept prototyping activities (e.g. NEC's WATMnet , ORL Cambridge's Radio ATM, Bell Labs' Bahama & SWAN , NTTs AWA) further confirmed the viability of wireless ATM technology for delivery of multimedia services to portable devices.

Wireless ATM technology is currently migrating from the research stage to standardization and early commercialization. The ATM Forum started a wireless ATM (WATM) working group in June 1996, with the objective of producing wireless and mobile extensions to the existing ATM UNI4.0 specification by the end of 1998. Significant R&D and trial product development efforts are now underway in various parts of the world, so that the technology should begin to reach the market over the next 2-3 years. The recent allocation of the so-called 5 GHz U-NII band by the U.S. Federal Communications Commission together with earlier ETSI rulings on HIPERLAN, represent a significant breakthrough at the regulatory level. This should result in increased commercial activity in this area. The authors visualize the emergence of public and private microcellular wireless networks which provide a variety of broadband mobile services including high-speed Internet access and audio/video delivery. The availability of such broadband wireless networks by the end of this decade should stimulate the evolution of a new generation of high-performance mobile computing devices in both vertical and horizontal markets worldwide. It is noted here that although wireless ATM is typically associated with mobile multimedia services, selected components of the technology can be applied to a variety of other application scenarios including PCS/cellular infrastructure, fixed/residential wireless access and microwave infrastructure links.

An overview of wireless ATM system architecture, design of key subsystems, technology development status and standardization is presented in the contribution entitled "Wireless ATM Technology: Present Status and Future Directions" by Dipankar Raychaudhuri. The paper concludes with a view of future directions in the area of wireless broadband technology.

In the paper entitled "A System for Wireless Data Services" by Krishan K. Sabnani et al and the authors present the design and implementation of a system for providing a selected set of wireless data services. The focus of this paper is in identifying the key issues and challenges, and providing design approaches and principles, at both protocol and system levels. As will become clear, the recurring theme in this paper is how to provide a maximal and interesting set of wireless data services while keeping the complexity of the end device and bandwidth usage (especially in the uplink direction) consistent with the wireless operating constraints. For concreteness, they have embodied their design and techniques in the construction of a set of servers for wireless data, which we called collectively a wireless data server (WDS). A prototype WDS has been implemented and is operational at Bell Laboratories.

During the last few years, the popularity of mobile cellular communications, the Internet, and laptop computing has resulted in development efforts to transmit data packets over the air, both for local area networks and for residential access. This development has the potential to enable efficient mobile computing, as well as to provide residential high speed data access capability without building new infrastructure. Although wireless packet radio technology was developed for the defense sector about a decade ago, the transmission efficiencies involved need to be significantly improved for a public service based on this technology. There exists an IEEE standard, 802.11, for wireless local area networks that operate at 2 Mbps. Another standard, HIPERLAN, was developed in ETSI for wireless local area networks operating at 20 Mbps (however, no prototype was built and there are no known product plans). Yet another standardization effort is ongoing at the ATM Forum and ETSI to develop “wireless ATM”. Cellular voice networks are beginning to support limited forms of data services. In addition, recently, a service offering for Internet access based on a proprietary transmission scheme has become popular in some cities and on some university campuses in the U.S. In spite of all this activity, indicative of the demand for wide-scale wireless packet transmission, it is widely recognized in the engineering community that technology development for packet data transmission over the air is not straightforward. Efficient transmission across unknown and time-varying channels in the burst mode requires significant complexity. Further, what is really needed today is the support of a variety of transmission formats: data, voice, and video, or integrated services. Efforts for networking standards to provide QoS support for integrated services over a variety of networks are under way at various standardization bodies. The issue is more complicated for the wireless channel, mainly because of its highly noisy nature. These issues in wireless packet networking are discussed by Ender Ayangoly in his contribution entitled "Wireless Packet and Wireless ATM Systems". The emphasis is on wireless ATM networking, although the general conclusions to be drawn are applicable to general wireless packet networks as well.

The need for wired-line quality voice and high speed data up to 2Mb/s is driving the international community to a third generation standard with one option based on Wideband-CDMA (WCDMA). In addition, the need for privacy and mobile fax or Internet hookups, as well as wired line voice quality, is driving cordless cellular telephones to third generation technology. Another application for WCDMA technology is the wireless local loop (WLL). The WLL permits a long distance carrier to bypass the local service provider and thereby reduce consumer costs while increasing profits. However, for this to be a successful venture it is required that the wireless technology yield the same quality as the wired technology. None of the first or second generation technologies can meet that objective. Another application of WLL is providing wireless telecommunication service to customers throughout the world who, today, do not have such wired service. However, even in this case of "catch-up" the local telephone service providers demand the highest quality. The demand for high quality can be met today only through the use of third generation WCDMA systems. These and number of other applications of WCDMA are reviewed in the paper entitled "Overview of Wideband CDMA" by Donald Schilling.

The impulse radio is described by Robert A. Scholtz and Moe Z. Win in the paper entitled "Impulse Radio". This form of ultra-wide band signaling, has properties that make it a viable candidate for short range communications in dense multipath environments. The paper describes the characteristics of impulse radio, gives analytical estimates of its multiple access capability, and presents propagation test results and their implications for the design of the radio receiver.

The overview of Wireless in Local Loop is presented by Ramjee Prasad and Chandan Kumar Chatterjee. This paper presents a bird's eye view on the hot topic of present day in telecommunications viz., Wireless in Local Loop (WILL). After laying down the description of WILL and stating its advantages and disadvantages, it goes on to discuss the details of different types of architecture. In brief it also presents the different standards in Europe, Japan and USA where all the development work has been done so far. The paper further dwells upon a few products relating to each standard and finally ends with a brief discussion on the air interface multiple access schemes. Mention or description of certain products in this paper should not, in any way, be construed by the reader as a recommendation for the product. These have been taken only as examples.

## 2.3 Antennas & Propagation

An adaptive array antenna can be considered as a software antenna because it can programmably form a desired beam pattern if an appropriate set of antenna weights is provided in software. It must be a typical tool for realizing a software radio. Ryuji Kohno in his contribution entitled "Spatial and Temporal Communication Theory Using Software Antennas for Wireless Communications" consider it as an adaptive filter in the space and time domains. In this way the communication theory can be generalized from a conventional time domain into both space and time domains. This paper introduces a spatial and temporal channel modeling, equalization, optimum detection for a single user and multiuser CDMA, precoding and joint optimization of both transmitter and receiver. Such spatial and temporal processing promises drastic improvement of performance as a practical countermeasure for multipath fading in mobile radio communications.

A review of modeling techniques that employ ray optics and the Uniform Theory of Diffraction (UTD) for site specific propagation prediction in the UHF band is given by Henry L. Bertoni and George Liang in the contribution entitled "Review Of Ray Modeling Techniques For Site Specific Propagation Prediction". A number of different methods are described which includes: a two dimensional ray trace in the horizontal plane, a method employing two 2D ray traces in orthogonal planes, ray tracing in full three dimensional space and the Vertical Plane Launch (VPL) technique which approximates a full 3D ray trace. When employed for microcells with ranges that are less than one kilometer, these techniques can provide accurate predictions of the propagation characteristics in a cluttered urban environment. The strengths and weaknesses of each type of ray tracing implementations is addressed along with direct comparisons between methods and with measurements in typical urban environments.

When developing standards for UMTS and FPLMTS, important issues include capacity and spectrum efficiency, which can be enhanced by antenna diversity techniques. The basic principles of these techniques are explained in a contribution entitled "Fundamentals of Smart Antennas for Mobile Radio Applications" by Paul Walter Baier et al.

As an example, these principles are applied to a specific UMTS air interface concept. This concept evolves from GSM by the introduction of a CDMA component in addition to FDMA and TDMA.

## 2.4 Advanced Systems & Technology

In Europe, Mobile and Personal Communications have always been considered a key driver for growth and innovation, as well as being a necessary building block of the Wireless Information Society. Since 1988, European Union (EU)-funded R&D projects have been working towards the development of the next generations of mobile communication concepts, systems and networks.

The ACTS programme, launched in 1995 and extending until 1998, provides a first opportunity to master and trial mobile and personal communications services and technologies, involving service providers, communications operators and equipment manufacturers.

From the user's perspective, the ACTS programme strives to ensure that current mobile services are extended to include multimedia and broadband services, that access to services is possible without regard to the underlying networks, and that convenient, light weight, compact, and power efficient terminals adapt automatically to whatever air-interface parameters are appropriate to the user's location, mobility, and desired services.

Well into the last half of the ACTS programme, it is time to make a first assessment of the progress relating to three aspects of R&D on Third Generation Mobile and Personal Communication Systems, namely future mobile/wireless services, mobile/wireless platforms, and enabling technologies.

At the same time, it becomes necessary to start looking ahead towards the next R&D programme, specially in the area of Mobile and Personal Communications. In this regard, the European Commission has recently put forward its position regarding the next Framework Programme, highlighting the need to continue if not intensify its support to Mobile and Personal Communications and the associated advanced technologies, services and applications. All these activities are summarised in the paper " Mobile and Personal Communications: ACTS and Beyond" by J. Schwartz da Silva et al. Since the work started in the standardization bodies ITU TG8/1 (International Telecommunication Union Task Group) committee for FPLMTS/IMT-2000 1986 and SMG5 (Special Group Mobile) subtechnical committee for UMTS (Universal Mobile Telecommunication System) 1991, the third generation activities have formed an umbrella for advanced radio system developments. During the recent years, standardization activities towards UMTS and FPLMTS/IMT-2000 have accelerated towards concrete specifications. According to present plans ETSI will select a UMTS Radio Access Concept by the end of 1997, and the main system parameters will be frozen by the end of 1998. Moreover, the FPLMTS/IMT-2000 radio transmission techniques evaluation process has started in ITU. Submissions of candidate technologies are expected latest by October 1998 and the completion of the ITU selection process by early 1999.

To support the standardization activities several research programs throughout the world have developed third generation air interface concepts. In addition, laboratory and field trials have been performed. In the contribution entitled "Overview of Research Activities for Third Generation Mobile Communications" by Tero Ojanperä the UMTS system concept and radio access system research activities throughout the world are described. The paper also discusses the FRAMES multiple access evaluation campaign, wideband CDMA based radio interfaces, TDMA and hybrid schemes, OFDM based air interface schemes, Time Division Duplex (TDD) concepts and related problems.

Some selected topics currently being subject of extensive research for applications in future CDMA network are summarized by Markku .Juntti and Savo Glisic in the contribution entitled "Advanced CDMA for Wireless Communications". Current research activities in this field cover a number of problems such as: modulation and error correcting coding, synchronization diversity, multiple access intracell interference (MAI), near-far problem and power control, intercell interference, overlay type internetwork interference, multiuser detectors and efficiency of interference cancellation, multipath propagation and adaptive antennas including space diversity, efficiency of RAKE receiver, inefficiency of signal and channel parameter estimation etc.All these efforts are directed towards increasing the system capacity and from that point of view the authors discuss some selected topics within this paper with clear emphasis on multiuser demodulators .

Masao Nagakawa and Lachlan B. Michael in their contribution entitled "Wireless Communication Technology in Intelligent Transport Systems" discuss applications of wireless communications in the emerging intelligent transport systems (ITS). Both applications , those already in use and those that are presently at the research or experimental level are surveyed. The use of ITS is expected to increase the capacities of highways and make road travel safer. In this environment the use of wireless communication is indispensable and new and different problems from cellular wireless communication must be overcome.

In this paper particular attention is paid to the uses of wireless communication in the vehicle-roadside link (where vehicles communicate in a two-way link with a roadside base station or information is received from broadcast beacons), and vehicle to vehicle wireless communication. The use of spread spectrum techniques to overcome problems of interference in ITS is examined and the problems of forming a distributed wireless network in an ever changing environment is discussed. Specific examples in this paper are taken mostly form Japanese research, though some recent research in Europe and America, and the differences in direction are examined.

Three basic topics should be addressed for the design of RF ICs targeting a specific system application: 1) transceiver architecture for optimum system and chip partitioning as well as optimum use of software techniques, 2) semiconductor technology for optimum performance, cost and integration path, and 3) circuit implementation. At this point in time, there is no practical universal RF IC transceiver architecture based on only one technology that can satisfy any system requirement. Every application has its own set of solutions (and sometimes several sets of solutions), but it is not inconceivable that "programmable" mobile radio transceivers serving multi-standard applications will emerge in the next few years. Christian Kermarrec in his contribution entitled "RFIC Design for Wireless Communications" addresses some of these issues and presents results that illustrate the trends and options in developing RF IC components for wireless communications.

Two main ingredients for the development of wireless communication VLSI circuits are signal processing technology and Digital CMOS technology. A review of the main characteristics and associated metrics is proposed by Frederic Boutaud, in his contribution entitled “VLSI Signal Processing Solutions for Wireless Communications: A Technology Overview”. Evolution and future trends are discussed. The impact on DSP and baseband processors is presented.

## References

[1] K. S. Gilhousen et al., "On the Capacity of a Cellular CDMA System," IEEE Trans. Vehic. Tech., vol. 40. no. 2, May 1991, pp. 303-312.

[2] A. M. Viterbi and A. J. Viterbi, "Erlang Capacity of a Power Controlled CDMA System," IEEE Journal on Selected Areas in Communications, vol. 11, no. 6, August 1993, pp. 892-899.

[3] J. C. Liberti, Jr. and T. S. Rappaport, "Analytical Results for Capacity Improvements in CDMA," IEEE Trans. Vehic. Tech., vol. 43. no. 3, August 1994, pp. 680-690.

[4] C. Y. Huang and D. G. Daut, "Evaluation of Capacity for CDMA Systems on Frequency-Selective Fading Channels," Proc. ICUPC ’96, pp. 975-979.

[5] A. F. Naguib, A. Paulraj and T. Kailath, "Capacity Improvement with Base-Station Antenna Arrays in Cellular CDMA," IEEE Trans. Vehic. Tech., vol. 43. no. 3, August 1994, pp. 691-698.

[6] R. Kohno and L. B. Milstein, "Spread Spectrum Access Methods for Wireless Communications," IEEE Communications Magazine, January 1995, pp. 58-67.

[7] W. C. Y. Lee, "Overview of Cellular CDMA," IEEE Trans. Vehic. Tech., vol. 40. no. 2, May 1991, pp. 291-301.

[8] M. Shimizu, Y. Asano and Y. Daido, "Reverse-Link Performance for Microcellular DS-CDMA Systems with Orthogonal Sequence Spreading," Proc. ICUPC '93, pp. 244-248.

[9] S. Kondo and L. B. Milstein, "Performance of Multicarrier DS CDMA Systems," IEEE Trans. Communications, vol. 44, no. 2, February 1996, pp. 238-246.

[10] A. Baier et al., "Design Study for a CDMA-Based Third-Generation Mobile Radio System," IEEE Journal on Selected Areas in Communications, vol. 12, no. 4, May 1994, pp. 733-743.

[11] C. L. Despins, G. Djelassem and V. Roy, "Comparative Evaluation of CDMA and FD-TDMA Cellular System Capacities with respect to Radio Link Capacity," Proc. PIMRC '96, pp. 387-391.

[12] M. Zorzi, "Improved analysis of the outage performance in cellular systems," Proc. ICUPC '96, pp. 250-254.

[13] V. Tralli and R. Verdone, "Bit error and outage probability for digital cellular systems with a small number of interferers," Proc. PIMRC '96, pp. 73-77.

[14] J-S. Wu, J-K. Chung and Y-C. Yang, "Co-channel Interference and Capacity for Two-Tier CDMA Cellular Systems," Proc. PIMRC '96, pp. 88-92.

[15] W. Huang and V. K. Bhargava, "Performance Evaluation of a DS/CDMA Cellular System with Voice and Data Services," Proc. PIMRC '96, pp. 588-592.

[16] C.-C. Lee and R. Steele, "Closed-loop power control in CDMA systems," IEE Proc.-Commun., Vol. 143, No. 4, August 1996, pp. 231-239.

[17] J. Blanz, A. Klein, M. Na(han and A. Steil, "Capacity of a cellular mobile radio system applying joint detection," COST 231 TD94 002, 18 pages.

[18] B. Gudmundson, J. Sköld and K. Ugland, "A Comparison of CDMA and TDMA Systems," Proc. VTC'92, pp. 732-735.

[19] M. R. Heath and P. Newson, "On the capacity of spread-spectrum CDMA for mobile radio," Proc. VTC'92, pp. 985-988.

[20] P. Jung, P. W. Baier and A. Steil, "Advantages of CDMA and Spread Spectrum Techniques over FDMA and TDMA in Cellular Mobile Radio Applications," IEEE Trans. Vehic. Tech., vol. 42. no. 3, August 1993, pp. 357-364.

[21] J. Blanz, A. Klein, M. Na(han and A. Steil, "Performance of a Cellular Hybrid C/TDMA Mobile Radio System Applying Joint Detection and Coherent Receiver Antenna Diversity," IEEE Journal on Selected Areas in Communications, vol. 12, no. 4, May 1994, pp. 568-579.

[22] A. J. Viterbi, A. M. Viterbi, K. S. Gilhousen and E. Zehavi, "Soft Handoff Extends CDMA Cell Coverage and Increases Reverse Link Capacity," IEEE

Journal on Selected Areas in Communications, vol. 12, no. 8, October 1994, pp. 1281-1288.

[23] P. Newson and M. R. Heath, "The Capacity of a Spread Spectrum CDMA System for Cellular Mobile Radio with Consideration of System Imperfections" IEEE Journal on Selected Areas in Communications, vol. 12, no. 4, May 1994, pp. 673-684.

[24] K. I. Kim, "CDMA Cellular Engineering Issues," IEEE Trans. Vehic. Tech., vol. 42. no. 3, August 1993, pp. 345-350.

[25] A. J. Viterbi, A. M. Viterbi and E. Zehavi, "Performance of Power-Controlled Wideband Terrestrial Digital Communication," IEEE Trans. Communications, vol. 41, no. 4, April 1993, pp. 559-569.

[26] A. Fukasawa et al., "Wideband CDMA System for Personal Radio Communications," IEEE Communications Magazine, October 1996, pp. 116-123.

[27] S. Moshavi, "Multi-User Detection for DS-CDMA Communications," IEEE Communications Magazine, October 1996, pp. 124-135.

[28] T. Eng and L. B. Milstein, "Comparison of Hybrid FDMA/CDMA Systems in Frequency Selective Rayleigh Fading," IEEE Journal on Selected Areas in Communications, vol. 12, no. 5, June 1994, pp. 938-951. Vol. E76-B, No. 8, August 1993, pp. 894-905.

[29] J.Jiménez et, al., "Preliminary Evaluation of ATDMA and CODIT System Concepts", SIG5 deliverable MPLA/TDE/SIG5/DS/P/002/b1, September 1995.

[30] H.Honkasalo, "The technical evolution of GSM", Proc. of Telecom 95, Geneva, October 1995.

[31] P.Ranta, A.Lappeteläinen, Z-C Honkasalo, "Interference cancellation by Joint Detection in Random Frequency Hopping TDMA Networks", Proceedings of ICUPC96 conference, Vol 1 , pp.428-432.

[32] M.Pukkila and P. Ranta, "Simultaneous Channel Estimation for Multiple Co-channel Signals in TDMA Mobile Systems", Proceedings of IEEE Nordic Signal Processing Symposium (NORSIG'96), Helsinki 24-26th September, 1996.

[33] T.Ojanperä et.al., "Design of a 3rd Generation Multirate CDMA System with Multiuser Detection, MUD-CDMA", Proceedings of ISSSTA96 conference, Vol 1 pp.334-338. , Mainz, Germany,1996.

[34] K.Pajukoski and J.Savusalo, "Wideband CDMA Test System", Proceedings of PIMRC97, Helsinki, September 1997.

[35] T.Ojanperä Tero et al., "FRAMES - Hybrid Multiple Access Technology ", Proceedings of ISSSTA96 conference, Vol 1, pp. 320 - 324, Mainz, Germany, 1996.

[36] T.Ojanperä et.al., "A Comparative Study of Hybrid Multiple Access Schemes for UMTS", Proceedings of ACTS Mobile Summit Conference, Vol 1., pp. 124-130, Granada, Spain, 1996.

[37] T.Ojanperä et.al, "Comparison of Multiple Access Schemes for UMTS", Proceedings of VTC97, Vol.2, pp. 490-494, Phoenix, U.S.A, May 1997.

[38] T. Ojanperä, A.Klein and P.O.Andersson, "FRAMES Multiple Access for UMTS", IEE Colloquium on CDMA Techniques and Applications for Third Generation Mobile Systems, London, May 1997.

[39] F.Ovesjö, E.Dahlman, T.Ojanperä, A.Toskala and A.Klein, "FRAMES Multiple Access Mode 2 - Wideband CDMA", Proceedings of PIMRC97, Helsinki, September 1997.

[40] A.Klein, R.Pirhonen, J.Sköld and R.Suoranta, "FRAMES Multiple Access Mode 1 - Wideband TDMA with and without Spreading" Proceedings of PIMRC97, Helsinki, September 1997.

[41] E.Nikula and E.Malkamäki, "High Bit Rate Services for UMTS using wideband TDMA carriers", Proceedings of ICUPC´96, Vol.2, pp. 562 - 566, Cambridge, Massachusetts, September/October 1996.

[42] ARIB FPLMTS Study Committee, "Report on FPLMTS Radio Transmission Technology SPECIAL GROUP, (Round 2 Activity Report)", Draft v.E1.1, January 1997.

[43] F.Muratore and V.Palestini, "Burst transients and channelization of a narrowband TDMA mobile radio system", Proceedings of the 38th IEEE Vehicular Technology Conference, 15-17 June, 1988, Philadelphia, Pennsylvania.

[44] K.Ohno et.al., "Wideband coherent DS-CDMA", Proc. IEEE VTC'95, pp.779 - 783, Chicago, U.S.A, July 1995.

[45] T.Dohi et.al., "Experiments on Coherent Multicode DS-CDMA:", Proc. IEEE VTC'96, pp.889 - 893, Atlanta GA, USA.

[46] F.Adachi et.al., "Coherent DS-CDMA: Promising Multiple Access for Wireless Multimedia Mobile Communications", Proc. IEEE ISSSTA'96, pp.351 - 358, Mainz, Germany, September 1996.

[47] S.Onoe et.al., "Wideband-CDMA Radio Control Techniques for Third Generation Mobile Communication Systems", Proceedings of VTC97, Vol.2, pp.835-839, Phoenix, USA, May 1997.

[48] K.Higuchi et.al., "Fast Cell Search Algorithm in DS-CDMA Mobile Radio Using Long Spreading Codes", Proceedings of VTC97, Vol.3, pp.1430-1434, Phoenix, USA, May 1997.

[49] F.Adachi, M.Sawahashi and K.Okawa, "Tree-Structured generation of orthogonal spreading codes with different lengths for forward link of DS-

CDMA mobile radio" , Electronics Letters, Vol.33, No.1., pp.27-28, January 1997.

[50] Baier et.al., "Design Study for a CDMA-Based Third Generation Mobile Radio System , IEEE JSAC Selected Areas in Communications, Vol.12, No.4, pp. 733 - 743, May 1994.

[51] P-G.Andermo and L-M.Ewerbring, "A CDMA-Based Radio Access Design for UMTS", IEEE Personal Communications, Vol.2, No.1, pp.48-53, February 1995.

[52] M.Ewerbring et.al., Performance Evaluation of Wideband Testbed based on CDMA", Proceedings of VTC'97, pp. 1009 - 1013, Phoenix, Arizona, USA.

[38] A.Urie et.al., "ATDMA System Definition", ATDMA deliverable R2084/AMCF/PM2/DS/R/044/b1, January 1995.

[53] A.Urie et.al., "An Advanced TDMA Mobile Access System for UMTS", IEEE Personal Communications, Vol.2, No.1, pp. 38-47, February 1995.

[54] A.Urie, "Advanced GSM: A Long Term Future Scenario for GSM", Proceedings of Telecom 95, Vol.2 pp.33-37, Geneva, October 1995.

[55] J.Jiménez et, al., "Preliminary Evaluation of ATDMA and CODIT System Concepts", SIG5 deliverable MPLA/TDE/SIG5/DS/P/002/b1, September 1995.

[56] J.Sköld et.al, "Cellular Evolution into Wideband Services", Proceedings of VTC97, Vol.2, pp. 485 - 489, Phoenix, USA, May 1997.

[57] K.Pehkonen et.al., "A Performance Analysis of TDMA and CDMA Based Air Interface Solutions for UMTS High Bit Rate Services", Proceedings of PIMRC97, Helsinki, September 1997.

[58] T.Ojanperä, P.Ranta, S.Hamalainen and A.Lappetelainen, "Analysis of CDMA and TDMA for 3rd Generation Mobile Radio Systems", Proceedings of VTC97, Vol.2, pp.840-844, Phoenix, USA, May 1997.

[59] M.Gustafsson. et.al, "Different Compressed Mode Techniques for Interfrequency Measurements in a Wide-band DS-CDMA System" submitted to PIMRC´97.

[60] M.Thornberg, "Quality based power control in the CODIT UMTS Concept", Proc. of RACE Mobile Telecommunications Summit Vol.2 pp.308-312, Cascais, Portugal November 1995.

[61] J.Mikkonen and J.Kruys, "The Magic WAND: a wireless ATM access system", Proceedings of ACTS Mobile Summit Conference, Vol 2., pp. 535 - 542, Granada, Spain, 1996.

[62] Y-W.Park et.al.,"Radio Characteristics of PCS using CDMA", Proc. IEEE VTC'96, pp.1661-1664, Atlanta GA, USA.

[63] E-K.Hong et.al., "Radio Interface Design for CDMA-Based PCS", Proceedings of ICUPC´96, pp.365-368, 1996.

[64] J.M.Koo et.al. , "Implementation of prototype wideband CDMA system", Proceedings of ICUPC´96, pp.797-800, 1996.

[65] J.M.Koo, "Wideband CDMA technology for FPLMTS", The 1st CDMA International Conference, Seoul Korea, November 1996.

[66] S.Bang, et al., Performance Analysis of Wideband CDMA System for FPLMTS, Proceedings of VTC'97, pp. 830 - 834, Phoenix, Arizona, USA.

[67] H-R.Park, "A Third Generation CDMA System for FPLMTS Application", The 1st CDMA International Conference, Seoul Korea, November 1996.

[68] A.Sasaki, "A perspective of Third Generation Mobile Systems in Japan", IIR Conference Third Generation Mobile Systems, The Route Towards UMTS, London, February 1997.

[69] S.C.Wales, "The U.K. LINK Personal Communications Programme: A DS-CDMA Air Interface for UMTS", Proceedings of RACE Mobile Telecommunications Summit, Cascais, Portugal, November 1995.

[70] B.Engström and C.Österberg, "A System for Test of Multi access Methods based on OFDM", Proc. IEEE VTC'94, Stockholm, Sweden, 1994.

[71] M.Ericson et.al., "Evaluation of the mixed service ability for competitive third generation multiple access technologies", Proceedings of VTC'97, Phoenix, Arizona, USA.

[72] M.Wahlqvist, R.Larsson and C.Österberg, "Time synchronization in the uplink of an OFDM system", Proc. IEEE VTC'96, pp.1569 - 1573, Atlanta GA, USA.

[73] R.Larsson, C.Österberg and M.Wahlqvist, "Mixed Traffic in a multicarrier System", Proc. IEEE VTC'96, pp.1259-1263, Atlanta GA, USA.

[74] ETSI SMG2, "Description of Telia's OFDM based proposal" TD 180/97 ETSI SMG2, May 1997.

[75] R.L.Peterson, R.E.Ziemer and D.E.Borth, "Introduction to Spread Spectrum Communications", Prentice Hall, 1995.

[76] A. Klein, P.W. Baier: Linear unbiased data estimation in mobile radio systems applying CDMA. IEEE Journal on Selected Areas in Communications, vol. SAC-11 (1993), pp. 1058-1066.)

[77] M.M.Naßhan, P. Jung, A. Steil, P.W. Baier, "On the effects of quantization, nonlinear amplification and band limitation in CDMA mobile radio systems using joint detection" Proceedings of the Fifth Annual International Conference on Wireless Communications WIRELESS'93, Calgary/Canada (1993), pp. 173-186.

[78] J.Yang et.al., "PN Offset Planning in IS-95 based CDMA system" Proceedings of VTC'97, Vol.3, pp.1435-1439, Phoenix, Arizona, USA.

# Part 1

# Communication Theory

# 1 CODING AND MODULATION FOR THE FADING CHANNEL

Ezio Biglieri

Giuseppe Caire

Giorgio Taricco

**Abstract:**

On fading channels the coding/modulation (C/M) paradigms developed for the Gaussian channel may not be valid anymore. For example, coded modulation schemes optimized for the independent Rayleigh channel do not attempt at maximizing the minimum Euclidean distance among error events, but rather their Hamming distance.

The channel model turns out to have a considerable impact on the choice of the preferred solution for C/M. Moreover, if the channel model is uncertain, or not stable enough in time to design a C/M scheme closely matched to it, then the best proposition may be that of a "robust" solution, that is, one that provides good performance in a wide variety of fading environments.

In this contribution we review a few important issues in C/M for the fading channel. By focusing our attention on the flat, independent Rayleigh fading channel, we discuss how some design criteria valid for the Gaussian channel should be modified.

## 1.1 INTRODUCTION: GENERAL CONSIDERATIONS

The increasing interest for digital mobile-radio transmission systems has led of late to the consideration of coding and modulation (C/M) for fading channels. For these channels the paradigms developed for the Gaussian channel may not be valid anymore, and a fresh look at the design philosophies is called for. Specifically, in the past system designer choices were driven by their knowledge of the behavior of modulation and coding over the Gaussian channel: consequently, they tried to apply to radio channels solutions that were far from optimum on channels where non-linearities, Doppler shifts, fading, shadowing, and interference from other users make the channel far from Gaussian.

A considerable body of work has been reversing this perspective, and it is now being widely recognized that C/M solutions for the fading channel should be selected by taking into account the basic features of the channel. One early example of this is the design of modulation schemes: starting from standard phase-shift keying, which has an excellent performance over the Gaussian channel, variations of its basic scheme were derived to cope with non-linear effects, tight bandwidth constraints, etc., and resulting into effective modulations like GMSK and $\pi/4$-QPSK. Another, more recent, example is the study of "fading codes," i.e., coded-modulation schemes that are specifically optimized for a Rayleigh channel, and hence do not attempt at maximizing the Euclidean distance between error events, but rather their Hamming distance.

As a result, the channel model turns out to have a considerable impact on the choice of the preferred solution of the C/M schemes. For example, the fading model may depend on the antenna system to be used. To be more specific, assume that intelligent antennas are available. Then the system may end up having radiation patterns which are so directive that a propagation model close to LOS will be a good model, possibly coupled with an "on-off channel," that is, a channel which for a percentage of time is a good (viz, Gaussian) channel, and for the remaining time is so bad that reliable transmission is not possible. In this situation, the standard solution of transmitting signals with the same energy all the time is often a bad choice.

If the channel model is uncertain, or not stable enough in time to design a C/M scheme closely matched to it, then the best proposition may be that of a "robust" solution, that is, a solution that provides suboptimum performance on a wide variety of channel models. As we shall see later, the use of antenna diversity with maximal-ratio combining provides good performance on a wide variety of fading environments. Another such solution is offered by bit-interleaved coded modulation (BICM). This is similar in spirit to the "pragmatic-TCM" solution [19], in the sense that it is based on off-the-shelf coder/decoder pairs, but it does not share with the latter the problem of having a small Hamming distance (which would impair the performance on a Rayleigh fading channel).

As another consideration, if Doppler shifts prove to be very relevant in determining the performance of the system, then double-differential demodulation (which is insensitive to Doppler) may provide a sensible choice [1].

### *1.1.1 Multi-user detection: the challenge*

The design of C/M schemes is further complicated when a multi-user environment is taken into account. The main problem here, and in general in communication systems that share channel resources, is the presence of multiple-access interference (MAI). This is generated by the fact that every user receives, besides the signal which is specifically directed to it, also some power from transmission to other users. This is not true only when CDMA is used, but also with space-division multiple access, in which intelligent antennas are directed toward the intended user. The earlier studies devoted to multi-user transmission simply neglected the presence of MAI. Typically, they were based on the naive assumption that, due to some version of the ubiquitous "Central Limit Theorem," signals adding up from a variety of users would coalesce to a process resembling Gaussian noise. Thus, the effect of MAI would be an increase of thermal noise, and any C/M scheme designed to cope with the latter would still be optimal, or at least near-optimal, for multiuser systems.

Of late, it was recognized that this assumption was groundless, and consequently several of the conclusions that it prompted were wrong. The central development of multi-user theory was the introduction of the optimum multi-user detector: rather than demodulating each user separately and independently, it demodulates all of them simultaneously. A simple example should suffice to appreciate the extent of the improvement that can be achieved by optimum detection: in the presence of vanishingly small thermal noise, optimum detection would provide error-free transmission, while standard ("single-user") detection is affected by an error probability floor which increases with the number of users. Multi-user detection was born in the context of terrestrial cellular communication, and hence implicitly assumed a MAI-limited environment where thermal noise is negligible with respect to MAI (high-SNR condition). For this reason coding was seldom considered, and hence almost all multiuser detection schemes known from the literature are concerned with symbol-by-symbol decisions.

### *1.1.2 Impact of decoding delay*

Another relevant factor in the choice of a C/M scheme is the decoding delay that one should allow: in fact, recently proposed, extremely powerful codes (the Turbo Codes) suffer from a considerable decoding delay, and hence their application might be useful for data transmission, but not for real-time speech. For real-time speech transmission, which imposes a strict decoding delay constraint, channel variations with time may be rather slow with respect to the maximum allowed delay. In this case the channel may be modeled as a "block-fading" channel, in which the fading is about constant for a number of symbol intervals. On such a channel, a single code word may be transmitted after being split into several blocks, each suffering from a different attenuation, thus realizing an effective way of achieving diversity.

### *1.1.3 Unequal error protection*

In some analog source coding applications, like speech or video compression, the sensitivity of the source decoder to errors in the coded symbols is typically not uniform: the quality of the reconstructed analog signal is rather insensitive to errors affecting certain classes of bits, while it degrades sharply when errors affect other classes. This happens, for example, when analog source coding is based on some form of hierarchical coding, where a relatively small number of bits carries the "fundamental information" and a larger number of bits carries the "details" like in the case of the MPEG2 standard.

Assuming that the source encoder produces frames of binary coded symbols, each frame can be partitioned into classes of symbols of different "importance" (i.e., of different sensitivity). Then, it is apparent that the best coding strategy aims at achieving lower BER levels for the important classes while admitting higher BER levels for the unimportant ones. This feature is referred to as unequal error protection (UEP). On the contrary, codes for which the BER is (almost) independent of the position of the information symbols are referred to as equal error protection (EEP) codes.

An efficient method for achieving UEP with Turbo Codes was recently studied in [7]. The key point is to match a non-uniform puncturing pattern to the interleaver of the Turbo-encoder in order to create locally low-rate Turbo Codes for the important symbols, and locally high-rate Turbo Codes for the unimportant symbols. In this way, we can achieve several protection levels while keeping constant the total code rate. On the decoding side, all what we need is to "depuncture" the received sequence by inserting zeroes at the punctured positions. Then, a single Turbo-decoder can handle different code rates, equal-error-protection Turbo Codes and UEP Turbo Codes.

### *1.1.4 Our review*

In this contribution we review a few important issues in coding and modulation for the fading channel. Here we focus our attention to the flat, independent

Rayleigh fading channel, and we discuss how three paradigms commonly accepted for the design of C/M for a Gaussian channel, which we call Euclid's, Lao Zi's, and Ungerboeck's paradigm, should be shifted when dealing with a fading channel.

"Euclid's paradigm" consists of designing a C/M scheme based on the maximization of the minimum Euclidean distance among signals. "Lao Zi's paradigm," named after the author of the classic book Tao Te Ching, consists of adapting oneself to the environment, a central Taoist tenet. In our parlance, it consists of "adapting the code to the channel." "Ungerboeck's paradigm" requires combining modulation and coding (and consequently demodulation and decoding) in a single entity.

Before discussing all this, we present in a tutorial fashion some results on the capacity of the Rayleigh fading channels: these results show the importance of coding on this channel, and the relevance of obtaining channel state information (CSI) in the demodulation process.

## 1.2 THE FREQUENCY-FLAT, SLOW RAYLEIGH-FADING CHANNEL

This channel model assumes that the duration of a modulated symbol is much greater than the delay spread caused by the multipath propagation. If this occurs, then all frequency components in the transmitted signal are affected by the same random attenuation and phase shift, and the channel is frequency-flat. If in addition the channel varies very slowly with respect the symbol duration, then the fading $R(t)\exp[j\Theta(t)]$ remains approximately constant during the transmission of one symbol (if this does not occur the fading process is called *fast*.)

The assumption of non-selectivity allows us to model the fading as a process affecting the transmitted signal in a multiplicative form. The assumption of a slow fading allows us to model this process as a constant random variable during each symbol interval. In conclusion, if $x(t)$ denotes the complex envelope of the modulated signal transmitted during the interval $(0, T)$, then the complex envelope of the signal received at the output of a channel affected by slow, flat fading and additive white Gaussian noise can be expressed in the form

$$r(t) = Re^{j\Theta}x(t) + n(t), \tag{1.1}$$

where $n(t)$ is a complex Gaussian noise, and $Re^{j\Theta}$ is a Gaussian random variable, with $R$ having a Rice or Rayleigh pdf and unit second moment, i.e., $E[R^2] = 1$.

If we can further assume that the fading is so slow that we can estimate the phase $\Theta$ with sufficient accuracy, and hence compensate for it, then coherent detection is feasible. Thus, model (1.1) can be further simplified to

$$r(t) = Rx(t) + n(t). \tag{1.2}$$

It should be immediately apparent that with this simple model of fading channel the only difference with respect to an AWGN channel resides in the fact that $R$, instead of being a constant attenuation, is now a random variable, whose value affects the amplitude, and hence the power, of the received signal. Assume finally that the value taken by $R$ is known at the receiver: we describe this situation by saying that we have *perfect* CSI. Channel state information can be obtained for example by inserting a pilot tone in a notch of the spectrum of the transmitted signal, and by assuming that the signal is faded exactly in the same way as this tone.

Detection with perfect CSI can be performed exactly in the same way as for the AWGN channel: in fact, the constellation shape is perfectly known, as is the attenuation incurred by the signal. The optimum decision rule in this case consists of minimizing the Euclidean distance

$$\int_0^T [r(t) - Rx(t)]^2 \, dt = |\mathbf{r} - R\mathbf{x}|^2 \tag{1.3}$$

with respect to the possible transmitted real signals $x(t)$ (or vectors $\mathbf{x}$).

A consequence of this fact is that the error probability with perfect CSI and coherent demodulation of signals affected by frequency-flat, slow fading can be evaluated as follows. We first compute the error probability $P(e \mid R)$ obtained by assuming $R$ constant in model (1.2), then we take the expectation of $P(e \mid R)$, with respect to the random variable $R$. The calculation of $P(e \mid R)$ is performed as if the channel were AWGN, but with the energy $\mathcal{E}$ changed into $R^2\mathcal{E}$. Notice finally that the assumptions of a noiseless channel-state information and a noiseless phase-shift estimate make the values of $P(e)$ thus obtained as representing a limiting performance.

Consider now the error probabilities that we would obtain with binary signals without coding (see [4] for a more general treatment).. For two signals with common energy $\mathcal{E}$ and correlation coefficient $\rho = (\mathbf{x}, \widehat{\mathbf{x}})/\mathcal{E}$ we have, for Rayleigh fading and perfect channel-state information,

$$P(e) = \frac{1}{2}\left(1 - \sqrt{\frac{(1-\rho)\mathcal{E}/2N_0}{1 + (1-\rho)\mathcal{E}/2N_0}}\right). \tag{1.4}$$

This can be upper bounded, for large signal-to-noise ratios, by

$$P(e) \leq \frac{2}{1-\rho}\,\frac{1}{\mathcal{E}/N_0}. \tag{1.5}$$

In the absence of CSI, one could take a decision rule consisting of minimizing

$$\int_0^T [r(t) - x(t)]^2 \, dt = |\mathbf{r} - \mathbf{x}|^2 \tag{1.6}$$

However, with constant envelope signals ($|\mathbf{x}|$ constant), the error probability obtained with (1.3) and (1.6) coincide because

$$\begin{aligned} P(\mathbf{x} \to \widehat{\mathbf{x}}) &= P(|\mathbf{r} - R\widehat{\mathbf{x}}|^2 < |\mathbf{r} - R\mathbf{x}|^2) \\ &= P(2R(\mathbf{r}, \mathbf{x} - \widehat{\mathbf{x}}) < 0) \\ &= P((\mathbf{r}, \mathbf{x} - \widehat{\mathbf{x}}) < 0) \; . \end{aligned}$$

Fig. 1.1 compares error probabilities over the Gaussian channel with those over the Rayleigh fading channel with perfect CSI. It is seen that the loss in error probability is considerable. As we shall see in a moment, coding can compensate for a substantial amount of this loss.

### 1.2.1 *Impact of coding*

The channel model is described in Fig. 1.2. Here, $x$ is the channel input, $a$ the channel state, $y$ the channel output, and $\widehat{a}$ the estimate of the channel state. We assume that $y$ depends only on $a$, $x$, and not on $\widehat{a}$.

The information we obtain on $x$ from the observation of $y$ and $\widehat{a}$ is given by

$$I(x;\, y, \widehat{a}) = \int_x \int_y \int_{\widehat{a}} p(x, y, \widehat{a}) \log_2 \frac{p(x \mid y, \widehat{a})}{p(x)} \, dx \, dy \, d\widehat{a},$$

where $p(\;\cdot\;)$ are probabilities or probability density functions. The channel capacity is given by

$$C = \max I(x; y, \widehat{a})$$

where the maximum is to be taken over the distribution of $x$. The calculation of this maximum can be avoided by choosing for $p(x)$ the uniform distribution $1/M$ ($M$ the number of channel symbols). Under suitable symmetry assumptions this choice provides the actual capacity. By making the simplifying assumption that transmission is binary, the symmetry condition is

$$p(y \mid a, -x) = p(-y \mid a, x)$$

which yields, with $p(x) = 1/2$,

$$C = 1 + \mathrm{E}_{\widehat{a}} \left[ \int_y p(y \mid \widehat{a}, 1) \log_2 \frac{p(y \mid \widehat{a}, 1)}{p(y \mid \widehat{a}, 1) + p(y \mid \widehat{a}, -1)} \, dy \right] \tag{1.7}$$

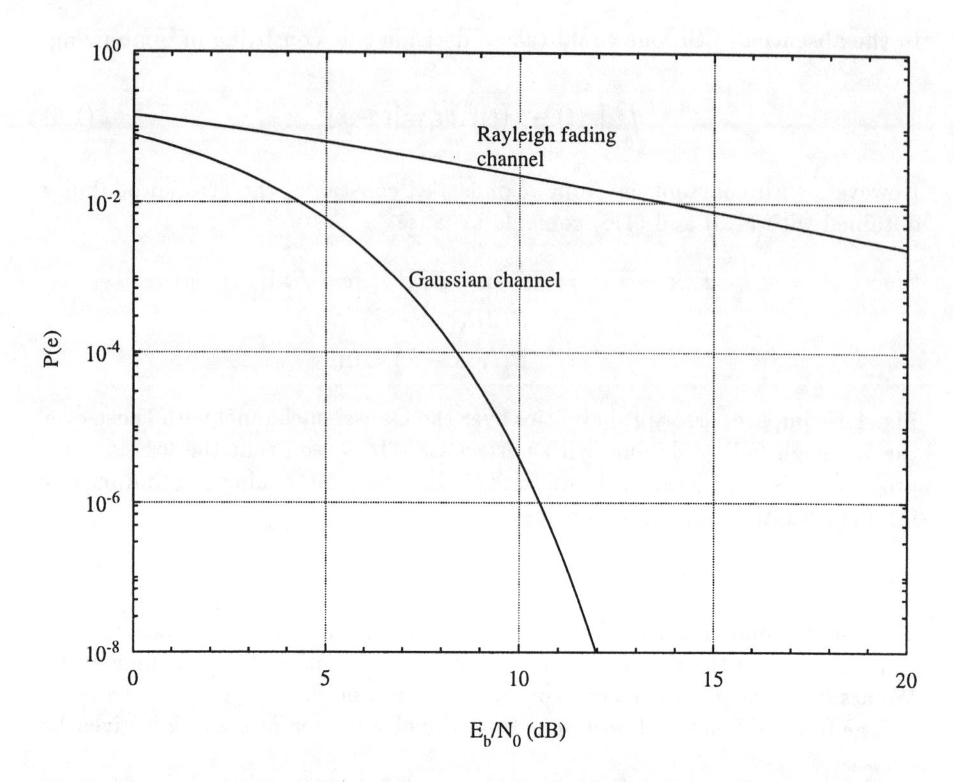


**Figure 1.1** Error probabilities of binary transmission over the Gaussian channel and over a Rayleigh fading channel with and without channel-state information.

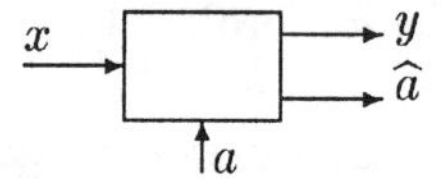


**Figure 1.2** Fading channel model: $x$ is the channel input, $a$ the channel state, $y$ the channel output, and $\widehat{a}$ the estimate of the channel state.

By observing that

$$p(\widehat{a}) = \int_a p(\widehat{a} \mid a) p(a)\, da$$

we have

$$p(y \mid \widehat{a}, x) = \int_a p(a \mid \widehat{a}) p(y \mid a, x)\, da. \tag{1.8}$$

We have two limiting special cases here:

**Perfect channel-state information..** This occurs when $\widehat{a} = a$, so that (1.7) yields

$$C = \mathrm{E}_a[C(a)] \tag{1.9}$$

where $C(a)$ is the capacity of the channel when the latter is in state $a$:

$$C(a) = 1 + \int_y p(y \mid a, 1) \log_2 \frac{p(y \mid a, 1)}{p(y \mid a, 1) + p(y \mid a, -1)}\, dy \tag{1.10}$$

**No channel-state information..** In this case

$$p(\widehat{a} \mid a) = p(\widehat{a}),$$

so that from (1.8) we have

$$C = 1 + \int_y p(y \mid 1) \log_2 \frac{p(y \mid 1)}{p(y \mid 1) + p(y \mid -1)}\, dy. \tag{1.11}$$

If we compare (Fig. 1.3) the channel capacities of a Rayleigh fading channel with and without CSI with that of a Gaussian channel, we see that the loss in capacity due to the fading is much smaller than the loss in terms of error probability. This shows that coding for the fading channel is actually highly beneficial.

## 1.3 EUCLID'S PARADIGM

A commonly accepted design criterion is to design coded schemes such that their minimum Euclidean distance is maximized. This is correct on the Gaussian channel with high SNR (although not when the SNR is very low: see [13]), and is often accepted, *faute de mieux*, on channels that deviate little from the Gaussian model (e.g., channels with a moderate amount of intersymbol interference). However, the Euclidean-distance criterion should be outright rejected over the Rayleigh fading channel. In fact, analysis of coding for the Rayleigh fading channel proves that Hamming distance (also called "code diversity" in this context) plays the central role here.

Assume transmission of a coded sequence $\mathcal{X} = (\mathbf{x}_1, \mathbf{x}_2, \ldots, \mathbf{x}_n)$ where the components of $\mathcal{X}$ are signal vectors selected from a constellation $\mathcal{S}$. We do not distinguish here among block or convolutional codes (with soft decoding), or block- or trellis-coded modulation. We also assume that, thanks to perfect (i.e., infinite-depth) interleaving the fading random variables affecting the various symbols $\mathbf{x}_k$ are independent. Hence we write, for the components of the received sequence $(\mathbf{r}_1, \mathbf{r}_2, \ldots, \mathbf{r}_n)$:

$$\mathbf{r}_k = R_k \mathbf{x}_k + \mathbf{n}_k, \tag{1.12}$$

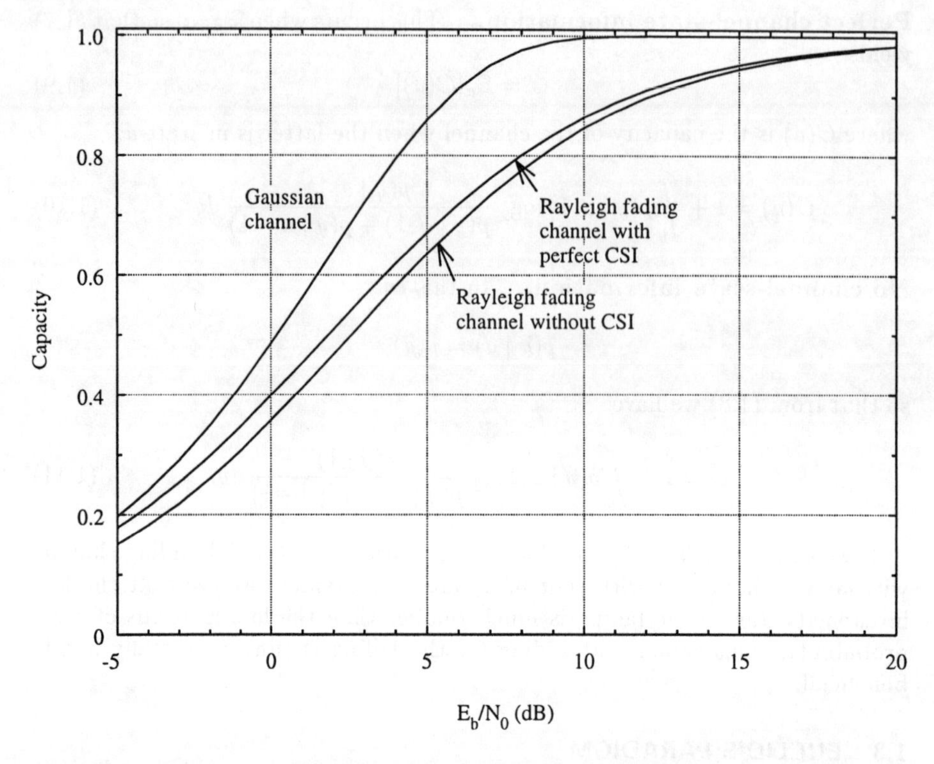


**Figure 1.3** Capacity of binary transmission over the Gaussian channel and over a Rayleigh fading channel with and without channel-state information.

where the $R_k$ are independent, and, under the assumption that the noise is white, the RV's $n_k$ are also independent.

Coherent detection of the coded sequence, with the assumption of perfect channel-state information, is based upon the search for the coded sequence $\mathcal{X}$ that minimizes the distance

$$\sum_{k=1}^{N} |\mathbf{r}_k - R_k \mathbf{x}_k|^2. \tag{1.13}$$

Thus, the pairwise error probability can be expressed in this case as

$$P\{\mathcal{X} \to \widehat{\mathcal{X}}\} \leq \prod_{k \in \mathcal{K}} \frac{1}{1 + |\mathbf{x}_k - \widehat{\mathbf{x}}_k|^2 / 4N_0} \tag{1.14}$$

where $\mathcal{K}$ is the set of indices $k$ such that $\mathbf{x}_k \neq \widehat{\mathbf{x}}_k$.

**An example..** For illustration purposes, let us compute the Chernoff upper bound to the word error probability of a block code with rate $R_c$. Assume that binary antipodal modulation is used, with waveforms of energies $\mathcal{E}$, and that the demodulation is coherent with perfect CSI. Observe that for $\widehat{\mathbf{x}}_k \neq \mathbf{x}_k$ we have

$$|\mathbf{x}_k - \widehat{\mathbf{x}}_k|^2 = 4\mathcal{E} = 4R_c\mathcal{E}_b,$$

where $\mathcal{E}_b$ denotes the average energy per bit. For two code words $\mathcal{X}$, $\widehat{\mathcal{X}}$ at Hamming distance $H(\mathcal{X}, \widehat{\mathcal{X}})$ we have

$$P\{\mathcal{X} \to \widehat{\mathcal{X}}\} \leq \left(\frac{1}{1 + R_c\mathcal{E}_b/N_0}\right)^{H(\mathcal{X}, \widehat{\mathcal{X}})}$$

and hence, for a linear code,

$$P(e) = P(e \mid \mathcal{X}) \leq \sum_{w \in \mathcal{W}} \left(\frac{1}{1 + R_c\mathcal{E}_b/N_0}\right)^w,$$

where $\mathcal{W}$ denotes the set of nonzero Hamming weights of the code, considered with their multiplicities. It can be seen that for high enough signal-to-noise ratio the dominant term in the expression of $P(e)$ is the one with exponent $d_{\text{min}}$, the minimum Hamming distance of the code. □

By recalling the above calculation, the fact that the probability of error decreases inversely with the signal-to-noise ratio raised to power $d_{\text{min}}$ can be expressed by saying that we have introduced a *code diversity* $d_{\text{min}}$.

We may further upper bound the pairwise error probability by writing

$$P\{\mathcal{X} \to \widehat{\mathcal{X}}\} \leq \prod_{k \in \mathcal{K}} \frac{1}{|\mathbf{x}_k - \widehat{\mathbf{x}}_k|^2/4N_0} = \frac{1}{[\delta^2(\mathcal{X}, \widehat{\mathcal{X}})/4N_0]^{H(\mathcal{X}, \widehat{\mathcal{X}})}} \tag{1.15}$$

(which is close to the true Chernoff bound for small enough $N_0$). Here

$$\delta^2(\mathcal{X}, \widehat{\mathcal{X}}) = \left[\prod_{k \in \mathcal{K}} |\mathbf{x}_k - \widehat{\mathbf{x}}_k|^2\right]^{H(\mathcal{X}, \widehat{\mathcal{X}})}$$

is the geometric mean of the non-zero squared Euclidean distances between the components of $\mathcal{X}$, $\widehat{\mathcal{X}}$. The latter result shows the important fact that the error probability is (approximately) inversely proportional to the *product* of the squared Euclidean distances between the components of $\mathbf{x}$, $\widehat{\mathbf{x}}$ that differ, and, to a more relevant extent, to a power of the signal-to-noise ratio whose exponent is the Hamming distance between $\mathcal{X}$ and $\widehat{\mathcal{X}}$.

We hasten to observe here that the expression obtained here for the pairwise error probability is an *upper bound*, rather than an exact expression. Thus, the results obtained should be interpreted with some care.

Further, we know from the results referring to block codes, convolutional codes, and coded modulation that the union bound to error probability for a coded system can be obtained by summing up the pairwise error probabilities associated with all the different "error events." For small noise spectral density $N_0$, i.e., for high signal-to-noise ratios, a few equal terms will dominate the union bound. These correspond to error events with the smallest value of the Hamming distance $H(\mathcal{X}, \widehat{\mathcal{X}})$. We denote this quantity by $L_c$ to stress the fact, to be discussed soon, that it reflects a diversity residing in the code. We have

$$P\{\mathcal{X} \to \widehat{\mathcal{X}}\} \underset{\leq}{\sim} \frac{\nu}{[\delta^2(\mathcal{X}, \widehat{\mathcal{X}})/4N_0]^{L_c}} \tag{1.16}$$

where $\nu$ is the number of dominant error events. For error events with the same Hamming distance, the values taken by $\delta^2(\mathcal{X}, \widehat{\mathcal{X}})$ and by $\nu$ are also of importance. This observation may be used to design coding schemes for the Rayleigh fading channel: here no role is played by the Euclidean distance, which is the central parameter used in the design of coding schemes for the AWGN channel.

For uncoded systems ($n = 1$), the results above hold with the positions $L_c = 1$ and $\delta^2(\mathcal{X}, \widehat{\mathcal{X}}) = |\mathbf{x} - \widehat{\mathbf{x}}|^2$, which shows that the error probability decreases as $N_0$. A similar result could be obtained for maximal-ratio combining in a system with diversity $L_c$. This explains the name of this parameter. In this context, the various diversity schemes may be seen as implementations of the simplest among the coding schemes, the repetition code, which provides a diversity equal to the number of diversity branches [18].

From the discussion above, we have learned that over the perfectly-interleaved Rayleigh fading channel the choice of a coding scheme should be based on the maximization of the code diversity, i.e., the minimum Hamming distance among pairs of error events. Since for the Gaussian channel code diversity does not play the same central role, coding schemes optimized for the Gaussian channel are likely to be suboptimum for the Rayleigh channel.

## 1.4 LAO ZI'S PARADIGM

The design procedure described in the section above, and consisting of adapting the C/M scheme to the channel, may suffer from a basic weakness. If the channel model is not stationary, as it is, for example, in a mobile-radio environment where it fluctuates in time between the extremes of Rayleigh and AWGN, then a code designed to be optimum for a fixed channel model might perform poorly when the channel varies. Therefore, a code optimal for the AWGN channel may be actually suboptimum for a substantial fraction of time. An alternative solution consists of doing the opposite, i.e., *matching the channel to the coding scheme*: the latter is still designed for a Gaussian channel, while the former is transformed from a Rayleigh-fading channel (say) into a Gaussian one, thanks to the introduction of antenna diversity and maximal-ratio combining.

The standard approach to antenna diversity is based on the fact that, with several diversity branches, the probability that the signal will be simultaneously faded on all branches can be made small. Another approach, which was investigated by the authors in [15, 16, 17], is philosophically different, as it is based upon the observation that, under fairly general conditions, a channel affected by fading can be turned into an additive white Gaussian noise (AWGN) channel by increasing the number of diversity branches. Consequently, it can be expected (and it was indeed verified by analyses and simulations) that a coded modulation scheme designed to be optimal for the AWGN channel will perform asymptotically well also on a fading channel with diversity, at the only cost of an increased receiver complexity. An advantage of this solution is its robustness, since changes in the physical channel affect the reception very little.

This allows us to argue that the use of "Gaussian" codes along with diversity reception provides indeed a solution to the problem of designing robust coding schemes for the mobile radio channel.

Fig. 1.4 shows the block diagram of the transmission scheme with fading and co-channel interference. Our initial assumptions, valid in the following unless otherwise stated, are [15, 16, 17]:

- PSK modulation
- $M$ independent diversity branches whose signal-to-noise ratio is inversely proportional to $M$ (this assumption is made in order to disregard the SNR increase that actually occurs when multiple receive elements are used).
- Flat, independent Rayleigh fading channel.
- Coherent detection with perfect channel-state information.
- Synchronous diversity branches.
- Independent co-channel interference, and a single interferer.

The codes examined are the following:

**J4**: 4-state, rate-2/3 TCM scheme based on 8-PSK and optimized for Rayleigh-fading channels [11].

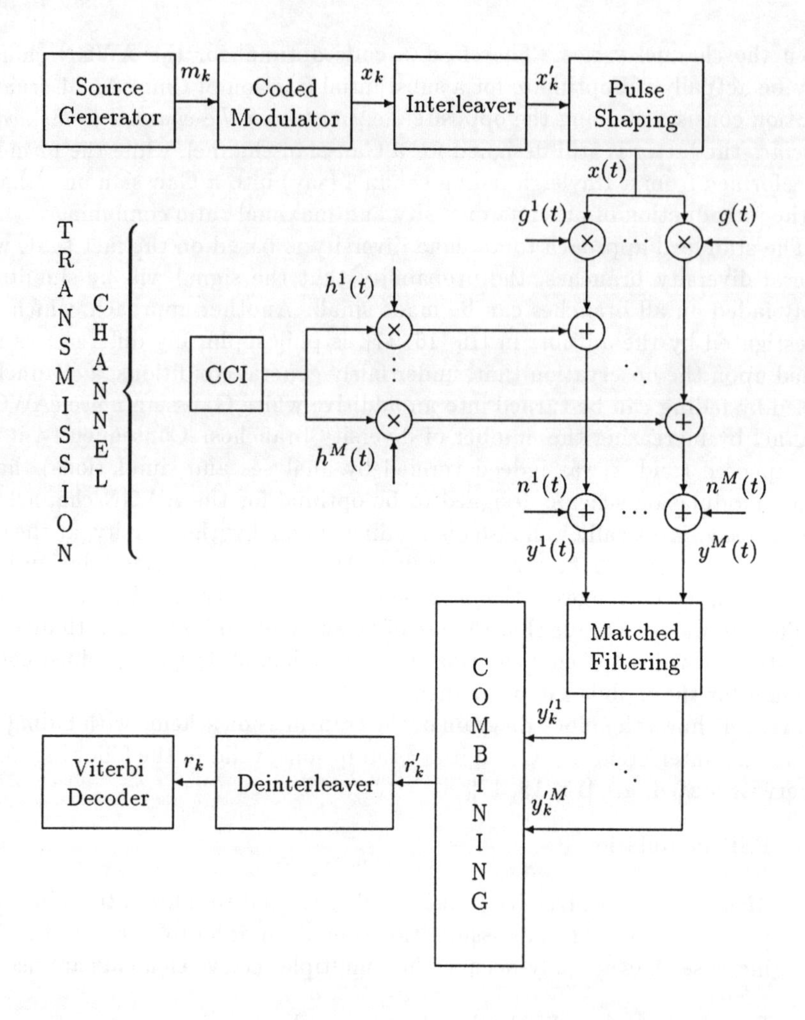


**Figure 1.4** Block diagram of the transmission scheme.

**U4:** 4-state rate-2/3 TCM scheme based on 8-PSK and optimized for the Gaussian channel.

**U8:** Same as above, with 8 states.

**Q64:** "Pragmatic" concatenation of the "best" rate-1/2 64-state convolutional code with 4-PSK modulator and Gray mapping [19].

Fig. 1.5 compares the performance of U4 and J4 (two TCM schemes with the same complexity) over a Rayleigh-fading channel with $M$-branch diversity. It is seen that, as $M$ increases, the performance of U4 comes closer and closer to

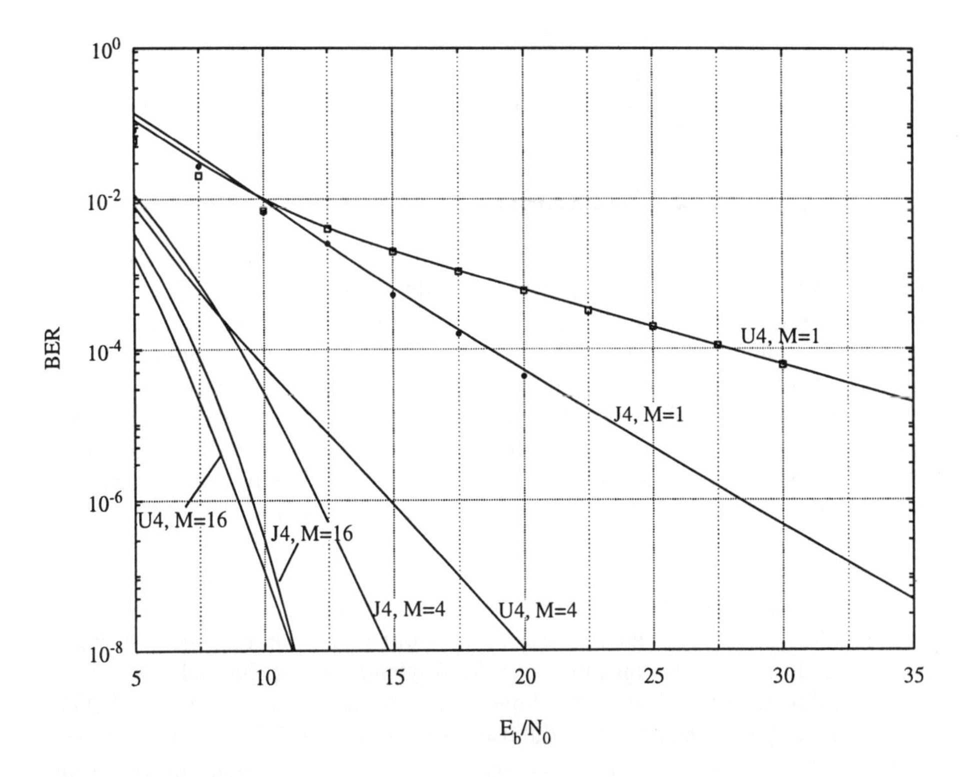


**Figure 1.5** Effect of antenna diversity on the performance of 4-state TCM schemes over the flat, independent Rayleigh-fading channel. J4 is optimum for the Rayleigh channel, while U4 is optimum for the Gaussian channel.

that of J4. Similar results hold for correlated fading: even for moderate correlation J4 loses its edge on U4, and for $M$ as low as 4 U4 performs better than J4 [15]. The effect of diversity is more marked when the code used is weaker. As an example, two-antenna diversity provides a gain of 10 dB at BER=$10^{-6}$ when U8 is used, and of 2.5 dB when Q64 is used [15]. The assumption of branch independence, although important, is not critical: in effect, [15] shows that branch correlations as large as .5 degrade system BER only slightly. The complexity introduced by diversity can be traded for delay: as shown in [15], in some cases diversity makes interleaving less necessary, so that a lower interleaving depth (and consequently a lower overall delay) can be compensated by an increase of $M$.

When differential or pilot-tone, rather than coherent, detection is used [16], a BER-floor occurs which can be reduced by introducing diversity. As for the effect of co-channel interference, even its BER-floor is reduced as $M$ increased (although for its elimination multi-user detectors should be employed). This shows that antenna diversity with maximal-ratio combining is highly instrumental in making the fading channel closer to Gaussian.

## 1.5 UNGERBOECK'S PARADIGM

Ever since 1982, when Ungerboeck published his landmark paper on trellis-coded modulation [14], it has been generally accepted that modulation and coding should be combined in a single entity for improved performance. Several results followed this line of thought, as documented by a considerable body of work aptly summarized and referenced in [11] (see also [5, Chap. 10]). Under the assumption that the symbols were interleaved with a depth exceeding the coherence time of the fading process, new codes were designed for the fading channel so as to maximize their diversity. This implied in particular that parallel transitions should be avoided in the code, and that any increase in diversity would be obtained by increasing the constraint length of the code. One should also observe that for non-Ungerboeck systems, i.e., those separating modulation and coding with binary modulation, Hamming distance is proportional to Euclidean distance, and hence a system optimized for the additive white Gaussian channel is also optimum for the Rayleigh fading channel.

A notable departure from Ungerboeck's paradigm was the core of [19]. Schemes were designed in which coded modulation is generated by pairing an $M$-ary signal set with a binary convolutional code with the largest minimum free Hamming distance. Decoding was achieved by designing a metric aimed at keeping as their basic engine an off-the-shelf Viterbi decoder for the *de facto* standard, 64-state rate-1/2 convolutional code. This implied giving up the joint decoder/ demodulator in favor of two separate entities.

Based on the latter concept, Zehavi [20] first recognized that the code diversity, and hence the reliability of coded modulation over a Rayleigh fading channel, could be further improved. Zehavi's idea was to make the code diversity equal to the smallest number of distinct *bits* (rather than *channel symbols*) along any error event. This is achieved by bit-wise interleaving at the encoder output, and by using an appropriate soft-decision bit metric as an input to the Viterbi decoder. For different approaches to the problem of designing coded modulation schemes for the fading channels see [6].

One of Zehavi's findings, rather surprising *a priori*, was that on some channels there is a downside to combining demodulation and decoding. This prompted the investigation whose results are presented in a comprehensive fashion in [2] (see also [3]).

An advantage of this solution is its robustness, since changes in the physical channel affect the reception very little. Thus, it provides good performance with a fading channel as well as with an AWGN channel (and, consequently, with a Rice fading channel, which can be seen as intermediate between the latter two).

## 1.6 CONCLUSIONS

This review was aimed at illustrating some concepts that make the design of codes for the fading channel differ markedly from the same task applied to the Gaussian channel. In particular, we have examined the design of "fading codes," i.e., C/M schemes which maximize the Hamming, rather than the Euclidean, distance, the interaction of antenna diversity with coding (which makes the channel more Gaussian), and the effect of separating coding from modulation in favor of a more robust C/M scheme. The issue of optimality as contrasted to robustness was also discussed to some extent.

### References

[1] E. Biglieri, M. Di Sciuva and V. Zingarelli, "Modulation and coding for mobile radio communications: Channel with correlated Rice fading and Doppler frequency shift," *IEEE Trans. Vehicular Technology*, to be published, 1997.

[2] G. Caire, G. Taricco, and E. Biglieri, "Bit-interleaved coded modulation," *IEEE Transactions on Information Theory*, to be published, 1997.

[3] S. A. Al-Semari and T. Fuja, "Bit interleaved I-Q TCM," *ISITA '96*, Victoria, B.C., September 17–20, 1996.

[4] E. Biglieri, G. Caire, and G. Taricco, "Error probability over fading channels: A unified approach," *European Transactions on Telecommunications*, 1997, to be published.

[5] E. Biglieri, D. Divsalar, P. J. McLane, and M. K. Simon, *Introduction to Trellis-Coded Modulation with Applications*, New York: MacMillan, 1991.

[6] J. Boutros, E. Viterbo, C. Rastello, and J.-C. Belfiore, "Good lattice constellations for both Rayleigh fading and Gaussian channels," *IEEE Trans. Inform. Theory*, vol. 42, no. 2, pp. 502–518, March 1996.

[7] G. Caire and G. Lechner, "Turbo-codes with Unequal Error Protection," *IEE Electronics Letters*, Vol. 32 No. 7, pp. 629, March 1996.

[8] G. Caire, J. Ventura, and E. Biglieri, "Coded and Pragmatic-Coded Orthogonal Modulation for the Fading Channel with Non-Coherent Detection," *Proceedings of the IEEE International Conference on Communications (ICC'95)*, Seattle (WA), June 18–22, 1995.

[9] T. Ericson, "A Gaussian channel with slow fading," *IEEE Trans. Inform. Theory*, Vol. 16, pp. 353–355, May 1970.

[10] J. Hagenauer, "Zur Kanalkapazität bei Nachrichtenkanälen mit Fading und gebündelten Fehlern," *AEÜ*, Vol. 34, No. 6, pp. 229–237, June 1980.

[11] S. H. Jamali and T. Le-Ngoc, *Coded-Modulation Techniques for Fading Channels*. New York: Kluwer Academic Publishers, 1994.

[12] C. Berrou and A. Glavieux, "Near optimum error correcting coding and decoding: Turbo-codes," *IEEE Trans. Communications*, Vol. 44, No. 10, pp. 1261–1271, October 1996.

[13] M. Steiner, "The strong simplex conjecture is false," *IEEE Trans. Inform. Theory*, pp. 721–731, May 1994.

[14] G. Ungerboeck, "Channel coding with multilevel/phase signals", *IEEE Trans. Inform. Th.*, vol. IT-28, pp. 56–67, Jan. 1982.

[15] J. Ventura-Traveset, G. Caire, E. Biglieri, and G. Taricco, "Impact of diversity reception on fading channels with coded modulation—Part I: Coherent detection," *IEEE Trans. Commun.*, Vol. 45, No. 5, May 1997.

[16] J. Ventura-Traveset, G. Caire, E. Biglieri, and G. Taricco, "Impact of diversity reception on fading channels with coded modulation—Part II: Differential block detection," *IEEE Trans. Commun.*, Vol. 45, No. 6, June 1997.

[17] J. Ventura-Traveset, G. Caire, E. Biglieri, and G. Taricco, "Impact of diversity reception on fading channels with coded modulation—Part III: Co-channel interference," *IEEE Trans. Commun.*, Vol. 45, No. 7, July 1997.

[18] N. Seshadri and C.-E. W. Sundberg, "Coded modulations for fading channels—An overview," *European Trans. Telecomm.*, Vol. ET-4, No. 3, pp. 309–324, May-June 1993.

[19] A. J. Viterbi, J. K. Wolf, E. Zehavi, and R. Padovani, "A pragmatic approach to trellis-coded modulation," *IEEE Communications Magazine*, vol. 27, n. 7, pp. 11–19, 1989.

[20] E. Zehavi, "8-PSK Trellis Codes for a Rayleigh Channel," *IEEE Trans. on Commun.*, vol. 40, no. 5, pp. 873-884, May 1992.

# 2 PER-SURVIVOR PROCESSING (PSP)

Andreas Polydoros
Keith M.Chugg

**Abstract:**
The principle of Per-Survivor Processing (PSP) provides a general framework for the approximation of likelihood-based data detection (search) algorithms whenever the presence of unknown quantities prevents the precise use of the classical Viterbi Algorithm (VA). The PSP concept is based on the realization that decision-directed estimation of the relevant unknown parameters should be embedded into the structure of the search algorithm itself, thus joining together the two main functions of the receiver. We argue that PSP not only offers superior and robust performance versus the classical segregated-task architectures in many challenging environments, but that it is also the natural way of approaching the decoding problem from first principles, thus making it the appropriate vehicle for integrated-task receiver design.

**keywords:** Per-Survivor Processing, Maximum Likelihood Sequence Detection, Parameter Estimation, Viterbi Algorithm, Adaptive Processing.

## 2.1 INTRODUCTION

In most situations encountered in practice, data decoding (referred to as "data detection" here) must be carried out in the presence of unknown, possibly time-varying channel parameters. When the prevailing signal-to-noise ratio (SNR) is not too large or is rapidly varying (as is the case in mobile wireless systems), *ad hoc* methods (also referred to as "conventional" or "classic") for data detection do not perform very well. There is, then, a need for more sophisticated solutions which recover much of the loss of these previously established techniques. Such solutions of higher complexity, which were perceived as practically infeasible until recently, are now becoming more attractive because of the increased computational power (speed and memory) of modern DSP's and IC's. Furthermore, the inherent structure of digital communication waveforms allows for accurate parametric modeling, meaning that the knowledge (or accurate estimation) of a few key waveform parameters suffices for reproducing the sample waveform of the transmitted random signal with high precision.

The above facts motivate the trend to "learn"[1] as much about the channel as possible and use that knowledge in the data detection process. It follows that the architecture employed for executing this learning process and its associated performance, measured either directly or indirectly through its impact on the deliverable Bit-Error Rate (BER) of the detected data, plays a crucial role in receiver (and overall system) performance. Although reducing the BER in the data detection process is a common and acceptable goal to all,[2] there is a multitude of approaches towards this goal in the presence of all these channel-induced obscuring factors. Now, there is no argument about the best thing to do IF everything about the channel were known, and the only randomness left was that of the data and the omnipresent thermal noise in the observation. That issue was settled more than 25 years ago [1] (although not necessarily implemented due to the complexity considerations mentioned before). But that "if" is a big IF: one never really *precisely* knows the channel, and in many challenging situations one may not even come close due to effects such as significant dynamics, lack of channel modeling resources, limited receiver resources (such as memory devoted to channel-state representation), quality of receiver electronic components, interference from other users due to the crowding of the radio spectrum, etc. In addition, the dual commercial needs for high-speed access and mobility/portability mean that the channels will cooperate less and less as time progresses, thus intensifying the search for the "right" receiver in ever more noncooperative environments. Within this framework, we can succinctly state that the architectures advocated by the PSP principle pertain to well-structured, **joint** data detection/parameter estimation algorithms that lead to performing the various receiver tasks in a **coupled** way, thus differing in an essential way from the individual-task-oriented (compartmentalized) receiver designs of the past; we elaborate on that below.

The specific philosophy behind the development of the broad family of approaches to which PSP belongs is to apply the *likelihood-functional* (LF) theory of inference upon stochastic signals in the presence of noise in order to (a) distinguish between various denumerable hypotheses[3]and (b) estimate other useful parameters. The notion of likelihood-based detection and estimation implies a probabilistic viewpoint which differs from the statistics-flavored approaches of non-parametric testing or moment-based processing (such as the property-restoral algorithms), or any other adaptive learning-system, AI or expert-system approach (such as neural nets or fuzzy logic).[4] It is, in other words, a mainstream approach, very familiar to the parametric-model-friendly communication community. The differences between the various algorithms in this family reside in the detailed probabilistic modeling assumptions as well as any externally imposed constraints on the processor structure.

Regarding the special member of this family which we call PSP, this paper will address the key issues of: what is it? (Section 2); how did it come about? (Section 3); where has it been applied so far and where is it headed? (Section 4); how do you benefit by using it and at what cost? (also in Section 4).

At this introductory point we may summarize our experience with PSP as follows:

- PSP is broad and general, in that it is not constrained to "convenient" types of stochastic channels, such as the complex Gaussian family (Rayleigh, Rician)[5]. Phase noise, Doppler shifts, random delays (epoch), non-Gaussian fading statistics with unknown correlation, like-user and unlike-user interference, uncertainty due to reduced-state description, etc., don't lend themselves to easy modeling and handling, yet they prevail in most real-world applications. It is thus desirable to have a tool that handles all of these things in a unified way (with proper modifications, of course).
- PSP is straightforward to understand, describe and implement (although hard to analyze!); it is also easy to identify the fundamental paradigm shift involved in PSP, away from the "classic" architectures of the past: it suggests that each hypothesized data sequence should have its own set of parameter (or even past data) estimates, as opposed to a single estimated parameter set, shared by all hypothesized data sequences.
- PSP is robust to modeling inaccuracies, imperfections, and to the need to fine-tune tracking parameters, as in the more traditional ("conventional") structures, with the corollary of enhanced performance (such as lowering error floors, smaller SNR required for a certain performance level, etc.).

These points are further elaborated upon below, starting with the description of the concept.

## 2.2 WHAT IS PSP?

PSP can be perceived and described as the common concept underpinning such seemingly disparate tasks as optimal data detection in randomly perturbed (and stochastically described) channels such as Rayleigh or Rician fading, digitized demodulation of analog (FM) signals, complexity reduction in data demodulation/decoding for a known channel with long dispersion, joint data detection and adaptive parameter filtering (acquisition and tracking) of unknown[6] and/or time-varying channel parameters, co-channel interference separation, etc. The embedded modulation format may be arbitrary: memoryless or with memory, constant-envelope or not, coded or uncoded, narrowband or spread, or no modulation at all (see the FM application); the channel may or may not have memory (it typically does but this feature is not, by itself, essential to the PSP concept), it may or may not include colored or other-user interference in addition to the omnipresent Gaussian white thermal noise, and, in general it may or may not suffer from all other typical parametric ailments described in the literature. The receiver may not even be interested in the data but, rather, in some other parameter of interest (angle-of-arrival, for instance, in a multisensor system). The two essential features that must be present, however, are:

- (a) Inter-Digit Memory (IDM), meaning that the combination of elements in the scenario is such that there exists parametric memory between adjacent digits[7] in the received waveform (which necessitates a path search for optimal or quasi-optimal decisions).

- (b) There is Residual Metric Information (RMI) that must be computed and provided to the likelihood branch metric computer of a classic path-search algorithm, information that essentially summarizes the impact of the past on each present digit (more generally, state) transition.

A detailed development of these concepts is contained in Appendix 2.5 along with several illustrative examples. To provide an introductory example, think of a simple convolutionally coded system: there exists for that a well-defined state-diagram and trellis description (satisfying the IDM requirement in the most straightforward way) plus a well-defined, finite-complexity (trellis size) decoding algorithm, the familiar Viterbi Algorithm (VA), that produces optimal sequence decisions in the presence of AWGN. Imagine that there is just a random phase rotation in the channel, a stochastic process essentially with some correlation time. The branch metrics in the VA must be provided with a piece of RMI, namely the (estimated) value of that rotation, for otherwise the metric computation is impossible. Clearly, the RMI conveys the impact of past accumulated knowledge (on the phase value, in this case) to the branch metric computer. Thus, both features are present and the stage is set for the possible application of PSP. Observe here that the channel phase could provide, by itself, both the IDM and the RMI features, implying that PSP would be applicable even if there is no coding or modulation memory whatsoever (more on that later).

It is straightforward to broaden both the application at hand and the set of associated parameters involved in order to realize how general the model can be, within which IDM and RMI appear and need to be handled. What is the essence of PSP, then, in such a broad setting? It is simply the notion that *the RMI must be provided separately to each such transition metric (branch) computation, and its value must be computed based on the individual survivor attached to the node from which each such transition starts.*

To contrast this PSP architecture with the past traditional structures, recall that the classical paradigm for detecting data in an ML framework, either on a symbol-by-symbol basis[8] or as a sequence, what is familiarly called Maximum Likelihood Sequence Detection (MLSD), involves performing the associated extraneous operations in a *segregated* way, namely in sub-systems distinct from the one performing the data decisions, the purpose of which is to estimate all these other parameters embedded in the received observed waveform in addition to the data sequence itself. These parameters are the familiar ones: waveform amplitudes (or powers), carrier phase, symbol or baud epoch, frequency offset (Doppler, if induced by dynamics), channel-dispersion coefficients (i.e., intersymbol interference—ISI), and the like. Accordingly, the traditional receiver includes sub-systems such as bit synchronizers, AGCs, suppressed-carrier loops, frequency trackers, channel equalizers (or estimators), all of which do the same thing: they extract values for the relevant parameters and provide them to the decoder box for appropriate data extraction in a resulting (approximately) AWGN environment. Clearly, errors in this parameter estimation (PE) procedure obviously affect performance (e.g., BER) negatively, and this effect should be factored in the link margin of the power budget computations. The fact that an alternative way of designing the receiver (i.e., PSP) can produce significant savings in the power budget, or even "close" the link in cases where that is impossible to achieve with the familiar structures is obviously of major significance to receiver designers.

It is best to represent the classical adaptive MLSD architecture in generic block-diagram form[9] as in Figure 2.1(a), which allows for two options: in the first, PE is based on the observed data only, and its output estimated vector $\hat{\mathbf{f}}$ is fed to the decoder; we may call this the "non-decision-directed (NDD)" structure.[10] In the second option, the alternative "decision-directed (DD)" structure is depicted, which includes a feedback link (the dashed line) from the data detector to the PE box. This link provides *tentative*[11] data decisions out of the detector, to be used by the external estimator as if they were the true data values for the purpose of removing the data uncertainty from the observation and thus concentrating on the remaining parameter uncertainty. It is intuitively clear that the quality of the estimated parameter will depend on the quality of the estimated data fed to it (i.e., its error rate); thus, a low input SNR (or other equivalent measure) will create a vicious cycle of poor tentative decisions driving the PE, whose deteriorated estimator worsens further the detector, etc. This error propagation, explicit or implicit, is the Achilles heel of any such DFE or adaptive MLSD structure.[12] Let us note, in passing, that these two structures in Figure 2.1(a) tend to perform similarly in terms of deliverable BER.

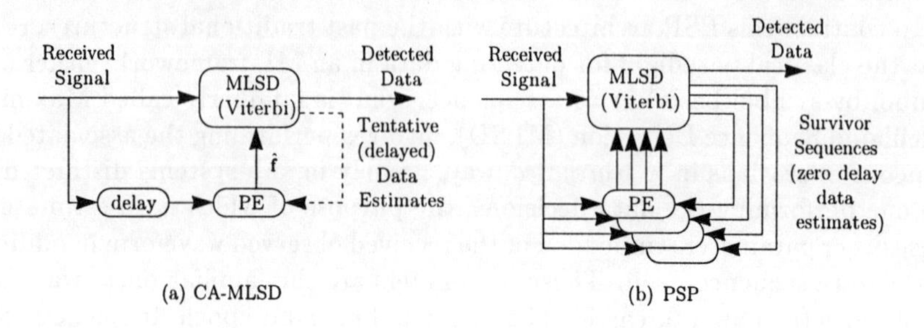


**Figure 2.1** The structure of (a) Conventional Adaptive (CA)-MLSD and (b) PSP

Another instructive way to view these classical "compartmentalized" designs is to assume a trellis search for the data detector (i.e., a VA-type detector), whose branch metric computations are affected by an external **global** estimated parameter vector, as shown in Figure 2.2(a): It is, again, implicitly assumed that the estimated vector $\hat{\mathbf{f}}$ is being updated every symbol, based on tentative decisions from the VA in the feedback sense described above.

It is now easy to succinctly explain the essence of PSP as encapsulated in Figures 2.1(b) and 2.2(b): each retained path in the search (a "survivor") keeps and updates *its own individual vector of estimated parameters, based on its own data sequence, namely its associated data history*. It follows that PE is performed in a **decentralized** manner, as opposed to the global manner of conventional receivers, which can also be interpreted as a zero-delay, **localized** decision feedback. Modifications of this fundamental new structure aside, this is basically all there is to PSP.

The possible variations on the PSP theme are only limited by our imagination. They include: the *search type* (trellis, sequential, symbol-by-symbol optimal, etc.); the *state/survivor structure* (full-state versus reduced-state or "mini-PSP", single versus multiple survivors per state, the M-algorithm, etc.); the *symbol/processing rate ratio* (i.e., number of samples and/or parameter updates per symbol); various *modeling options* (as in co-channel interference (CCI) problems); and so on. A particular combination will fit any given prob-

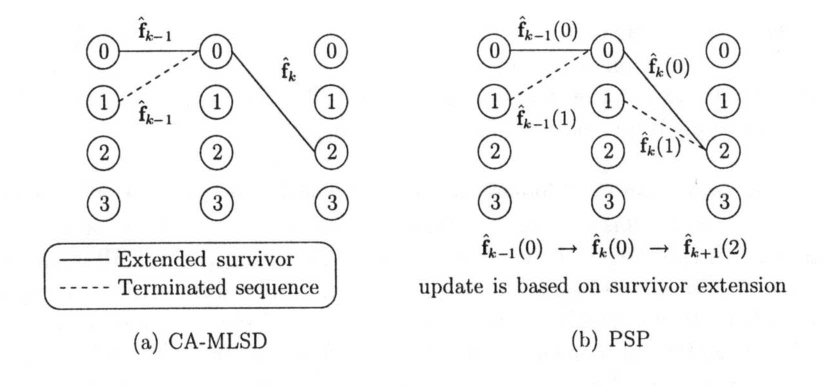


**Figure 2.2** The trellis structure of (a) CA-MLSD and (b) PSP

lem better than others[13], and the right pairing in various scenarios is the topic of much current research activity.

A final note on the trellis structures above: although the associated state space is typically created by the memory of the modulator/coder plus any finite-state memory of the channel (say, ISI), or a reduced sub-space of that combination, that does not have to be the case. One can create an *apparent memory* on the data (i.e., construct a hypothetical trellis of arbitrary size), simply for the purposes of joint data demodulation and parameter tracking via PSP. In other words, we may not only reduce memory, we may also deliberately increase it upwards beyond the length of the modulation memory (e.g., see Section A-.0.2). The receiver selected memory length, therefore, can be set to an arbitrary value subject to the memory and computation limitations of the receiver. The relationship between the receiver memory and the physical memory of the process is a topic requiring further exploration.

## 2.3 A BRIEF HISTORY OF PSP

Like many other interesting ideas, PSP has been re-invented independently by many people over the last 35 years or so, in different guise each time. There appears to be a dichotomy between the stochastic-signal literature and the data demod./parameter estimation literature[14], with a certain convergence now slowly taking place.

Building upon, re-interpreting and simplifying the seminal work of R. Price [4] on the LF of a *Gaussian* signal in Gaussian noise (such as reception through a Rayleigh or Rician complex Gaussian channel), Kailath [5] provided multiple versions of the classic quadratic formula for the log-likelihood functional of the observation (conditioned on each hypothesized version of the data path, namely what could be called the *per-path* likelihood). In a subsequent article [6], he offered an explicit per-path *channel* estimator/correlator version in addition to the classic *total-output-signal* estimator/correlator structure, always for the Gaussian-signal case (i.e., linear processing for either estimator). If *survivors* are just *retained paths*, where the path pruning is performed by some appropriate rule (including, in the unrealistic case, no pruning), then PSP is, indeed, synonymous with per-retained path-processing, a procedure of elective suboptimality (due to complexity considerations). The subsequent generalization of the MMSE-estimator/correlator log-likelihood formula for *any* diffusion process of the Ito type (see [7] for details) implies also a broader appeal for the stochastic version of PSP, although the cost of path pruning in such environments has not been examined yet.

In the non-random, dynamic-parameter sphere, [8] is the earliest allusion to a PSP-like structure applied to Doppler or FM phase tracking via an ingeniously constructed finite-hypothesis test on a quantized variable, obviously motivated[15] by the then recent VA. The classic trio of articles [9, 1, 10] on the VA had provided a major impetus to research on optimal trellis search techniques for a variety of problems, but that optimality did require a known channel plus Gaussian noise. Since the assumption of a precisely known channel is utopian, the most straightforward variation on the VA architecture was proposed[16] soon thereafter, namely to augment it with an external channel estimator (our previous PE) as per Figure 2.1(a); the conventional adaptive MLSD was born.

What happens when the ISI is too long (say, infinite, as in an ARMA channel), in which case the VA cannot be applied? Besides other ISI-truncating solutions, an explicit PSP solution along a MA-defined data trellis was proposed in [14]. A generalization of the concept was independently proposed later in [15] with a broader definition of the reduced-memory trellis states. Reduced-State-Sequence Estimation (RSSE) for coded signals was also independently proposed in [16, 17], all of which involved imaginative state-space reduction techniques and PSP computation of the created RMI (i.e., the remainder of the full-state description).

All the above work involved a known channel which was somehow difficult to include in a straightforward version of MLSD, hence the reduction. The early 90's saw the introduction of novel concepts for dealing with unknown (and possibly time-varying) channels, which we can roughly group into three very related problems[17]: (a) unknown fixed channel, which is dealt with via joint ML data and channel estimation schemes (time recursive or batch) [18, 19, 20, 21] (b) time-varying but statistically known channel, typically of the complex

Gaussian family [22, 23, 24] and (c) time-varying purely unknown channel, for which adaptive PSP was first explicitly (and independently) proposed either as a special technique to a given problem [25, 26, 27, 28, 29, 30, 31], or as an umbrella concept and general principle, applicable to any such environment [18, 32]. Since then, PSP has flourished in various directions; (see [33, 34, 35, 36] as a sample), having recently found a firm theoretical basis in the non-stochastic environment also [37, 38, 39].

## 2.4 APPLICATIONS

We describe in this section certain applications where PSP algorithms have been applied successfully and we provide some illustrative results. The comparisons are typically versus the conventional MLSD structure for the same problem (for hard decisions), which is viewed as the benchmark design. Care has been taken to optimize both algorithms (step sizes for both and tentative decision delay for the conventional), in order to make the comparison equitable. The list of such applications is obviously not exhaustive, and it only pertains to results obtained by the authors and their associates; the references cited in the previous section should be consulted for a more thorough variety of applications and results.

### *2.4.1 MLSD multipath fading ISI channels*

Multipath fading, the classic phenomenon in wireless transmission, has been viewed as both a curse (we would rather receive a single unfaded path of a given total energy!) and a blessing (given that we have to have fades in such environments, it is preferable to have and benefit from the inherent diversity provided by the multipath-induced ISI than not to have it). It is by now well documented that linear and DFE-based equalization are both inadequate in terms of performance in such fading links [26], and the Viterbi equalizer of GSM-type receivers (another name for CA-MLSD) has been diagnosed with faulty performance in rapid fading. As a result, MLSD with distributed, per-hypothesized-sequence processing (i.e, PSP) has been a popular topic in the literature, as discussed in the previous section, and it has been tackled from multiple angles (deterministic and stochastic models). The advantage of PSP (simplest flavor of the VA-PSP with a single survivor per node) against CA-MLSD is demonstrated below in Figure 2.3, assuming a healthy degree of dynamics (symbol-rate-normalized Doppler bandwidth of $2 \cdot 10^{-3}$), for a three-tap, independent Rayleigh fading channel [32].

The Figure above exhibits the key benefit of employing PSP in all fading channel conditions: it lowers the error floor encountered at high SNR where the primary source of performance degradation is due to the channel estimation

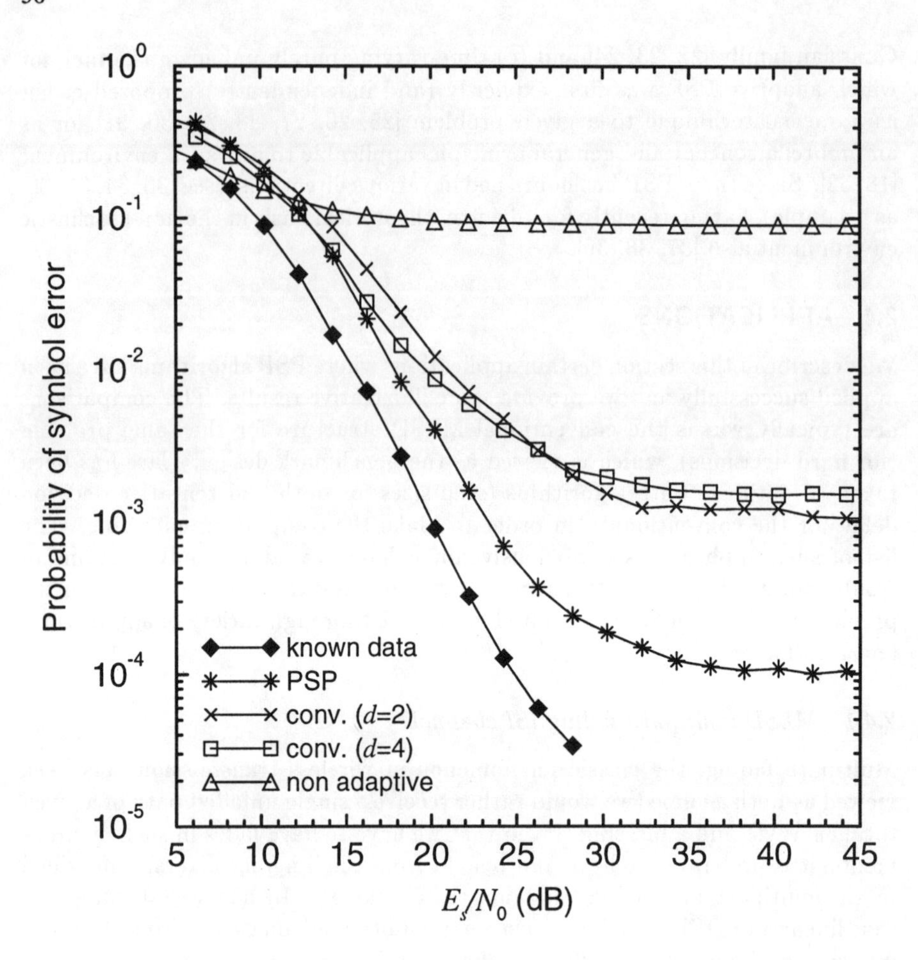


**Figure 2.3** Comparison of PSP vs. conventional MLSD for a frequency selective fading channel

error. When the dynamics are high (as in this Figure), it becomes difficult to track the channel with a single external estimator, whereas it is more effective with PSP. Conversely, when there are no significant dynamics and/or thermal noise dominates performance, there is no justification for the higher complexity of PSP and any version of the single-estimator architecture would do [42]. We should note here that the LMS version employed in the present example does not capitalize on the stochastic nature of the channel (i.e., Rayleigh) or its statistics; it just treats it as unknown but time-varying channel. A PSP armed with that specific statistical knowledge would track the channel with per-survivor Kalman filters (or its reduced-complexity version [43]), providing further performance improvement, although the issue of robustness of such statistics-based algorithms in the absence of exact such information is still largely open.

### 2.4.2 *Soft-decision algorithms in multipath fading channels*

Since channel coding is extensively employed to combat very noisy or time-varying fading channels, an appropriate and efficient receiver would have to jointly estimate the channel, equalize the received sequence and decode it. This task is many times complicated by the presence of an interleaver that decorrelates the channel errors input to the decoder. If one MLSD processor is used for channel equalization ("the Viterbi equalizer") and another for the decoding process ("the Viterbi decoder"), then only hard decisions can be fed from the equalizer, via the de-interleaver, to the decoder, thus limiting the achievable overall performance. To counter that, several schemes have been introduced which provide soft decisions at the output of the equalizer (see [45] for an extensive bibliography on the subject). Among them, the symbol-by-symbol Maximum A-Posteriori (MAP) probability algorithm developed in [46], which minimizes the symbol-error probability under the constraint of a fixed delay, is well suited for real-time applications since it only requires forward processing. Unfortunately, the number of variables that need to be stored and updated increases exponentially with the delay (in the form that was described in [46]) making it rather impractical. Recently, [47] reformulated the Abend & Fritchman algorithm; its new realization, named Optimum Soft-output Algorithm (OSA), has a complexity which is only linearly increasing with the delay constraint. Moreover, a Suboptimum Soft-output Algorithm (SSA) was also introduced in [47], and results therein indicated a small degradation versus the OSA. All of the above, however, assumed perfect channel state information. Let us note here that there also exist alternative architectures to the soft-decisions philosophy, pertaining to super-trellis state definitions that include the de-interleaver in their formation [33], and which also apply PSP for complexity-reduction purposes.

In a recent article [48], several adaptive extensions of the OSA and SSA algorithms were proposed. The Conventional Adaptive OSA/SSA (CA-OSA/CA-SSA) receivers were introduced as the soft-decision counterparts of the hard-decision CA-MLSD, and similarly the soft-decision adaptive version of PSP. Their performance (of the soft-decision group) was demonstrated to be greatly superior versus the standard (baseline) configuration, which consists of a hard-decision inner equalizer followed by convolutional decoder, as expected.

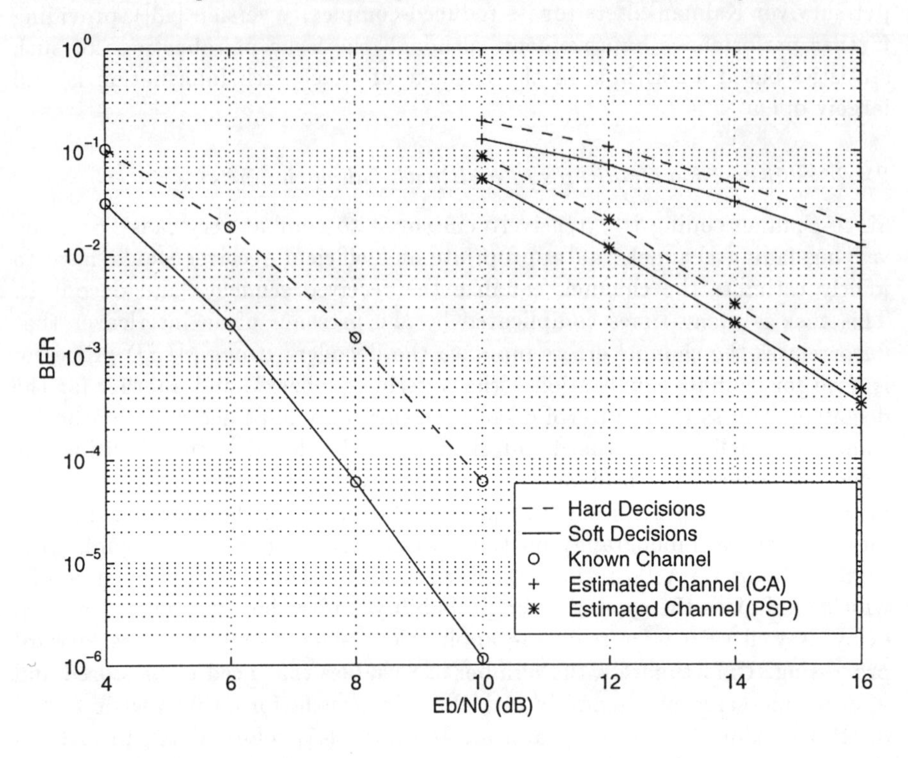


**Figure 2.4** Performance of the decoded CA-MLSD, CA-SSA, PSP-MLSD and PSP-SSA algorithms for a 4-tap channel

In terms of the comparative performance of PSP versus CA soft-decision algorithms, a coded system was configured as follows [49]: A zero-terminated sequence of 1710 bits is encoded by a rate-1/2 convolutional encoder resulting in 1710 quaternary symbols, which are then interleaved using a 57 × 30 block interleaver. The interleaved sequence is QPSK modulated, formatted into 30 bursts together with a training sequence of 26 symbols, and sent over an 4-tap Rayleigh fading channel with Doppler spread $f_d$. The transmission scheme and signal format is similar to GSM, except for the simpler interleaving and modulation (QPSK instead of GMSK). The receiver consists of the equalizer (CA-MLSD, PSP-MLSD, CA-SSA or PSP-SSA), the de-interleaver and the decoder for the convolutional code. The RLS algorithm was used and the forgetting factor $\mu$ was optimized for each channel. The results in Figure 2.4 arc for $f_dT = 0.005$, for which there exists a 3-4 dB gain of PSP versus CA .

### *2.4.3 CPM with various channel phase and timing impairments*

CPM is a popular power- and bandwidth-efficient scheme [51] which generates constant-envelope waveforms and is therefore particularly useful in wireless channels with nonlinear amplifiers. Applications of CPM are, however, still encumbered by implementation complexity and synchronization problems, as amply described in [29]: Complexity arises from the large number of states (trellis size) needed to describe the CPM signal and search over [52], but it is now known that only a small part of the trellis needs attention, leading to either a "reduced state" decoder [53] (part of the trellis is deleted and the algorithm searches over the rest) or to a "reduced-search" decoder [54].

Synchronization in CPM is challenging due to the smoothness of the resulting waveforms. All three classes of proposed techniques (non-linear tone-regenerating circuits [55], joint maximum likelihood estimation via a large super-trellis of data sequence and carrier phase modeled as a discrete random walk [56, 57], decision-directed phase lock loops [58, 59]) perform well, provided that the carrier phase varies very slowly in time. In practice this is not always true and substantial performance degradation may occur with respect to the jitter-free case.

Again, the PSP technique can be employed to jointly estimate these parameters and decode the CPM signal. Results indicate that PSP-based demodulation can recover a significant share of the aforementioned loss, due primarily to the fact that PSP provides the decoder with high-reliability, zero-delay preliminary decisions (that is, for the survivor closest to the actual transmitted sequence, which is the one that counts). Figure 2.5 below demonstrates the performance advantage of PSP versus CA-MLSD for a Doppler-impaired CPM demodulation task, under the following parameters: normalized Doppler offset $f_dT = 0.01$, a 3-symbol raised cosine pulse and phase unwrapping for the CA-MLSD, meaning that the $d$-lag phase estimates get projected ("unwrapped") to the present phase based on the frequency-offset estimate.

We conclude from Figure 2.5 that significant performance improvement is observed under Doppler shift conditions when the PSP technique is used. As indicated, PSP can approach the coherent reception symbol error probability curve, while the conventional methods fail.

### *2.4.4 TCM with phase noise tracking*

It is well known [60] that the memory embedded in TCM can produce error correction gains inherent in classical forward error correction (FEC) mecha-

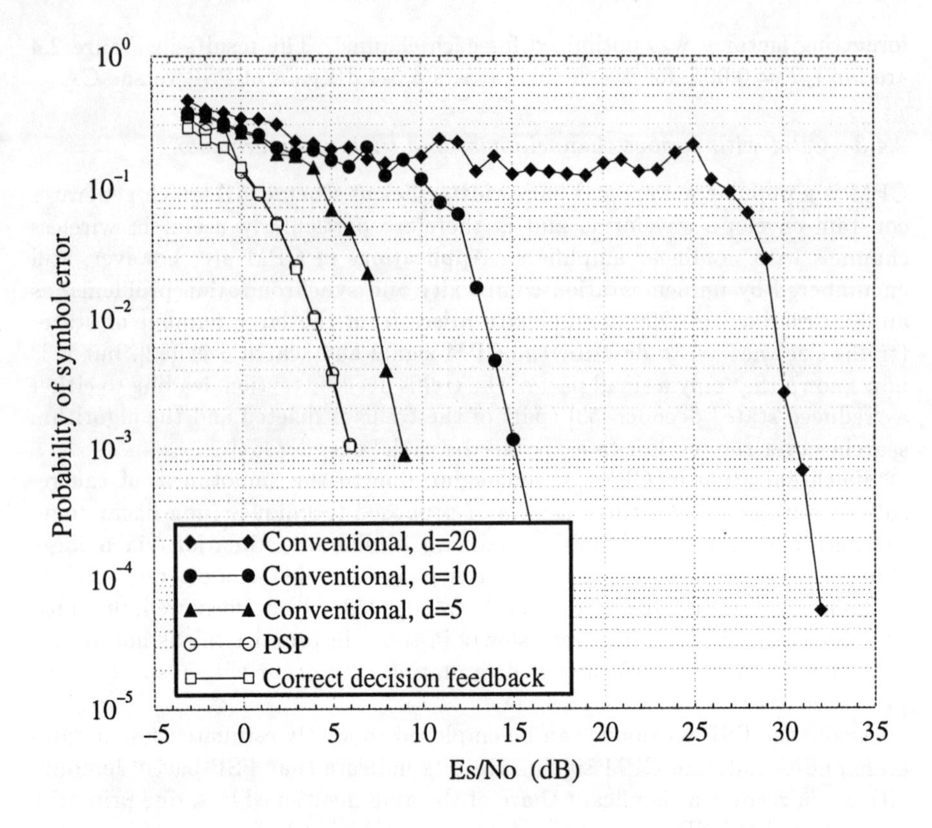


**Figure 2.5** Demodulation of partial response CPM in Doppler offset

nisms for the AWGN channel, but without the attendant bandwidth expansion of the latter case. It has been recently established that TCM, properly combined with interleaving techniques, can provide powerful performance gains in either flat [61] or frequency-selective [62] fading channels. However, a near-perfect knowledge or estimation of the carrier phase is required for the decoder to operate properly, and significant performance degradation is observed in the "partially coherent" case where the phase estimation is noisy [63, 64]. Counteracting such phase-noise-induced impairments is a major challenge, especially for lower-rate applications where the bit-to-bit phase change due to the longer bit duration may result in an unacceptable BER floor. This again raises the option of using PSP in such a TCM system, as demonstrated in Figure 2.6 below, for the particular environment of a Wiener phase noise process, where the

Gaussian increment variance = 0.00125 rad$^2$. Note that the tentative decision delay choice $d = 1$ is the best possible for CA-MLSD, in view of the one-step optimal prediction of a discrete-time, Wiener-increment phase process.

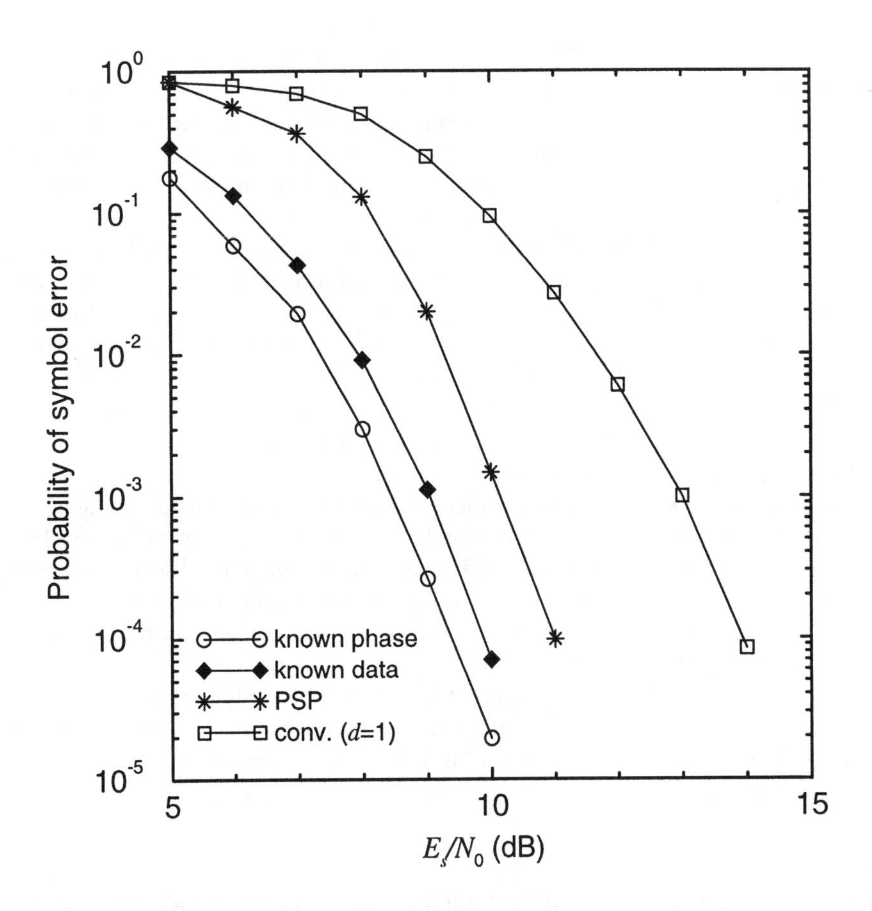


**Figure 2.6** PSP versus conventional MLSD for 8PSK-TCM with phase noise

An SNR performance gain of 3 dB can be discerned between PSP and CA-MLSD,which is significant in view of the fact that phase noise (produced from oscillators and other electronic parts) may be the limiting factor in such links.

### 2.4.5 Array processing against multipath

When demodulating data in a system equipped with antenna diversity, such as a cellular base station with multiple antenna elements arranged in a proper geometry (not necessarily linear), the modeling considerations are as in the above-mentioned applications, but in addition now there is the novel feature of inter-element correlation. That can range from very high (leading to a model of phase-shifted copies of the same waveform being received in independent noise per sensor, the basis of interferometric methods) to very low (as modeled in classic diversity combining of statistically independent multiple paths), depending on the actual physical environment and the nature of the reflected multipath rays. Accordingly, two distinct approaches to array-based MLSD can be developed [65, 66, 67]: (i) the physical channel approach (which we may call $(\mathbf{c},\theta)$–based), and (ii) the overall channel approach (the $\mathbf{H}$–based). In the former, the particular gain $c_i$ and angle-of-arrival $\theta_i$ for each multipath signal is estimated, whereas in the latter the overall effect of the transmit and receive filtering and the physical channel is estimated as a composite parameter. The method eventually selected should be robust to various mismatched or unexpected channel conditions (e.g., Doppler spectrum, correlation between sensor measurements, co-channel interference, etc.).

In addition to this modeling choice, the receiver must decide on how this parameter estimation is to be combined with the detection of unknown data, namely in a PSP or conventional fashion. A comparison of methods (CA-MLSD versus PSP) is provided in Figure 2.7 below for a four-path channel spread over two data symbols with fractionally-spaced processing, parameterized by the number of antennas, $M$.

The advantage of PSP in terms of a lowered BER floor is again evident, although increasing the number of elements tends to lower the floor for CA-MLSD, hence diminishing the need for PSP (depending, of course, on the desired operating point).

### 2.4.6 Combating CCI in digital TDMA transmission

Current narrowband TDMA digital cellular standards (GSM, DCS1800, DECT, IS-54) are primarily Co-Channel-Interference (CCI) limited, hence the need for elaborate frequency planning and reuse. The ability of a TDMA system to withstand higher levels of such interference will obviously translate to higher packing of cells and an increase in system capacity. Under CCI conditions, the statistical properties of the sum of interference plus additive noise can be quite different from the Gaussian model, especially when there are only a few dominant interferers. It should therefore be expected that significant performance improvement can be achieved by designing a receiver which recognizes and

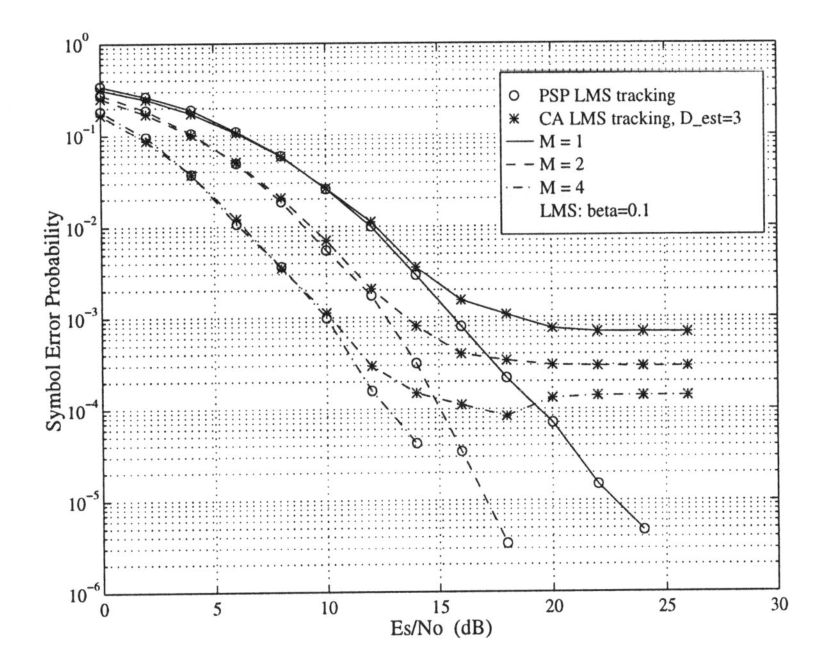


**Figure 2.7** $\mathbf{H}$-PSP vs. $\mathbf{H}$-CA-MLSD

exploits the intrinsic statistical properties of the interference; the fractionally spaced equalizers in [68, 69] demonstrate just that.

As an alternative method for even better performance, the PSP technique can be used either with a single antenna element [50, 70], or with multiple elements. There is a modeling degree of freedom in terms of how the CCI is treated: either its data stream is jointly modeled and estimated along with the intended-user, or it is treated as colored noise and adaptively mitigated by the trellis search for the main user.

To demonstrate the idea, let us assume one dominant interferer (the technique extends to multiple interferers, but at a concomitant increase in complexity) and perform joint, forward, recursive data and channel estimation for both users in a PSP fashion [50]. One typically assumes independent fading channels for the intended and the CCI signal, respectively. Furthermore, four sub-cases must be distinguished regarding the amount of data aiding available for either channel to be acquire. Although the case of data-aided intended-user channel acquisition and blind CCI channel acquisition appears to be the most reasonable for the commercial cellular scenario, one can easily envision applications for the other three. For the example demonstrated in Figure 2.8 below, joint channel acquisition is performed for the intended channel plus the co-channel, but it is data-aided for the intended user (known data stream during an epoch-known preamble or midamble) and blind for the CCI (unknown data that is jointly estimated with the channel) during the same period. Following the preamble period, joint intended-user and CCI data estimation is performed, and tracking of both channels takes place continuously. Modulation is QPSK with a quasi-static, 3-tap delay line Rayleigh fading channel. Comparison is with a DFE-based adaptive receiver structure that nulls out the CCI. No thermal noise is assumed.

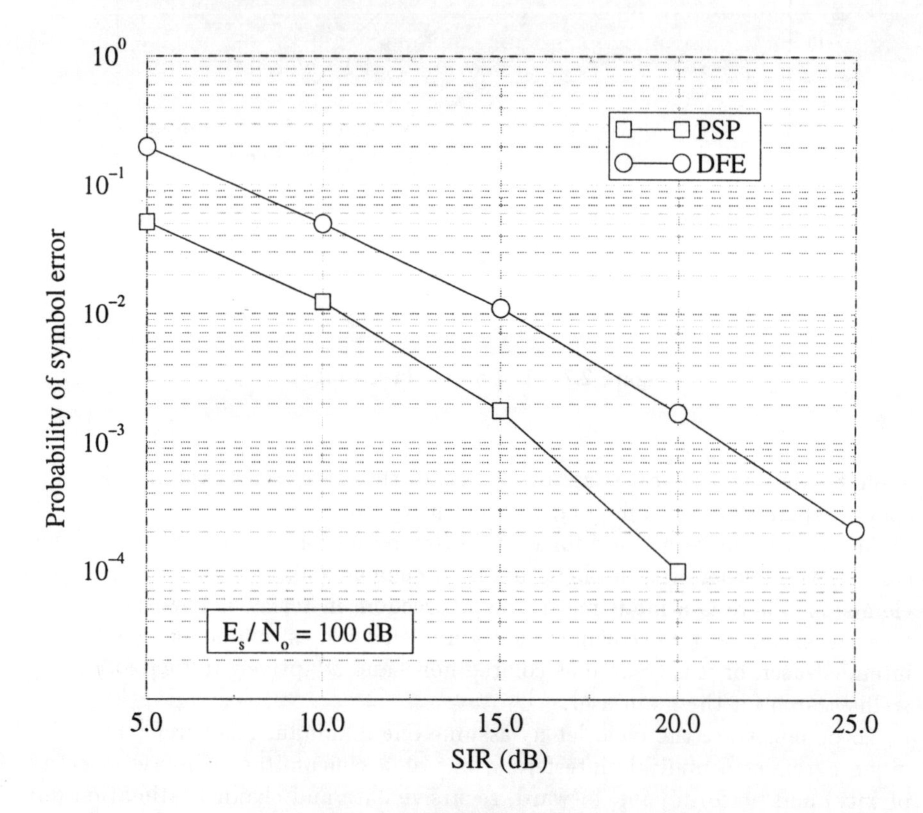


**Figure 2.8** CCI mitigation via PSP and DFE

There is approximately a 2-dB Signal-to-Interference-Ratio (SIR) gain of PSP versus DFE, which could be expected to increase in a time-varying environment. Furthermore, the robustness of a PSP-based array processor to unmodeled (i.e, unexpected) CCI was demonstrated in [66] in a time-varying environment for SIR as low as 10 dB.

### 2.4.7 *Rapid Blind Channel Acquisition*

Most of the applications described above have focused on the *tracking mode*, where the processor has a relatively good initial estimate of the RMI and must simply track any variations in this quantity while detecting subsequent data. The performance characteristics during the purely blind *acquisition mode* are significantly different. This has been demonstrated in [37, 71]. In particular, the case of an unknown ISI signal in AWGN was considered and it was shown that there exist equivalence classes of data sequences. Equivalent sequences are those which have a joint maximum likelihood metric which is the same regardless of the received signal (i.e., the the sequence convolved with its best channel estimate yields a signal which is identical to another sequence convolved with its own best channel estimate). These equivalence classes are the primary reason for misacquisition, which leads to a poor channel estimate and erroneous data detection. When a misacquisition does not occur the BER is quite good and is converted to quite rapidly. The importance of maintaining more paths in the exhaustive tree-search is amplified in the acquisition mode relative to the tracking mode. This is illustrated in Figure 2.9 where a BPSK signal was distorted by an ISI channel with taps $\mathbf{f} = (1\ 2\ 1)$ and AWGN. The curves in Figure 2.9 show the BER plotted against the time after starting with an all zero channel estimate for various RLS-PSP-based algorithms. The parameter $L_t$ is the effective ISI length assumed for the purposes of trellis construction (e.g., the trellis has $2^{L_t}$ states). It can be seen that the additional complexity yields very rapid acquisition to the lower limit of the known channel MLSD performance.

The example illustrated in Figure 2.9 assumes a purely blind (possibly non-cooperative) signaling environment. This may be a valid assumption for broadcast channels or for some military applications. However, for many commercial applications one has control of the signaling. Traditionally, a known training sequence is inserted into a data packet to allow for channel identification and a graceful transition to the tracking mode. This training affects the system throughput directly since no information is conveyed. The notion of Coded Training and Modulation (CTM), which was introduced in [72], is to use joint channel and data estimation algorithms in conjunction with some sequence purging in order to provide data throughput and channel identification simultaneously. The basic idea is to select a group of sequences such that the misacquisition probability is minimized and the data throughput is maximized.

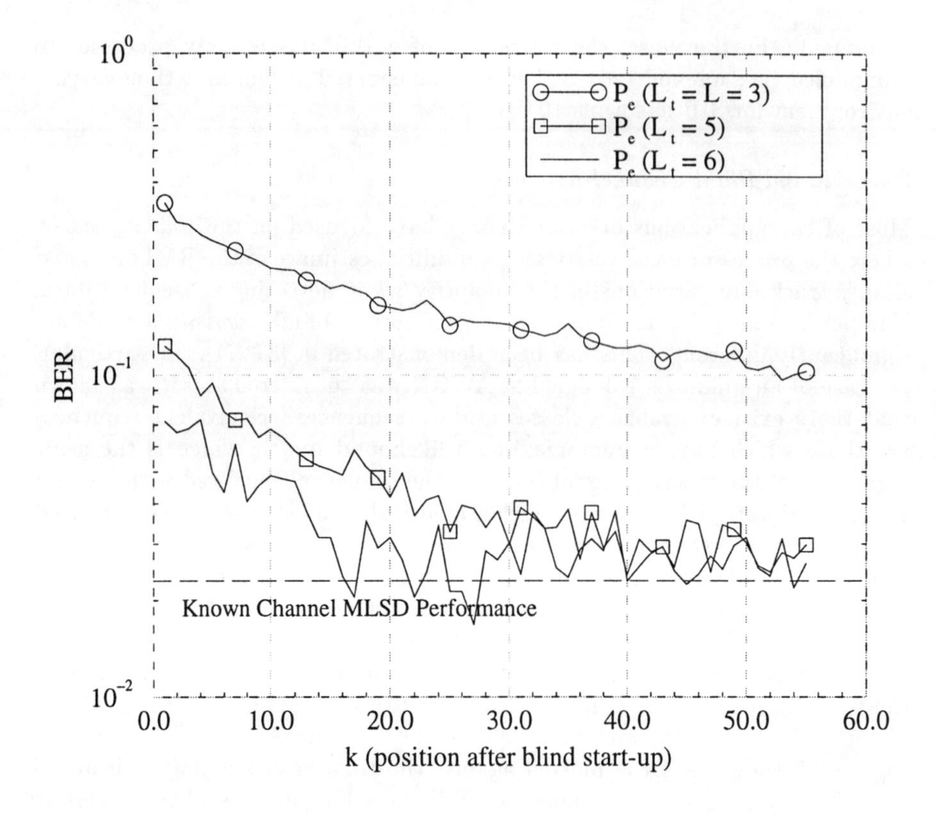


**Figure 2.9** The performance of multiple state PSP with various trellis sizes and SNR = 10 dB.

This selection process, which is essentially equivalent to designing a high-rate, non-Euclidean distance, coding and modulation scheme, is the topic of ongoing research.

## 2.5 CONCLUSIONS

This paper has discussed in some depth the notion of Per-Survivor Processing. The main message to be conveyed is that PSP provides an intuitively-satisfying and theoretically-justifiable approach to approximating likelihood computations (e.g., MLSD) when the complexity of the prevailing environment or of the required optimal processing prohibits exact likelihood processing. A large variety of applications have been demonstrated in the open literature and were summarized in this paper.

Perhaps the most promising applications of PSP in the future are those in which the signaling environment is either extremely noncooperative, or its structure is too complex to model precisely. In other words, the increased robustness associated with the PSP approach may prove to be its greatest asset. Examples of these types of applications are reduced-complexity blind channel estimation (mini-PSP) [40], modulation classification in ISI [41], co-channel interference mitigation for TDMA mobile channels [50], etc. As with any new concept, it is expected that it will require some time before the larger engineering community understands the capabilities of PSP and begins to factor these capabilities into system designs. Some examples along these lines include the possibility of more efficient frequency reuse patterns enabled by PSP for array processing and CCI mitigation, lower overhead for training utilizing CTM concepts, higher data rates via joint PSP-based, adaptive diversity combining, and, eventually, cheaper transceivers using components taking advantage of the robustness of PSP. Also, we emphasize that PSP is a general concept, much as likelihood detection itself, and may eventually find widest application in nonobvious arenas (e.g., twist-pair or cable modems, storage channels, etc.). Algorithm extraction, performance analysis, complexity evaluation, and implementation architectures are the obvious PSP research fronts in all of the aforementioned applications.

An overall conclusion from our common experience thus far is that the class of PSP decoding algorithms can achieve significant performance improvement over conventional techniques whenever the prevailing dynamics or overall uncertainty is significant, at a complexity level that is higher than that of conventional algorithms, but not dramatically so.

*Acknowledgment*

The authors would like to thank Jim Mulligan and Mike Dillon for their support and encouragement and the following current and former USC Ph.D. students who have contributed significantly to our understanding of PSP: Dr. Ching-Kae Tzou, Dr. Norman Lay, Achilleas Anastasopoulos, and Gent Paparisto.

## Notes

1. "estimate", "acquire", "track" are all related notions.

2. Other goals are occasionally desirable, pertaining to the quality of the estimates of the parameters themselves such as, for instance, angle-of-arrival estimation in a multisensor system, Doppler estimation for velocity estimation, etc.

3. Fitting examples are (a) choosing a winning Maximum Likelihood (ML) data sequence out of a finite total set of candidates within a given observation period, (b) classify a digital signal within a given menu, (c) detect a digitally emitting target in noise, etc.

4. This is not to say that such tools may not be brought to bear in the *implementation* of the resultant LF algorithms.

5. For such channels, propagating the channel estimates in time through Kalman-filter algorithms is familiar, easier to formulate and optimal at the same time.

6. Unknown both in the sense of their actual values in a particular observation segment, as well as in their probabilistic description.

7. "Digits" here stands for bits, encoded data symbols or digitized (quantized) representations of a stochastic process.

8. Symbol-by-symbol is meant here either in the optimal MAP sense or in a simple memoryless way, assuming that the modulation permits it.

9. The external PE block represents in a generic way the collection of all required external estimators; in other words, it may consist of many independent sub-systems.

10. Costas and power loops for phase tracking of suppressed carriers would fall in this category, for instance.

11. "Tentative" means allowing the decisions destined for the PE box to be extracted from the VA at a smaller delay than those destined for the deliverable output, a necessary compromise in rapidly varying environments.

12. See [2] and [3] for lucid explanations and rich bibliography on such structures.

13. Each combination will also induce different levels of required memory and computation, which should be kept in mind when various software/hardware constraints in speed, data throughput, power consumption, silicon area and the like, are factored in.

14. A reasonable explanation for this lack of conceptual cross-pollination might be that the random-process literature deals either with fixed parameters of known probability density, or restricted types of dynamic processes such as Gauss-Markov, for which optimal filtering (Kalman) is well defined and relatively easy to implement (linear in the observation); on the other hand, tracking of general time-varying parameters which are nonlinearly embedded in the observation (the more realistic practical case) lends itself easier to the latter formulation.

15. It is noteworthy that the particular problem involved no digital data sequence!

16. [11, 12, 13] were some of the seminal publications that established this type of adaptive MLSD.

17. The list of references following is not exhaustive due to space limitations.

## References

[1] G. Forney, "MLSE of Digital Sequences in the Presence of Intersymbol Interference," *IEEE Trans. Inform. Th.*,, IT-18,pp.363-378, May 1972.

[2] J. Proakis, *Digital Communications*, 2nd Ed., Mc-Graw Hill, 1989.

[3] S. Benedetto, E. Biglieri and V. Castellani, *Digital Transmission Theory*, Prentice-Hall, 1987.

[4] R. Price, "Optimum Detection of Random Signals in Noise, with Application to Scatter-Multipath Communication, I," *IRE Trans. Inform. Th.*, IT-2, pp.125-135, Dec. 1956.

[5] T. Kailath, "Correlation Detection of Signals Perturbed by a Random Channel," *IRE Trans. Inform. Th.*, IT-6, pp.361-366, June 1960.

[6] T. Kailath, "Optimum Receivers for Randomly Varying Channels," *Proc. 4th Symposium Inform. Th.*, Butterworth Scientific Press, London, pp.109-122, 1961.

[7] T. Kailath, "A General Likelihood-Ratio Formula for Random Signals in Gaussian Noise," *IEEE Trans. Inform. Th.*, IT-15, pp.350-361, May 1969.

[8] C. Cahn, "Phase Tracking and Demodulation with Delay," *IEEE Trans. Inform. Th.*, IT-20, pp.50-58, Jan. 1974.

[9] A. Viterbi, "Error Bounds for Convolutional Codes and an Asymptotically Optimum Decoding Algorithm,"*IEEE Trans. Inform. Th.*, IT-13, pp.259-260, Apr. 1967.

[10] G. Forney, "The Viterbi Algorithm," *Proc. of IEEE,* vol. 61, pp.268-278, March 1973.

[11] H. Kobayashi, "Simultaneous Adaptive Estimation and Decision Algorithm for Carrier Modulated Data Transmission Systems," *IEEE Trans. Commun.*, COM-19, pp.268-280, June 1971.

[12] F. Magee and J. Proakis, "Adaptive MLSE for Digital Signaling in the Presence of Intersymbol Interference," *IEEE Trans. Inform. Th.*, IT-19, pp.120-124, Jan. 1973.

[13] G. Ungerboeck, "Adaptive ML Receiver for Carrier-Modulated Data-Transmission Systems," *IEEE Trans. Commun.*, COM-22, pp.624-636, May 1974.

[14] A. Polydoros, "MLSE in the Presence of Infinite ISI," *Masters Thesis,* SUNY Buffalo, Dec. 1978; also: A. Polydoros and D. Kazakos, "MLSE in the Presence of Infinite ISI," *Proc. ICC*, Boston, MA, pp. 25.2.1-25.2.5, June 1979.

[15] A. Duel-Hallen and C. Heegard, "Delayed Decision Feedback Estimation," *IEEE Trans. Commun.*, COM-37, pp.428-436, May 1989.

[16] M. Eyuboğlu and S. Qureshi, "Reduced- State Sequence Estimation with Set Partitioning and Decision Feedback," *IEEE Trans. Commun.*, COM-36, pp.13-20, Jan. 1988.

[17] P. Chevillat and E. Eleftheriou, "Decoding of Trellis-Encoded Signals in the Presence of ISI and Noise," *IEEE Trans. Commun.*, COM-36, pp.669-676, July 1989.

[18] A. Polydoros and R. Raheli, "The Principle of PSP: A General Approach to Approximate and Adaptive MLSE," *Communication Sciences Institue, University of Southern California,* Report no. CSI-90-07-05, July 1990.

[19] N. Seshadri, "Joint Data and Channel Equalization Using Fast Blind Trellis Search Techniques," *Proc. Globecom*, pp.1659-1663, Dec.1990; also:*IEEE Trans. Commun.*, COM-42, pp.1000-1016, No.2/3/4 1994.

[20] M. Ghosh, "An Optimal Approach to Blind Equalization," *Ph.D. Thesis, USC*, Dec. 1991.

[21] R. Iltis, "A Bayesian MLSE Algorithm for a priori Unknown Channels and Symbol Timing," *IEEE Journ. Sel. Areas Commun.*, JSAC-10, pp.579-588, April 1992.

[22] R. E. Morley, Jr. and D. L. Snyder, "Maximum Likelihood Sequence Estimation for Randomly Dispersive Channels," *IEEE Trans. Commun.*, vol. COM-27, No. 6, pp. 833-839, June 1979.

[23] J. Lodge and M. Moher, "ML Estimation of CPM Signals Transmitted over Rayleigh Flat Fading Channels," *IEEE Trans. Commun.*, COM-38, pp.787-794, June 1990.

[24] X. Yu and S. Pasupathy, "Innovations-Based MLSE for Rayleigh Fading Channels," *IEEE Trans. Commun.*, COM-43, pp.1534-1544, No. 2/3/4 1995.

[25] A. Reichman and R. Scholtz, "Joint Phase Estimation and Data Decoding for TCM Systems," *Proc. FISCTA*, Scottland, Sept. 1991.

[26] J. Lin, F. Ling and J. Proakis, "Joint Data and Channel Estimation for TDMA Mobile Channels," *Proc. PIMRC92*, pp.235-239, Oct.1992; also: *Intern. Journal Wireless Inform. Networks,* Vol. 1, 1994.

[27] Z. Xie, C. Rushforth, R. Short and T. Moon,"Joint Signal Detection and Parameter Estimation in Multiuser Communications," *IEEE Trans. Commun.*, COM-41, pp.1208-1216, Aug. 1993.

[28] C-K. Tzou, R. Raheli and A. Polydoros, "Applications of PSP to Mobile Digital Communications," *Proc. Globecom,* Dec. 1993.

[29] A. D'Andrea, U. Mengali and G. Vitetta, "Multiple Phase Synchronizartion in Continuous Phase Modulation," *Digital Signal Processing: A Review Journal,* A. Polydoros, Ed.,Vol. 3, pp. 188-198, July 1993.

[30] A. D'Andrea, U. Mengali and G. Vitetta, "Approximate ML Decoding of Coded PSK with No Explicit Carrier Phase Reference," *IEEE Trans. Commun.*, COM-42, pp.1033-1040, No. 2/3/4 1994.

[31] H. Kubo, K. Murakami and T. Fujino, "An Adaptive MLSE for Fast Time-Varying ISI Channels," *IEEE Trans. Commun.*, pp.1872-1880, No. 2/3/4 1994.

[32] R. Raheli, A. Polydoros and C-K. Tzou, "PSP: A General Approach to MLSE in Uncertain Environments," *IEEE Trans. Commun.*, COM-43, pp.354-364, No. 2/3/4 1995.

[33] S. Fechtel, "Equalization for Coded Modulation on Fading Channels," *Proc. IEEE Benelux Workshop*, Belgium, May 1995.

[34] K. Hamied and G. Stuber, "An Adaptive Truncated MLSE Receiver for Japanese Personal Digital Cellular," *IEEE Trans. Veh. Techn.*, VT-45, pp.41-50, Feb. 1996.

[35] R. Raheli, G. Marino and P. Castoldi, "PSP and Tentative Decisions: What is in Between?," *IEEE Trans. Commun.*, COM-44, pp.127-129, Feb. 1996.

[36] G. Vitetta, D. Taylor and U. Mengali, "Double Filtering Receivers for PSK Signals Transmitted Over Rayleigh Frequency-Flat Fading Channels," *IEEE Trans. Commun.*, COM-44, pp.686-695, June 1996.

[37] K. M. Chugg and A. Polydoros, "On the Existence and Uniqueness of Joint Channel and Data Estimates," *IEEE IT Symposium*, Sept. 1995.

[38] K. M. Chugg and A. Polydoros,"MLSE for an Unknown Channel-Part I: Optimality Considerations," *IEEE Trans. Commun.*, COM-44, pp.836-846, July 1996

[39] K. M. Chugg and A. Polydoros, "MLSE for an Unknown Channel – Part II: Tracking Performance," *IEEE Trans. Commun.*, vol. 44, pp. 949-958, August 1996.

[40] A. Polydoros and N. Lay, "Channel Estimation and Blind Equalization Using Minimum State PSP," *Proc.MILCOM'95*, pp.993-997, Nov. 1995.

[41] N. Lay and A. Polydoros, "Modulation Classification of Signals in Unknown ISI Environments" *Proc.MILCOM'95*, pp.170-174, Nov. 1995.

[42] G. Castellini, F. Conti, E. DelRe and L. Pierucci, "A Continuously Adaptive MLSE Receiver for Mobile Communications: Algorithm and Performance",*IEEE Trans. Commun.*, COM-45, pp.80-88, January 1997.

[43] M. Rollins and S. Simmons,"Per-Survivor Kalman Processing in Fast Frequency-Selective Channels", *Proc. ICC'94*, New Orleans, May 1994; also, to appear in *IEEE Trans. Commun.*, 1997.

[44] J. Seymour and M. Fitz, "Near-Optimal Symbol-by-Symbol Detection Schemes for Flat Rayleigh Fading", *IEEE Trans. Commun.*, COM-43, pp.1525-1533, No. 2/3/4 1995.

[45] U. Hanson and T. Aulin, "Theoretical Performance Evaluation of Different Soft-Decision Algorithms," *Proc. of ISITA '94,* pp. 875-880, Nov. 1994, Sydney, Australia.

[46] K. Abend and B. D. Fritchman, "Statistical Detection for Communication Channels with Intersymbol Interference," *Proc. IEEE,* vol. 58, No. 5, May 1970, pp. 779-785.

[47] Y. Li, B. Vucetic and Y. Sato, "Optimum Soft-Output Detection for Channels with Intersymbol Interference," *IEEE Trans. Inf. Theory,* vol. IT-41, No. 3, pp. 704-713, May 1995.

[48] A. Anastasopoulos and A. Polydoros, "Soft-Decisions Per-Survivor Processing for Mobile Fading Channels," *Proc. VTC'97*, Phoenix, AZ, May 4-7, 1997.

[49] A. Anastasopoulos and A. Polydoros, "Adaptive Soft-Decision Algorithms for Mobile Fading Channels," submitted to *European Trans. Telecomm.*, April 1997.

[50] A. Polydoros and G. Paparisto, "PSP for Joint Data and Channel Estimation in Multipath Fading and Co Channel Interference Channels,"*Proc.MILCOM'95 (Classified)*, Nov. 1995.

[51] J.B. Anderson and C-E.W. Sundberg, "Advances in Constant Envelope Coded Modulation", *IEEE Commun. Mag.*, vol. 29, no. 12, Dec. 1991, pp. 36-45.

[52] J.B. Anderson, T. Aulin, and C-E.W. Sundberg, *Digital Phase Modulation*, Plenum Press, New York, 1986.

[53] A. Svensson,"Reduced State Sequence Detection of Partial Response Continuous Phase Modulation", *IEE Proc., I*, 138, 4, pp. 256-268.

[54] S.J. Simmons and P.H. Wittke, "Low Complexity Decoders for Constant Envelope Digital Modulations", *IEEE Trans. Commun.*, pp. 1273-1280, Dec. 1983.

[55] A.N. D'Andrea, U. Mengali and R. Reggiannini, "Carrier Phase and Clock Recovery for Continuous Phase Modulated Signals", *IEEE Trans. Commun.*, pp. 1095-1101 , Oct. 1987.

[56] S.J. Simmons and P.J. McLane, "Low-complexity Carrier Tracking Decoders for Continuous Phase Modulations", *IEEE Trans. Commun.*, pp. 1285-1290, Dec. 1985.

[57] J.M. Liebetreu,"Joint Carrier Phase Estimation and Data Detection Algorithms for Multi-h CPM Data Transmission", *IEEE Trans. Commun.*, pp. 873-881, Sept. 1986.

[58] A-N. Premji and D.P. Taylor, "Receiver Structures for Multi-h Signalling Formats", *IEEE Trans. Commun.*, vol. COM-35, pp. 439-451, Apr. 1987.

[59] A.J. MacDonald and J.B. Anderson, "PLL Synchronization for Coded Modulation',' *Conf. Rec. ICC ' 91*, pp. 52.6.1-52.6.5., June 1991.

[60] G. Ungerboeck, "Channel Coding with Multilevel/Phase Signals", *IEEE Trans. on Inform. Theory*, vol. IT-28, pp. 55-67, Jan. 1982.

[61] E. Biglieri, D. Divsalar, P.J. McLane and M.K. Simon, *Introduction to Trellis-Coded Modulation with Applications*, MacMillan, New York, 1991.

[62] A. Anastasopoulos and K. M. Chugg, "TCM for Frequency-Selective, Interleaved Fading Channels Using Joint Diversity Combining," *submitted to CTMC, Globecom'97.*

[63] E. Zehavi and G. Kaplan, "Phase Noise Effects on M-ary PSK Trellis Codes", *IEEE Trans. Commun.*, vol. COM-39, , March 1991.

[64] C.-K. Tzou, "Per-Survivor Processing: A General Approach to MLSE in Uncertain Environments," *Ph.D. Dissertation, University of Southern California,* December 1993.

[65] G. Paparisto, K. M. Chugg, N. E. Lay and A. Polydoros, "A PSP Antenna Array Receiver for Joint Angle, Multipath and Data Estimation," submitted to Globecom'97.

[66] G. Paparisto and K. M. Chugg, "PSP Array Processing for Multipath Fading Channels", submitted to *IEEE Trans.Commun.*, June 1997.

[67] S. Diggavi and A. Paulraj, "Signal Detection for Time-Varying Vector Channels", *29th Asilomar Conference on Signals, Systems and Computers*, pp. 152-156, Pacific Grove, Oct.30 - Nov.2, 1995.

[68] C.L. Despins, D.D. Falconer and S.A. Mahmoud, "Compound Strategies of Coding, Equalization and Space Diversity for Wide-band TDMA Indoor Wireless Channels", *IEEE Trans. Vehic. Tech.*, vol. VT-41, no. 4, Nov. 1992.

[69] W.A. Gardner and S. Venkataraman, "Performance of Optimum and Adaptive Frequency-shift Filters for Co-channel Interference and Fading ", *Proc. 24th Asilomar Conf. Signals, Syst., Comput.*, Pacific Grove, CA, Nov.1990.

[70] J. Hamkins, E. Satorius, G. Paparisto and A. Polydoros, "A Comparative Study of Co-Channel Interference Suppression Techniques," Proceedings of IMSC'97, Pasadena, CA, June 1997.

[71] K. M. Chugg, "Acquisition Performance of Blind Sequence Detectors Using Per-Survivor Processing," *Proc. Veh. Tech. Conf. 1997,* Phoenix, AZ, May 1997, pp. 539-543.

[72] K. M. Chugg, "PSP: An Important Component of Integrated Receiver Processing," *26-th Annual IEEE Communication Theory Workshop 1997,* Tucson, AZ, April 1997.

[73] K. M. Chugg, "The Suboptimality of the Viterbi Algorithm for Random Channels," (in preparation).

## Appendix: Likelihood development and PSP details

The description of PSP is formalized in this appendix, first by setting up the general concepts of likelihood computation, then by applying these concepts to a few simple, specific examples.

Most of the applications of interest fit into the general (complex baseband) model of

$$r(t) = y(t; \mathbf{a}) + n(t) \tag{A.1}$$

where $n(t)$ is the complex baseband equivalent of AWGN and $\mathbf{a}$ is the underlying digital sequence to be detected. We use boldface to denote a vector, with the convention that

$$\mathbf{a}_{k_1}^{k_2} = \begin{bmatrix} a_{k_2} & a_{k_2-1} & \cdots & a_{k_1} \end{bmatrix}^T \tag{A.2}$$

where $(\cdot)^T$ represents the transpose operation and the absence of the subscript and superscript implies the entire sequence. The actual structure of $\mathbf{a}$ and the mapping to $y(\cdot)$ may be quite general. For example, $\mathbf{a}$ could represent a digitized version of an analog quantity and $y(\cdot)$ may contain the effects of interference or random distortions. We assume this continuous-time signal is converted to a discrete-time, symbol-spaced signal of the form

$$z_k = x_k(\mathbf{a}) + w_k \qquad k = 0, 1, \ldots K-1 \tag{A.3}$$

where $w_k$ is discrete time circular complex AWGN with variance $\sigma^2$. The details of this conversion (i.e., the receiver front-end processing) are not discussed in detail here. However, we note in passing that the signals in (A.3) may represent vectors as the result of oversampling, which in most cases can be justified as approximately information loseless [21, 38]. We also make the assumption that $x_k(\mathbf{a}) = x_k(\mathbf{a}_0^k)$ – i.e., that the mapping from the data $\mathbf{a}$ to noise-free signal $\mathbf{x}$ is causal. The data sequence is assumed to be independent with each element uniformly distributed over an $M$–ary alphabet $\mathcal{A}$.

Virtually all likelihood computations of interest involve computation of the conditional density $f(\mathbf{z}|\tilde{\mathbf{a}})$, where $\tilde{\mathbf{a}}$ denotes a hypothesized or conditional realization of the transmitted data sequence $\mathbf{a}$. Obviously, this is the basis for MLSD. Similar terms are also computed and averaged over in order to compute the a-posteriori probabilities (APPs) $p(\tilde{a}_k|\mathbf{z})$, although we do not describe this in detail. Thus, it is desired to compute this conditional density recursively at the symbol rate. This can be accomplished via Bayes' rule

$$\begin{aligned} f(\mathbf{z}_0^k|\tilde{\mathbf{a}}_0^k) &= f(z_k|\mathbf{z}_0^{k-1}, \tilde{\mathbf{a}}_0^k) f(\mathbf{z}_0^{k-1}|\tilde{\mathbf{a}}_0^k) && \text{(A.4)} \\ &= \prod_{i=1}^{k} f(z_i|\mathbf{z}_0^{i-1}, \tilde{\mathbf{a}}_0^i), && \text{(A.5)} \end{aligned}$$

where the causality assumption has been used in the last equality.

Consider performing MLSD based on (A.5). If it is possible to compute $f(z_i|\mathbf{z}_0^{i-1}, \tilde{\mathbf{a}}_0^i)$ recursively from $f(z_{i-1}|\mathbf{z}_0^{i-2}, \tilde{\mathbf{a}}_0^{i-1})$, then one may view the MLSD problem as a general $M$-ary tree-search problem. Unfortunately, the complexity of this search grows exponentially with the data record $k$, limiting its utility for even short data packets. However, if there exists a finite integer value $L$ with the property that[1]

$$f(z_i|\mathbf{z}_0^{i-1}, \tilde{\mathbf{a}}_0^i) = f(z_i|\mathbf{z}_0^{i-1}, \tilde{\mathbf{a}}_{i-(L-1)}^i) \tag{A.6}$$

then this exhaustive tree-search folds into an $M^{L-1}$-state trellis search with complexity growing only linearly in the data record. This is a familiar concept when one starts with an finite-state Markov process observed in AWGN [10]. The subtle distinction here being that we define $L$ and hence the state structure by the statistical properties of the observation as they relate to conditioning on the data (see also [22]). In fact we define the *augmented state* (state transition) at time $i$ as $\mathbf{a}_{i-(L-1)}^i$ associated with the minimum value of $L$ satisfying the property in (A.6). With this condition satisfied, it is clear that the VA finds the MLSD solution via a trellis search.

In some cases the value of $L$ required to fold the tree into a trellis without loss of optimality is too large to be considered practical for implementation. Alternatively, there may be no such value for $L$ – i.e., for many applications the exhaustive search is required for optimality. In such cases, one may force a folding on the tree with a trellis based on a memory of $L_t$. In this case, the conditional value of $\tilde{\mathbf{a}}_0^{i-L_t}$ is not available at time $i$ (i.e., not all hypotheses are retained). In PSP this is handled by summarizing the relevant information in $\tilde{\mathbf{a}}_0^{i-L_t}$ in the RMI, denoted by $\Theta(\check{\mathbf{a}}_0^{i-L_t}; \tilde{\mathbf{a}}_{i-(L_t-1)}^i)$. This notation is intended to convey the fact that all possible values of $\tilde{\mathbf{a}}_{i-(L_t-1)}^i$ are considered, but that for each of these hypothesized state transitions, only one[2] *survivor sequence* $\check{\mathbf{a}}_0^{i-L_t}$ is maintained. This is clarified via some well-known applications.

### *A-.0.1 Example: MLSD for a known ISI channel*

In the well-known case of finite ISI

$$x_k(\mathbf{a}) = x_k(\mathbf{a}_{k-(L-1)}^k) = \sum_{m=0}^{L_{MA}-1} f_m a_{k-m}, \tag{A.7}$$

where $\{f_m\}$ are the known equivalent channel coefficients, the condition in (A.6) holds via

$$f(z_i|\mathbf{z}_0^{i-1}, \tilde{\mathbf{a}}_0^i) = f(z_i|\tilde{\mathbf{a}}_{i-(L_{MA}-1)}^i) = \mathcal{N}(z_i; x_i(\tilde{\mathbf{a}}_{i-(L_{MA}-1)}^i); \sigma^2). \tag{A.8}$$

Here, the Gaussian noise density function with mean $m$ and variance $\sigma^2$ is denoted by $f(w;m;\sigma^2)$. Thus, conditioned on the state transition, the observation is independent of all other observations. This case was described in detail in [10]. Notice that there is no RMI required in this case.

The number of states in the VA-based processor is $M^{L_{MA}-1}$. If this is too large, a reduced state approximation may be used. A simple approach was described in [15] in which a trellis based on $L_t < L_{MA}$ is used to force the folding. In this case the value of $x_i(\tilde{\mathbf{a}}^i_{i-(L_{MA}-1)})$ in (A.8) is approximated by

$$x_i(\tilde{\mathbf{a}}^i_{i-(L_t-1)};\Theta(\breve{\mathbf{a}}_0^{i-L_t})) = \sum_{m=0}^{L_t-1} f_m\tilde{a}_{k-m} + \sum_{m=L_t}^{L_{MA}-1} f_m\breve{a}_{k-m}. \tag{A.9}$$

In this case of reduced state approximate MLSD, the RMI is $\breve{\mathbf{a}}^{i-L_t}_{i-(L_{MA}-1)}$.

In the case where the channel impulse response is extremely long (e.g., high data-rate twisted pair subscriber lines), one may choose to model the channel as an autoregressive/moving-average channel (ARMA)

$$x_k(\mathbf{a}_0^k) = \sum_{m=0}^{L_{MA}-1} f_m a_{k-m} + \sum_{m=1}^{N_{AR}} g_m x_{k-m}(\mathbf{a}_0^{k-m}). \tag{A.10}$$

There is no $L$ which can be defined in the sense of satisfying the folding condition in this case. However, as suggested in [14, 15], we may force a folding of the tree with any value of $L_t$ which suits our complexity and performance requirements. A natural choice is to take $L_t = L_{MA}$, in which case $x_k(\mathbf{a}_0^k)$ is approximated by

$$x_k(\breve{\mathbf{a}}_0^{k-L_{MA}};\tilde{\mathbf{a}}^k_{k-(L_{MA}-1)}) = \sum_{m=0}^{L_{MA}-1} f_m\tilde{a}_{k-m}+\sum_{m=1}^{N_{AR}} g_m x_{k-m}(\breve{\mathbf{a}}_0^{k-L_{MA}};\tilde{\mathbf{a}}^k_{k-(L_{MA}-1)}). \tag{A.11}$$

In words, the AR portion of the noise free-channel output at time $k$ is based on the per-survivor noise free outputs – i.e., the RMI is $x_{k-m}(\breve{\mathbf{a}}_0^{k-L_{MA}};\tilde{\mathbf{a}}^k_{k-(L_{MA}-1)})$ for $m = 1,2\ldots N_{AR}$.

Notice that in the last two examples, PSP is utilized for complexity reduction even when there is no uncertainty regarding the channel impulse response.

### *A-.0.2 Example: Memoryless modulation with random parameters*

This example illustrates the fact that the IDM need not come from come dispersion in the channel or coding/modulation in order for the PSP concept to be applicable. For a simple example consider the flat fading signal given by

$$x_k(\mathbf{a}) = x_k(a_k) = c_k a_k, \tag{A.12}$$

where $c_k$ is a complex Gaussian process. It follows that

$$f(z_i|\mathbf{z}_0^{i-1}, \tilde{\mathbf{a}}_0^i) = \mathcal{N}(z_i; \hat{z}_{i|i-1}(\tilde{\mathbf{a}}_0^i); \hat{\sigma}_{i|i-1}^2(\tilde{\mathbf{a}}_0^i)). \tag{A.13}$$

The conditional mean and variance of $z_i$ given all past observations and current and past data are given via affine MMSE estimates. In general, these conditional second moments can be computed via per-survivor Kalman Filtering (MMSE estimation). It has been noted in [23] that, whenever $z_i$ is an $L$-th order AR process, the required conditional mean and variances depend only on $\mathbf{a}_{i-(L-1)}^i$ via a finite linear predictor. Thus, under this assumption the folding condition in (A.6) is met and the VA provides the optimal processing. We note in passing that if the fading process $c_i$ is AR and the AWGN is in fact present, then $z_i$ is *not* AR and the folding does not occur (i.e., the VA is not not an optimal processor). It has also been claimed that if $c_i$ is a Moving Average process (which is a special case of a finite support correlation process) then the folding condition occurs [22], an issue which is the topic of current research [73].

In any case, the RMI in the computation of (A.13) is the required Kalman Filter recursion parameters, including the conditional mean and variance. In general the folding condition is not met and the forced folding solution is an $M^{L_t-1}$–state PSP-KF processor. The value of $L_t$ is a design choice and may be selected to trade complexity and performance. This concept generalizes directly to the frequency-selective case (i.e., where IDM is present) as described in [21].

# 3 COMMUNICATION OVER MULTIPATH FADING CHANNELS: A TIME-FREQUENCY PERSPECTIVE

Akbar M. Sayeed
Behnaam Aazhang

**Abstract:**
Dynamics of multipath fading have a major effect on the performance of mobile wireless communication systems. The inherently time-varying nature of the mobile wireless channel makes nonstationary signal processing techniques particularly attractive for system design. Time-frequency representations are powerful tools for time-varying signal processing, and in this paper, we present a time-frequency view of wireless communication over multipath channels. Our discussion is anchored on a fundamental finite-dimensional time-frequency representation of the wireless channel that facilitates diversity signaling by exploiting multipath and Doppler shifts. The substantially higher level of diversity afforded by time-frequency processing over conventional techniques translates into significant gains in virtually all aspects of system performance. We illustrate the utility of the time-frequency framework via novel signaling and receiver structures, and multiuser acquisition and interference-suppression algorithms.

## 3.1 INTRODUCTION

The wireless channel has a major affect on the performance of mobile communication systems. The channel is an inherently time-varying system and, for optimal performance, the characteristics of the channel must be incorporated in system design. The time-varying nature of the channel dictates that time-varying signal processing techniques be made to bear on the problem.

Time-frequency representations (TFRs) are powerful tools for the representation and processing of time-varying signals and systems. They are two-dimensional (2D) signal representations jointly parameterized in terms of both time and frequency. As such, TFRs provide a natural framework for representing the time-varying mobile wireless channel, and the optimal signaling and receiver structures for communications over such channels.

In this paper, we present a time-frequency perspective of wireless communications over time-varying multipath channels. Our discussion is primarily in the context of code-division multiple access (CDMA) systems because of their well-known ability to combat multipath fading. Starting with a time-frequency description of the mobile wireless channel, we arrive at a canonical finite-dimensional time-frequency representation of the channel that will serve as the backbone of our treatment.

The canonical time-frequency-based channel representation shows that spread-spectrum signaling over time-varying multipath channels possesses additional degrees of freedom that are not exploited by existing communications systems. Essentially, CDMA systems possess a large time-bandwidth product (TBP) that can be exploited to provide diversity against channel fading. The state-of-the-art RAKE receiver that achieves multipath diversity exploits only the large bandwidth but not the large TBP of CDMA systems. The time-frequency channel representation identifies Doppler as another dimension for diversity, and facilitates the exploitation of joint multipath-Doppler diversity by fully utilizing the available TBP in CDMA systems [1, 2]. Thus, a time-frequency approach to communication over multipath channels has the potential of delivering substantial gains over conventional techniques in virtually all aspects of system performance.

The fundamental time-frequency channel representation can be exploited in a variety of aspects of communication system design and analysis ranging from new signaling and receiver structures to multiuser detection, timing-acquisition and interference-suppression to information-theoretic issues related to multipath fading channels. We illustrate the power of the time-frequency paradigm by focusing on two main themes: 1) novel signaling and receiver structures, and 2) a new approach to multiuser timing-acquisition and interference-suppression that fully incorporates the multipath channel. Our objective is not to give a detailed description of the techniques but to give a flavor for the advantages of the time-frequency approach in the context of the two themes.

The next section briefly introduces TFRs that we use throughout the paper. In Section 3.3, we provide a natural time-frequency-based description of the mobile wireless channel in the context of CDMA systems, and introduce the canonical time-frequency channel representation. Section 3.4 describes the basic TFR-based receiver structures that exploit joint multipath-Doppler diversity. In Section 3.5, we discuss some nontraditional signaling and receiver structures that maximally exploit Doppler diversity. Section 3.6 leverages the time-frequency receiver structure to propose a near-far resistant framework for multiuser timing acquisition and interference-suppression. Some concluding remarks and avenues for future research are discussed in Section 3.7.

## 3.2 TIME-FREQUENCY PRELIMINARIES

TFRs are 2D signal representations in terms of both time and frequency, and are powerful tools for representing time-varying signals and systems. As we will see, TFRs are ideally suited for processing signals transmitted over the inherently time-varying mobile wireless channel.

The concepts of time and frequency shifts play a fundamental role in the theory of TFRs and we will use them throughout the paper. We will denote time and frequency shifts by operators $\mathbf{T}_\tau$ and $\mathbf{F}_\theta$, respectively, defined as

$$(\mathbf{T}_\tau s)(t) \stackrel{def}{=} s(t-\tau) \ , \ (\mathbf{F}_\theta s)(t) \stackrel{def}{=} s(t)e^{j2\pi\theta t} \ ,$$

where the parameter $\tau$ denotes the value of the time-shift, and $\theta$ denotes the value of the frequency-shift introduced in the signal.

The TFR of primary interest to us in this paper is the short-time Fourier transform (STFT) which is defined for a signal $s(t)$ as

$$\mathbf{STFT}_s(\theta,\tau;g) \stackrel{def}{=} \int s(t)g^*(t-\tau)e^{-j2\pi\theta t}dt$$

for a given window function $g(t)$. It is a projection of the signal onto a family of functions $\psi_{(\theta,\tau)}$ that are time- and frequency-shifted versions of $g(t)$:

$$\mathbf{STFT}_s(\theta,\tau;g) \stackrel{def}{=} \langle s, \psi_{(\theta,\tau)} \rangle \tag{3.1}$$

$$\psi_{(\theta,\tau)}(t) \stackrel{def}{=} (\mathbf{F}_\theta \mathbf{T}_\tau g)(t) = g(t-\tau)e^{j2\pi\theta t} \ , \tag{3.2}$$

where $\langle \cdot, \cdot \rangle$ denotes the innerproduct. STFT can also be interpreted as a *time-frequency correlation* function. In fact, for each fixed $(\theta,\tau)$, $\mathbf{STFT}_s(\theta,\tau;g)$ is the matched-correlator output for optimally detecting the time-frequency-shifted signal, $\psi_{(\theta,\tau)} = g(t-\tau)e^{j2\pi\theta t}$, in the presence of additive white Gaussian

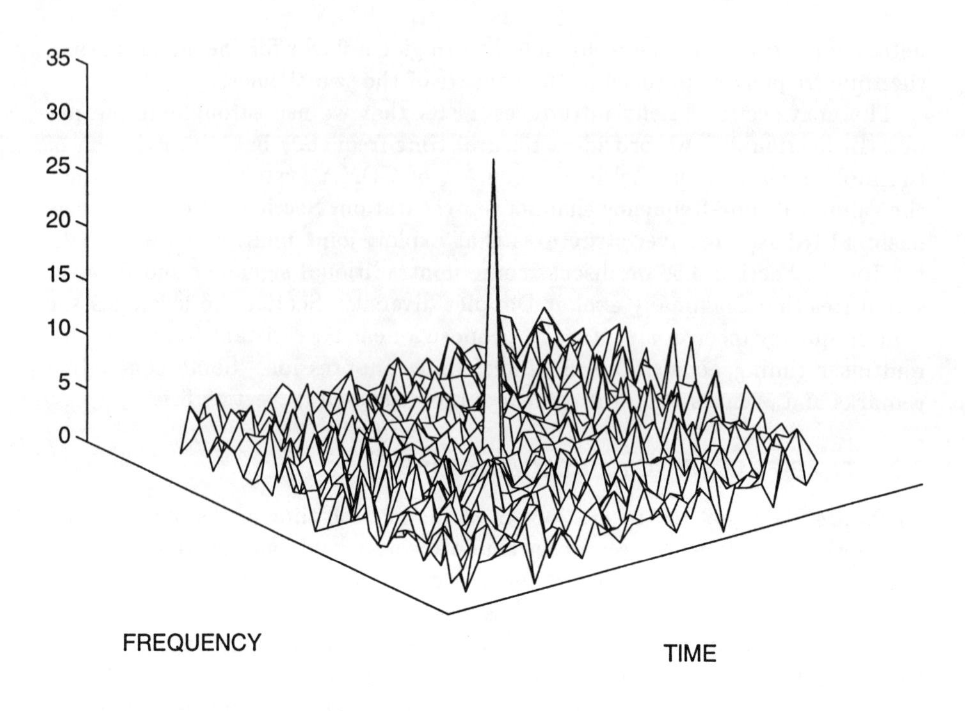


**Figure 3.1** Mobile wireless channel: Linear time-varying system.

noise (AWGN); that is, $s(t) = \psi_{(\theta,\tau)}(t) + w(t)$, where $w(t)$ is AWGN. The STFT is an integral part of the time-frequency receivers presented in this paper.

## 3.3 THE FUNDAMENTAL CHANNEL REPRESENTATION

In this section, we introduce the canonical channel representation that is central to our time-frequency framework for CDMA systems. For clarity of exposition, we present the channel representation in terms of the channel-response to a generic spread-spectrum signal. Different signaling and modulation schemes will be introduced in later sections as needed.

Figure 3.1 shows are schematic of the wireless channel. The baseband signal at the receiver $r(t)$ is given by

$$r(t) = s(t) + w(t)$$

where $w(t)$ is the AWGN with power spectral density $\mathcal{N}_0/2$, and $s(t)$ is channel-transformed version of the transmitted baseband signal $x(t)$

$$s(t) = \int_0^\infty h(t,\tau)x(t-\tau) .$$

We are interested in a representation for the time-varying channel [3], described by the kernel $h(t,\tau)$, that maps $x(t)$ into $s(t)$.

It suffices to consider the unmodulated signal to characterize the effects of the channel. Thus, for simplicity of exposition, we consider $x(t) = q(t)$, where $q(t)$ is a spread-spectrum signal of duration $T$ given by

$$q(t) = \sum_{n=0}^{N-1} a[n]v(t - nT_c) ,$$

where $v(t) = I_{[0,T_c)}(t)$, the chip waveform, $T_c$ is the chip interval, $I_{[a,b)}(t)$ is the indicator function of $[a,b)$, $T = NT_c$, and $a[n]$ is the spreading code (sequence) corresponding to $q(t)$ [4]. The parameter $N \gg 1$ is called the spreading gain of the spread spectrum system. An equivalent representation of the channel which is central to our discussion is in terms of the *spreading function* defined as [3, 4]

$$H(\theta,\tau) \stackrel{def}{=} \int h(t,\tau)e^{-j2\pi\theta t}dt .$$

The corresponding representation of $s(t)$ is

$$s(t) = \iint H(\theta,\tau)(\mathbf{F}_\theta \mathbf{T}_\tau q)(t)d\theta d\tau = \iint H(\theta,\tau)q(t-\tau)e^{j2\pi\theta t}d\theta d\tau ,$$

and thus $\theta$ corresponds to the Doppler shifts (frequency-shifts) introduced by the channel (temporal variations), and $\tau$ corresponds to the multipath delays (time-shifts). The spreading function $H(\theta,\tau)$ quantifies the time-frequency spreading introduced by the channel. Note from (3.1) that if we represent the channel by an operator $\mathbf{H}$, it is composed of a linear combination of time- and frequency-shift operators

$$\mathbf{H} = \iint H(\theta,\tau)\mathbf{F}_\theta \mathbf{T}_\tau d\theta d\tau .$$

It is well-known that an arbitrary time-varying linear system admits such a representation in terms of time and frequency shifts [5, 6].[1]

### 3.3.1 *Statistical Channel Parameters*

The time-variant channel impulse response $h(t, \tau)$ is best modeled as a stochastic process and a realistic model in many situations is the wide-sense stationary uncorrelated scatterer (WSSUS) model [4, 3] in which the temporal variations in $h(t, \tau)$ are represented as a stationary Gaussian process, and the channel responses at different lags (different scatterers) are uncorrelated (independent). Thus, the channel is characterized by second-order statistics which are given by[2]

$$\mathrm{E}\{H(\theta_1, \tau_1)H^*(\theta_2, \tau_2)\} = \Psi(\theta_1, \tau_1)\delta(\theta_1 - \theta_2)\delta(\tau_1 - \tau_2) \ ,$$

where

$$\Psi(\theta, \tau) \stackrel{def}{=} \mathrm{E}\left\{|H(\theta, \tau)|^2\right\} \ ,$$

and $\delta(t)$ denotes the Dirac delta function. The function $\Psi(\theta, \tau) \geq 0$ is called the *scattering function* and denotes the power in the different multipath-delayed and Doppler-shifted signal components. It is particularly useful for describing some salient characteristics of the channel [4, 3]. The maximum range of the values of $\tau$ over which $\Psi(\theta, \tau)$ is essentially nonzero is called the *multipath spread* of the channel and is denoted by $T_m$. Similarly, the maximum range of the $\theta$ values over which $\Psi(\theta, \tau)$ is essentially nonzero is called the *Doppler spread* of the channel and is denoted by $B_d$. In terms of these channel parameters (3.1) becomes

$$s(t) = \int_0^{T_m}\int_0^{B_d} H(\theta, \tau)(\mathbf{F}_\theta \mathbf{T}_\tau q)(t) d\theta d\tau = \int_0^{T_m}\int_0^{B_d} H(\theta, \tau) q(t - \tau) e^{j2\pi\theta\tau} d\theta d\tau \ .$$

Thus, $T_m$ signifies the maximum multipath delay introduced by the channel, and $B_d$ corresponds to the maximum Doppler shift produced by the channel. Note that if there is no time-variation in the channel, then $B_d = 0$; that is, $h(t, \tau) = h(\tau)$, a time-invariant channel.

### 3.3.2 *The Canonical Channel Representation*

The following result describes the fundamental channel representation, obtained by a $(\theta, \tau)$-sampling of (3.3.1), that plays a central role in our discussion [2, 1].

**Theorem 1.** If the multipath spread is greater than or equal to the chip period, $T_m \geq T_c$, and the Doppler spread is greater than or equal to the reciprocal to the waveform duration, $B_d \geq 1/T$, then the received signal $s(t)$ in (3.3.1) admits the finite-dimensional representation

$$s(t) \approx \frac{T_c}{T} \sum_{l=0}^{L-1} \sum_{p=0}^{P-1} \widehat{H}\left(\frac{p}{T}, lT_c\right) u_{p,l}(t)$$

where $L = \lceil T_m/T_c \rceil$, $P = \lceil B_d T \rceil$, and $\widehat{H}(\theta,\tau)$ is a bandlimited approximation of $H(\theta,\tau)$. The waveforms $u_{p,l}(t)$'s are defined as

$$u_{p,l}(t) \stackrel{def}{=} (\mathbf{F}_{\frac{p}{T}} \mathbf{T}_{lT_c} q)(t) = q(t - lT_c) e^{j\frac{2\pi p t}{T}}$$

and are approximately orthogonal

$$\langle u_{p,l}, u_{p',l'} \rangle \approx \|q\|^2 \delta_{p-p'} \delta_{l-l'} ,$$

where $\delta_m$ denotes the Kronecker delta function. □

The $(\theta,\tau)$-sampling in (3.3.2) is depicted in Figure 3.2. It is worth noting

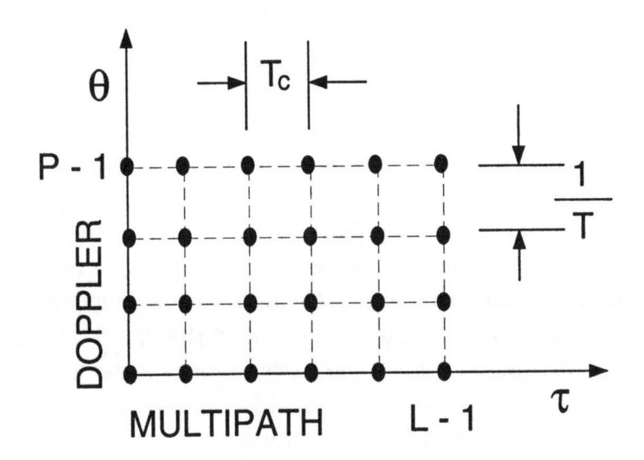


**Figure 3.2** Sampling of the time-frequency plane inherent in the canonical channel representation.

that there is virtually no loss of information in the approximate representation (3.3.2) due to the $(\theta,\tau)$-sampling. The reason is that due to the time- and band-limited nature of the signaling waveform $q(t)$, the receiver only "sees" a corresponding time- and band-limited version $(\widehat{H}(\theta,\tau))$ of the channel $(H(\theta,\tau))$, which in turn justifies the sampling of the spreading function in (3.3.2). Note that (3.3.2) also implies the following operator representation for the channel as a finite linear combination of time-frequency-shift operators (compare with (3.1)):

$$\widehat{\mathbf{H}} \approx \frac{T_c}{T} \sum_{l=0}^{L-1} \sum_{p=0}^{P-1} \widehat{H}\left(\frac{p}{T}, lT_c\right) \mathbf{F}_{\frac{p}{T}} \mathbf{T}_{lT_c} .$$

The above channel approximation depends on the TBP of the spread-spectrum signal $q(t)$: the larger the TBP, the finer the $(\theta,\tau)$-sampling as depicted in

Figure 3.2. We now discuss some useful interpretations and implications of the above representation.

### 3.3.3 *Discussion*

The channel transforms the deterministic (1D) signaling waveform $q(t)$ into a (PL-dimensional) stochastic signal, $s(t)$, as described by (3.3.1), and (3.3.2) is a canonical representation of $s(t)$. In fact, (3.3.2) is a Karhunen-Loève-like expansion of the received signal $s(t)$: the $\widehat{H}(p/T, lT_c)$'s are uncorrelated random variables and the waveforms $u_{p,l}(t)$'s are (roughly) orthogonal. The approximate orthogonality of $\{u_{p,l}(t)\}$ is illustrated in Figure 3.3 which show a 3D plot of the time-frequency correlation function

$$\mathbf{STFT}_q\left(\frac{p}{T}, lT_c; q\right) = \left\langle q, \mathbf{F}_{\frac{p}{T}} \mathbf{T}_{lT_c} q \right\rangle = \int q(t) q^*(t - lT_c) e^{-j\frac{2\pi pt}{T}} dt$$

of a direct sequence spread-spectrum waveform generated from a length 31 M-sequence [4].

The signal representation (3.3.2) also facilitates *diversity signaling* [4] that can be exploited by an appropriate receiver structure (See Sections 3.4 and 3.5). Essentially, the orthogonality of the $u_{p,l}(t)$'s implies that they can be processed as separate "channels," and the independence of the $\widehat{H}\left(\frac{p}{T}, lT_c\right)$ implies that those channels can be processed independently to provide diversity [1, 2]. Thus, as depicted in Figure 3.2, the received signal is resolved into $PL$ independent multipath-Doppler components (diversity channels) by appropriate sampling of multipath-Doppler (time-frequency) plane.

The diversity signaling is possible due to the large TBP of the spread-spectrum signaling waveform $q(t)$, as is easily evident from Figure 3.2. The larger the TBP product, the smaller the product $T_c \times \frac{1}{T}$, and the larger the number of delay-Doppler diversity channels. It is worth noting that samples on the multipath axis are due to the large bandwidth and correspond to multipath diversity that is exploited by the conventional RAKE receiver [1, 2, 4]. The samples on the Doppler axis are due to the processing time $T$ and correspond to "Doppler diversity" that is exploited by the "Doppler RAKE" receiver [1, 2]. For maximal channel exploitation both multipath and Doppler diversity should be jointly exploited as is done by the time-frequency-based receiver structures described in Sections 3.4 and 3.5.

The canonical signal representation (3.3.2) is based on the time-frequency sampling grid depicted in Figure 3.2. By *fixing* the relative location of the multipath-Doppler components, the grid provides a more parsimonious channel model: assuming perfect synchronization, the channel is completely described by the *number* of multipath-Doppler components and the corresponding channel

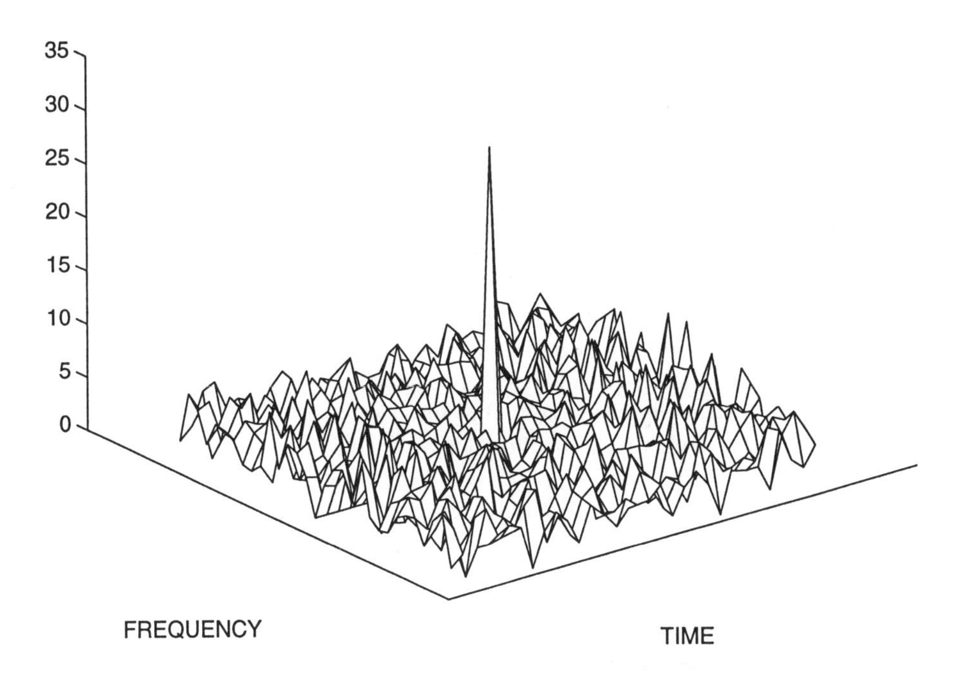


**Figure 3.3** Time-frequency correlation function of a spread spectrum signaling waveform (length-31 M-sequence).

coefficients — the locations of the different multipath-Doppler components are all known once one is known. Such structure in the channel model is very useful both from a receiver-implementation viewpoint and from an estimation viewpoint. In particular, we will see how this structure yields efficient receiver and signaling structures (Sections 3.4 and 3.5), and can be exploited to generate powerful multiuser timing-acquisition and interference-suppression algorithms for multipath channels (Section 3.6).

## 3.4 BASIC DIVERSITY SIGNALING AND RECEPTION

In this section, we describe the basic form of the time-frequency receivers that exploit the canonical representation (3.3.2) to achieve joint multipath-Doppler diversity in CDMA systems. We restrict our discussion to binary signaling and negligible intersymbol interference. Extensions to $M$-ary PSK signaling are straightforward, and the effects of intersymbol interference will be discussed in the next section in the context of a signaling scheme employing long overlapping codes.

The complex baseband transmitted signal $x(t)$ in Figure 3.1 for a continuous data stream can be represented as

$$x(t) = \sum_i q_i(t - iT)$$

where $T$ is the symbol period which is also the duration of the signaling waveform $q_i(t)$, $q_i(t) \in \{q^1(t), q^0(t)\}$, that could either correspond to antipodal signaling ($q^1(t) = -q^0(t)$) or orthogonal signaling ($\langle q^1, q^0 \rangle = 0$). The waveforms $q^m(t)$, $m = 0, 1$ are spread-spectrum waveforms of the form (3.1). Under our assumption of negligible intersymbol interference ($T_m \ll T$), "one-shot" analysis suffices in which each symbol can be treated independently;[3] that is, the $i$-th symbol is processes using $r(t), (i-1)T \le t < iT$, only. Thus, without loss of generality, we base our discussion on the $i = 0$ symbol:

$$r(t) = s(t) + n(t) \quad , \quad t \in [0, T)$$

where

The representation (3.3.2) clearly identifies the matched-filter waveforms needed to extract the independent multipath-Doppler components:

$$u_{p,l}^m(t) \stackrel{def}{=} \left( \mathbf{F}_{\frac{p}{T}} \mathbf{T}_{lT_c} q^m \right)(t) = q^m(t - lT_c) e^{j\frac{2\pi p t}{T}} \quad , \quad p = 0, 1, \cdots, P-1 \ , \quad l = 0, 1, \cdots, L-1 \ . \tag{3.3}$$

Recalling the definition of the STFT in (3.2), we note that the matched filter outputs can be precisely computed via *sampled* STFTs; that is, the sufficient statistics are given by

$$\langle r, u_{p,l}^m \rangle \stackrel{def}{=} \mathbf{STFT}_r\left(\frac{p}{T}, lT_c; q^m\right) \ .$$

Figure 3.4 illustrates the simple implementation of the sampled STFT via a bank of *identical* matched filters determined by the signaling waveforms. The STFT outputs in (3.4) can be combined to yield optimal coherent and noncoherent receiver structures as described next.

### 3.4.1 *Coherent Processing*

If estimates of the channel coefficients $\widehat{H}(p/T, lT_c)$ are available, such as through a pilot-based channel estimation, coherent processing can be used. The optimal

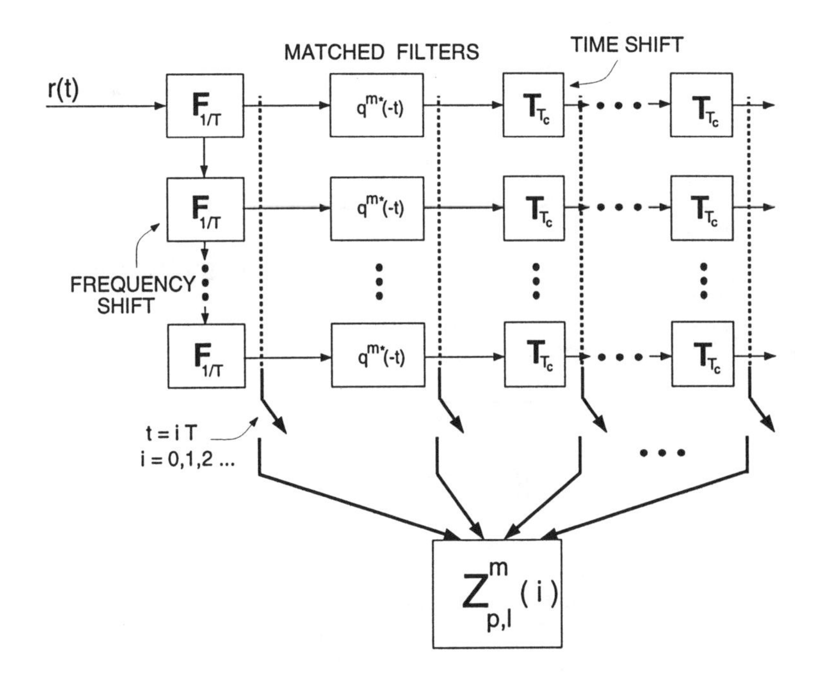


**Figure 3.4** Implementation of sampled STFT using a bank of identical matched filters (signaling waveform). $Z^m_{p,l}(i)$ denotes the sufficient statistics for the $i$-the symbol; that is, $Z^m_{p,l}(i) = \left\langle r_i, u^m_{p,l} \right\rangle$, where $r_i(t) = r(t + iT)$.

test statistic, which is applicable to both antipodal or orthogonal signaling, is given by

$$\widehat{m}_C = \operatorname{sgn}\left( \operatorname{real}\left\{ \sum_{l=0}^{L-1} \sum_{p=0}^{P-1} \widehat{H}\left(\frac{p}{T}, lT_c\right) \left[ \langle r, u^1_{p,l} \rangle - \langle r, u^0_{p,l} \rangle \right] \right\} \right)$$

where the $u_{p,l}(t)$'s are defined in (3.3). Figure 3.5 illustrates the detector structure for coherent processing.

### *3.4.2 Noncoherent (Quadratic) Processing*

If the actual values of the channel coefficients are not available, noncoherent processing can be used. If the second-order channel statistics ($\Psi(\theta, \tau)$) are known, the optimal noncoherent (quadratic) test statistic can be used which is

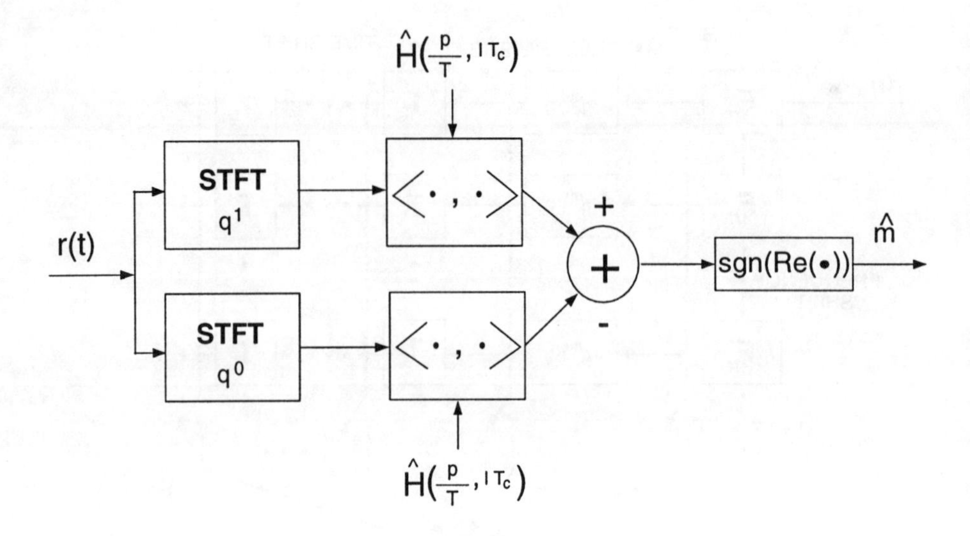


**Figure 3.5** STFT-based coherent detector structure for exploiting joint multipath-Doppler diversity.

given by

$$\widehat{m}_{NC} = \text{sgn}\left(\sum_{l=0}^{L-1}\sum_{p=0}^{P-1} \widehat{\Psi}\left(\frac{p}{T}, lT_c\right)\left[\left|\langle r, u^1_{p,l}\rangle\right|^2 - \left|\langle r, u^0_{p,l}\rangle\right|^2\right]\right)$$

where

$$\widehat{\Psi}(\theta,\tau) \stackrel{def}{=} \frac{\|q\|^2 \mathrm{E}\left\{\widehat{H}(\theta,\tau)|^2\right\}}{\|q\|^2 \mathrm{E}\left\{\widehat{H}(\theta,\tau)|^2\right\} + \mathcal{N}_0}.$$

If the channel statistics are not known, the equal-gain noncoherent combiner can be used which assumes uniform power in the different multipath-Doppler components; that is, $\widehat{\Psi}(\theta,\tau) = 1$. Figure 3.6 illustrates the quadratic detector structure which is, of course, only applicable with orthogonal signaling.

### *3.4.3 Potential Performance Gains*

In this section, we demonstrate the potential performance gains of the time-frequency receivers relative to the conventional RAKE receiver that is the state-of-the-art in CDMA systems. The time-frequency receivers are applicable in fast-fading, frequency-selective scenarios and can deliver significant

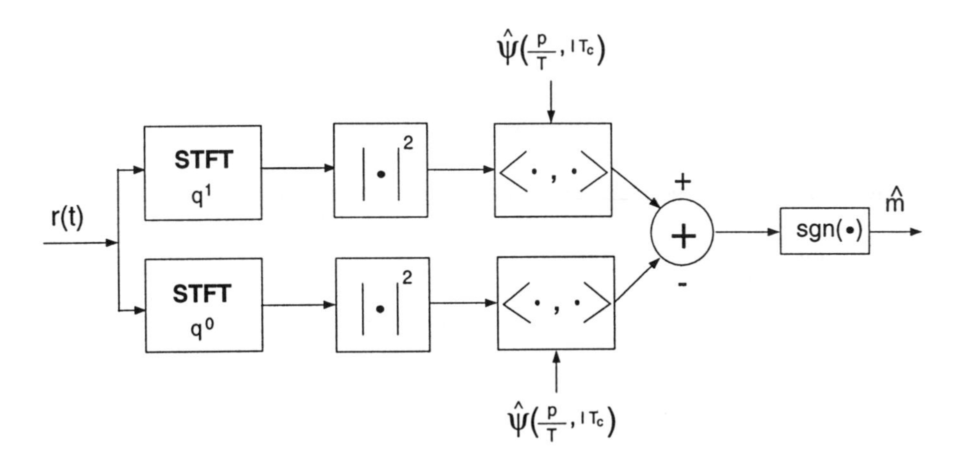


**Figure 3.6** STFT-based quadratic detector structure for exploiting joint multipath-Doppler diversity.

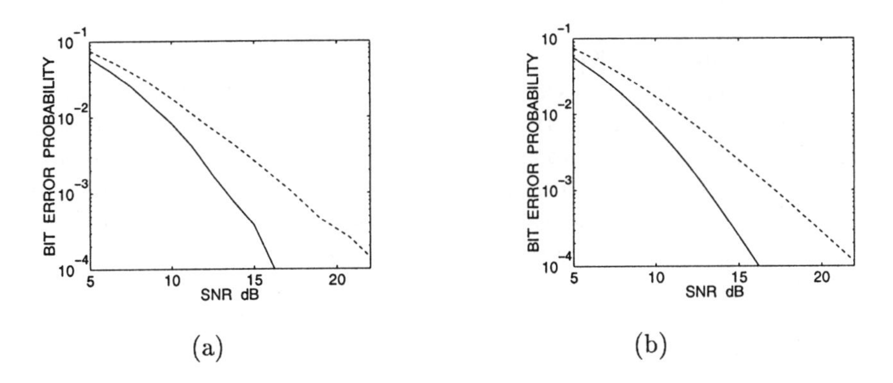


**Figure 3.7** Performance comparison between conventional RAKE (dashed) versus multipath-Doppler RAKE (solid) for coherent processing. (a) Theoretical. (b) Simulated.

performance gains by exploiting joint multipath-Doppler diversity. In fact, all time-frequency techniques that are based on the canonical channel representation (3.3.2) share the higher level of diversity inherent in the joint multipath-Doppler processing, and are thus capable of exhibiting similar performance gains.

Figure 3.7 compares the performance of the conventional RAKE receiver to that of time-frequency receiver for $L = 2$ and $P = 2$; that is, one multipath component and one Doppler component are resolvable. All components are assumed to have the same power. Thus, the RAKE corresponds to 2-point diversity, whereas the time-frequency RAKE corresponds to 4-point diversity. The plots in Figure 3.7 are based on coherent orthogonal signaling and show the bit error probability as a function of SNR for the two receivers.[4] The theoretical curves are based on analytic expressions for probability of bit error [2, 4] and the simulation results are based on 100000 independent experiments that use signaling waveforms based on length-31 M-sequences waveforms[1, 2]. Figure 3.7 clearly show the substantial performance gains achievable with joint time-frequency processing. For example, at $P_e \approx 10^{-4}$, there is a 6dB gain in performance which means that the transmitted power can be reduced by more than a factor of 4 in the case of time-frequency detectors relative to the conventional RAKE.

## 3.5 MAXIMALLY TIME-SELECTIVE SIGNALING AND RECEPTION

In this section, we use the fundamental channel representation (3.3.2) to propose a new signaling and receiver structure that extends the receivers of previous section to maximally exploit Doppler diversity. We present an overview of the techniques to provide a flavor of our time-frequency approach. A detailed discussion is provided in [7].

Time-frequency receivers exploit joint multipath-Doppler diversity by utilizing the large TBP of spectrum-spectrum signaling waveforms via (3.3.2). Larger diversity translates into improved receiver performance, and as we noted earlier, the larger the TBP, the larger the number of achievable diversity components. Clearly, the TBP product can be increased by increasing the bandwidth or the signaling duration or both. The larger bandwidth (smaller $T_c$) translates into larger diversity due to the larger number of resolvable independent multipath components (see Figure 3.2). The larger signaling duration ($T$) achieves higher diversity by resolving a larger number of independent Doppler-shifted signal components (see Figure 3.2).

In this section, we propose an approach for achieving maximal Doppler diversity by using long signaling waveforms. Consequently, successive symbol waveforms necessarily overlap in time in order to keep the data rate constant, resulting in intersymbol interference. Our methodology shares several features with the approach presented in [8, 9]. However, in contrast to [8, 9], our techniques clearly identify the mechanism — Doppler diversity — by which improved performance is achieved. As such, our framework also provides an alternative, simpler, and intuitively appealing explanation to the ideas presented in [8, 9]. A comparison of our framework with the techniques in [8, 9] is provided in [7].

For simplicity of exposition, we focus on the single-user case using BPSK signaling, and assume perfect synchronization. The results can be extended to the multiuser case [10], and multiuser synchronization issues will be discussed in the next section.

To be consistent with our previous notation, let $T$ denote the duration of the spread-spectrum signaling waveform $q$, and let $T_d$ denote the symbol period which determines the data rate. Note that in this case $T \gg T_d$, and let $T = DT_d$ for some integer $D \gg 1$, which we refer to as the *overlap factor.* The transmitted baseband signal can be represented as

$$x(t) = \sum_i b(i)q(t - iT_d)$$

which consists of overlapping symbols since $T_d \ll T$. Due to the intersymbol interference, the one-shot detector is no longer sufficient.[5] In fact, it is easy to see that for optimal detection all symbols need to be decoded jointly, since at the very least, each symbol overlaps with the preceding and succeeding symbol. Mathematically, this scenario is analogous to asynchronous multiuser detection [11], and intersymbol-interference (ISI) channels [12, 4].

From a practical viewpoint, this entails that a block of symbols has to be decoded jointly. Without loss of generality, assume that we are interested in decoding the 0-th symbol, and consider a block of $2I + 1$ symbols centered around it. Thus, the observation interval is $[-IT_d, IT_d + T)$ corresponding to the transmitted waveform

$$x_0(t) = \sum_{i=-I}^{I} b(i)q(t - iT_d) \ , \quad -IT_d \leq t < IT_d + T \ .$$

Similarly define $s_0(t)$, $w_0(t)$ and $r_0(t) = s_0(t) + w_0(t)$. Using the canonical representation (3.3.2), $s_0(t)$ is given by

$$s_0(t) \approx \frac{T_c}{T} \sum_{i=-I}^{I} b(i) \sum_{l=0}^{L-1} \sum_{p=0}^{P-1} \widehat{H}^i_{p,l} u^i_{p,l}(t)$$

where $\widehat{H}^i_{p,l} \stackrel{def}{=} \widehat{H}^i\left(\frac{p}{T}, lT_c\right)$ are the channel coefficients corresponding to the $i$-th symbol, and

$$u^i_{p,l}(t) \stackrel{def}{=} (\mathbf{F}_{\frac{p}{T}} \mathbf{T}_{(lT_c+iT_d)} q)(t) = q(t - lT_c - iT_d)e^{j\frac{2\pi pt}{T}} \ .$$

It is evident from (3.5) that given (estimates of) the channel coefficients, $\widehat{H}^i_{p,l}$, the correlator outputs corresponding to the $u^i_{p,l}(t)$'s are sufficient statistics for the given block of symbols.

It is convenient to introduce a vector notation for the sufficient statistics. Let the correlator outputs be denoted by

$$z^i_{p,l} \stackrel{def}{=} \langle r_0, u^i_{p,l} \rangle = \langle s_0, u^i_{p,l} \rangle + \langle w_0, u^i_{p,l} \rangle \ ,$$

where $i = -I, \cdots I$, $p = 0, \cdots P-1$, and $l = 0, \cdots L-1$. By concatenating the $z^i_{p,l}$'s for the different values of the indices $l$, $p$, and $i$ (in that order), we can represent the correlator outputs by a $(2I+1)PL \times 1$ vector which we denote by $\mathbf{z}$. Similarly, define the vectors $\mathbf{u}(t)$, $\mathbf{u}^i(t)$ and $\mathbf{u}^i_p(t)$ in terms of the $u^i_{p,l}(t)$'s, and the vectors $\mathbf{h}^i$ and $\mathbf{h}^i_p$ in terms of the scaled channel coefficients $\frac{T_c}{T}\widehat{H}^i_{p,l}$. Finally, define the $(2I+1)PL \times (2I+1)$ matrix $\mathbf{H}$ of channel coefficients as

$$\mathbf{H} \stackrel{def}{=} \begin{bmatrix} \mathbf{h}^{-I} & \mathbf{0} & \cdots & \mathbf{0} \\ \mathbf{0} & \mathbf{h}^{-I+1} & \mathbf{0} & \cdots \\ \vdots & \vdots & \ddots & \vdots \\ \mathbf{0} & \cdots & \mathbf{0} & \mathbf{h}^{I} \end{bmatrix}$$

With the established notation, the received signal $s_0(t)$ can be represented as

$$s_0(t) = \mathbf{u}^T(t)\mathbf{H}\mathbf{b}$$

where $\mathbf{b}$ is the $(2I+1) \times 1$ vector composed of the bits $b(i)$. Recall that $\mathbf{u}(t)$ is a $(2I+1)PL \times 1$ vector and $\mathbf{H}$ is a $(2I+1)PL \times (2I+1)$ matrix. It follows that the vector $\mathbf{z}$ of correlator outputs can be expressed as

$$\mathbf{z} = \mathbf{R}\mathbf{H}\mathbf{b} + \mathbf{w}$$

where $\mathbf{R}$ is a $(2I+1)PL \times (2I+1)PL$ signal correlation matrix defined as

$$\mathbf{R} \stackrel{def}{=} \int_{-IT_d}^{IT_d+T} \mathbf{u}^*(t)\mathbf{u}(t)^T dt = \begin{bmatrix} \mathbf{R}_{-I,-I} & \mathbf{R}_{-I,-I+1} & \cdots & \mathbf{R}_{-I,I} \\ \mathbf{R}_{-I+1,-I} & \mathbf{R}_{-I+1,-I+1} & \cdots & \mathbf{R}_{-I+1,I} \\ \vdots & \vdots & \vdots & \vdots \\ \mathbf{R}_{I,-I} & \mathbf{R}_{I,-I+1} & \cdots & \mathbf{R}_{I,I} \end{bmatrix}$$

where the $PL \times PL$ matrices $\mathbf{R}_{i,j}$ are defined as

$$\mathbf{R}_{i,j} \stackrel{def}{=} \int_{-IT_d}^{IT_d+T} \mathbf{u}^{i*}(t)\mathbf{u}^j(t)^T \ ,$$

and $\mathbf{w}$ is the noise component of $\mathbf{z}$ corresponding to $w_0(t)$ in (3.5). Evidently, $\mathbf{w}$ is zero-mean, Gaussian-distributed with correlation matrix $\mathcal{N}_0\mathbf{R}$.

The representation of the correlator outputs (sufficient statistics) in (3.5) is of the same form as in the case of multiuser detection [13, 11] or ISI channels

[12, 4]. The optimal receiver [11, 12] requires sequence decoding over the block and suffers from high complexity. For practical implementation, a variety of suboptimal techniques that have been developed for multiuser detection (or ISI channels) may be adapted to this problem [13, 14, 15, 16, 4]. For example, a decorrelating stage [13] (which is equivalent to the zero-forcing solution in ISI channels [4]) may be applied first before coherently combining the various multipath-Doppler components to achieve diversity [10]. A detailed discussion of this approach appears elsewhere [7].

We note that the signaling based on long, overlapping codes proposed above crucially depends on the time-frequency channel representation which is encapsulated in the matrices **H** and **R**. Moreover, since the matrix **R** denotes the correlation between the different time-frequency shifted versions of the *same* underlying spreading code, the correlations are not very significant because of the approximate orthogonality property (see Theorem 1). However, for a large value of the overlap factor $D$, the contributions due to other symbols become significant and must be taken into account[7]. In this context, the approximately "lossless" codes of [8, 9] may be useful as well.

## 3.6 MULTIUSER TIMING ACQUISITION

Thus far, we have explored the time-frequency framework in the single-user case with perfect timing information. As mentioned earlier, the time-frequency approach can extended to incorporate multiuser detection [10]. In this section, we leverage the fundamental channel representation (3.3.2) to address the important problem of multiuser timing acquisition [17]. Our approach offers a powerful alternative to existing multiuser acquisition techniques [18, 19, 20, 21, 22, 23] in that it fully incorporates the multipath channel effects and does not require knowledge of actual channel coefficients thereby resulting in a dramatic reduction in complexity. Moreover, the proposed approach is near-far resistant and is blind in that no training sequence is required. Again, our emphasis is not on details but on how the time-frequency approach can be exploited in the context of acquisition.

### *3.6.1 Formulation*

Assume BPSK signaling and consider a system of $K$ users so that the received waveform can be expressed as

$$r(t) = \sum_{k=1}^{K} s_k(t - \tau_k) + w(t)$$

where $s_k(t-\tau_k)$ is the received signal corresponding to the $k$-th user, and $\tau_k$ is the corresponding unknown delay relative to some fixed reference. Without loss of generality, we assume that[6]

$$0 \leq \tau_1 \leq \tau_2 \cdots \leq \tau_K \leq T .$$

In terms of our channel representation, the signal corresponding to the $k$-th user can be expressed as

$$s_k(t) \approx \frac{T_c}{T} \sum_i b_k(i) \sum_{l=0}^{L-1} \sum_{p=0}^{P-1} \widehat{H}_{p,l}^{k,i} q_k(t - iT - lT_c) e^{j\frac{2\pi p t}{P}}$$

where $b_k(i)$ denotes the $i$-th symbol of the $k$-th user, $\widehat{H}_{p,l}^{k,i} \stackrel{def}{=} \widehat{H}^{k,i}\left(\frac{p}{T}, lT_c\right)$ denote the channel coefficients corresponding to the $i$-th symbol of the $k$-th user, and $q_k(t)$ is the spreading waveform of the $k$-th user. Evidently, we need an observation interval of duration at least $2T$ in order to observe one complete symbol waveform for each user. Thus, without loss of generality, we consider the observation interval $[0, 2T]$ which includes the $i = -1, 0, 1$ symbols for each user.[7]

Conditioned on the bit sequence $b_k(i)$, the received signal $r(t)$ is a zero-mean Gaussian process. Moreover, the correlation function of $r(t)$ is independent of the PSK bit sequence. Thus, the timing acquisition problem can be considered as a problem of estimating the parameters $(\tau_k)$ of a Gaussian process.

The correlation function of the received signal $r(t)$ is given by

$$R(t_1, t_2) \stackrel{def}{=} \mathrm{E}\{r(t_1) r^*(t_2)\} = \sum_{k=1}^{K} R_k(t_1 - \tau_k, t_2 - \tau_k) + \mathcal{N}_0 \delta(t_1 - t_2) \quad ,$$

where $R_k(t_1, t_2)$ is the correlation function of $s_k(t)$ given by

$$R_k(t_1, t_2) \approx \frac{T_c^2}{T^2} \sum_i \sum_{p=0}^{P-1} \sum_{l=0}^{L-1} \widehat{\Psi}_{p,l}^{k} q_k(t_1 - iT - lT_c) q_k^*(t_2 - iT - lT_c) e^{j\frac{2\pi p}{T}(t_1 - t_2)}$$

which follows from (3.6.1) where $\widehat{\Psi}_{p,l}^{k} \stackrel{def}{=} \widehat{\Psi}^{k}\left(\frac{p}{T}, lT_c\right)$ are the samples of the scattering function for the $k$-th user.[8] On the space of functions on $[0, 2T)$, denote by $\mathbf{R}_k(\tau_k)$ the operator defined by $R_k(t_1 - \tau_k, t_2 - \tau_k)$:

$$(\mathbf{R}_k(\tau_k) x)(t) \stackrel{def}{=} \int_0^{2T} R_k(t - \tau_k, u - \tau_k) x(u) du \, .$$

It follows that on the observation interval $[0, 2T)$, the correlation function $R(t_1, t_2)$ admits the operator representation

$$\mathbf{R}(\tau) \stackrel{def}{=} \sum_{k=1}^{K} \mathbf{R}_k(\tau_k) + \mathcal{N}_0 \mathbf{I}$$

where $\mathbf{I}$ denotes the identity operator, and $\tau \stackrel{def}{=} [\tau_1, \tau_2, \cdots \tau_k] \in [0, T)^K$ denotes the unknown vector of timing offsets.[9]

### 3.6.2 *Maximum Likelihood Approach*

The log-likelihood function for the observation waveform $r(t)$, $0 \leq t < 2T$, is given by [24]

$$L^{\tau}(r) = \langle \mathbf{R}^{-1}(\tau) r, r \rangle - \log\left(\det\left(\mathbf{R}(\tau)\right)\right)$$

where $\det(\cdot)$ denotes the determinant of the operator (product of the eigenvalues).[10] Thus, the maximum-likelihood (ML) estimate of the timing vector $\tau$ is given by

$$\tau^{ML} = \arg \max_{\tau \in [0,T)^K} L^{\tau}(r) .$$

From (3.6.2), it is clear that finding $\tau^{ML}$ is a complicated problem since the nonlinear dependence of $\mathbf{R}^{-1}(\tau)$ on $\tau$ cannot be functionally characterized in general, and the likelihood surface will be plagued by many local maxima. Thus, a direct ML approach is not feasible.

Under a weak signal assumption,[11] we can simplify the problem by considering the locally-optimal likelihood function[12] which results in the local ML estimates [24]

$$\tau_k^{LO} = \arg \max_{\tau_k \in [0,T)} \langle \mathbf{R}_k(\tau_k) r, r \rangle \ , \quad k = 1, 2, \cdots K \ .$$

The estimator in (3.6.2) is simple: compute a sequence of quadratic forms for different (quantized) values of $\tau_k \in [0, T)$, and choose the $\tau_k$ corresponding to largest value. However, this "decoupled" estimator is clearly not near-far resistant since it does not account for the contribution of the other users in $r(t)$.

### 3.6.3 *Interference-Suppression-Based Approach*

Both the estimators (3.6.2) and (3.6.2) are *quadratic* in the observed signal $r(t)$ and represent two extremes: (3.6.2) is optimal, requires statistics for all users ($\mathbf{R}(\tau)$) and is computationally intractable, whereas (3.6.2) is simple, requires the statistics of only the desired user ($\mathbf{R}_k(\tau_k)$), but is not near-far resistant. Our approach exploits the canonical channel model and strikes a balance between this complexity versus performance trade-off by modifying (3.6.2) to be near-far resistant via interference suppression techniques.

For $k$-th user, the estimator (3.6.2) consists of an array of quadratic processors, $\langle \mathbf{R}_k(\tau_k)r, r\rangle$, each "matched" to particular value of the delay $\tau_k \in [0, T)$. To suppress the contribution due to other users, we define the optimal quadratic processor for a particular delay $\tau_k$ of $k$-th user as the solution to the following constrained optimization problem

$$\begin{aligned} \mathbf{Q}_k(\tau_k) &= \arg\min_{\mathbf{Q}} \mathrm{E}\{\langle \mathbf{Q}r, r\rangle\} = \arg\max_{\mathbf{Q}} \mathrm{Tr}\,(\mathbf{Q}\mathbf{R}(\tau)) \\ & \quad \text{subject to } \mathrm{Tr}\,(\mathbf{Q}\mathbf{R}_k(\tau_k)) = 1 \,. \end{aligned} \tag{3.4}$$

The intuitive motivation for (3.4) is that the optimal quadratic processor $\mathbf{Q}_k(\tau_k)$ should pass the signal components in the "direction" of the delay $\tau_k$ of the $k$-th user at a fixed gain, while minimizing the contribution due to other signal components (interference). Note that the above optimization problem is similar in spirit to the concept of beamforming in array processing [25], and the linear processor design in [14, 26].

Our algorithm is based on the following characterization of the solution to (3.4) at the true underlying delay $\tau_k$ and under the assumption of no noise. We state the result without proof.

**Theorem 2.** Consider the noise-free situation; that is, $\mathcal{N}_0 = 0$, $\mathbf{R}(\tau) = \sum_k \mathbf{R}_k(\tau_k)$. At the true underlying delay $\tau_k$ of $k$-th user, the optimal quadratic processor $\mathbf{Q}_k(\tau_k)$ solving (3.4) is given by[13]

$$\mathbf{Q}_k(\tau_k) = \alpha \sum_{n=1}^{L_k P_k} \mathbf{c}_n \otimes \mathbf{c}_n$$

where $\{c_n(t) : \; t \in [0, 2T), n = 1, 2, \cdots L_k P_k\}$ are the $L_k P_k$ generalized eigenvectors corresponding to the unit (largest) generalized eigenvalue of the eigenequation

$$\mathbf{R}(\tau)\mathbf{c} = \lambda \mathbf{R}_k(\tau_k)\mathbf{c} \,,$$

$L_k$ and $P_k$ are the number of multipath and Doppler components, respectively, in the channel representation (3.3.2) for user $k$, and $\alpha$ is chosen to satisfy the constraint in (3.4). □

Inspired by the above result, the basic idea behind our algorithm is to quantize the interval $[0, T)$ into intervals of length $T_c/2$, as done in [23], and set up a multihypothesis testing problem corresponding to the different quantized values of the timing-offset $\tau_k$ of the desired user $k$. Once $\tau_k$ has been determined to within half a chip interval, finer quantization can be used to improve the estimate. The essential generic steps are outlined below.

**Timing Acquisition Pseudo-Algorithm:**
**Step 1.** Let $I = 2N + 1$, where $T = NT_c$. Define

$$\tau_k(i) = \frac{T_c}{2} i \, , \quad i = 0, 1, \cdots I - 1 \, .$$

**Step 2.** For each $i$, compute $\mathbf{Q}_k(\tau_k(i))$ via (3.6.3) corresponding to the $L_k P_k$ largest generalized eigenvalues of

$$[\mathbf{R}(\tau) - \mathcal{N}_0 \mathbf{I}] \, \mathbf{c} = \lambda \mathbf{R}_k(\tau_k(i)) \mathbf{c} \, .$$

**Step 3.** Estimate the delay $\tau_k$ as

$$\widehat{\tau}_k = \tau_k(i_{max}) = \arg \max_{\tau_k(i)} \mathrm{Tr} \left( \mathbf{Q}_k(\tau_k(i)) [\mathbf{R}(\tau) - \mathcal{N}_0 \mathbf{I}] \right) \, ,$$

or as based on a hypothesis test on the closeness of the $L_k P_k$ largest generalized eigenvalues of (3.6.3) to unity (see Theorem 2).
**Step 4.** Get a refined estimate by further quantizing the $T_c/2$ interval determined by $i_{max}$ and the neighboring index yielding the larger (smaller) metric in Step 3. □

### *3.6.4 Discussion*

To apply the acquisition algorithm in practice, $\mathbf{R}(\tau)$ and $\mathcal{N}_0$ can be estimated from received data. The correlation function $\mathbf{R}_k(\tau_k)$ corresponding to the desired user $k$ is determined by the canonical channel representation and is given by (3.6.1). In the absence of knowledge about the channel statistics, $\Psi^k_{p,l} = 1$ may be used.

The most attractive feature of the proposed approach is that it incorporates the canonical time-frequency channel representation in the acquisition algorithm without having to estimate the channel coefficients, thereby dramatically reducing the complexity. Due to the quadratic processing, the channel is represented via its second-order statistics which can be reliably estimated or approximated. The near-far resistance of the algorithm essentially stems from Theorem 2 which states that after accounting for the effects of noise, the best quadratic processor for the desired user (corresponding to the correct delay) is independent of the power of other users. Details of this approach appear elsewhere [27].

We note that our acquisition approach is equally applicable in slow fading situations in which we can ignore the Doppler aspects. Moreover, once the timing has been acquired, a near-far resistant linear or quadratic time-frequency equalizer can be designed along the lines of [10] or [14, 26]. Finally, even though we have discussed timing acquisition, our framework can be readily extended to incorporate (carrier) frequency acquisition as well.

## 3.7 CONCLUSION

Central to modern mobile communication networks is the fast-fading multipath wireless channel that is one of the single most important factors affecting the system performance. For optimal processing, time-varying signal processing techniques are needed to account for the inherently time-varying dynamics of the channel. We have provided a time-frequency framework for mobile wireless communications that emphasizes this time-varying perspective. The signal processing tools underlying our framework are TFRs that perform joint time-frequency processing.

At the heart of our framework is a canonical finite-dimensional time-frequency representation of the channel in terms of time (multipath) and frequency (Doppler) shifts of the transmitted signal. The fundamental channel representation not only provides new insights but also leads to new design strategies that can deliver significantly improved performance due to joint multipath-Doppler diversity signaling.

The multipath-Doppler diversity inherent in our time-frequency framework is achieved by exploiting the large TBP of spread-spectrum signaling waveforms in CDMA systems. The conventional RAKE receiver only utilizes the large bandwidth to achieve multipath diversity. The additional Doppler diversity afforded by joint time-frequency processing translates into substantial performance gains in virtually every aspect of system design.

Doppler diversity is essentially achieved by doing processing over sufficiently long time intervals so that the effective channel becomes time-selective. In Section 3.5 we presented the basic ideas behind a novel signaling scheme in which long overlapping spreading waveforms are used to achieve long processing times. A promising direction for future research is the use of block processing to achieve Doppler diversity via some form of sequence decoding. Such a scheme could be applicable in existing systems such as the IS-95 standard in which a block of 200 symbols is processed together for the purpose of interleaving [28].

In addition to novel signaling and receiver structures, the canonical time-frequency channel representation can be leveraged in several other aspects of system design. In particular, in Section 3.6 we developed a framework for multiuser timing-acquisition based on quadratic processing. The acquisition framework incorporates the multipath fading channel and fully exploits the signal structure provided by the canonical time-frequency representation. Moreover, the approach can be extended to provide near-far resistant algorithms for joint multiuser acquisition and demodulation over multipath fading channels [10, 27].

The fundamental time-frequency-based signal representation also has a curious connection to the "spreading versus coding" ideas of Massey [29]. As such, our representation suggests that the channel provides a form of coding via multipath-Doppler signaling. An interesting research issue is the exploration of this connection to quantify the advantages of CDMA systems in the context of multipath fading.

In short, our time-frequency approach provides a unified signal processing framework for communication over multipath fading channels. It ties in with several existing approaches, suggests new design strategies, provides intuitively appealing alternatives to others, and leads to several new insights and directions for future research that could prove quite fruitful.

## Notes

1. The special class of linear time-invariant systems can be represented solely in terms of the time-shift operator which results in the well-known convolution integral.

2. We assume a zero-mean channel (Rayleigh fading). Extension to non-zero mean situations (Rician fading) is straightforward.

3. In conjunction with the assumption of negligible intersymbol interference, the conditions of Theorem 1 become: $T_c \leq T_m \ll T$ and $1/T \leq B_d \ll W$, where $W \approx 1/(2T_c)$ is the bandwidth of the signaling waveforms. In Section 3.5, we relax the condition $B_d \geq 1/T$ by considering signaling based on long, overlapping codes in which block-processing becomes necessary due to intersymbol interference.

4. Similar performance gains are obtained for noncoherent processing [1, 2].

5. However, it may perform satisfactorily under certain conditions. See the discussion later in this section.

6. In this section, for simplicity of exposition, we restrict our discussion to nonoverlapping codes of duration $T$, in contrast to the long overlapping codes of last section.

7. Longer intervals can be incorporated straightforwardly.

8. Note that under the WSSUS assumption, the channel *statistics* do not change over time.

9. After appropriate sampling, these operators are represented by matrices over finite dimensional vector spaces.

10. See also footnote 3.7.

11. That is, the SNR for each user is sufficiently low.

12. The first term in the Taylor expansion of the likelihood function as a function of signal power.

13. The notation $\mathbf{c}_n \otimes \mathbf{c}_n$ denotes a rank-1 projection operator defined as $((\mathbf{c}_n \otimes \mathbf{c}_n)x)(t) \stackrel{def}{=} \int c(t)c^*(u)x(u)du$.

## References

[1] A. M. Sayeed and B. Aazhang, "Exploiting Doppler diversity in mobile wireless communications," in *Proc. 1997 Conf. Inf. Sci. Syst. (CISS'97)*, 1997.

[2] A. M. Sayeed and B. Aazhang, "Joint multipath-Doppler diversity in mobile wireless communications," *submitted to IEEE Trans. Commun.*, May 1997.

[3] P. A. Bello, "Characterization of randomly time-variant linear channels," *IEEE Trans. Commun. Syst.*, vol. CS-11, pp. 360–393, 1963.

[4] J. G. Proakis, *Digitial Communications.* New York: McGraw Hill, 3rd ed., 1995.

[5] R. G. Shenoy and T. W. Parks, "The Weyl correspondence and time-frequency analysis," *IEEE Trans. Signal Processing*, vol. 42, pp. 318–332, Feb. 1994.

[6] W. Kozek, "Time-frequency signal processing based on the Wigner-Weyl framework," *Signal Processing*, vol. 29, pp. 77–92, Oct. 1992.

[7] S. Bhasyam, A. M. Sayeed, and B. Aazhang, "Time-selective signaling and reception in multiple-access communication over fading channels," *in preparation.*

[8] G. W. Wornell, "Spread-signature CDMA: Efficient multiuser communications in the presence of fading," *IEEE Trans. Inform. Theory*, vol. IT-41, pp. 1418–1438, Sep. 1995.

[9] G. W. Wornell, "Spread-response precoding for communication over fading channels," *IEEE Trans. Inform. Theory*, vol. IT-42, pp. 488–501, Mar. 1996.

[10] A. M. Sayeed, A. Sendonaris, and B. Aazhang, "Multiuser detectors for fast-fading multipath channels," in *Proc. 31st Asilomar Conf. Signals, Syst., and Computers*, November 1997.

[11] S. Verdu, "Minimum probability of error for asynchronous Gaussian multiple-access channels," *IEEE Trans. Inform. Theory*, vol. IT-32, pp. 85–96, Jan. 1986.

[12] G. D. Forney, "Maximum likelihood sequence estimation of digital sequences in the presence of intersymbol interference," *IEEE Trans. Inform. Theory*, vol. IT-18, pp. 363–378, May 1972.

[13] R. Lupas and S. Verdu, "Near-far resistance of multiuser detectors in asynchronous channels," *IEEE Trans. Commun.*, vol. 38, pp. 496–508, Apr. 1990.

[14] U. Madhow and M. L. Honig, "MMSE interference suppression for direct-sequence spread-spectrum CDMA," *IEEE Trans. Commun.*, vol. 42, pp. 3178–3188, Dec. 1994.

[15] A. Duel-Hallen, "Decorrelating decision-feedback multiuser detector for synchronous code-division multiple-access channel," *IEEE Trans. Commun.*, vol. 41, pp. 285–290, Feb. 1993.

[16] M. K. Varanasi and B. Aazhang, "Multistage detection in asynchronous code-division multiple-access communications," *IEEE Trans. Commun.*, vol. 38, pp. 509–519, Apr. 1990.

[17] U. Madhow and M. Pursley, "Acquisition in direct-sequence spread-spectrum communication networks: an asymptotic analysis," *IEEE Trans. Inform. Theory*, vol. IT-39, pp. 903–912, May 1993.

[18] S. Y. Miller and S. C. Schwartz, "Parameter estimation for asynchronous multiuser communincation," in *Proc. 1989 Conf. Inform. Sci. Syst. (CISS'89)*, pp. 294–299, 1989.

[19] S. E. Bensley and B. Aazhang, "Subspace-based channel estimation for code division multiple access communication systems," *IEEE Trans. Commun.*, vol. 44, pp. 1009–1020, Aug. 1996.

[20] A. Radovic and B. Aazhang, "Iterative algorithms for joint data detection and delay estimation for code division multiple access communication systems," in *Proc. 31st Ann. Allerton Conf. Commun., Contr., Computing, Monticello, IL*, pp. 1–10, 1993.

[21] R. F. Smith and S. L. Miller, "Code timing estimation in a near-far environment for direct-sequence code-division multiple-access," in *Proc. Milcom '94*, pp. 47–51, 1994.

[22] E. G. Strom, S. Parkvall, S. L. Miller, and B. E. Ottersten, "Propagation delay estimation in asynchronous direct-sequence code-division multiple access systems," *IEEE Trans. Commun.*, vol. 44, pp. 84–93, Jan. 1996.

[23] U. Madhow, "Blind adaptive interference suppression for the near-far resistant acquisition and demodulation of direct-sequence CDMA systems," *IEEE Trans. Signal Processing*, vol. 45, pp. 124–136, Jan. 1997.

[24] H. V. Poor, *An Introduction to Signal Detection and Estimation*. Springer-Verlag, 1988.

[25] B. D. Vanveen and K. M. Buckley, "Beamforming: A versatile approach to spatial filtering," *IEEE Signal Processing Magazine*, pp. 4–24, April 1988.

[26] M. L. Honig, U. Madhow, and S. Verdu, "Blind adaptive multiuser detection," *IEEE Trans. Inform. Theory*, vol. IT-41, pp. 944–960, Jul. 1995.

[27] A. M. Sayeed and B. Aazhang, "Noncoherent multiuser timing acquisition over multipath fading channels," *in preparation*.

[28] T. S. Rappaport, *Wireless Communications*. New Jersey: Prentice Hall, 1996.

[29] J. L. Massey, "Towards an information theory of spread-spectrum systems," in *Code Division Multiple Access Communications* (S. G. Glisic and P. A. Lippänen, eds.), pp. 29–46, Kluwer Academic Publishers, 1995.

# 4 ITERATIVE DECODING ALGORITHMS

Branka Vucetic

**Abstract:**

In this tutorial paper we present iterative algorithms which can be used for decoding of concatenated codes. The decoding operation is based on either a maximum a posteriori (MAP) algorithm or a Viterbi algorithm generating a weighted soft estimate of the input sequence. The iterative algorithm performs the information exchange between the two component decoders. The performance gain of the MAP algorithm over the Viterbi algorithm at low SNR leads to a slight performance advantage.

The MAP algorithm is computationally much more complex than the Viterbi algorithm. The operations in the MAP algorithm are multiplications and exponentiations while in the Viterbi algorithm they are simple add, compare and select operations.

As an example these algorithms are applied to decoding of turbo codes and their performance is compared on a Gaussian channel.

## 4.1 INTRODUCTION

Iterative decoding algorithms has been used in decoding of the class of parallel concatenated codes [12] first introduced by Berrou et al in 1993. An iterative decoder for concatenated codes consists of two soft output decoders for the component codes separated by the interleaver. The component decoders are based on either a maximum a posteriori (MAP) algorithm or a Viterbi algorithm generating a weighted soft estimate of the input sequence. The iterative algorithm performs the information exchange between the two component decoders.

The iterative decoding of turbo codes in [12] is based on the soft output maximum a posteriory probability (MAP) algorithm.

The MAP algorithm was first introduced by Chang and Hancock in 1966 [51] and was proposed for signal detection on channels with memory. The same algorithm was used by Bahl et al in 1974 [50] for decoding of linear codes.

The Viterbi algorithm can also be modified to produce soft outputs [1]–[11].

The MAP algorithm minimises the symbol (or bit) error probability, while the Viterbi algorithm minimises the sequence error probability. The performance of the MAP and Viterbi algorithms are identical at high SNR. The performance gain of the MAP algorithm over the Viterbi algorithm at low SNR leads to a performance advantage in iterative decoding.

The MAP algorithm is computationally much more complex than the Viterbi algorithm. The operations in the MAP algorithm are multiplications and exponentiations while in the Viterbi algorithm they are simple add, compare and select operations.

## 4.2 THE MAP DECODING ALGORITHM

The MAP algorithm outputs a hard decision and a real number, which is the posteriory probability (APP). The APP is a measure of the probability of the hard output decision.

The encoding process of a linear convolutional code can be described as a discrete time finite-state Markov process. This process can be graphically represented by state and trellis diagrams.

The considered system model is shown in Fig. 4.1.

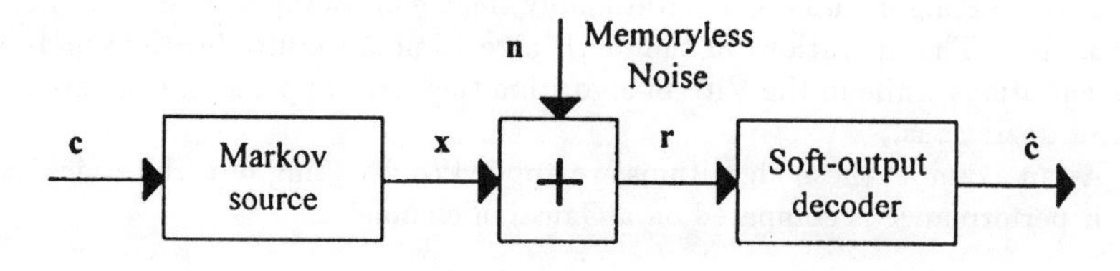


**Figure 4.1** System Model

The encoded block at the output of the encoder, modelled as a Markov source, denoted by $\mathbf{x}_t$, is transmitted through a discrete memoryless channel. $\mathbf{r}_t$ is the noise corrupted version of $\mathbf{x}_t$.

The decoder gives an estimate of the input to the Markov source, by examining $\mathbf{r}_t$.

In general the input bits, $c_t$, $t = 1, 2, \cdots N$, where $N$ is an integer, can be nonbinary, but for simplicity we assume that they are independently generated binary symbols and have equal a priori probabilities.

The bit sequence

$$\mathbf{c}_t = (c_1, c_2, \cdots c_N)$$

is encoded by an $(n, 1, m)$ recursive systematic convolutional encoder. The code trellis has a total number of $M = 2^m$ distinct states, indexed by the integer $l$, $l = 0, 1, \cdots M - 1$. The state of the trellis at time $t$ is denoted by $S_t$. The state sequence from time $t$ to $t'$ is denoted by $S_t^{t'}$ and it is given by

$$\mathbf{S}_t^{t'} = (S_t, S_{t+1}, \cdots, S_{t'})$$

The corresponding encoder output sequence is

$$\mathbf{v}_t^{t'} = (\mathbf{v}_t, \mathbf{v}_{t+1}, \cdots, \mathbf{v}_{t'})$$

where

$$\mathbf{v}_t = (v_{t,0}, v_{t,1}, \cdots, v_{t,n-1})$$

and are defined by $n$ encoder generator sequences.

The coded sequence $\mathbf{v}_t^{t'}$ is modulated by a BPSK modulator. The modulated sequence is denoted by $\mathbf{x}_t^{t'}$ and given by

$$\mathbf{x}_t^{t'} = (\mathbf{x}_t, \mathbf{x}_{t+1}, \cdots, \mathbf{x}_{t'})$$

where

$$\mathbf{x}_t = (x_{t,0}, x_{t,1}, \cdots, x_{t,n-1})$$

and

$$x_{t,i} = 2v_{t,i} - 1, \quad i = 0, \cdots, n-1 \tag{4.1}$$

The modulated sequence $\mathbf{x}_t^{t'}$ is corrupted by additive with Gaussian noise during transmission, ending up with the received sequence

$$\mathbf{r}_t^{t'} = (\mathbf{r}_t, \mathbf{r}_{t+1}, \cdots, \mathbf{r}_{t'})$$

where

$$\mathbf{r}_t = (r_{t,0}, r_{t,1}, \cdots, r_{t,n-1})$$

and

$$r_{t,i} = x_{t,i} + n_{t,i} \qquad i = 0, 1, \cdots, n-1$$

where $n_{t,i}$ is a zero-mean Gaussian noise random variable with variance $\sigma^2$. Each noise sample is assumed to be independent from each other.

The content of the shift register in the encoder at time $t$ represents $S_t$ and it transits into $S_{t+1}$ in response to the input $c_{t+1}$ giving as output the coded block $\mathbf{v}_{t+1}$. The state transition of the encoder is shown in the state diagram.

Let $c_t$ be the information bit associated with the transition $S_{t-1}$ to $S_t$, producing as output $\mathbf{v}_t$. The decoder gives an estimate of the input to the Markov source, by examining $\mathbf{r}_t$. The MAP algorithm provides the log likelihood denoted by $\Lambda(c_t)$, given the received sequence $\mathbf{r}_1^\tau$

$$\Lambda(c_t) = log \frac{Pr\{c_t = 1 \mid \mathbf{r}_1^\tau\}}{Pr\{c_t = 0 \mid \mathbf{r}_1^\tau\}} \tag{4.2}$$

where $Pr\{c_t = i \mid \mathbf{r}_1^\tau\}$, $i = 0, 1$, is the APP of the data bit $c_t$.

The decoder makes a decision by comparing $\Lambda(c_t)$ to a threshold equal to zero

$$c_t = \begin{cases} 1 & \text{if } \Lambda(c_t) \geq 0 \\ 0 & \text{otherwise} \end{cases} \tag{4.3}$$

We can compute the APPs in (4.2) as

$$Pr\{c_t = 0 \mid \mathbf{r}_1^\tau\} = \sum_{(l',l)\epsilon B_t^0} Pr\{S_{t-1} = l', S_t = l | \mathbf{r}_1^\tau\} \tag{4.4}$$

where $B_t^0$ is the set of transitions $S_{t-1} = l' \rightarrow S_t = l$ that are caused by the input bit $c_t = 0$.

Also

$$Pr\{c_t = 1 \mid \mathbf{r}_1^\tau\} = \sum_{(l',l)\in B_t^1} Pr\{S_{t-1} = l', S_t = l | \mathbf{r}_1^\tau\} \tag{4.5}$$

where $B_t^1$ is the set of transitions $S_{t-1} = l' \rightarrow S_t = l$ that are caused by the input bit $c_t = 1$.

Eq. (4.4) can be written as

$$Pr\{c_t = 0 | \mathbf{r}_1^\tau\} = \sum_{(l',l)\in B_t^0} \frac{Pr\{S_{t-1} = l'; S_t = l; \mathbf{r}_1^\tau\}}{Pr\{\mathbf{r}_1^\tau\}} \tag{4.6}$$

The APP of the decoded data bit $c_t$ can be derived from the joint probability defined as

$$\sigma_t(l',l) = Pr\left\{S_{t-1} = l'; S_t = l, \mathbf{r}_1^\tau\right\}, \quad l = 0,1,\cdots,M-1 \tag{4.7}$$

Eq. (4.6) can be written as

$$Pr\left\{c_t = 0|\mathbf{r}_1^\tau\right\} = \sum_{(l',l)\in B_t^0} \frac{\sigma_t(l',l)}{Pr\left\{\mathbf{r}_1^\tau\right\}} \tag{4.8}$$

Similarly the APP for $c_t = 1$ is given by

$$Pr\left\{c_t = 1|\mathbf{r}_1^\tau\right\} = \sum_{(l',l)\in B_t^1} \frac{\sigma_t(l',l)}{Pr\left\{\mathbf{r}_1^\tau\right\}} \tag{4.9}$$

The log-likelihood $\Lambda(c_t)$ is then

$$\Lambda(c_t) = log\frac{\sum_{(l',l)\in B_t^1} \sigma_t(l',l)}{\sum_{(l',l)\in B_t^0} \sigma_t(l',l)} \tag{4.10}$$

The log-likelihood $\Lambda(c_t)$ represents the soft output of the MAP decoder. It can be used as an input to another decoder in a concatenated scheme or in the next iteration in an iterative decoder. Finally, the decoder can make a hard decision by comparing $\Lambda(c_t)$ to a threshold equal to zero.

The joint probability $\sigma_t(l',l)$ can be computed recursively [50].

## 4.3 ITERATIVE MAP DECODING

A turbo code encoder based on a (2,1,4) recursive convolutional code [12] is shown in Fig. 4.2.

The first encoder in Fig. 4.2 has the information sequence $\mathbf{c}$ of length $N$ as its input

$$\mathbf{c} = (c_1, c_2, \cdots c_N)$$

It produces two output sequences denoted by $\mathbf{v}_o$ and $\mathbf{v}_1$. The second encoder has as its input the interleaved information sequence, denoted by $\tilde{\mathbf{c}}$. It generates two output sequences $\tilde{\mathbf{v}}_0$ and $\tilde{\mathbf{v}}_1$. However, the output sequences $\tilde{\mathbf{v}}_0$ is not transmitted because it can be recovered by deinterleaving from the sequence $\mathbf{v}_0$.

The first role of the interleaver is to generate a long code. Second, it decorrelates the inputs to the two decoders so that an iterative suboptimum decoding algorithm based on information exchange between the two component decoders

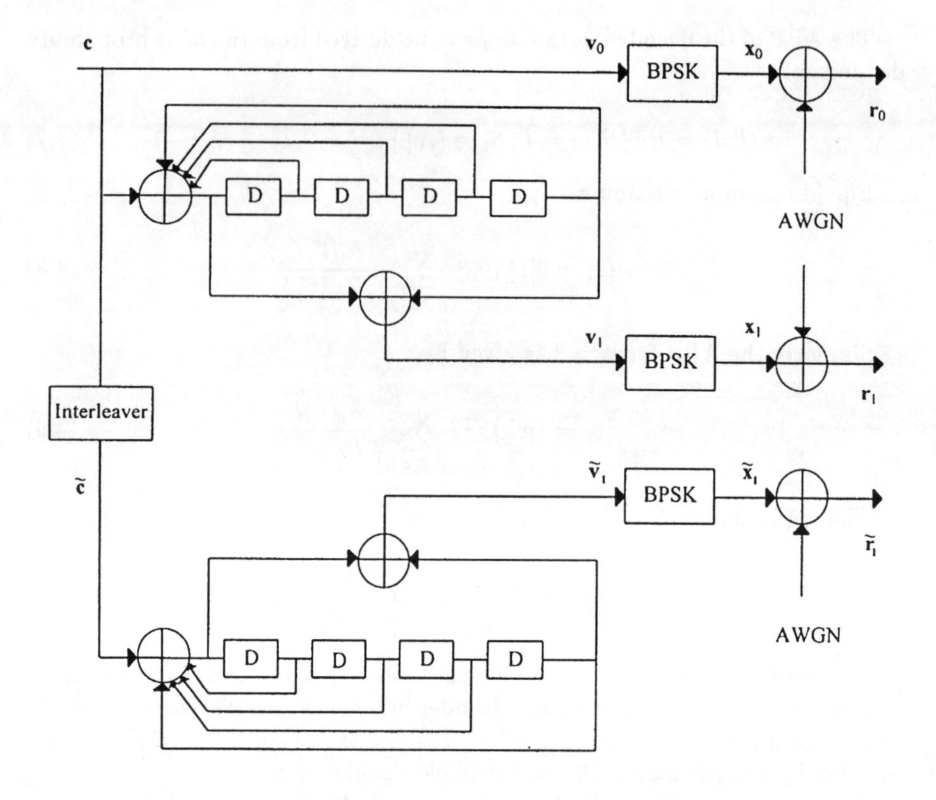


**Figure 4.2** A turbo encoder

can be applied. If the input sequences to the two component decoders are decorrelated there is a high probability that after the correction in one decoder the remaining errors should become correctable in the second decoder.

In the pseudo-random interleaver a block of $N$ input bits is read into the interleaver and read out pseudo-randomly. The pseudo-random pattern must be available at the decoder as well.

A turbo encoder generates an $(n\,(N+m)\,,N)$ block code where $n$ is the number of encoded sequences which is 3 in this example.

In general, the two component codes and their rates can be different.

Various overall turbo code rates such as 2/3, 3/4, 5/6 and so on, can be obtained by puncturing the 1/3 turbo code for the encoder shown in Fig. 4.2.

Another variation of the above turbo code construction is to use more than two component codes.

Decoding algorithms with soft output information, MAP or soft output Viterbi algorithms (SOVA) can be used for decoding of turbo codes.

An iterative turbo decoder is shown in Fig. 4.3.

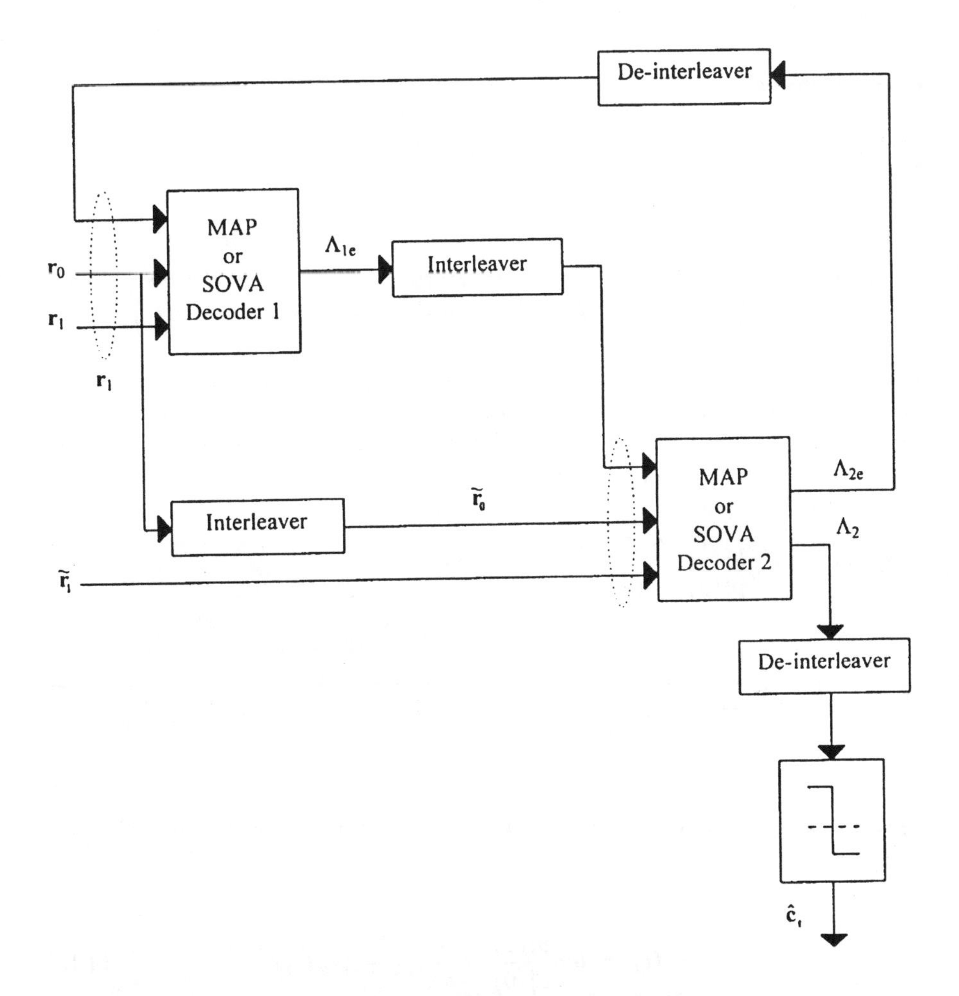


**Figure 4.3** An iterative turbo code decoder

In the MAP based iterative scheme, the first MAP decoder takes as input the received information sequence $\mathbf{r}_0$ and the received parity sequence generated by the first encoder $\mathbf{r}_1$, in Fig. 4.2. It then produces a soft output which is interleaved and used by the second decoder together with the interleaved received information sequence $\tilde{\mathbf{r}}_0$ and the received parity sequence produced by the second encoder, $\tilde{\mathbf{r}}_1$. The second decoder also produces a soft output which can be again used in the first decoder to further improve the soft output and so on.

After a certain number of iterations the soft outputs of both MAP decoders stop to produce further improvements. Then the last stage of decoding makes a hard decision.

Let us examine the first MAP decoder in the first iteration. The log-likelihood of the first MAP decoder is given by

$$\Lambda_1(c_t) = log\frac{\sum_{l=0}^{M-1}\alpha_t(l)p_t^1(1)exp\left(-\frac{\sum_{j=0}^{n-1}\left(r_{t,j}-x_{b,j}^1(l)\right)^2}{2\sigma^2}\right)\cdot\beta_t\left(S_f^1(l)\right)}{\sum_{l=0}^{M-1}\alpha_t(l)p_t^1(0)exp\left(-\frac{\sum_{j=0}^{n-1}\left(r_{t,j}-x_{b,j}^0(l)\right)^2}{2\sigma^2}\right)\cdot\beta_t\left(S_f^0(l)\right)} \tag{4.11}$$

where we introduce the notation $p_t^1(1)$ and $p_t^1(0)$ for the a priori probabilities for 1 and 0 at the input of the first encoder, respectively, since they will differ from the correspondong a priori probabilities at the input of the second decoder denoted by $p_t^2(1)$ and $p_t^2(0)$.

Rewrite $\Lambda_1(c_t)$ as

$$\begin{aligned}\Lambda_1(c_t) &= log\frac{p_t^1(1)}{p_t^1(0)}\\ &+ log\frac{\sum_{l=0}^{M-1}\alpha_t(l)exp\left(-\frac{\left(r_{t,0}-x_{b,0}^1\right)^2+\sum_{j=1}^{n-1}\left(r_{t,j}-x_{b,j}^1(l)\right)^2}{2\delta^2}\right)\cdot\beta_t\left(S_f^1(l)\right)}{\sum_{l=0}^{M-1}\alpha_t(l)exp\left(-\frac{\left(r_{t,0}-x_{b,0}^0\right)^2+\sum_{j=1}^{n-1}\left(r_{t,j}-x_{b,j}^0(l)\right)^2}{2\sigma^2}\right)\beta_t\left(S_f^0(l)\right)}\end{aligned} \tag{4.12}$$

since the code is systematic $x_{b,0}^i$, $i = 0, 1$ are independent of the trellis and state $l$.

$\Lambda_1(c_t)$ could be further decomposed into

$$\Lambda_1(c_t) = log\frac{p_t^1(1)}{p_t^1(0)} + \frac{2}{\sigma^2}r_{t,0} + \Lambda_{1e}(c_t) \tag{4.13}$$

where

$$\Lambda_{1e}(c_t) = log\frac{\sum_{l=0}^{M-1}\alpha_t(l)exp\left(-\frac{\sum_{j=1}^{n-1}\left(r_{t,j}-x_{b,j}^1(l)\right)^2}{2\sigma^2}\right)\beta_t\left(S_f^1(l)\right)}{\sum_{l=0}^{M-1}\alpha_t(l)exp-\left(\frac{\sum_{j=1}^{n-1}\left(r_{t,j}-x_{b,j}^0(l)\right)^2}{2\sigma^2}\right)\beta_t\left(S_f^0(l)\right)} \tag{4.14}$$

$\Lambda_{1e}(c_t)$ is called the *extrinsic information.* It is a function of the redundant information introduced by the encoder and supplied by the decoder.

Let us observe the input to the second MAP decoder. Since the input to the second decoder $\mathbf{r}_1^T$ includes the interleaved version of $\mathbf{r}_0$, denoted by $\tilde{\mathbf{r}}_0$, the received information signal $\tilde{r}_{t,0}$ correlates with the interleaved soft output $\tilde{\Lambda}_1(c_t)$. Therefore, the contribution due to $r_{t,0}$ must be taken out from $\Lambda_1(c_t)$.

However, $\Lambda_{1e}(c_t)$ does not contain $r_{t,0}$ and it can be used as the a priori probability for decoding in the second stage.

In the first iteration, in the first MAP decoder it is assumed that $p_t^1(1) = p_t^1(0) = 0.5$.

In the second iteration we use the extrinsic information from the second MAP decoder to compute a priori probability $p_t^1(1)$ and $p_t^1(0)$ in the following way

$$\Lambda_{2e}(c_t) = log\frac{p_t^1(1)}{p_t^1(0)} \tag{4.15}$$

From (4.15) and the relationship

$$p_t^1(1) = 1 - p_t^1(0)$$

we can write for the a priori probabilities

$$p_t^1(1) = \frac{e^{\tilde{\Lambda}_{2e}(c_t)}}{1 + e^{\tilde{\Lambda}_{2e}(c_t)}} \tag{4.16}$$

$$p_t^1(0) = 1 - \frac{e^{\tilde{\Lambda}_{2e}(c_t)}}{1 + e^{\tilde{\Lambda}_{2e}(c_t)}} \tag{4.17}$$

Substituting (4.16) and (4.17) into (4.13) we get

$$\Lambda_1(c_t) = \Lambda_{1e}(c_t) + \frac{2}{\sigma^2}r_{t,0} + \tilde{\Lambda}_{2e}(c_t) \tag{4.18}$$

Similarly

$$\Lambda_2(c_t) = \Lambda_{2e}(c_t) + \frac{2}{\sigma^2}\tilde{r}_{t,0} + \tilde{\Lambda}_{1e}(c_t) \tag{4.19}$$

## 4.4 SUMMARY OF THE ITERATIVE MAP DECODING METHOD

1. Initialise $\Lambda_{2e}^{(0)}(c_t) = 0$.

- For iterations $r = 1, 2, \cdots I$ where $I$ is the total number of iterations, compute $\Lambda_1^{(r)}(c_t)$ and $\Lambda_2^{(r)}(c_t)$ by using Eq. (4.11).

- Compute $\Lambda_{1e}^{(r)}(c_t)$ as

$$\Lambda_{1e}^{(r)}(c_t) = \Lambda_1^{(r)}(c_t) - \frac{2}{\sigma^2} r_{t,0} - \tilde{\Lambda}_{2e}^{(r-1)}(c_t) \tag{4.20}$$

- Compute $\Lambda_{2e}^{(r)}$ as

$$\Lambda_{2e}^{(r)}(c_t) = \Lambda_2^{(r)}(c_t) - \frac{2}{\sigma^2} \tilde{r}_{t,0} - \tilde{\Lambda}_{1e}^{(r)}(c_t) \tag{4.21}$$

2. After $I$ iterations make a hard decision on $c_t$ based on $\Lambda_2(c_t)$.

## 4.5 SOFT OUTPUT VITERBI ALGORITHM

The MAP algorithm gives minimum symbol error probability and provides optimum soft outputs as symbol-by-symbol a posteriori probabilities (SAPs). The MAP algorithm needs the knowledge of noise variance and performs computations in the probabilistic domain. We proposed an asymptotically optimum soft output algorithm for multilevel modulations that stores a soft survivor for each state and updates it at each recursion step. It has distance metric [4].

We modified the soft output Viterbi algorithm for block codes so that it has the complexity of only 1.5 the complexity of the standard Viterbi algorithm, does not require noise variance estimation and operates in the distance domain where only simple operations of addition and multiplication are performed [5] [6].

The algorithm can be successfully used to iteratively decode turbo codes with a slight loss of 0.6dB relative to the MAP algorithm.

We consider the same system as shown in Fig. 4.1.

The standard Viterbi decoding algorithm finds an information sequence $\mathbf{c} = \mathbf{c}_1^t$ that corresponds to the modulated sequence $\mathbf{x} = \mathbf{x}_1^t$ in the trellis diagram such that the probability $Pr(\mathbf{c}, \mathbf{r}_1^t)$ is maximised. This probability is given by

$$\begin{aligned} Pr(\mathbf{c}, \mathbf{r}_1^t) &= Pr(\mathbf{c})\, Pr(\mathbf{r}_1^t|\mathbf{c}) \\ &= Pr(\mathbf{c})\, Pr(\mathbf{r}_1^t|\mathbf{x}_1^t) \end{aligned} \tag{4.22}$$

since there is one to one correspondence between $\mathbf{c}$ and $\mathbf{x}$.

In order to simplify the operations we introduce the log function $logPr(\mathbf{c}, \mathbf{r}_1^t)$. For the Gaussian channel $logPr(\mathbf{c}, \mathbf{r}_1^t)$ becomes

$$\begin{aligned} logPr(\mathbf{c}, \mathbf{r}_1^t) &= \sum_{t=1}^N logP(c_t) + \sum_{t=1}^N log \prod_{i=0}^{n-1} \frac{1}{\sqrt{2\pi}\sigma} e^{-\frac{(r_{t,i}-x_{t,i})^2}{2\sigma^2}} \\ &= \sum_{t=1}^N logP(c_t) + \frac{nN}{\sqrt{2\pi}\sigma} - \sum_{t=1}^N \sum_{i=0}^{n-1} \frac{(r_{t,i}-x_{t,i})^2}{2\sigma^2} \end{aligned} \tag{4.23}$$

where $P(c_t)$ is the a priori probability of a binary symbol $c_t$.

Expression (4.23) shows that maximising $Pr(\mathbf{c}, \mathbf{r}_1^t)$ is equivalent to minimising the Euclidean distance $\sum_{t=1}^{N} \sum_{i=0}^{n-1} (r_{t,i} - x_{t,i})^2$ between the received sequence $\mathbf{r}_1^t$ and modulated sequences $\mathbf{x}_1^t$ in the trellis since $P(c_t)$ is constant and equal 0.5 for binary independently generated symbols.

We assign to each branch on the path $\mathbf{x}$ in the trellis the Euclidean distance, called *branch metric* and denoted by $\nu_t$, as follows

$$\nu_t^{(\mathbf{x})} = \sum_{i=0}^{n-1} (r_{t,i} - x_{t,i})^2 \tag{4.24}$$

Then the path metric corresponding to the path $\mathbf{x}$, denoted by $\mu_t^{(\mathbf{x})}$ is given by

$$\mu_t^{(\mathbf{x})} = \sum_{t=1}^{N} \nu_t^{(\mathbf{x})} \tag{4.25}$$

The standard Viterbi decoder selects a path in the trellis with the minimum path metric.

The soft output Viterbi algorithm (SOVA) estimates the soft output information in the form of the log likelihood function $\Lambda(c_t)$, given the received sequence $\mathbf{r}_1^\tau$

$$\Lambda(c_t) = log \frac{Pr\{c_t = 1 \mid \mathbf{r}_1^\tau\}}{Pr\{c_t = 0 \mid \mathbf{r}_1^\tau\}} \tag{4.26}$$

where $Pr\{c_t = i \mid \mathbf{r}_1^\tau\}$, $i = 0, 1$, is the a priori probability of the data bit $c_t$.

The SOVA decoder makes a hard decision by comparing $\Lambda(c_t)$ to a threshold equal to zero

$$c_t = \begin{cases} 1 & \text{if } \Lambda(c_t) \geq 0 \\ 0 & \text{otherwise} \end{cases} \tag{4.27}$$

The SOVA decoder selects the path $\mathbf{x}$ with the minimum path metric $\mu_{\tau,min}$ as the maximum likelihood path in the same way as the standard Viterbi algorithm. The probability of selecting this path is proportional to

$$Pr\{\mathbf{x}|\mathbf{r}_1^\tau\} \sim e^{-\mu_{\tau,min}} \tag{4.28}$$

Let us denote by $\mu_{\tau,c}$ the metric of the strongest competitor of the maximum likelihood path, which is the minimum path metric for the path obtained when the trellis symbol on the maximum likelihood path at time $t$ is replaced by its complementary symbol.

If the hard estimate of the maximum likelihood path at time $t$ is 1, then the estimate at time $t$ of the strongest competitor is 0. Therefore

$$Pr\{c_t = 1|\mathbf{r}_1^\tau\} \sim e^{-\mu_{\tau,min}} \quad (4.29)$$

$$Pr\{c_t = 0|\mathbf{r}_1^\tau\} \sim e^{-\mu_{\tau,c}} \quad (4.30)$$

The logarithm of the ratio of the above two probabilties is

$$\log \frac{Pr\{c_t = 1|\mathbf{r}_1^\tau\}}{P\{c_t = 0|\mathbf{r}_1^\tau\}} = \log \frac{e^{-\mu_{\tau,min}}}{e^{-\mu_{\tau,c}}} = \log e^{\mu_{\tau,c}-\mu_{\tau,min}} = \mu_{\tau,c} - \mu_{\tau,min} \quad (4.31)$$

Let us denote by $\mu_\tau^1$ the minimum path metric if $c_t$ is 1 and $\mu_\tau^0$ the minimum path metric if $c_t$ is 0.

The soft output of the SOVA decoder can in general be expressed as

$$\begin{aligned}\Lambda(c_t) &= log \frac{P\{c_t = 1 \mid \mathbf{r}_1^\tau\}}{P\{c_t = 0 \mid \mathbf{r}_1^\tau\}} \\ &= \mu_\tau^0 - \mu_\tau^1 \end{aligned} \quad (4.32)$$

If the decision is made on a finite length block, as in block codes, turbo codes or convolutional codes in TDMA systems, the proposed algorithm can be implemented in as a bidirectional recursive method with forward and backward recursions. As an example, the following simple two-state trellis diagram, shown in Fig. 4.4 is used to illustrate the algorithm. The trellis has a total of $W+1 =$

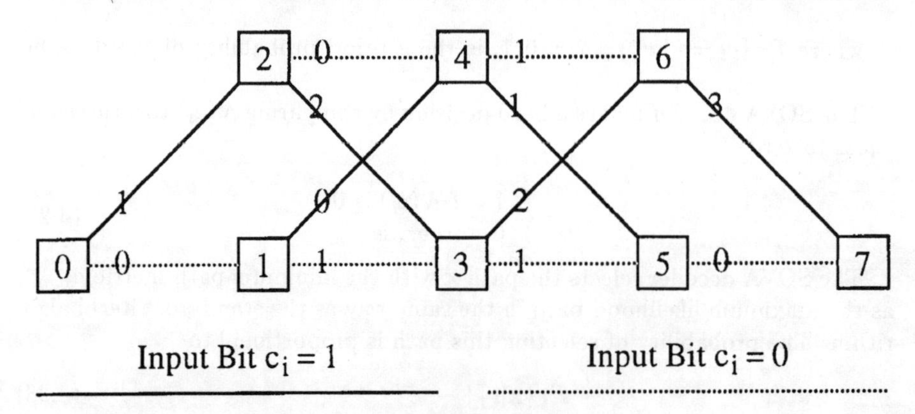


**Figure 4.4** An example of a trellis diagram

$M\dot{\tau} = 8$ nodes, which are denoted by squares and labeled by the numbers $0, 1, \ldots, 7$. The trellis starts with node 0 and ends at node $W = 7$. Each branch is labeled by its branch metric, which can be computed from the received sequence $\mathbf{r}$. The forward recursion uses the well known operation of Add-Compare-Select (ACS) to find:

1. The $W$ minimum path metrics, denoted by $\mu_{f,i}$, from node 0 to the other $i = 1, 2, ..., W = 7$ nodes.

2. The maximum-likelihood path metric, denoted by $\mu_{min}$, from node 0 to node $W = 7$.

The result of the forward recursion is shown in Fig. 4.5 where the nodes are labeled with the minimum metrics $\mu_{f,\tau}$, for $\tau = 1, 2, ..7$ from node 0. The concatenated branches show the maximum-likelihood path wich has the path metric equal to the sum of the branch metrics $\mu_{min} = 1$. The backward recur-

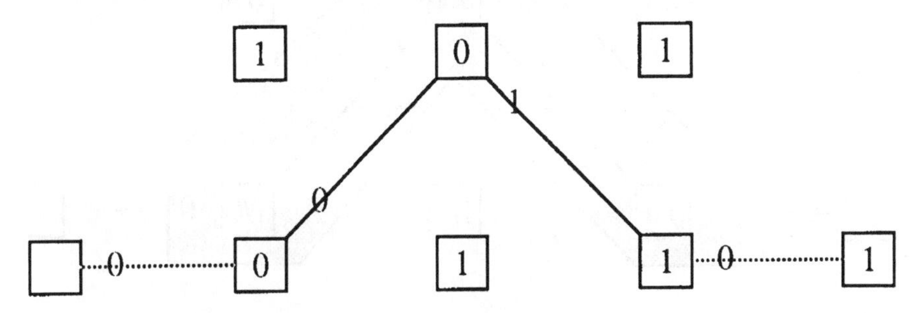


**Figure 4.5** Forward recursion

sion also uses the operation of ACS to find:

1. The $W - 1$ minimum path metrics, denoted by $\mu_{b,i}$, from node $W = 7$ to nodes $i = W - 1, W - 2, \ldots, 1$. These minimum metrics together with those from the forward recursion allow the computation of the strongest competitor path metric $\mu_c$ as

$$\mu_c = \mu_{f,i} + \mu_{b,i}, \tag{4.33}$$

2. The soft output value is then from Eq. (4.32)

$$\Lambda(c_t) = \mu^0 - \mu^1 \tag{4.34}$$

For example, in Fig. 4.5 to the hard estimate at time $t = 1$ is read from the maximum likelihood path and it is 0. Therefore, at time $t = 1$

$$\mu_{min} = \mu^0 = 1$$

The metric of the strongest competitor at time $t = 1$ $\mu_c$ corresponds to the complementary symbol which is 1. Thus

$$\mu_c = \mu^1$$

The results for both forward and backward recursion are shown in Fig. 4.6 The first number in the nodes in Fig. 4.6 represents the minimum path metrics for the forward recursion and the second number represents the minimum path metrics for the backward recursion. For example, at time $t = 1$, the hard

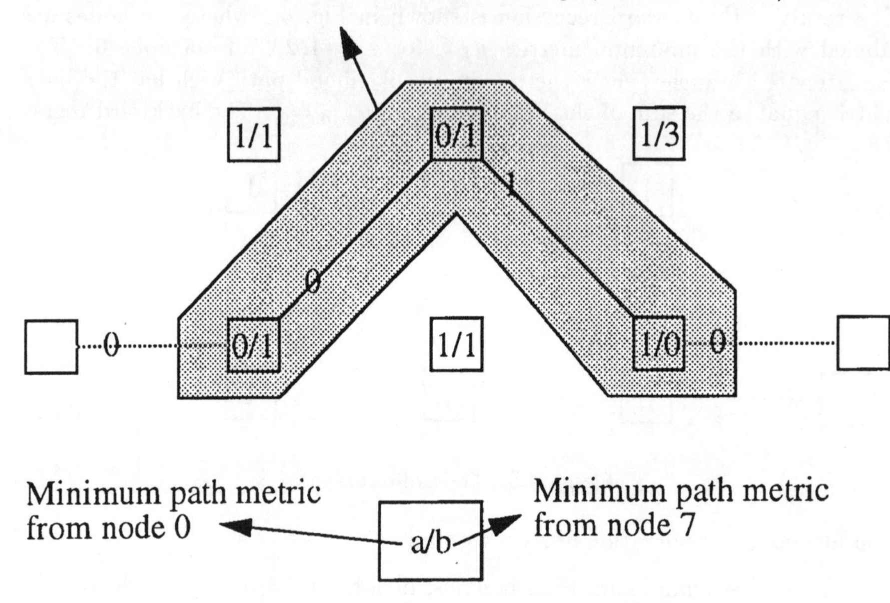


**Figure 4.6** Forward-backward recursion

estimate of the transmitted symbol can be read from the maximum likelihood path in Fig. 4.5 and it is $\hat{c}_1 = 0$. If we remove the path which corresponds to binary 0 at time 1, we obtain the path which corresponds to $c_1 = 1$. This path ends in node 2 at time 1. For this path the minimum path metric is obtained from Eq. (4.33) as the sum of the feedforward path metric $\mu_{f,1} = 1$, shown by the first number, and the backward path metric $\mu_{b,1} = 1$, shown by the second number in the node, as follows

$$\mu_c = \mu^1 = \mu_{f,i} + \mu_{b,i} = 1 + 1 = 2$$

The soft output value is then from Eq. (4.34) computed as

$$\Lambda(c_t) = \mu^0 - \mu^1 = 1 - 2 = -1$$

Note that the soft-output decision can be made after each backward recursion step. Because of this, no substantial extra memory is required for backward recursion. The computational complexity of the forward recursion is equivalent to that of the Viterbi algorithm. The computational complexity of the backward recursion is usually less than that of the Viterbi algorithm since only part of the trellis need to be searched. Therefore, the computational complexity is upper bounded by two times that of the Viterbi algorithm. With binary input bits $c_t$, the computational complexity is about 1.5 times that of the Viterbi algorithm.

## 4.6 ITERATIVE SOFT OUTPUT VITERBI ALGORITHM

Let us consider again a turbo encoder shown in Fig. 4.2 The block diagram of the iterative soft output Viterbi algorithm is shown in Fig. 4.3. The first soft output Viterbi algorithm (SOVA) decoder produces the soft output $\Lambda_1(c_t)$ as where $t$ is the sampling instant. The soft output is interleaved and used by the second SOVA decoder. The second SOVA decoder also produces its soft output $\Lambda_2(c_t)$ which is used in the next iteration in the first SOVA decoder.

The soft output of the $i-th$ SOVA decoder, where $i = 1, 2$, is obtained from Eq. (4.32) as

$$\Lambda_1(c_t) = \mu_\tau^0 - \mu_\tau^1 \tag{4.35}$$

In iterative decoding the a priori probability $c_t$ in general is different from 0.5, as the previous decoding stage will provide its improved estimate relative to the one in an independent binary source. Therefore the path metric in Eq. (4.35) for the first SOVA decoder can be computed, from Eqs. (4.23), (4.25) and (4.24) as

$$\mu_t^{(x)} = \mu_{t-1}^{(x)} + \sum_{i=0}^{n-1} (r_{t,i} - x_{t,i})^2 - c_t log\left[\frac{P(c_t = 1)}{P(c_t = 0)}\right] \tag{4.36}$$

where $c_t log\left[\frac{P(c_t=1)}{P(c_t=0)}\right]$ is provided by the previous decoding stage. As with the MAP algorithm, the soft output information $\Lambda(c_t)$ from the SOVA decoder is decomposed into two parts, the intrinsic information $\Lambda_i(c_t)$ and the extrinsic information $\Lambda_e(c_t)$. The extrinsic information $\Lambda_e(c_t)$ can be used as the a priori information in the next stage decoding after interleaving or deinterleaving. Let us denote by $\Lambda_1(c_t)$ the output of the first decoder. Its intrinsic information is $\Lambda_{1i}(c_t)$ and the extrinsic information is $\Lambda_{1e}(c_t)$ . By combining Eqs. (4.32) and (4.36) the output of the first SOVA decoder can be expressed as

$$\begin{aligned}
\Lambda_1(c_t) &= \mu_\tau^0 - \mu_\tau^1 \\
&= \left\{ \mu_{\tau-1}^0 + \sum_{i=0}^{n-1} \left(r_{t,i} - x_{t,i}^0\right)^2 \right\} \\
&\quad - \left\{ \mu_{\tau-1}^1 + \sum_{i=0}^{n-1} \left(r_{t,i} - x_{t,i}^1\right)^2 - \log\left[\frac{P(c_t=1)}{P(c_t=0)}\right] \right\} \\
&= \log\left[\frac{P(c_t=1)}{P(c_t=0)}\right] + 4r_{t,0} + 2\sum_{i=1}^{n-1}(x_{t,i}^1 - x_{t,i}^0)r_{t,i} \\
&\quad + \mu_{\tau-1}^0 - \mu_{\tau-1}^1 \\
&= \Lambda_{1i}(c_t) + \Lambda_{1e}(c_t) \qquad (4.37)
\end{aligned}$$

where

$$\Lambda_{1i}(c_t) = \log\left[\frac{P(c_t=1)}{P(c_t=0)}\right] + 4r_{t,0} \qquad (4.38)$$

and

$$\Lambda_{1e}(c_t) = 2\sum_{i=1}^{n-1} \left(x_{t,i}^1 - x_{t,i}^0\right) r_{t,i} + \mu_{\tau-1}^0 - \mu_{\tau-1}^1 \qquad (4.39)$$

So, the extrinsic information $\Lambda_{1e}(c_t)$ can be obtained from (4.37) and (4.38) as

$$\begin{aligned}
\Lambda_{1e} &= \Lambda_1(c_t) - \Lambda_{1i}(c_t) \\
&= \Lambda_1(c_t) - \log\left[\frac{P(c_t=1)}{P(c_t=0)}\right] - 4r_{t,0} \qquad (4.40)
\end{aligned}$$

where $\log\left[\frac{P(c_t=1)}{P(c_t=0)}\right]$ is the loglikelihood ratio of the a priori probabilities equal to the deinterleaved extrinsic information of the second decoder in the iterative decoding. Let us denote by $\tilde{\Lambda}_{2e}(c_t)$ the deinterleaved extrinsic information of the second decoder. Then Expression (4.40) becomes

$$\begin{aligned}
\Lambda_{1e}(c_t) &= \Lambda_1(c_t) - \Lambda_{1i}(c_t) \\
&= \Lambda_1(c_t) - \tilde{\Lambda}_{2e}(c_t) - 4r_{t,0} \qquad (4.41)
\end{aligned}$$

Similarly, let us denote by $\Lambda_2(c_t)$ the output of the second decoder with its intrinsic information and extrinsic information $\Lambda_{2i}(c_t)$ and $\Lambda_{2e}(c_t)$, respectively. Then as in (4.40) we get

$$\Lambda_{2e}(c_t) = \Lambda_2(c_t) - \Lambda_{2i}(c_t)$$

$$= \Lambda_2(c_t) - \log\left[\frac{P(c_t=1)}{P(c_t=0)}\right] - 4r_{t,0}$$
$$= \Lambda_2(c_t) - \tilde{\Lambda}_{1e}(c_t) - 4r_{t,0} \quad (4.42)$$

where $\tilde{\Lambda}_{1e}(c_t)$ is the interleaved extrinsic information from the first decoder.

## 4.7 SUMMARY OF THE ITERATIVE SOVA DECODING ALGORITHM

1. Initialise $\Lambda_{2e}^{(0)}(c_t) = 0$.
2. For iterations $r = 1, 2, \cdots I$ where $I$ is the total number of iterations, compute $\Lambda_1^{(r)}(c_t)$ and $\Lambda_2^{(r)}$ by using (4.32) and (4.36).
3. After $I$ iterations make a hard decision on $c_t$ from $\Lambda_2^{(I)}(c_t)$

## 4.8 MAP AND SOVA PERFORMANCE COMPARISON

The bit error probability for the 16-state rate 1/3 turbo code for the MAP and SOVA decoding algorithms with the same number of iterations of 18 and interleaver size of 4096 are shown in Fig. 4.7. The SOVA decoding results in a performance degradation of 0.7dB at the bit error probability of $10^{-5}$ but it is significantly less complex. The simulation running time of the SOVA decoder is about one half of the running time of the MAP decoder.

## 4.9 CONCLUSIONS

The iterative MAP and bidirectional soft output Viterbi algorithms are presented and applied to decoding of turbo codes. The MAP algorithm is computationally much more complex than the Viterbi algorithm. The operations in the MAP algorithm are multiplications and exponentiations while in the Viterbi algorithm they are simple add, compare and select operations.

As an example these algorithms are applied to decoding of turbo codes and their performance is compared on a Gaussian channel. The MAP iterative algorithm has a performance advantage on the Gaussian channel of about 0.5dB at the bit error probability of $10^{-4}$ relative to the SOVA iterative decoding method.

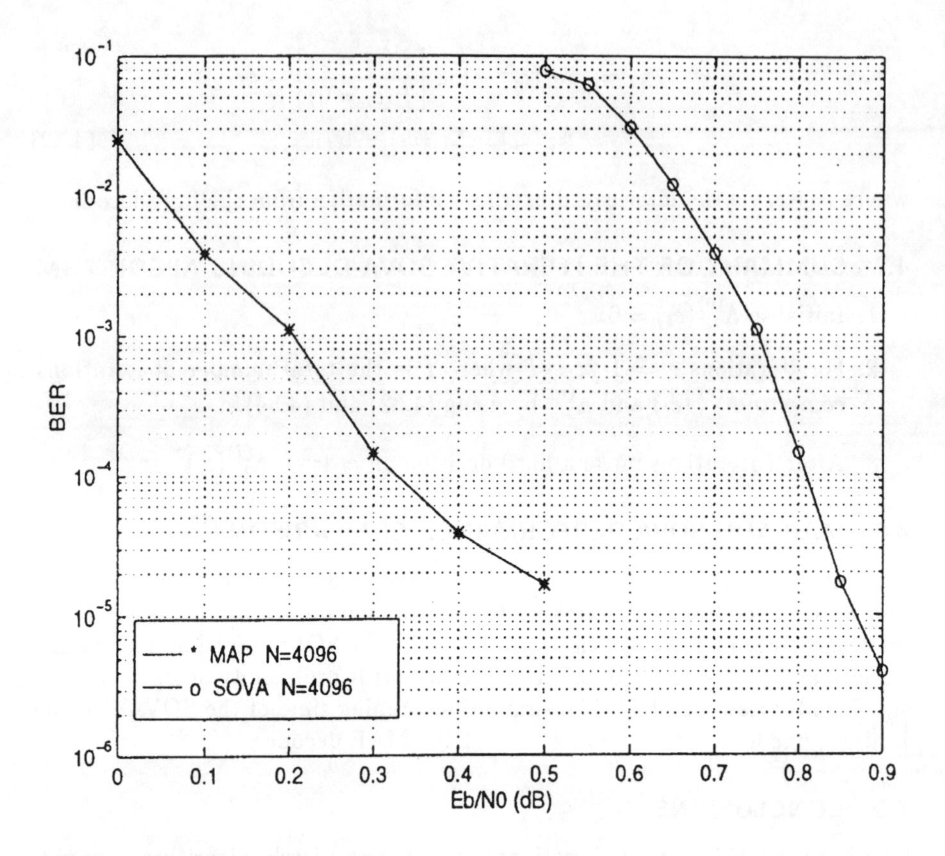


**Figure 4.7** Bit error performance of a 1/3 rate turbo code on a Gaussian channel for various decoding algorithms

## References


[1] G. D. Forney, JR. "The Viterbi Algorithm", Proceeding of IEEE, Vol. 61, No. 3, March 1973.

[2] Yamamoto H. and Itoh K., "Viterbi Decoding for Convolutional Codes with Repeat Request", IEEE on Inform. Theory, vol. IT-26, pp. 540-547, Sept. 1980.

[3] J. Hagenauer, P. Hoeher, "A Viterbi Algorithm with Soft-Decision Outputs and its Applications," Conf. Rec. *GLOBECOM'89*, Dallas, Texas, Vol. 3, pp. 47.1.1–47.1.7, Nov. 1989.

[4] Y. Li, B. Vucetic and Y. Sato, "Optimum Soft Output Detection for Channels with Intersymbol Interference", IEEE Trans. Inform. Theory, Vol-41, No-3, May 1995, p. 704-713.

[5] Yunxin Li and Branka Vucetic, "A Low-complexity Soft-output TDMA Receiver", TELFOR '95, Belgrade, 3-8 Dec. 1995, Yugoslavia.

[6] Yunxin Li and Branka Vucetic, "A Genmeralized MLSE Algorithm", INNSP '95, China, December 10-13, 1995.

[7] T. Hashimoto "A list-Type Reduced-Constraint Generalization of the Viterbi Algorithm," *IEEE Trans. Inform. Theory*, Vol. IT-33, No. 6, pp. 866–876, Nov. 1987.

[8] B. Vucetic, "Bandwidth Efficient Concatenated Coding Schemes on Fading Channels" IEEE Trans. Commun., Jan. 1993.

[9] T. Schaub and J.W. Modestino "An Erasure Declaring Viterbi Decoder and its Application to Concatenated Coding Systems," *ICC'86*, IEEE Cat. No. CH23143/86, pp. 1612-1616, 1986

[10] Branka Vucetic, Elvio Leonardo and Lin Zhang,"Soft Output Multistage Decoding of Multilevel Block Codes", IEEE ITW'95 Proceedings, Rydzyna, Poland, June 15-19 1995.

[11] N. Seshadri and C-E.W. Sundberg, "Generalized Viterbi Algorithms for Error Detection with Convolutional Codes," Conf. Rec. *GLOBECOM'89*, Dallas, Texas, Vol. 3, pp. 43.3.1–43.3.5, Nov. 1989.

[12] C. Berrou, A. Glavieux and P. Thitimajshima, Near Shannon Limit Error-Correcting Coding and Decoding Turbo Codes (1), Proc. ICC'93, Geneva, Switzerland, pp. 1064-1070, May 1993.

[13] S. Hirasawa et al, Modified Product Codes, IEEE Trans. Inform. Theory, Vol. IT-30, March 1984, pp. 299-306.

[14] G.D. Forney, Concatenated Codes, MIT, Cambridge, MA, USA, 1966.

[15] P. Elias, Error-free Coding, IEEE Trans. Inform. Theory, Vol. IT-4, pp. 29-37, Sept. 1954.

[16] H. Imai and S. Hirakawa, A New Multilevel Coding Method Using Error-Correcting Codes", IEEE Trans. Inform. Theory, Vol. IT-23, May 1975.

[17] J. Hagenauer and P. Robertson, Iterative (Turbo) Decoding of Systematic Convolutional Codes with the MAP and SOVA Algorithms", Proc. ITG Conf. Frankfurt, Germany, Oct. 1994.

[18] S. Le Goff et al, Turbo Codes and High Spectral Efficiency Modulation, Proc. Globecom'94, San Francisco, California, USA, pp. 645-649, Dec. 1994.

[19] P. Robertson, Illuminating the Structure of Code and Decoder of Parallel Concatenated Recursive Systematic (Turbo) Codes, Proc. Globecom'94, San Francisco, California, USA, pp. 1298-1303, Dec. 1994.

[20] A.J. Viterbi and J.K. Omura, *Principles of Digital Communications and Coding*, New York McGraw Hill , 1979.

[21] Claude Berrou and Alain Glavieux, Turbo-codes: General Principles and Applications, Audio and Video Digital Radio Broadcasing System and Techniques 1994, pp.215-226,

[22] P. Jung, Novel Low Complexity Decoder for Turbo-Codes, Electronics Letters, 1995, Jan., Vol. 31, No. 2, pp. 86–87,

[23] A.S. Barbulescu and S.S. Pietrobon, Terminating the Trellis of Turbo-Codes in the Same State, Electronics Letters, 1995, Jan., Vol. 31, No. 1, pp. 22-23

[24] P. Jung and M. Nabhan Performance Evaluation of Turbo Codes for Short Frame Transmission Systems, Electronics Letters, 1994, Jan., Vol. 30, No. 2, pp. 111-113

[25] Joachim Hagenauer and Lutz Papke, Iterative Decoding of Binary Block and Convolutional Codes, IEEE Trans. Inform. Theory, 1996, March, Vol. 42, No. 2, pp. 429-445.

[26] S. Benedetto and G. Montorsi, Average Performance of Parallel Concatenated Block Codes, Electronics Letters, 1995, Feb., Vol. 31, No. 3, pp. 156-158.

[27] S. Benedetto and G. Montorsi, Performance Evaluation fo Turbo Codes, Electronics Letters, 1995, Feb., Vol. 31, No. 3, pp. 163-165.

[28] S. Benedetto and G. Montorsi, Design of Parallel Concatenated Convolutional Codes, IEEE Trans. Commun. 1996, May, Vol. 44, No. 5, pp. 591-600.

[29] S. Benedetto, G. Montorsi, D. Divsalar and F. Pollara, Serial Concatenation of Interleaved Codes: Performance Analysis, Design, and Iterative Decoding, TDA Progress Report 42-126, Aug. 1996.

[30] S. Benedetto and G. Montorsi, Unveiling Turbo Codes: Some Results on Parallel Concatenated Coding Schemes, IEEE Trans. Inform. Theory, Vol. 42, No. 2, March 1996, pp. 409-428.

[31] Claude Berrou, Patrick Adde, Ettiboua Angui and Stephane Faudeil, A Low Complexity Soft-Output Viterbi Decoder Architecture, ICC93, 1993, pp. 737-740.

[32] J. Lodge, R.Young, P. Hoeher, J. Hagenauer, Separable MAP "Filters" for the Decoding of Product and Concatenated Codes, ICC93, 1993, pp. 1740-1745.

[33] Robert J. McEliece, Eugene R. Rodemich, Jung-Fu Cheng, The Turbo Decision Algorithm, The 33rd Alerton Conference on Communications, Computing and Control, October 1995, pp. 1-11.

[34] K.Fazel, L.Papke, Combined Multilevel Turbo-code With 8PSK Modulation, GLOBECOM95. 1995, pp. 649-653.

[35] Stephane Le Goff, Alain Glavieux and Claude Berrou, Turbo-Codes and High Spectral Efficiency Modulation, GLOBECOM94, 1994, pp. 645-649.

[36] Lin Zhang, Weimin Zhang, Jeff T. Ball and Martin C. Gill, MILCOM96, An Extremely Robust Tutbo Coded HF Modem, 1996.

[37] W.J. Blackert and S.G. Wilson, Turbo Trellis Coded Modulation, Proc. CISS'96, 1996, Princeton, NJ, USA.

[38] W.J. Blackert, E.K. Hall and S.G. Wilson, An Upper Bound on Turbo Code Free Distance, ICC96, 1996, pp. 957-961.

[39] Patrick Robertson and Thomas Worz, A Novel Bandwidth Efficient Coding Scheme Employing Turbo Codes, ICC96, 1996, pp. 962-967.

[40] Ramesh Pyndiah, Annie Picart and Alain Glavieux, Performance of Block Coded 16-QAM and 64-QAM Modulations, GLOBECOM95, 1995, pp. 1039-1043

[41] Joachim Hagenauer and Peter Hoeher, A Viterbi Algorithm with Soft-Decision Outputs and its Applications, GLOBECOM89, 1989, pp. 1680-1686.

[42] U. C. G. Fiebig and P. Robertson, Soft Decision Decoding in Fast Frequency Hopping Systems with Convolutional Codes and Turbo Codes, ICC96, 1996, pp. 1064-1070.

[43] Stefan Kaiser and Lutz Papke, Optimal Detection when Combining OFDM-CDMA with Convolutional and Turbo Channel Coding, ICC96, 1996, pp. 343-348.

[44] Gerard Battail, Claude Berrou and Alain Glavieux, Pseudo-Random Recusive Convolutional Coding for Near-Capacity Performance, GLOBECOM93, 1993.

[45] D. Divsalar, S. Dolinar, R. J. Mceliece and F. Pollara, Transfer Function Bounds on the Performance of Turbo Codes, JPL TDA Progress Report 42-122, Aug. 1995.

[46] D. Divsalar and F. Pollara, Turbo Codes for Deep-Space Communications, JPL TDA Progress Report 42-120, Feb. 1995

[47] D. Divsalar and F. Pollara, Mutiple Turbo Codes for Deep-Space Communications, JPL TDA Progress Report 42-121, May. 1995

[48] K. Abend and B.D. Fritchman, "Statistical Detection for Communication Channels with Intersymbol Interference," *Proc. IEEE* , Vol. 58, No. 5, pp. 779–785, May 1970.

[49] J. Raviv, " Decision Making in Markov Chains Applied to the Problem of Pattern Recognition," *IEEE Trans. Inform. Theory*, Vol. IT-13, pp. 536-551, Oct. 1967

[50] L.R. Bahl, J. Cocke, F. Jelinek and J. Raviv, "Optimal Decoding of Linear Codes for Minimizing Symbol Error Rate," *IEEE Trans. Inform. Theory* , Vol. IT-20, pp. 284–287, Mar. 1974.

[51] R.W. Chang and J.C. Hancock "On Receiver Structures for Channels Having Memory," *IEEE Trans. Inform. Theory*, Vol. IT-12, pp. 463-468, Oct. 1966.

[52] P.R. Chevillat and E. Eleftheriou, "Decoding of Trellis-Encoded Signals in the Presence of Intersymbol Interference and Noise," *IEEE Trans. Commun.*, Vol. 37, No. 7, pp. 669–676, Jul. 1989.

# 5 ACQUISITION OF DIRECT-SEQUENCE SPREAD-SPECTRUM SIGNALS

Dilip V. Sarwate

**Abstract:** In order to successfully demodulate a digitally modulated signal, the receiver must first synchronize its oscillators and clocks to the received signal. If direct-sequence spread-spectrum modulation is used, then it is also necessary to synchronize to the code sequence that was used to produce the spectral spreading. Methods used to achieve coarse synchronization to the code sequence, also called acquisition of the code sequence, will be discussed in this chapter. Serial search methods require comparatively little hardware but can take a long time to achieve acquisition. On the other hand, parallel search methods can acquire the code sequence very quickly, but can be somewhat expensive to implement. A survey of some of these basic ideas is presented and some improvements on known methods are suggested.

## 5.1 INTRODUCTION

In order to function successfully, a receiver in a digital communication system must first synchronize itself with the incoming signal. There are many facets to this synchronization process. The receiver must achieve *carrier synchronization* by synchronizing its oscillator to the radio-frequency (RF) carrier signal at its input. Synchronization in both frequency and phase is needed for coherent demodulation while only frequency synchronization suffices for noncoherent demodulation. Once carrier synchronization has been achieved, the receiver attempts *symbol synchronization* – that is, the receiver determines when each symbol starts and ends so that it can sample the demodulated baseband waveform at the appropriate time instants. Since the symbols are usually grouped into codewords or frames or packets etc., the receiver must achieve *frame synchronization* as well – that is, the receiver must determine where each such group starts and ends. Such tasks are common to all digital communication systems, and there is an extensive literature on these subjects. The subject of this chapter, however, is an additional task that must be performed by the receiver in a *direct-sequence spread-spectrum* (DS/SS) communication system, and that is the achievement of *code synchronization* whereby the receiver synchronizes its locally generated code sequence (also known as the signature sequence) to the code sequence in the received signal.

Generally speaking, a DS/SS receiver must achieve code synchronization before it can attempt carrier synchronization. It is, of course, possible (in theory) to remove binary modulation by squaring (and quaternary modulation by taking the fourth power of) the incoming signal, and then to synchronize to the unmodulated double frequency (or quadruple frequency) signal via a phase-lock loop, just as one does in narrowband communication systems. However, such methods often fail in practice for direct-sequence modulation for the following reason. The signal energy is spread over a large bandwidth, and attempts to recover a substantial fraction of this energy cause far too much noise to be injected into the phase-lock loop. Consequently, the loop has difficulty in locking to and tracking the signal. On the other hand, narrowband filtering of the signal prior to squaring reduces the noise but it also reduces the signal energy that is captured. Thus, the phase-lock loop still has a very poor signal-to-noise ratio and cannot achieve lock easily. On the other hand, once code synchronization has been achieved, multiplying the incoming signal with the local (synchronized) code replica strips off the direct-sequence modulation, leaving an "ordinary" narrowband digitally modulated signal that can be synchronized to and demodulated via standard techniques.

Not only is code synchronization a necessary first step for successful data transmission over a link in a DS/SS system, but the failure to achieve code synchronization also limits the capacity of DS/SS networks. A typical measure of system capacity of a DS/SS network is the number of simultaneous transmissions that can be successfully supported subject to a performance measure such as the error probability or the signal-to-noise ratio satisfying specified criteria. It is known that this capacity is proportional to the bandwidth expansion factor (also called processing gain) (see e.g. [13], [15]), that is, the capacity is asymptotically $O(W)$ where $W$ is the bandwidth expansion factor. Of course, the $O(W)$ signals can be demodulated with the specified error probability or signal-to-noise ratio only if code synchronization has been achieved already! On the other hand, if one wants the probability of failure to achieve code synchronization to be no larger than a specified value, then the number of simultaneous transmissions that can be supported is asymptotically $O(L/\log L)$ where $L$ is the size of the matched filter used in the code synchronization process [10]. Since $L$ is typically comparable to or smaller than $W$, it might be expected that the capacity of DS/SS networks will be more severely limited by failure to achieve code synchronization on the individual links rather than poor performance as measured by error probability or signal-to-noise ratio. In fact, numerical evaluation of these results on capacity indicates that the effect is not merely an asymptotic artifact that is of no practical concern in real-world systems, but occurs at relatively small values of the bandwidth expansion factor [10]. We remark that these results are for a simple system in which there is no aid to synchronization, that is, the receiver must acquire code synchronization from the spread-spectrum signal itself without the help of auxiliary timing signals such as pilot tones. Performance is better if additional timing information is available, or if the initial uncertainty in the time difference between the received code sequence and the locally generated code sequence is small, or if two-stage acquisition schemes are used, but nonetheless, such results indicate the significant effect of acquisition performance on the overall operation of DS/SS systems [11].

In spite of the fundamental importance of code synchronization in direct-sequence spread-spectrum systems, it has received comparatively little attention in the literature. In contrast, there is an extensive literature on demodulation methods, receiver design, and coding for spread-spectrum communication. Some of the system models and some of the methods used in the latter studies have been applied occasionally to the study of code synchronization schemes. Unfortunately, such applications are sometimes not fully justifiable, though, of course, it may well be that the numerical results do not differ significantly from those obtained by more careful modeling. Such matters are also discussed in the remainder of this chapter.

## 5.2 SYSTEM MODEL

### *5.2.1 The Transmitted Signal*

Let $\{c\} = (\ldots, c_0, c_1, c_2, \ldots)$ denote a binary $\{+1, -1\}$ sequence of period $N$ and let

$$c(t) = \sum_{j=-\infty}^{\infty} c_j \Pi_{T_c}(t - jT_c)$$

denote the corresponding pulse train of period $NT_c$. Here, $\Pi_{T_c}(\cdot)$ denotes the rectangular pulse function of duration $T_c$ centered at $T_c/2$, with each such pulse being referred to as a *chip* of duration $T_c$. Both $\{c\}$ and $c(t)$ are referred to as a *code sequence* or signal or *signature sequence* or signal. Notice that $c^2(t) = 1$ for all $t$. The pulse train $c(t)$ has bandwidth proportional to $T_c^{-1}$, and antipodal phase modulation of an RF carrier signal $\sin(\omega_0 t + \phi)$ by $c(t)$ produces the *phase-coded* carrier signal

$$\sin(\omega_0 t + \phi + (\pi/2)c(t)) = c(t)\cos(\omega_0 t + \phi)$$

whose bandwidth is also proportional to $T_c^{-1}$.

Let $W \leq N$ be an integer, and let $T = WT_c$. The binary data signal

$$b(t) = \sum_{i=-\infty}^{\infty} b_i \Pi_T(t - iT)$$

of rate $T^{-1}$ bits/s has bandwidth proportional to $T^{-1}$. When this data signal modulates the phase-coded carrier described above, the resulting signal $b(t)c(t)\cos(\omega_i t + \phi)$ still has bandwidth proportional to $T_c^{-1} = WT^{-1}$ which is larger than the data signal bandwidth by the factor $W$. For this reason, $W$ is called the bandwidth expansion factor. Note that $W$ chips are transmitted during each data bit interval of duration $T$ and that the (baseband) signaling scheme is just antipodal signaling with the pulse-coded signal

$$\pm \sum_{j=Wi}^{W(i+1)-1} c_j \Pi_{T_c}(t - jT_c)$$

being used to transmit the $i$-th data bit instead of the usual rectangular pulse $\pm\Pi_T(t - iT)$. If $W = N$, then the same baseband waveform is used for all the data bits, whereas if $W < N$, then successive data bits are transmitted using different signaling waveforms.

The presence of data modulation in the received signal complicates the code synchronization process at the receiver in various ways. Therefore, in many spread-spectrum systems, the transmitter aids the code synchronization process by transmitting only the phase-coded carrier signal (without data modulation) at the beginning of each transmission. Such aids to synchronization may also be repeated at periodic intervals to help receivers that may need re-syncronization. Thus, in what follows, we shall often assume for simplicity that there is no data modulation present, and that the code synchronization process is attempting to synchronize to the signature sequence in the phase-coded carrier only.

### 5.2.2 *The Received Signal*

In describing the received signal, let us assume the receiver clock as the time reference, so that $c(t)$ is the locally generated code sequence, and cos $\hat{\omega}t$ the locally generated carrier signal. Then, in its simplest form, the receiver input corresponding to the transmitted signal described above can be expressed as

$$r(t) = \sqrt{2V}b(t + \delta T_c)c(t + \delta T_c)\cos(\omega_i t + \theta) + n(t), \tag{5.1}$$

where $V$ is the received signal power, $\delta T_c$ is the time difference between the two code signals, $\theta$ is the RF phase of the received carrier, and $n(t)$ is additive white Gaussian noise with two-sided power spectral density $\eta_0/2$. Note that we have written the carrier radian frequency as $\omega_i$ in (1) to account for the possibility of Doppler shifts as well the uncertainty in the receiver's knowledge of the transmitter's exact carrier frequency. In actuality, because of multipath propagation and fading, several such signals with different amplitudes, delays, phase-shifts, and Doppler shifts might be received from the same transmitter, and in case of multiple-access communications, signals from other transmitters (with different signature sequences) may also be present at the receiver input. We shall consider the effect of these additional signals as well as the effect of the various impairments caused by the channel later. For the moment, let us assume that the receiver input is exactly as in (1).

### 5.2.3 *Code Acquisition*

The code synchronization process must estimate the value of $\delta$ which the time difference, measured in units of the chip duration $T_c$, between the received code signal and the locally generated code signal. If the estimate $\delta_{est}$ equals $\delta$ exactly, then, since $c^2(t) = 1$ for all $t$, multiplication of the received signal of (1) by the local code signal $c(t + \delta_{est}T_c) = c(t + \delta T_c)$ strips off the direct-sequence modulation, that is, despreads the wideband signal into the conventional narrowband PSK signal $\sqrt{2V}b(t + \delta T_c)\cos(\omega_i t + \theta)$ with baseband signals that are the usual rectangular pulses of duration $T$. This narrowband signal can be demodulated via a conventional receiver. Even if $\delta_{est}$ is nearly (but not exactly) equal to $\delta$, a substantial portion of the received signal energy is despread. Thus, it is possible to attempt carrier synchronization (and even demodulation) once $\delta$ has been estimated approximately. Of course, lack of perfect code synchronization reduces the signal-to-noise ratio for these latter processes. For this reason, it is convenient to divide the code synchronization process into the two stages of *acquisition* and *tracking* [14]. Acquisition refers to the coarse synchronization of the received sequence and locally generated sequence to within some fraction of the chip duration of the code sequence. Once acquisition has been accomplished, a code tracking loop is employed to achieve fine alignment of the two sequences and maintain that alignment. In this chapter, we consider only the acquisition problem.

The acquisition problem is that of finding an estimate $\delta_{est}T_c$ of the unknown time shift $\delta T_c$ such that $\delta T_c - \delta_{est}T_c$ is within the pull-in range of the code tracking loop. Since the spreading code has period $NT_c$, we can assume that $\delta \in [0, N)$ and thus the estimate $\delta_{est} \in [0, N)$ also. The code signal is said to be acquired if $\delta - \delta_{est}$ is within the pull-in range of the loop. However, because we are dealing with periodic signals, acquisition also occurs if $\delta$ is nearly 0 and $\delta_{est}$ is nearly $N$ (or vice versa). Hence, the following definition describes precisely when acquisition has occurred:

$$\min\{|\delta - \delta_{est}|, N - |\delta - \delta_{est}|\} \leq \zeta \tag{5.2}$$

for some specified $\zeta$ corresponding to the pull-in range of the code tracking loop. However, for simplicity of notation in what follows, we shall assume that the difference $\delta - \delta_{est}$ is reduced modulo $N$ and is always in the range $[-N/2, N/2)$. With this simplified notation, we can state that acquisition occurs whenever $|\delta - \delta_{est}| \leq \zeta$. Also, the pull-in range of a code tracking loop is typically $\pm T_c/2$, so we restrict our attention to this case only and set $\zeta = \frac{1}{2}$. It is often not much more difficult to obtain results for smaller values of $\zeta$, but, as noted in [1], the various algorithms for acquisition can be more complicated. Thus, throughout this chapter, we assume that acquisition has occurred whenever

$$|\delta - \delta_{est}| \leq \frac{1}{2} \tag{5.3}$$

The code signal is not acquired whenever this condition does not hold. Thus, the error probability or failure probability of an acquisition scheme $S$ is defined to be

$$P_e[S] = P[|\delta - \delta_{est}| > \frac{1}{2}]. \tag{5.4}$$

Acquisition methods can be broadly classified as *serial* search methods and *parallel* search methods. In a serial search scheme, the receiver compares $r(t)$ with a locally generated signal of the form $\hat{r}(t) = c(t + \hat{\delta}t)\cos(\hat{\omega}t + \hat{\theta})$ over some time interval. Here $\hat{\delta}, \hat{\omega}$ and $\hat{\theta}$ denote the receiver's current estimates of the time difference $\delta$, the input carrier frequency $\omega_i$, and phase $\theta$. On the basis of the comparison, the receiver either decides that $r(t)$ and $\hat{r}(t)$ are nearly identical, that is, $r(t)$ has been successfully acquired, or (more likely) that the two signals are quite different. In the latter case, the receiver changes one or more of its estimates $\hat{\delta}, \hat{\omega}$ and $\hat{\theta}$ by a small amount, and repeats the comparison. In this fashion, the range of possible values of $\delta$ and $\omega_i$ is searched in discrete steps, until such time that the receiver acquires $r(t)$ Obviously, very little hardware is needed to implement a serial search scheme, but the total time required for successful acquisition can be quite large. In contrast, parallel search schemes are expensive to implement, but acquire very rapidly. In a parallel search scheme, $r(t)$ is compared *simultaneously,* that is, in parallel, to "all possible" signals of the form $\hat{r}(t)$. Of course, in practice, the various signals $\hat{r}(t)$ differ from each other in discrete steps of delay, frequency, and phase just as with serial search methods. The results of these simultaneous comparisons are then processed to decide which of the $\hat{r}(t)$'s most closely resembles $r(t)$. Obviously, parallel search methods can acquire the signal in little more than the time for one signal comparison, but the numerous simultaneous comparisons require considerable hardware. However, as will be discussed later, signal processing techniques can reduce the equipment requirements considerably and parallel search schemes are likely to become increasingly viable alternatives in the future.

The simplest method of determining whether the received code signal $c(t+\delta T_c)$ is nearly synchronized with the locally generated code signal $c(t+\hat{\delta}T_c)$ is to multiply $r(t)$ by $c(t+\hat{\delta}T_c)$, filter the product through a bandpass filter with center frequency $\hat{\omega}$, and measure the signal energy at the output of the filter. If condition (3) is satisfied, the product $c(t+\delta T_c)c(t+\hat{\delta}T_c)$ equals 1 over an interval of $T_c/2$ or more in each chip interval of length $T_c$. As a result, the product $r(t)c(t+\hat{\delta}T_c)$ has substantial signal energy at frequency $\omega_i$. Except when $\omega_i$ and $\hat{\omega}$ are quite different due to a large Doppler shift or uncertainty about the transmitter's carrier frequency, this signal energy is detectable at the output of the bandpass filter. If large frequency uncertainties exist, a (tunable) filter can be swept across the range of frequency uncertainty to capture this energy. If (3) is not satisfied or the frequency error is large, there is relatively little energy in the passband. Consequently, the receiver can declare that the signal has been acquired whenever it detects substantial signal energy at the output of bandpass filter.

Instead of simple bandpass filtering, it is often preferable to use *matched* filtering (or correlation techniques) for detecting near-synchronism. The receiver correlates $r(t)$ with the in-phase and quadrature phase-coded signals $\sqrt{2}c(t+\hat{\delta}T_c)\cos\hat{\omega}t$ and $-\sqrt{2}c(t+\hat{\delta}T_c)\sin\hat{\omega}t$ over an interval of length $LT_c$. Thus, the receiver measures

$$X = \int_0^{LT_c} r(t)\sqrt{2}c(t+\hat{\delta}T_c)\cos\hat{\omega}t\,dt, \tag{5.5}$$

and

$$Y = \int_0^{LT_c} -r(t)\sqrt{2}c(t+\hat{\delta}T_c)\sin\hat{\omega}t\,dt, \tag{5.6}$$

where $X$ and $Y$ can be modeled as independent Gaussian random variables whose means are approximately $\sqrt{V}L(1-|\delta-\hat{\delta}|)T_c\cos\theta$ and $\sqrt{V}L(1-|\delta-$

$\hat{\delta}|)T_c \sin\theta$ if (3) holds. It follows that the signal amplitude $\sqrt{X^2+Y^2}$ at the correlator output is a Ricean random variable when (3) holds. On the other hand, when (3) does not hold, $X$ and $Y$ are approximately zero-mean, and thus the signal amplitude is usually modeled as a Rayleigh random variable. As before, signal acquisition can be declared whenever the amplitude exceeds a threshold. Notice that if $\hat{\delta} = \delta$, and $\hat{\omega} = \omega_i$, then the correlators are exactly equivalent to the noncoherent matched filter for the detection of the received signal.

If we assume without loss of generality that $\hat{\delta}$ is an integer $k$, or equivalently, that the integrations in (5) and (6) begin at a time t such that $t+\hat{\delta}T_c = kT_c$ and continue over the next $L$ chip intervals, then we can use the fact that $c(t+\hat{\delta}T_c)$ is a constant over each chip interval $iT_c \leq t < (i+1)T_c$ to write

$$X = \sum_{i=k}^{k+L-1} c_i \int_{iT_c}^{(i+1)T_c} r(t)\cos\hat{\omega}t\,dt = \sum_{i=k}^{k+L-1} c_i \cdot x_i, \tag{5.7}$$

and

$$Y = \sum_{i=k}^{k+L-1} c_i \int_{iT_c}^{(i+1)T_c} -r(t)\sin\hat{\omega}t\,dt = \sum_{i=k}^{k+L-1} c_i \cdot y_i, \tag{5.8}$$

where the integrals can be viewed as matched filtering of the chips via integrate-and-dump correlators whose output is sampled (and the integrator dumped) every $T_c$ seconds. The matched filtering is followed by digital (i.e., discrete-time) correlations that compute the inner products shown in (7) and (8). In fact, since each $c_i$ is either +1 or −1, the digital correlators need not do any multiplications – additions and subtractions suffice to compute the correlations. Furthermore, the receiver can compute the inner product of $(x_k, x_{k+1}, \ldots, x_{k+L-1})$ and $(y_k, y_{k+1}, \ldots, y_{k+L-1})$ not just with $(c_k, c_{k+1}, \ldots, c_{k+L-1})$ but also with *any* $(c_i, c_{i+1}, \ldots, c_{i+L-1})$. Thus, parallel acquisition techniques need not use multiple correlators to compute quantities such as in (5) and (6). Instead, a pair of chip matched filters followed by multiple digital correlators can be used. Another simplification results from the following observation. If we wish to compute the correlations (that is, inner products) of the sequence $(x_k, x_{k+1}, \ldots, x_{k+L-1})$ with the $K$ sequences $(c_i, c_{i+1}, \ldots, c_{i+L-1})$, $j \leq i \leq j+K-1$, we can instead compute the cyclic convolution of the sequence $(x_k, x_{k+1}, \ldots, x_{k+L-1}, 0, 0, \ldots, 0)$ of length $K+L-1$ with the sequence $(c_{j+K+L-2}, c_{j+K+L-3}, \ldots, c_j)$. As is well known, we can compute this in $O((K+L)\log(K+L))$ steps via fast Fourier transform (FFT) techniques. Of course, FFT computations require complex multiplications whereas the brute-force method uses only additions and subtractions, and this should be taken into account in choosing the computational approach.

### 5.2.4 *Channel Impairments*

The channel changes the transmitted signal in various ways and the effect of these channel impairments must be mitigated during the acquisition process. Consider first the problem of Doppler shifts in frequency caused by relative motion between the transmitter and receiver. If the length of the transmission path is changing at the rate of 30 m/s, roughly automobile speed on highways, the Doppler shift in carrier frequency is approximately 0.1 part in $10^6$ which amounts to 100 Hz at a 1 GHz carrier frequency. If this frequency difference is not tracked and compensated for by the receiver, the locally generated carrier in a *coherent* demodulator will fluctuate from being perfectly in phase with the received signal to being completely opposite in phase just 5 milliseconds later to being perfectly in phase again another 5 milliseconds later. However, unless the data rate on the channel is very low, such fluctuations will not affect a *noncoherent* or differentially coherent demodulator. For example, if the data rate is $10^4$ bits/s, the phase difference between the received carrier signal $\cos(\omega_i t + \theta)$ and the locally generated carrier signal $\cos \hat{\omega} t$ will slowly change from $\theta$ to the negligibly different $\theta \pm \pi/50$ during the transmission of one data bit. In particular, the effect on the computations (5) and (6) is very small, and the net reduction in performance can be accounted for quite adequately by reducing the signal-to-noise ratio slightly. In short, Doppler shifts in frequency can almost be ignored for the purposes of analysis of acquisition in land mobile communications. In contrast, the receiver may be more uncertain about the transmitted frequency because of the instability of the transmitter oscillator, because manufacturing variations might cause the natural frequency of the transmitter oscillator to be different from its nominal value $\omega_0$, because of the effect of ambient conditions on the transmitter oscillatir, and the like. This uncertainty can be a more serious problem because the difference between the nominal transmitter carrier frequency and the actual transmitter carrier frequency $\omega_i$ can be much larger than that expected from Doppler shifts caused by relative motion. For example, a 1 GHz oscillator rated as being stable to 5 parts in $10^6$ can have frequency changes as large as 5 kHz. Such frequency variations are accommodated by varying the locally generated carrier frequency $\hat{\omega}$ in steps during the acquisition process. As should be obvious from the above discussion, reducing the frequency error to less than 200 Hz (say) should be adequate for code acquisition, and once the code has been acquired, the code-tracking loop and the carrier phase-lock loop can be called upon to fine-tune the local phase-coded carrier signal.

Multipath propagation is a channel impairment which results in signal fading. The rays that traverse different paths can interfere constructively or destructively, and the resulting interference pattern can have peaks and nulls separated spatially by distances on the order of a wavelength. Temporal variations in signal strength can also occur in a few milliseconds. A simple model for such fading is as follows. The signal is assumed to have been received via a large number of paths of nearly identical length. The central limit theorem is then invoked to model $\sqrt{V}$ and $\theta$ in (1) as independent random variables with $\sqrt{V}$ having a Rayleigh distribution and $\theta$ a uniform distribution on $[0, 2\pi)$. If one of the paths is a direct (line-of-sight) path or via a specular reflection, $\sqrt{V}$ is often taken to be a Ricean random variable. The corresponding channel models are called the (frequency nonselective or flat fading) Rayleigh and Ricean channel models respectively. These simple fading channel models will suffice for the purposes of this chapter, but more elaborate models have also been used in the literature. The reader is referred to Chapter 8 of [14] for an excellent tutorial introduction to these models for fading channels. Before considering the effect on acquisition in spread-spectrum communications, however, we briefly consider some standard results on digital communication over fading channels.

To account for the temporal variations in signal strength, the fading channel model includes the implicit assumption that $\sqrt{V}$ and $\theta$ are actually sample functions from *very* slowly varying random processes. Thus, over time intervals of length $T$ or less, $\sqrt{V}$ and $\theta$ can be assumed to be constant. The error probability for a given bit thus depends on the signal strength during that bit interval, and this error probability can be quite small or quite large depending on the value of $\sqrt{V}$. Furthermore, since $\sqrt{V}$ and $\theta$ are assumed to be essentially constant over a bit interval, their values during the next bit interval can be only slightly different from their values during the current bit interval. Hence, the error probability for any given bit is only very slightly different from the error probability of the previous bit or the next bit. However, the bit error probability does vary considerably over time as the signal slowly fades in and out. Assuming that the appropriate ergodic condition holds, the time average of the bit error probability is the same as the ensemble average, that is, the expected value of the error probability as a function of $\sqrt{V}$ and $\theta$. It is well known [14] that on a Rayleigh fading channel, this average probability of error decreases only as an inverse function of the average signal-to-noise ratio. Heuristically, this is because during deep fades, when $\sqrt{V}$ is very small, the bit error probability is very large, and these large values dominate the average. In contrast, the error probability on the classical additive white Gaussian noise channel decreases exponentially with increasing signal to noise ratio.

Error-control coding with interleaving can be used to improve performance on a fading channel. In fact, interleaving is essential. Without interleaving, deep fades will cause most of the bits in a codeword to have very large error probability. Thus, unless very powerful codes are used, the large decoder error probability during deep fades will still dominate the average error probability performance. On the other hand, when the signal is not so deeply faded, the low rate of the powerful code will decrease the throughput and thus reduce the performance. Thus, the interleaver must have sufficient depth to spread out the codeword symbols in time so that not all of the symbols are affected by deep fades. Unfortunately, large interleaver depth increases the hardware costs as well as the decoding delay, and the system designer must take these factors into account. Nevertheless, if the interleaving depth is adequate, it is reasonable to assume, as is commonly done, that the random variables representing the signal amplitude and phase of "successive" bits in a codeword are independent because these bits are actually received sufficiently apart in time to justify this assumption. However, the assumption of independent fading cannot be justified for bits that are actually received in temporal succession over the channel.[1] For the nonselective fading processes that we consider in this chapter, successive bits should be assumed to have identical fading levels, or if some degree of time selectivity is to be incorporated, the fading levels of successive bits should be correlated in accordance with the autocorrelation function of the fading process [19]. As a final comment, we note that interleaved error-control coding is essentially a form of diversity transmission since the information in a codeword is spread over time, and successful decoding (that is, error-free transmission) requires that most of the symbols be received when the signal is not too deeply faded for correct demodulation.

For the acquisition problem, fading is a deleterious condition that cannot be easily ameliorated by interleaving, coding, or diversity. First, notice that interleaving and diversity transmission are not particularly useful in improving performance. The receiver needs to estimate the channel conditions that exist *now,* and not the average of the channel conditions over some time period corresponding to the interleaving depth, or the average of the instantaneous channel conditions in several virtual channels corresponding to the diversity transmissions. Thus, the chip measurements $x_i$ and $y_i$ exhibited in (7) and (8) are made in strict temporal succession, and all of them should be assumed to be faded identically. In fact, it is reasonable to assume that measurements $X$ and $Y$ in (5) and (6) over $[0, LT_c)$ and similar measurements $\tilde{X}$ and $\tilde{Y}$ over $[LT_c, 2LT_c)$ are also essentially faded identically, etc. It follows that during deep fades, error-control coding will not help either, because all the codeword symbols will be demodulated with high error probability. The truth of the matter is that acquisition is difficult during deep fades and communication is difficult during deep fades. This is, of course, well known to the everyday user of cellular telephones – sometimes, it is virtually impossible to establish a circuit from where one is, and it is necessary to move to a "better" location where there is more signal strength to obtain service. Thus, the (conditional) probability of failing to acquire the received signal is dependent on the instantaneous level of the fading process, and it is this probability that is more important than its time average or ensemble average. These averages usually exhibit a depressingly low rate of decrease with increasing signal power for the same reason as that noted earlier, namely that the poor performance during deep fades tends to dominate the average error probability. Even more depressing is the fact that, as discussed above, the commonly used tools in the communications engineer's toolkit do not lead to significant improvements in performance. For these reasons, in our study of acquisition, we shall ignore the statistical nature of $\sqrt{V}$ and state results as if it were a constant. The results can be interpreted as being conditional distributions, probabilities, etc. conditioned on the random variable $\sqrt{V}$ having a specific value. The interested reader can average these quantities with respect to whatever distribution is deemed appropriate for $\sqrt{V}$.

When the signal is received via paths that differ considerably in length, the signal model in (1) can be modified to include several terms of the form exhibited each with different delays, phase shifts, and fading levels. Since the difference in delays leads to constructive interference at some frequencies and destructive interference at other frequencies, such channels are said to exhibit frequency-selective fading. Since the signal bandwidth is proportional to $T_c^{-1}$, a rule of thumb is that signals with differential delays on the order of $T_c$ or more are distinguishable as arising from different paths, while smaller delays are modeled as Rayleigh or Ricean fading. Depending on the strength of the signal in each distinguishable path, it can be detected and acquired, and in some systems such as the North American Digital Cellular System (IS-95), rake receivers are used to combine these multipath signals in order to increase the effective received signal power [14]. From the point of view of the acquisition system trying to acquire one of these signals, however, the other (as yet unacquired) multipath signals are just interference. Similarly, any signals received from many transmitters (usually with different code sequences and possibly in many versions due to multipath), also cause interference in the acquisition process. The exact analysis of the effect of all these interfering signals on the performance of the acquisition system is difficult, and very often, the central limit theorem is used to model the net effect of the interference as an increase in the effective value of the noise power. With appropriate modifications, then, simple results on the performance of acquisition in additive white Gaussian noise can be used to predict the results in the presence of multiple-access interference and multipath interference.

In the remainder of this chapter, we consider simple serial and parallel search schemes for acquisition of DS/SS signals in additive white Gaussian noise. No auxiliary timing signals are used; the receiver must estimate $\delta$ from the received signal itself. The results presented give the conditional probability of failing to acquire given the level of fading of the (nonselective) fading process, and can be averaged over the fading statistics if needed. Selective fading processes are not considered at all. The results can also be used to determine the performance in the presence of multipath and multiple-access signals provided that the net effect of these interfering signals is treated as effectively increasing the noise power level.

## 5.3 SERIAL ACQUISITION SCHEMES

An excellent description of serial search schemes and the system issues and trade-offs involved is given in [14]. In the simplest form of a serial scheme, the receiver computes $X$ and $Y$ as in (5) and (6) and compares $X^2 + Y^2$ to a threshold. If $X^2 + Y^2$ exceeds the threshold, the signal is declared to have been acquired. If not, the parameters of the local code generator and oscillator are changed and the correlations computed again. The parameters of interest include the length of the integration time $LT_c$; the threshold value which affects the false alarm probability (the probability of declaring the signal to be acquired when in fact it is not) and the detection probability (the probability of correctly declaring the signal to be acquired); and the step size in changing the estimated carrier frequency $\hat{\omega}$ or the estimated delay $\hat{\delta}$. The latter step size is usually taken to be $\frac{1}{2}T_c$ or $\frac{1}{4}T_c$. All of these parameters affect the mean time to acquire the signal. If the step size is too small, the search necessarily takes longer. If the integration time is too small, the false alarm probability is too high, and so on.

In practice, most serial search schemes are two-stage schemes. When $X^2+Y^2$ exceeds the threshold, instead of declaring that the signal has been acquired, the system enters a verification stage in which the correlations of (5) and (6) are carried out over a much longer time interval. The result is again tested (against a different threshold) to see if the signal has indeed been acquired. The length of the verification phase also affects the mean time to acquisition because each time a false alarm occurs, the system enters the verification stage, and then usually exits it to resume the search. It is typically assumed that verification is perfect and that the system is always able to distinguish between false alarms and true acquisition in the verification stage. Of course, this is only approximately true in practice, and the time for verification can be quite long if very small probabilities of false verification are desired.

One problem that is usually ignored in the analysis of serial search schemes is the exact description of the statistics of the signal amplitude $\sqrt{X^2 + Y^2}$ when (3) does not hold. Generally, it is assumed that this is a Rayleigh density corresponding to no signal output being present in $X$ and $Y$. This is not quite correct. Usually, there is a small signal component because the correlations in (7) and (8) are never exactly zero. Unfortunately, it is difficult to get an accurate model for this signal component. Its value varies with the position in the code sequence in a manner that is not amenable to analysis. Even useful upper bounds on these correlation values are difficult to obtain. One method that has been tried is to treat these as random. The point to be made, though, is that the threshold is often chosen on the basis of the signal component being exactly zero. Hence, because of these nonzero correlations, the false alarm probability (that is, the probability that the system enters the verification stage) is usually larger than is indicated by the analysis. Furthermore, this causes the mean time to acquisition to be larger than the value predicted by the analyses.

One difficulty with serial search schemes is that they are fixed-length tests. The length of the test required to achieve a given false alarm probability and detection probability depends on the signal level. If the signal level can be expected to be changing, we either need to estimate it and set the length of the test appropriately or make the test long enough (that is, $LT_c$ large enough) that the given performance can usually be achieved. An alternative scheme that avoids this problem in part is based on the sequential probability ratio test (SPRT). Once again, we refer the reader to [14] for details of the scheme. Here, successive chip matched filter outputs are observed and the likelihood ratio is compared to two thresholds $A$ and $B$ where $A < 1 < B$. If the likelihood ratio exceeds $B$, successful acquisition is declared and the system proceeds to the verification stage. If the likelihood ratio is smaller than A, it is decided that the local code signal and the received code signal are not in synchronism; the delay estimate is adjusted and the test is restarted. Finally, if the likelihood ratio lies between $A$ and $B$, another chip observation is made and the likelihood ratio is recomputed. When the signal-to-noise ratio is large, one or the other threshold is crossed after only a few chip observations while if the signal-to-noise ratio is low, a larger number of chip observations are required.

The performance of a SPRT scheme depends critically on the likelihood ratio being used. In [14], the standard assumption is made that this is the ratio of a Ricean density to a Rayleigh density. It is easily shown that neither of these assumptions is fully valid. In the case of near synchronism, the parameters of the density of the Ricean random variable depend on whether or not the chip interval in question contains a transition from $+1$ to $-1$ or vice versa. Similarly, under the assumption that the code sequences are not in synchronism, it is still possible for the chip observation to have a Ricean density. The proper model for this latter case is what is called the *random sequence* model rather than the more popular *zero sequence* model [1], [5], [6] [11]. In [1], [5], [6], it is shown that if an SPRT is designed with the zero sequence model, then its false alarm probability is an *increasing* function of the signal-to-noise ratio! Heuristically, the effect becomes more apparent at high signal-to-noise ratio because the mismatch between the statistical description and the actual signal distribution is more severe.

Finally, a novel scheme is presented in [5] and [7] in which the chip observations are divided into two subsets of nearly equal size. In one set, there is no transition from +1 to −1 or vice versa during the corresponding chip interval whereas in the other set there is such a chip transition. Of course, this assumes that the signals are in near synchronism. The SPRT formulation in [14] applies to the first set. On the other hand, since the other set always contains chip transitions, these can be used to estimate the fractional part of the delay. In other words, the delay $\delta$ can be expressed as $k + \epsilon$ where $k$ is an integer and $0 \leq \epsilon < 1$. The integer part $k$ is estimated by the SPRT exactly as in [2] or [14] while the fractional part is estimated separately and independently. Of course, most of the SPRT tests end in a declaration that the two signals are not in synchronism, in which case the estimate of the fractional part is discarded. But when the SPRT does end up accepting the current settings, an exceedingly accurate estimate of the fractional part of the delay can be supplied to the code-tracking loop. Additional details about this scheme can be found in [6]. For other schemes using SPRT methods, see [8] [9].

## 5.4 PARALLEL ACQUISITION SCHEMES

In a parallel acquisition scheme, the receiver computes, in parallel, the correlation of the received signal with in-phase and quadrature RF carriers that have been modulated with all possible delayed versions of the code sequence. Here, of course, the delays are all integer multiples of $T_c$. Suppose that $K$ delayed versions of the code sequence are possible, where we assume without loss of generality, that the possible delays are $\hat{\delta} = i,\ 0 \leq i \leq K-1$. For convenience in analysis, we modify the definitions of (5) and (6) slightly so as to obtain various parameters in suitable form. Thus, for $0 \leq i \leq K-1$, the $i$-th in-phase correlation is

$$X_i = \frac{2}{\sqrt{\eta_0 T_c}} \int_0^{LT_c} r(t)c(t + iT_c)\cos\omega_0 t\, dt, \tag{5.9}$$

whereas the $i$-th quadrature correlation is

$$Y_i = \frac{-2}{\sqrt{\eta_0 T_c}} \int_0^{LT_c} r(t)c(t + iT_c)\sin\omega_0 t\, dt. \tag{5.10}$$

These correlation values are sufficient statistics for estimating the delay $\delta$ given that the data preprocessing consists of chip matched filtering and sampling at the chip rate. In fact, as noted in Section II.C, only one pair of chip correlators followed by $K$ digital correlators are needed, or in same cases, FFT methods can be used to advantage.

In the remainder of this chapter, it will be assumed that the code sequence is a maximal-length binary linear feedback shift register sequence (also called a PN sequence or $m$-sequence) of period $N = 2^n - 1$ and that $K = N$. In this case, the FFT methods referred to above can be replaced by fast Hadamard transform algorithms that require no multiplications [1]. However, regardless of the method used to compute the $X_i$'s and $Y_i$'s, their distribution is as follows. If $\delta = k + \epsilon$, where $k \in \{0, \dots, K-1\}$ and $\epsilon \in [0, 1)$, we have

$$X_i = \begin{cases} p[(1-\epsilon)N - \epsilon]\cos\theta + V_i, & \text{if } i = k \\ p[\epsilon - 1 + \epsilon N]\cos\theta + V_i, & \text{if } i = k+1 \\ -p\cos\theta + V_i, & \text{otherwise} \end{cases} \tag{5.11}$$

and

$$Y_i = \begin{cases} p[(1-\epsilon)N - \epsilon]\sin\theta + W_i, & \text{if } i = k \\ p[\epsilon - 1 + \epsilon N]\sin\theta + W_i, & \text{if } i = k+1 \\ -p\sin\theta + W_i, & \text{otherwise.} \end{cases} \tag{5.12}$$

where $p = [2VT_c/\eta_0]^{1/2}$ is a measure of the chip SNR, and $V_i$ and $W_i$ are independent zero-mean Gaussian random variables with variance $N$. Conditioned on $\delta$ and $\theta$, $X_i$ and $Y_i$ are independent Gaussians with variance $N$, and $Cov[X_i, Y_j] = -Cov[X_j, Y_i] = 0$. On the other hand, $Cov[X_i, X_j] = Cov[Y_i, Y_j] = -1$ if $i \neq j$. Using these facts, it can be shown that if $\theta$ is assumed to be uniformly distributed on $[0, 2\pi)$, the conditional joint density of the complex vector $\mathbf{Z} = \mathbf{X} + j\mathbf{Y}$ given $\delta = k + \epsilon$ is

$$\begin{aligned} & h_{\mathbf{Z}|\delta}(\mathbf{z}|k+\epsilon) = \\ & \quad C \exp\left[-\frac{1}{2(N+1)}\left(\sum_{i=0}^{N-1}|z_i|^2 + \left|\sum_{i=0}^{N-1} z_i\right|^2\right)\right] \\ & \quad \times \exp[-p^2 q(\epsilon)] I_0(p\, r(\mathbf{z}, k, \epsilon)), \end{aligned} \tag{5.13}$$

where $C = [(2\pi)^N (N+1)^{N-1}]^{-1} \exp(-p^2 N/2)$, $q(\epsilon) = (N+1)(\epsilon^2 - \epsilon)$, and $r(\mathbf{z}, k, \epsilon)) = |(1-\epsilon)z_k + \epsilon z_{k+1}|$.

We will now present the optimal estimator as developed in [18]. The optimal estimator maximizes the *a posteriori* probability that $\delta$ lies in an interval of unit length centered at $\hat{\delta}$. Under the assumption that $\delta$ is uniformly distributed over $[0, N)$, the noncoherent optimal estimate is given by

$$\delta_{ncopt} = \arg \max_{\hat{\delta} \in [0,N)} \int_{\hat{\delta}-1/2}^{\hat{\delta}+1/2} h_{\mathbf{Z}|\delta}(\mathbf{z}|u)\, du. \tag{5.14}$$

As shown in [18], after eliminating factors of $h_{\mathbf{Z}|\delta}(\mathbf{z}|k+\epsilon)$ that do not depend on $k$ or $\epsilon$, the integral in (5.14) is proportional to

$$I_{ncopt}(\hat{\delta}) = \int_{\hat{\delta}-l+1/2}^{1} e^{-p^2 q(\epsilon)} I_0(p\, r(\mathbf{z}, l-1, \epsilon))d\epsilon + \int_{0}^{\hat{\delta}-l+1/2} e^{-p^2 q(\epsilon)} I_0(p\, r(\mathbf{z}, l, \epsilon))d\epsilon, \quad (5.15)$$

for $l - 1/2 \leq \hat{\delta} \leq l + 1/2$. $I_{ncopt}(\hat{\delta})$ is differentiable everywhere, so the extrema may be found using standard calculus techniques as for the coherent optimal scheme. The algorithm $\mathcal{S}_{ncopt}$ for finding $\delta_{ncopt}$ is as follows:

For each $0 \leq l \leq N-1$, if $|z_l| > \min\{|z_{l-1}|, |z_{l+1}|\}$, there are local extrema of the function $I_{ncopt}(\hat{\delta})$ at

$$\hat{\delta}_l = l + \frac{-B_l \pm \sqrt{B_l^2 - 4A_l C_l}}{2A_l}, \quad (5.16)$$

where

$$\begin{aligned} A_l &= |z_{l+1} - z_l|^2 - |z_l - z_{l-1}|^2 \\ B_l &= |z_{l+1}|^2 + |z_{l-1}|^2 - 2|z_l|^2 \\ C_l &= (|z_{l+1} + z_l|^2 - |z_l + z_{l-1}|^2)/4. \end{aligned} \quad (5.17)$$

For each such $\hat{\delta}_l \in [l-1/2, l+1/2]$, calculate $I_{ncopt}(\hat{\delta})$ using (5.15) and find the global maximum of these values. $\delta_{ncopt}$ is the location of this global maximum.

It is believed, but has not yet been proved, that the global maximum of $I_{ncopt}(\hat{\delta})$ must be in an interval $(l-1/2, l+1/2)$ such that $|z_l| > \max\{|z_{l-1}|, |z_{l+1}|\}$, in which case there is only one candidate local maximum in each such interval. In any case, the optimal noncoherent estimator is quite difficult to implement, even more so than the corresponding coherent scheme. Evaluation of the decision statistic $I_{ncopt}(\hat{\delta})$ requires numerical integration of exponential and Bessel functions. Furthermore, this calculation must be performed roughly $N/3$ times, for each local maximum. As in the case of the optimal coherent parallel acquisition scheme, the error probability is very difficult to find analytically. We will next present several suboptimal estimators which are easier to implement.

## 5.5 SUBOPTIMAL PARALLEL ESTIMATORS

In this section we present three suboptimal noncoherent parallel acquisition schemes as detailed in [21] and [23]. This work closely follows the approach taken in the development of the suboptimal coherent schemes in [22]. The schemes that we present here are of varying complexities, but all are comparable to $\mathcal{S}_{ncopt}$ in performance.

### 5.5.1 *Locally Optimal Estimator*

The locally optimal estimator $\mathcal{S}_{ncloe}$ is defined as the limiting form of the optimal estimator as $p \to 0$ (recall $p$ is a measure of the chip SNR). The decision statistic for $\mathcal{S}_{ncloe}$ is found by evaluating the derivative of $I_{ncopt}(\hat{\delta})$ at $p = 0$. This process yields the decision statistic

$$\begin{aligned} I_{ncloe}(\hat{\delta}) &= \frac{1}{6}\left(|z_l|^2 + |z_{l-1}|^2 + \mathrm{Re}(z_{l-1}z_l^*)\right) \\ &+\frac{1}{2}\left(|z_l|^2 - |z_{l-1}|^2\right)(\hat{\delta} - l + 1/2) \\ &+\frac{1}{2}\left(|z_{l-1}|^2 - |z_l|^2 + \mathrm{Re}(z_l(z_{l+1} - z_{l-1})^*)\right)(\hat{\delta} - l + 1/2)^2 \\ &+\frac{1}{6}\left(|z_{l+1} - z_l|^2 - |z_l - z_{l-1}|^2\right)(\hat{\delta} - l + 1/2)^3. \end{aligned} \tag{5.18}$$

This is differentiable everywhere, so its extrema can be found easily, and the scheme $\mathcal{S}_{ncloe}$ is

> For each $0 \le l \le N-1$, if $|z_l| > \min\{|z_{l-1}|, |z_{l+1}|\}$, there are local extrema of the function $I_{ncopt}(\hat{\delta})$ at
>
> $$\hat{\delta}_l = l + \frac{-B_l \pm \sqrt{B_l^2 - 4A_lC_l}}{2A_l}, \tag{5.19}$$
>
> where $A_l$, $B_l$, and $C_l$ are given by (5.17). For each such $\hat{\delta}_l \in [l - 1/2, l + 1/2]$, calculate $I_{ncloe}(\hat{\delta})$ using (5.18) and find the global maximum of these values. $\delta_{ncloe}$ is the location of this global maximum.

Note that the the locations of the local maxima of $I_{ncloe}(\hat{\delta})$ are precisely the same as those given by $I_{ncopt}(\hat{\delta})$. Therefore, $\delta_{ncloe}$, the global maximum of $I_{ncloe}(\hat{\delta})$, is one of the local maxima of $I_{ncopt}(\hat{\delta})$, and hopefully it is also the global maximum $\delta_{ncopt}$ of $I_{ncopt}(\hat{\delta})$. Simulation results show that $\mathcal{S}_{ncloe}$ performs very well, and approximates the performance of the optimal scheme for a wide range of SNR.

$\mathcal{S}_{ncloe}$ is certainly easier to implement than $\mathcal{S}_{opt}$, as its decision statistic is simply the cubic equation (5.18) rather than the integral expression (5.15). This decision statistic must be evaluated at a large number of points, however, and the overall scheme is still difficult to analyze.

### 5.5.2 *Hybrid Estimator*

The second suboptimal noncoherent acquisition scheme that we present, $\mathcal{S}_{ncmo}$, is a hybrid of the optimal estimator and the chip synchronous maximum likelihood estimator. The noncoherent maximum likelihood estimator $\mathcal{S}_{ncmle}$ maximizes the density function (5.13). When $\epsilon = 0$, $\delta_{ncmle} = l^* = \arg\max\{|z_l|\}$.

We use this same decision statistic in $\mathcal{S}_{ncmo}$, but let the actual estimate be the local maxima in the interval $[l-1/2, l+1/2]$ as given by $\mathcal{S}_{ncopt}$. Our scheme $\mathcal{S}_{ncmo}$ is then

Let $l^* = \arg\max_{0 \leq l \leq N-1}\{|z_l|\}$. The the estimate of $\delta$ is

$$\delta_{ncmo} = l^* + \frac{-B_{l^*} - \sqrt{B_{l^*}^2 - 4A_{l^*}C_{l^*}}}{2A_{l^*}}, \tag{5.20}$$

where $A_l$, $B_l$ and $C_l$ are given by (5.17).

Note that only one of the roots is given in (5.20). This is because in the case that $|z_l| > \max\{|z_{l-1}|, z_{l+1}|\}$, only one local maxima of (5.15) can occur in $[l-1/2, l+1/2]$, and it can easily be shown that the root given in (5.20) is the correct one.

$\mathcal{S}_{ncmo}$ requires very little calculation compared to either $\mathcal{S}_{ncopt}$ or $\mathcal{S}_{ncloe}$. One does not need to find all the candidate local maxima via (5.16) and (5.17) so the search is simplified. The performance of $\mathcal{S}_{ncmo}$ is quite good as well, although not as good as $\mathcal{S}_{ncloe}$. The only drawback of this scheme is the fact that the fractional part of the estimate (5.20) presents difficulties when attempting to evaluate the scheme analytically.

### 5.5.3 *Simplified Hybrid Estimator*

The final noncoherent parallel acquisition scheme that we present approximates the performance of $\mathcal{S}_{ncmo}$. This scheme is a noncoherent analogue of the coherent scheme $\mathcal{S}_{mo}$ presented in [22], in the sense that it uses the square envelope data $|z_i|^2$ in place of the in-phase data $x_i$ that are used in the coherent receiver. This approximating version of $\mathcal{S}_{ncmo}$ is denoted $\mathcal{S}_{ncmoa}$, and is as follows:

Let $l^* = \arg\max_{0 \leq l \leq N-1}\{|z_l|\}$. The the estimate of $\delta$ is

$$\delta_{ncmo} = l^* + \frac{|z_{l^*+1}|^2 - |z_{l^*-1}|^2}{2(2|z_l^*|^2 - |z_{l^*-1}|^2 - |z_{l^*+1}|^2)}. \tag{5.21}$$

$\mathcal{S}_{ncmoa}$ is easier to implement than $\mathcal{S}_{ncmo}$ (there are even fewer calculations involved in forming the estimate), and simulation results show that the performance of $\mathcal{S}_{ncmoa}$ is practically identical to that of $\mathcal{S}_{ncmo}$. Furthermore, we can derive some analytical bounds on the performance of this scheme.

## 5.6 PERFORMANCE ANALYSIS

Exact expressions for the probability of error for any of the noncoherent parallel acquisition schemes are difficult to derive. The statistics $|Z_i|$ are correlated Ricean random variables, and the decision statistics and/or estimates given by $\mathcal{S}_{ncopt}$, $\mathcal{S}_{ncloe}$, and $\mathcal{S}_{ncmo}$ are not easily analyzed. On the other hand, we can provide some approximations and bounds on the conditional probability of error for $\mathcal{S}_{ncmoa}$.

When $\epsilon = 0$, we have the following formula for $P_{e|\epsilon}[\mathcal{S}_{ncmoa}]$:

$$\begin{aligned} &P_{e|0}[\mathcal{S}_{ncmoa}] = \\ &(1+\frac{1}{N})e^{-M^2N/2}\int_0^\infty\int_0^\infty uze^{-\frac{1}{2}(u^2+z^2)} \\ &\times J_0\left(uz\sqrt{1/N}\right) I_0\left(pz\sqrt{N+1}\right) \\ &\times \left(\int_0^z xe^{-\frac{1}{2}(x^2-u^2/N)} J_0(xu\sqrt{1/N})dx\right)^{N-1} dudz. \end{aligned} \tag{5.22}$$

The right side of (5.22) is the error probability for noncoherent $N$-ary simplex signaling [12]. We believe that $P_{e|\epsilon}[\mathcal{S}_{ncmoa}]$ and $P_{e|\epsilon}[\mathcal{S}_{ncmo}]$ are minimum when $\epsilon = 0$, so (5.22) is a conjectured lower bound on the average error probabilities of $\mathcal{S}_{ncmoa}$ and $\mathcal{S}_{ncmoa}$.

Given that $\delta = k + \epsilon$, it can be easily shown that with $i \neq k, k+1$, an upper bound on $P_{e|\epsilon}[\mathcal{S}_{ncmoa}]$ is given by:

$$\begin{aligned} &P_{e|\epsilon}[\mathcal{S}_{ncmoa}] \leq \\ &(N-2)\min\{P[|Z_i| > |Z_k|], P[|Z_i| > |Z_{k+1}|]\} \\ &+P[(2\epsilon-1)|Z_k|^2 - \epsilon|Z_{k+1}|^2 + (1-\epsilon)|Z_{k-1}|^2 > 0] \\ &+P[(1-2\epsilon)|Z_{k+1}|^2 - (1-\epsilon)|Z_k|^2 + \epsilon|Z_{k+2}|^2 > 0]. \end{aligned} \tag{5.23}$$

For $i \neq k, k+1$, the various terms in (5.23) are given as follows:

$$\begin{aligned} &P[|Z_i| > |Z_k|] = \\ &\quad M(\nu_1, \nu_2) - \frac{1}{2}\exp\left(-\frac{\nu_1^2+\nu_2^2}{2}\right) I_0(\nu_1\nu_2), \\ &\quad \triangleq P_1(p, N, \epsilon), \end{aligned} \tag{5.24}$$

and

$$P[|Z_i| > |Z_{k+1}|] = P_1(p, N, 1-\epsilon), \tag{5.25}$$

where

$$\begin{aligned} \nu_1 &= p\sqrt{(N-\sqrt{N^2-1})/8}\left(1+\epsilon\sqrt{\frac{N+1}{N-1}}\right), \\ \nu_2 &= p\sqrt{(N+\sqrt{N^2-1})/8}\left(1-\epsilon\sqrt{\frac{N+1}{N-1}}\right). \end{aligned} \tag{5.26}$$

In (5.24), $M(\cdot,\cdot)$ denotes the Marcum Q function [4]

$$M(a,b) = \int_b^\infty x \exp\left(-\frac{x^2+a^2}{2}\right) I_0(ax)dx.$$

The second term in (5.23) can be found by noting that the probability that $F_1 = (1-2\epsilon)|Z_{k+1}|^2 - (1-\epsilon)|Z_k|^2 + \epsilon|Z_{k+2}|^2$ is greater than zero is

$$\begin{aligned} P[F_1 > 0] &= \frac{1}{2\pi}\int_0^\infty \int_{-\infty}^\infty \phi_{F_1}(-jt)e^{-jft}dtdf \\ &= \frac{1}{2} + \frac{1}{2\pi j}\int_{-\infty}^\infty \frac{\phi_{F_1}(-jt)}{t}dt, \end{aligned} \tag{5.27}$$

where $\phi_{F_1}(-jt)$ is the characteristic function of the Hermitian quadratic form $F_1$, which has a well-known form [4, 24]. After some simplification, this technique yields the integral formulas

$$\begin{aligned} & P[(2\epsilon-1)|Z_k|^2 - \epsilon|Z_{k+1}|^2 + (1-\epsilon)|Z_{k-1}|^2 > 0] = \\ & \frac{1}{2} - \frac{1}{\pi}\int_0^\infty \frac{\exp\left(\frac{p^2(N+1)t^2A_2(\epsilon,N,t)}{A_1(\epsilon,N,t)}\right)}{tA_1(\epsilon,N,t)} \\ & \times \left((1+a_2(\epsilon,N)t^2)\sin\left(\frac{p^2(N+1)tA_3(\epsilon,N,t)}{A_1(\epsilon,N,t)}\right)\right. \\ & \left. + a_3(\epsilon,N)t^3\cos\left(\frac{p^2(N+1)tA_3(\epsilon,N,t)}{A_1(\epsilon,N,t)}\right)\right) dt \\ & \triangleq P_2(p,N,\epsilon), \end{aligned} \tag{5.28}$$

where $A_1$, $A_2$, and $A_3$ are polynomial functions of $\epsilon$, $N$, and $t$, and $a_2$ and $a_3$ depend on $\epsilon$ and $N$. The last term in (5.23) is

$$\begin{aligned} P[(1-2\epsilon)|Z_{k+1}|^2 - (1-\epsilon)|Z_k|^2 + \epsilon|Z_{k+2}|^2 > 0] = \\ P_2(p,N,1-\epsilon). \end{aligned} \tag{5.29}$$

The integrals in (5.28) and (5.29) are quite complicated, but nonetheless provide, in conjunction with equations (5.24) and (5.25), an analytical expression for an upper bound on the conditional error probability of the scheme $S_{ncmoa}$. We believe, however, that $P_2(p,N,\epsilon) + P_2(p,N,1-\epsilon)$ is maximum at $\epsilon = 0.5$. If we also let $N$ become very large, so that $\nu_1 \approx 0$ and $\nu_2 \approx p(1-\epsilon)\sqrt{N}/2$, then by noting that $M(0,b) = e^{-b^2/2}$, we may obtain from (5.23) the following conjectured asymptotic upper bound on the conditional error probability:

$$\begin{aligned} P_{e|\epsilon}[S_{ncmoa}] \le \frac{1}{2}(N-2)\min\{e^{-p^2N(1-\epsilon)^2/8}, e^{-p^2N\epsilon^2/8}\} \\ + e^{-p^2N/32}. \end{aligned} \tag{5.30}$$

We may then average this over $\epsilon \in [0, 1)$ to obtain the following conjectured asymptotic upper bound on the average error probability:

$$P_e[\mathcal{S}_{ncmoa}] \leq (N-2)\frac{\sqrt{2\pi}}{p\sqrt{N}}\left(Q(p\sqrt{N}/4) - Q(p\sqrt{N}/2)\right)$$
$$+e^{-p^2 N/32}. \qquad (5.31)$$

We can in fact prove that the average error probability of $\mathcal{S}_{ncmoa}$ decreases exponentially with increasing SNR by observing that for every $0 \leq \epsilon < 1$, the terms in (5.23) can be bounded by exponentially decreasing functions of $p$. Note that $M(a,b) - \frac{1}{2}\exp\left(-\frac{a^2+b^2}{2}\right) I_0(ab) \leq \frac{b}{b-a} exp(-\frac{(b-a)^2}{2}) - \frac{1}{2}\exp\left(-\frac{a^2+b^2}{2}\right)$ if $a < b$, so the first term in (5.23) decreases exponentially with increasing $p$. We may also upper bound the second term in (5.23) by noting that $P[F_1 > 0] \leq e^{-yF_1} = h_{F_1}(y)$, $y \leq 0$, where $h_{F_1}(y)$ is the moment-generating function of $F_1$, if it exists. It can then be shown that for every $\epsilon$ there does exist some $y_0 < 0$ such that $h_{F_1}(y_0)$ exists and decreases exponentially with increasing $p$. A similar argument is used to upper bound the third term in (5.23). Therefore we know that $P_e[\mathcal{S}_{ncmoa}]$ decreases exponentially with increasing SNR, which also implies that $P_e[\mathcal{S}_{ncopt}]$ decreases exponentially with increasing SNR. For further details, consult [23].

## 5.7 CONCLUDING REMARKS

Synchronization of code sequences in direct-sequence spread-spectrum communication systems is a fundamental task that has not been studied as extensively as it deserves. This chapter has presented an idiosyncratic survey of some topics in the coarse synchronization, that is, the acquisition of code sequences. It is hoped that the reader will be stimulated to study this fascinating and sometimes frustrating subject in more detal.

## 5.8 ACKNOWLEDMENT

Some of the work presented here was done in conjunction with Richard Korkosz, Andrew Slonneger, and Meera Srinivasan, one-time graduate students and advisees, from whom I learned more than they learned from me. Nonetheless, the opinions expressed herein, as well as the errors of interpretation, are solely my own.

## Notes

1. One exception would be when the fading is so highly time-selective that the correlation time of the fading process is much smaller than the bit duration. In this case, however, the assumption of the fading level being essentially constant over the bit duration is not untenable.

## References

[1] K. K. Chawla and D. V. Sarwate, "Parallel acquisition of PN sequences in DS/SS systems," *IEEE Transactions on Communications*, vol. COM-42, pp. 2155–2164, May 1994.

[2] K. K. Chawla and D. V. Sarwate, "Acquisition of PN sequences in chip synchronous DS/SS systems using a random sequence model and the SPRT," *IEEE Transactions on Communications*, vol. COM-42, pp. 2324-2334, June 1994.

[3] M. Cohn and A. Lempel, "On fast M-Sequence transforms," *IEEE Transactions on Information Theory,* vol. IT-23, pp. 135-137, January 1977.

[4] C. W. Helstrom, *Statistical Theory of Signal Detection,* Pergamon Press, 1968.

[5] R. A. Korkosz and D. V. Sarwate, "Serial acquisition of PN sequences in chip asynchronous DS/SS systems," in *Proc. MILCOM Conference,* vol. 3, pp. 789-793, October 1993.

[6] R. A. Korkosz, "Serial acquisition of PN sequences in direct-sequence spread-spectrum communication systems," Ph.D. thesis, Departmen t of Electrical and Computer Engineering, University of Illinois at Urbana-Champaign, Urbana, Illinois, 1994.

[7] R. A. Korkosz and D. V. Sarwate, "Estimation of delay of sgnature sequences in chip asynchronous DS/SS systems," in *IEEE International Conference on Communications Record,* vol. 3, pp. 1696–1700, May 1994.

[8] S. Tantaratana and A. W. Lam, "Noncoherent sequential acquisition for DS/SS systems," in *Proceedings of the 29th Annual Allerton Conference on Communication, Control, and Computing,* pp. 370-379, October 1991.

[9] S. Tantaratana and A. W. Lam, "Mean acquisition time for noncoherent PN sequence sequential acquisition schemes," in *Proc. MILCOM Conference,* vol. 3, pp. 784-788, October 1993.

[10] U. Madhow and M. B. Pursley, "Acquisition in direct-sequence spread-spectrum communication networks: An asymptotic analysis," *IEEE Transactions on Information Theory,* vol. 39, pp. 903–912, May 1993.

[11] U. Madhow and M. B. Pursley, "Mathematical modeling and performance analysis for a two-stage acquisition scheme for direct-sequence spread-spectrum CDMA," *IEEE Transactions on Communications,* vol. 43, pp. 2511–2520, September 1995.

[12] A. H. Nuttall, "Error probabilities for equicorrelated M-ary signals under phase-coherent and phase-incoherent reception," *IRE Transactions on Information Theory,* vol. IT-8, pp. 305-314, July 1962.

[13] R. L. Pickholtz, L. B. Milstein, and D. L. Schilling, "Spread spectrum for mobile communications," *IEEE Transactions on Vehicular Technology,* vol. 40, pp. 312–322, May 1991.

[14] R. L. Peterson, R. E. Ziemer, and D. E. Borth, *Introduction to Spread Spectrum Communications*, Prentice Hall, 1995.

[15] M. B. Pursley, "The role of spread spectrum in packet radio networks," *Proceedings of the IEEE,* vol. 75, pp. 116–134, January 1987.

[16] D. V. Sarwate and M. B. Pursley, "Crosscorrelation properties of pseudorandom and related sequences," *Proceeding of the IEEE,* vol. 68, pp. 593-619, May 1980.

[17] J. L. Shanks, "Computation of the fast Walsh-Fourier transform" *IEEE Transactions on Computers,* pp. 457-459, May 1969.

[18] A. M. Slonneger and D. V. Sarwate, "Noncoherent parallel acquisition of PN Sequences in direct-sequence spread-spectrum systems," *Proceedings of the Second IEEE International Symposium on Spread-Spectrum Techniques and Applications,* 31-34, November 1992.

[19] E. A. Sourour and S. C. Gupta, "Direct-sequence spread-spectrum parallel acquisition in a fading mobile channel," *IEEE Transactions on Communications,* vol. 38, pp. 992-998, July 1990.

[20] M. Srinivasan and D. V. Sarwate, "Parallel acquisition of spreading sequences in DS/SS systems," *Proceedings of the Conference on Information Sciences and Systems*, vol. II, pp. 1163–1168, 1994.

[21] M. Srinivasan and D. V. Sarwate, "Noncoherent parallel acquisition of spreading sequences in DS/SS systems," *Proceedings of the Conference on Information Sciences and Systems*, vol. II, pp. 858–863, 1996.

[22] M. Srinivasan and D. V. Sarwate, "Simple schemes for parallel acquisition of spreading sequences in DS/SS systems," *IEEE Transactions on Vehicular Technology,* vol. 45, pp. 593–598, August 1996.

[23] M. Srinivasan, "Parallel acquisition of spreading sequences in direct-sequence spread-spectrum communication systems," Ph.D. thesis, Department of Electrical and Computer Engineering, University of Illinois at Urbana-Champaign, Urbana, Illinois, 1996.

[24] G. L. Turin, "The characteristic function of Hermetian quadratic forms in complex normal variables," *Biometrika,* vol. 47, pp. 199-201, June 1960.

# Part 2

# Wireless Communications

# 1 EVOLVING WIRELESS LAN INDUSTRY - PRODUCTS AND STANDARDS

Kaveh Pahlavan,
Ali Zahedi
Prashant Krishnamurthy

**Abstract**

This paper provides an overview of the existing wideband wireless local communication industry. First the evolution of the wireless LAN industry from concept to product development is discussed. Then an overview of the existing and emerging standards and international activities in US, EC, and Japan is provided. An update and comparison of the existing standards covers IEEE 802.11, HIPERLAN, and Wireless ATM. The activities in the US will cover WINForum and U-NII. The EC activities under ACTS/RACE all addressing WLAN services for the third generation wireless networks. The Japanese activities includes lower speed services for indoor/outdoor as well as high speed indoor applications perceived for the future of this industry in Japan.

## 1.1 INTRODUCTION

We are emerging to a new and exciting era for the wireless LANs. After a decade of self realization for the WLAN industry, today markets for WLAN and inter-LAN bridges are finding their way in the Health Care, Manufacturing, Finance, and Education. The IEEE 802.11 standard for WLANs, operating in the unlicensed ISM bands, is emerging as a mature standard presenting a well defined technology that is adopted by the manufacturers and accepted by the users. Chipsets are developed for the 802.11 that makes it available for further software creativity to develop new applications and expand the market. ETSI's RES-10 has defined another alternative technology, HIPERLAN I, that is more focused toward the ad-hoc networking applications and supports higher data rates up to 23Mbps. An ad-hoc network provides an environment to set up a temporally wireless network among a group of users equipped with WLANs. In the last year, research around wireless ATM soared like an epidemic engaging numerous companies in examining the suitability of yet another alternative technology for the implementation of wireless LANs. The continual demand of the wireless LAN industry for more unlicensed bands in useful spectrums, initiated by WINForum, resulted in the release of 20 MHz unlicensed band around 2GHz for asynchronous and isochronous applications and the U-NII (formerly SUPERNet) won 300 MHz of unlicensed bands in 5GHz earlier this year. On the other hand, the pan-European third generation cellular service UMTS is considering connectionless packet switched networks as class D bearer services and under its research arm ACTS the MEDIAN, WAND, SAMBA, and AWACS projects are addressing WLAN services. The third generation systems will incorporate WLAN standards into the universal mobile services.

This paper provides an updated overview of all the activities around the WLAN industry.

## 1.2 THE BIRTH OF THE WLAN INDUSTRY

The concept of wireless LAN (WLAN) was first introduced at the IBM's Research Laboratories in Switzerland in late 1970's [GFE79]. The diffused Infrared technology was examined for the implementation of a 1 Mbps WLAN for manufacturing floors where the installation and maintenance of wiring are extremely expensive. This project was shortly followed by an experiment in the HP laboratories in California [FER80] using spread spectrum technology in a typical HP office setting. The architecture of the HP offices is open and there is no wall separating working areas. Wiring is snaked through the walls and lack of walls increases the cost of wiring. The IR experiments did not achieve more than a few hundred Kbps. The spread spectrum project needed a band and a petition was made to FCC for frequency bands. Both projects were abandoned.

In the early 1980's Codex/Motorola in Massachusetts started the development of a WLAN product using traditional radio modems at 1.71 GHz. They were unsuccessful in obtaining these bands from the FCC and this development project was abandoned as well. The only successful product in this period was the so called Data Radio. A group of computer hackers in British Columbia, Canada developed a wireless local network connecting computer terminals using standard voice band modems operating over walkie-talkie radios employing a CSMA type access protocol. In the mid-1980's the gross income of this company exceeded a few tens of millions of dollars per year. However, these networks were operating at 9600 bps and by no means they could be considered a WLAN. The minimum data rate for the IEEE 802 standard committee for LANs is 1 Mbps. This experiment showed that for the wireless local networks mobility alone can attract substantial new applications and consequently a new market.

The major break through for the WLAN industry was the release of the unlicensed ISM bands at 0.9, 2.4, and 5.7 GHz in the May of 1985 [MAR85]. There were two important aspects for these bands (1) they were unlicensed (2) the use of spread spectrum technology was mandated for operation. The relation between a licensed band and an unlicensed band is similar to relation between a private backyard and a public park. If you can afford a backyard you can barbecue there; otherwise you may go to the public park and, space available, do the same thing. The spread spectrum technology was assumed to provide privacy (resistance to interference from others) in the public park so that everyone can enjoy the free public yard without interference from others. The release of ISM bands coincided with some publications [PAH85] underlining the importance of wireless office information networks and suggesting alternative technologies for their implementation. Immediately after these events a momentum for the design of WLAN and wireless PBX products started in several companies. WLANs were new products developed by small companies or small groups in large companies, non of them affording backyards (licensed frequency bands) in expensive neighborhoods of 1GHz where income-rich voice applications were operating. These companies had three choices [PAH85] for implementation of WLANs (1) use the spread spectrum technology in the unlicensed ISM bands (2) migrate to higher frequencies (10's of GHz) where more bands are available but the development of the technology is challenging (3) use the IR technology.

## 1.3 EVOLUTION OF WLAN PRODUCTS AND APPLICATIONS

In the late 1980's the three technologies mentioned before were examined by variety of manufacturers for product development and around 1990 these products appeared in the market. NCR, Netherlands introduced the WaveLAN operating at 900MHz ISM bands; Motorola introduced the Altair operating in licensed 18GHz bands, and Photonics introduced the Photolink that was using the diffused IR technology. These efforts were initiated in mid to late 1980's when coaxial cable was still the dominant medium for wired LANs and the installation and relocation costs of the LANs constituted a major share of the total costs of the LAN industry that was around a couple of tens of billions of dollars per year. Wireless LANs at that time were considered to be an alternative that avoids the expensive and troublesome installation and relocation of the coaxial cabled LANs. Assuming that WLANs will capture 10-15% of the coaxial cable LANs, the early market predictions for wireless LANs were around $0.5-2 billions for the mid 1990's. However, by the time that WLAN products appeared in the market the less troublesome twisted pair wiring (TP) technology, similar to the existing telephone wiring, had already replaced the coaxial cabled LANs. The TP technology had considerably reduced the installation and relocation difficulties and and on top of that it had provided an opportunity for another jump in data rate to 100Mbps using the fast Ethernet technology. As a result, the first generation Indoor Wire-Replacement WLAN products that were mostly designed in a box consuming around 20W power (not supportable by batteries and therefore not suitable for laptops) did not met the market predictions. This was perhaps the end of the first generation WLAN products. In this period we learned that the solution to the cabling problem is twisted pair wiring not wireless medium. The twisted pair wiring is available in most buildings, telephone companies have a century of expertise in how to install them, and using hub based architecture they can be scaled up to provide bandwidths up to Gbps range.

Another event that shaped the direction of the WLAN industry was an enormous demand for small portable terminals to be used in warehouses for inventory check, in stock exchange halls to connect the brokers to the networks, in and around building to check in rental cars, in the supermarkets to check the price tags, and even in the restaurants to take the orders. This applications indicated that there is another important angle to wireless local networks: the “mobility”. Wireless local networks had found its own applications, evolving around mobility of the terminal, that was independent from high bandwidth that was the main drive for wired LANs. There are numerous WLAN applications demanding mobility that does not necessarily need 1 Mbps but they need local mobility for the terminal. With the appearance of the first generation WLANs in the market people started to discover these new applications.

The second generation wireless LANs evolved in two directions. One group developed PCMCIA card WLANs for laptops toaddress the demand for local mobility and its associated applications. The other group added directional antennas to the first generation shoe-box type WLAN products and marketed them as inter-LAN bridges for outdoor applications. Development of PCMCIA WLAN cards resulted in the first WLAN market based on "mobility" which provides a complementary feature to the wired LANs - rather than replacing them. With the availability of the PCMCIA cards for the laptops the concept of a wireless campus area network (CAN) network attracted significant attention. From marketing point of view, rather than selling WLANs one-by-one to overcome wiring difficulties, a CAN that should cover a campus will sell hundreds of PCMCIA card WLANs and many access points to cover a campus. A CAN does not replace the wired LANs but it complements the access to the backbone by providing mobility to the terminals. Leading universities started installation of these experimental networks and the National Science Foundation in the US supported the Center for Wireless Information Studies at WPI to implement a testbed for performance monitoring of the CANs. Massive market for a CAN demands interoperability among the terminals and a unified standard to support that interoperability. Otherwise, these application should aim at the narrower vertical markets for specific applications.

As we mentioned before, manufacturers of the first generation WLANs who had designed a shoe-box WLAN found a new market niche that we refer to as wireless inter-LAN bridges. The same WLAN that covers less than 100 meter in the indoor areas can use a directional antenna to cover a few kilometers of line of sight radio connection in the outdoor areas. With the data rates around 2 Mbps, these products can easily replace the need for wiring or leasing traditional wiring connections such as T1 lines. The cost of a wireless T1 equivalent will go around $10K that is close to the cost of leasing a short distance T1 line for one year. Therefore, WLANs that were designed for indoor wire replacement and were not successful found their market as the outdoor point-to-point wire replacement to connect two LANs that are several tens of kilometers apart from one another. Since the length of the wire to be replaced by n inter-LAN bridge is much longer and the outdoor wiring in the streets is much more difficult and expensive, this application turned out to meet an unexpected and growing market for the WLAN industry. The income from inter-LAN bridges turned out to be the bread and butter of several companies in the past few years. An inter-LAN bridge is a wireless local loops for data applications. However, in terms of the market size, wireless local loops can be used for every home but inter-LAN bridges have much more limited number of applications. The income from these products was enough to encourage most investors to continue the operation of several manufacturers of the first generation WLAN products and wait for the market hike of the wireless wideband local networks.

Today's WLAN market aims at four category of applications [WOZ96]

- Healthcare Industry
- Factory floors
- Banking industry
- Educational institutions

In the healthcare, in addition to the traditional equipment such as laptops, notebooks, and hand-held terminals, special wireless services such as electronic thermometer and blood pressure monitoring devices are expected to be involved in the wireless local communications. These devices are used to provide mobile access to the clinical and pharmaceutical data bases for the physician as well as personal and health data entering. In the manufacturing floors and factory environment, in addition to accessing and entering data to the database, wireless networks enables rapid modification of the assembly lines and instant network access to the delivery trucks at the dock stations. In the banking industry, WLANs facilitate extending the network facilities to other braches, upgrading the systems without disrupting banking operation, reorganizing and rearranging the branches, and the access to Internet. In the educational environments WLANs facilitates distant learning using wireless classrooms, provides access to the Internet, computational facilities, and databases servers to the students using notebooks.

All wireless LAN products available on PCMCIA cards are either direct sequence spread spectrum (DSSS) or frequency hoping spread spectrum (FHSS) operating in ISM bands. The diffused IR technology is used for nomadic access in shorter distances for applications such as nomadic access for laptops to printers or in specific areas within the hospitals, such as radiology departments, where using radio signal is not encouraged. In addition to spread spectrum technology, other technologies such as direct beam IR (DBIR) and traditional radios are used for inter-LAN bridge applications. According to January 1997 Frost & Sullivan report on the North American Wireless Office Hardware market, the total 1996 revenue for wireless offices was $390 million from which $218 million belonged to wireless LANs.

## 1.4 PROSPECTS OF WORLD WIDE WLAN INDUSTRY

Today's world wide market for the WLAN primarily belongs to the US. For the manufacturers aiming at this market the two most important issues are the availability of unlicensed bands for future product development and development of an spectrum etiquette for the unlicensed bands that secure a minimum availability for the users in different scenarios. In 1994, the FCC released 20 MHz of spectrum between two parts of licensed bands in the 1.9 GHz for operation of *unlicensed personal communications services* (U-PCS). The unlicensed nature of the frequency bands necessitated a ***"coexistence etiquette"*** which was developed by the WINTech, the technical subcommittee of the WINForum industry association. This spectrum etiquette forms the basis for the rules adopted by the FCC for operation in the U-PCS bands.

On January 9$^{th}$ 1997, the FCC released 300 MHz of unlicensed spectrum at 5.15-5.35 GHz and 5.725-5.825 GHz for use by a new category of radio equipment called "Unlicensed National Information Infrastructure" or U-NII devices. This decision of the FCC makes a large amount of spectrum available for wireless LAN applications which can provide data rates of upto several tens of Mbps needed for multimedia type of applications. The frequencies allocated are compatible with those specified earlier by the CEPT for HIPERLAN. The WINForum spectrum etiquette was also suggested by the FCC for U-NII devices, but was not adopted because of concerns about its suitability to accommodate WATM services.

Although WLAN market is primarily in the US, there are significant WLAN related activities in the EC and Japan. In the EC under the ACTS/RACE projects several research programs are devoted to examine variety of technologies for wideband local applications. The intention is to employ these technologies into the evolving third generation cellular systems. Several Japanese companies are developing WLAN products for the US market and a vision of the future for the Japanese market is evolving. In the rest of this section we address these activities with some additional details.

### *1.4.1 Spectrum Etiquette for U-PCS Bands*

This section briefly describes the spectrum etiquette specified by the FCC on the basis of WINForum's work for U-PCS devices, further details about this protocol is available in [STE94]. The interesting point about the WINForum etiquette is the assignment of two separate bands for the asynchronous and isochronous transmissions. This separation implies that, in the view of WINForum, voice and bursty data transmissions are incompatible enough to use separate channels in the air. This separation is in contrast with the view of those working in wired communications (such as ISDN and ATM) in which the trend is toward the integration of the facilities to carry the voice and bursty data.

The 20 MHz of spectrum allocated for U-PCS is shown in Figure 1.

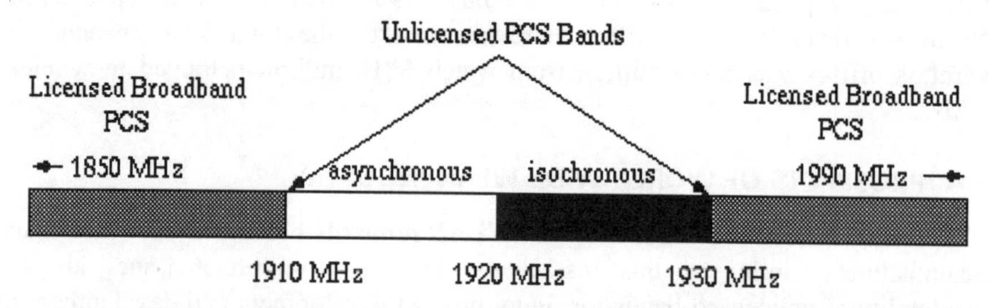


**Figure 1** Spectrum allocated for Unlicensed PCS [STE94]

The spectrum between 1.91 and 1.92 GHz is allocated for asynchronous type of services, mainly high speed bursts of information like one way data communications employing carrier sense multiple access (CSMA) for channel access. The 1.92-1.93 GHz band is allocated for isochronous devices using regular repetitive, delay intolerant transmissions with two-way communications accomplished through time-division duplex techniques. The maximum transmitter power in each case is restricted to $0.1\sqrt{B}$ Watts where B is the emission bandwidth in MHz. This power requirement is assumed to be suitable for indoor radio communications for up to 100 ft.

There are three basic principles adopted for the spectrum etiquette in the U-PCS bands: (a) listen before talk (or transmit) (LBT) protocol (b) low transmitter power and (c) restricted duration of transmissions. The spectrum is divided into several time-frequency slots which can be referred to as channels. Each device senses the channel it intends to use before transmission and shifts to another channel if the one it senses is already occupied. A low transmitter power ensures a short operation range and thus limits the geographic extent of the interference enabling reuse of the same time-frequency slots. The transmitter has to give up the channel after a certain maximum period of time to provide other transmitters an opportunity to use the channel. No monitoring of the channel is required during transmissions. The requirements on timing accuracy is quite rigid making this protocol unsuitable for ATM transport.

## 1.5 U-NII DEVICES

The U-NII initiative started with WINForum filing a petition before the FCC requesting allocation of 250 MHz of spectrum for high speed devices providing 20 Mbps in May 1995. This was followed by a petition by Apple Computer requesting allocation of 300 MHz of spectrum in the same frequency bands for promoting full deployment of the so-called National Information Infrastructure (NII)[1]. These bands and devices were referred to as "Shared Unlicensed PErsonal Radio Network" (SUPERNet) bands and devices. The FCC believes that this spectrum would enable opportunities for providing advanced telecommunications services to educational institutions, health care providers, libraries etc [FCC96] thereby significantly assisting in meeting universal service goals [HUD94] as set forth in the Telecommunications act of 1996 [TEL96]. In its report, the FCC adopted the term U-NII for these devices that will support the creation of new wireless local area networks, facilitate wireless access to the NII and meet multimedia communications requirements by providing spectrum for a wide range of devices and service offering such as enabling broadband wireless computer network (and Internet) connection to all classrooms, retrieving patient data such as X-Rays and medical charts from databases, etc.

The FCC has allocated 300 MHz of spectrum for U-NII devices. The frequency bands and their respective technical restrictions are shown in Table 1. To encourage maximum flexibility in developing new technologies minimum technical restrictions have been adopted. Detailed restrictions or standards can only delay the implementation of the appropriate products that are able to attract commercial applications.

**Table 1** FCC Requirements for the U-NII Frequency Bands

| Band of operation | Maximum Tx Power | Max. Power with antenna gain of 6 dBi | Maximum PSD | Applications: suggested and/or mandated | Other Remarks |
|---|---|---|---|---|---|
| 5.15 - 5.25 GHz | 50 mW | 200 mW | 2.5 mW/MHz | Restricted to indoor applications | Antenna must be an integral part of the device |
| 5.25 - 5.35 GHz | 250 mW | 1000 mW | 12.5 mW/MHz | Campus LANs | Compatible with HIPERLAN |
| 5.725-5.825 GHz | 1000 mW | 4000 mW | 50 mW/MHz | Community networks | Longer range in low-interference (rural) environs. |

[1] The FCC Describes the NII as a group of networks including the public switched telephone network (PSTN), radio and television networks and other networks yet to be built, which will together serve the communications and information processing needs of the people of the US in the future.

Although no devices have been developed yet for U-NII bands, several activities for developing standards and products in these bands have already started. The Mobile Information Infrastructure (MII), an ongoing research project at Bell Labs which is partially funded by the National Institute of Standards and Technology is considering wireless ATM applications in this band [AYA96]. The IEEE 802.11 has set up a study group for higher speed physical layer standards development in the 5 GHz band with the same MAC protocol used currently for the 2 Mb/s standard using spread spectrum technology in the ISM bands. Data rates beyond 20 Mbps as well as capability of the MAC layer to support data, voice and video services are examined in this study group. Since in the U-NII, as opposed to ISM bands, there is no restriction to use spread spectrum technology, other modulation techniques such as OFDM and GMSK have been suggested. Because of the increase in the frequency of operation, in the U-NII the coverage of an access point (AP) is expected to reduce by about 40% as it is compared by the coverage of an 802.11 AP. This coverage could be reduced additionally if the spread spectrum technology is not adopted. At higher U-NII frequencies the LAN adapter is expected to be more expensive, but the cost/performance ratio is expected to be improved.

### *1.5.1 European and Japanese Wideband Wireless Activities*

The ACTS (Advanced Communications Technologies and Services) program in Europe is involved in several wideband wireless activities which focus mainly on two different areas: UMTS and MBS. The FRAMES (Future Radio Wideband Multiple Access System) and RAINBOW (Radio Access Independednt Broadband on Wireless) projects are evaluating air interface standards, network functions, mobility, location, call and connection control for the UMTS to connect to the B-ISDN infrastructure. Under Wireless Customer Premises Network / Mobile Broadband System (WCPN/MBS) there are four major wideband projects [Agh96,ACT96]. The MEDIAN project is evaluating and implementing a wireless LAN capable of operating at 60 GHz and providing 155 Mbps. The prototype will demonstrate one base station at 155 Mbps and two portables, one at 34 Mbps and the other at 155 Mbps. The Magic WAND project, see Table 2, demonstrates wireless ATM transmission at 5 GHz. The *System for Advanced Mobile Broadband Applications* (SAMBA) develops an MBS trial of two base and two mobile stations operating at 40 GHz. The *Advanced Wireless ATM Communications Systems* (AWACS) is based on a testbed already available through one of the project partners of ACTS and targets the development and demonstration of wireless access to B-ISDN services through this testbed. The terminals are of low mobility and operate at 19 GHz with data rates of up to 34 Mbps.

In addition to these projects, ACTS is also involved in satellite based wideband wireless projects like SECOM (Satellite EHF Communications for Mobile Multimedia Services) and SINUS (Satellite Integration into Networks for UMTS Services).

In Japan [Mor96,MPT96] the Ministry of Post and Telecommunications (MPT) is the primary agency involved in future wideband wireless networks. Radio regulation was updated to provide 26 MHz of bandwidth in the 2.4 GHz band for small power data communications that fits the IEEE 802.11. The major problem with this band appears to be the small bandwidth of 26 MHz as it is compared with

84 MHz available in the ISM bands in the US. There is another 80 MHz of spectrum available for licensed "local area radio stations" at 19 GHz that is similar to the frequencies used by Motorola's ALTAIR product.

The are more recent activity in Japan in local wideband wireless networks. Multimedia Mobile Access Communication (MMAC) committee has proposed two systems for a mobile communication infrastructure that works seamlessly with fibre optic networks. Terminals that can operate both indoors and outdoors at 6-10 Mbps (high-speed wireless access) and those that operate only indoors without handoffs (ultra high-speed radio LAN) systems, both operating in the millimeter wave frequencies, are investigated by MMAC. Equipment development in system design and fundamental RF technologies and standardization is expected to be completed by 2000 A.D and commercial services are expected to start by the year 2002. Another wideband project in Japan is the ATM Wireless Access (**AWA**) project that is carried out by NTT. The shortage of spectrum in the lower frequencies has resulted in focusing the activity in the *super high frequency (SHF)* band (10-30 GHz). TDMA with TDD will be the air interface with 20 to 80 Mbps. Each user terminal is expected to provide up to 12 Mbps [AWA96].

## 1.6 OVERVIEW OF WLAN STANDARDS

There are three standardization activities for wireless LANs. The IEEE standard that started in IEEE 802.4 Token Bus wired LAN as IEEE 802.L in 1988. In 1990 the IEEE 802.4L changed its name to IEEE 802.11 to form a stand alone WLAN standard in the IEEE 802 LAN standards organization. The IEEE 802.11 is a mature standard adopted by several manufacturers for product development. The HIPERLAN, developed by RES-10 of the ETSI as a pan-European standard for high speed wireless local networks. The so called HIPERLAN I, the first defined technology by this standard group, is completed and other versions using wireless ATM and other alternatives is under consideration [Wil96]. Although a couple of prototypes were developed for the HIPERLAN I but it has not yet been adopted by any manufacturer for product development. The bands assigned for HIPERLAN in the EC was one of the motives for the FCC to release the U-NII bands discussed in the last section. The third standard for WLANs is under study by the Wireless ATM (WATM) working group of the ATM Forum that started last year. This group will develop specifications for radio access, MAC layer and mobility support for WATM. In the past couple of years several companies have developed prototypes to implement the so called wirless ATM technology. More recently, these activities merged under the WATM Forum. The major incentive for this technology is to provide an end to end ATM solution for the mobile terminals of the future. The rest of this section intend to provides a comparative evaluation of these evolving standards.

The 802.11 standard [IEE96] supports infrastructure based and ad-hoc topologies. In a infrastructure network, shown in Fig. 2.a, mobile terminals communicate with the backbone network through an Access Point (AP). The AP is a bridge interconnecting the 802.11 network to the backbone wired infrastructure. In this configuration, distribution system interconnects multiple Basic Service Sets (BSS) through access points to form a single infrastructure network named Extended Service Set (ESS). A mobile terminal can roam among different BSSs in one ESS without losing the connectivity to the backbone. In ad-hoc configuration, shown in Fig. 2.b, the mobile terminals communicate with each other in an independent BSS without connectivity to the wired backbone network.. In this case some of the functions of the AP, such as release of a beacon with a defined ID and timing reference, that are needed to form and maintain a BSS are provided by one of the mobile terminals. The cells in both configuration shown in Fig. 2 can overlap with one another. In addition to the contention based CSMA/CA access method suited to asynchronous data applications, the IEEE 802.11 also supports the contention free prioritized point coordination function (PCF) mechanism to support time bounded isochronous applications. The MAC services in the IEEE 802.11 supports authentication, encryption, and power conservation mechanism that are not available in the MAC layer of other 802 standards such as 802.3.

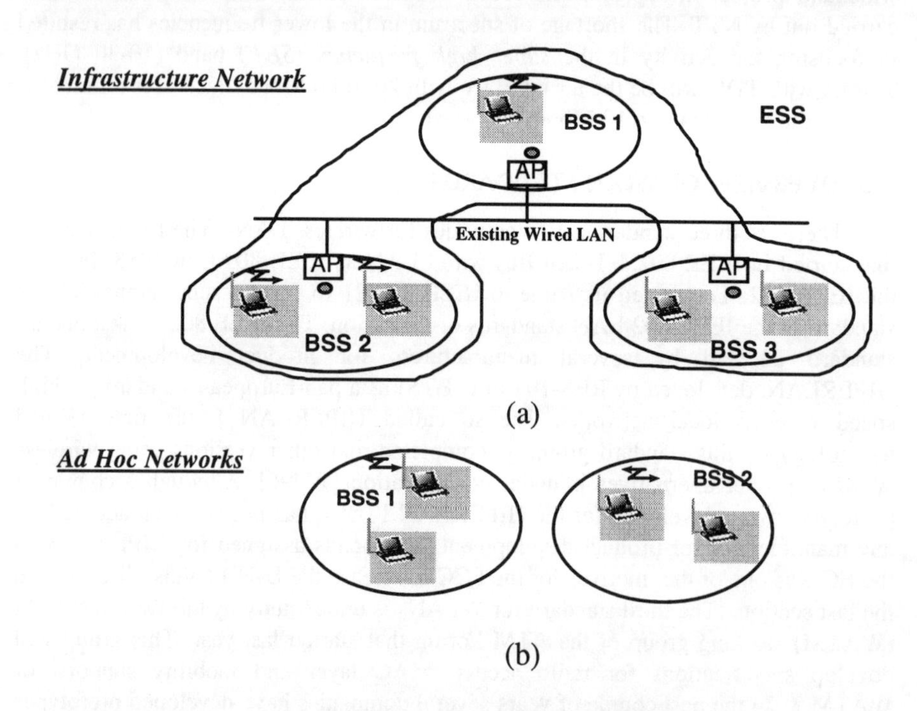


**Figure 2** Two Methods for IEEE 802.11 Configuration

The HIPERLAN network [ETS96, ETS94] supports multi-hop ad-hoc configuration shown in Fig. 3. Each node in HIPERLAN network has a node identifier (NID) and a HIPERLAN network identifier (HID) used to differentiate HIPERLANs from each other. The nodes holding the same HID are the members of a HIPERLAN. A nodes may be a member of several HIPERLANs simultaneously. The multi-hop routing extends the HIPERLAN communication beyond the radio range of a single node. Each HIPERLAN node is either a forwarder or a non-forwarder. A non-forwarder node accepts the packet that it has received. A forwarder node retransmits the received packets, if it does not have its own node address, to other terminals in its neighborhood. Each non-forwarder node should select at least one of its neighbors as forwarder. Inter-HIPERLAN forwarding needs bilateral cooperation and agreement between two HIPERLANs. To support routing and maintain the operation of a HIPERLAN the forwarder and non-forwarder nodes need to periodically update six and four data bases respectively. These data bases are identified in the Figure 3. The Non-Preemptive Multiple Access (NPMA) protocol used in HIPERLAN is a listen before talk protocol that supports both asynchronous and isochronous transmissions. The HIPERLAN defines a lifetime for each packet; after reaching the lifetime, if the packet has not arrived to the destination it is removed from the system. The packet lifetime mechanism and the priority scheme in the HIPERLAN facilitates the control of the quality of service. In addition to encryption and power conservation the MAC layer also handles the routing.

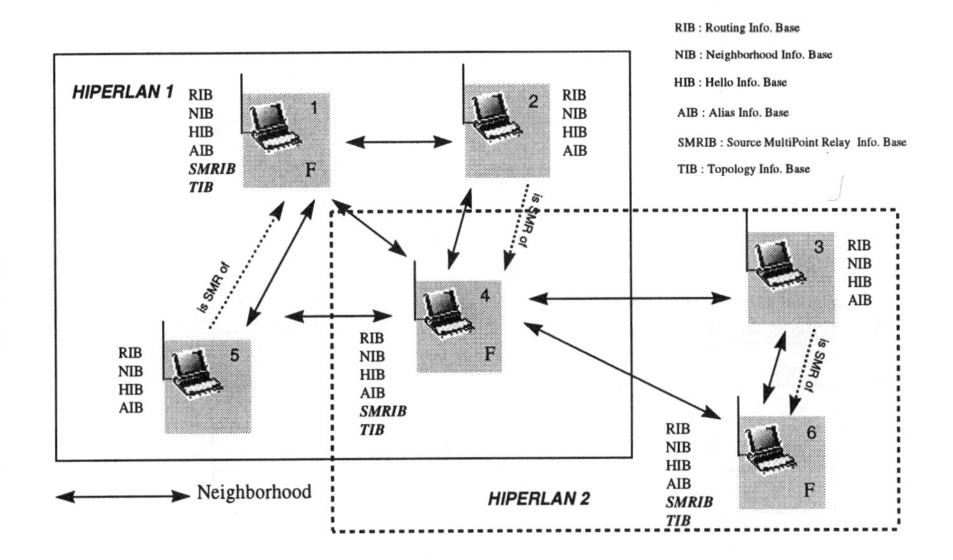


**Figure 3** HIPERLAN ad-hoc network configuration

All the existing WATM prototypes support only infrastructure based topology. In the infrastructure topology, considered in IEEE 802.11 and WATM prototypes, the mobile terminal should connect to the wired backbone using an AP that is connected either to a LAN or an ATM network. There are four options to interconnect the two air interfaces (802.11 WLAN and WATM) to the two wired backbones (legacy LANs and evolving ATM networks) namely WLAN-LAN, WATM-ATM, WLAN-ATM, and WATM-LAN. The protocol stack needed for the implementation of any of these options is different. Figure 4 a,b,c shows the WLAN-LAN, WLAN-ATM, and the WATM-ATM protocol stacks respectively. The first option is addressed in the IEEE 802.11 community, the other two options are considered in the WATM community [AYA96]. The fourth option WATM-LAN is not considered because if we go to WATM air interface we assume that the backbone network employs ATM switches.

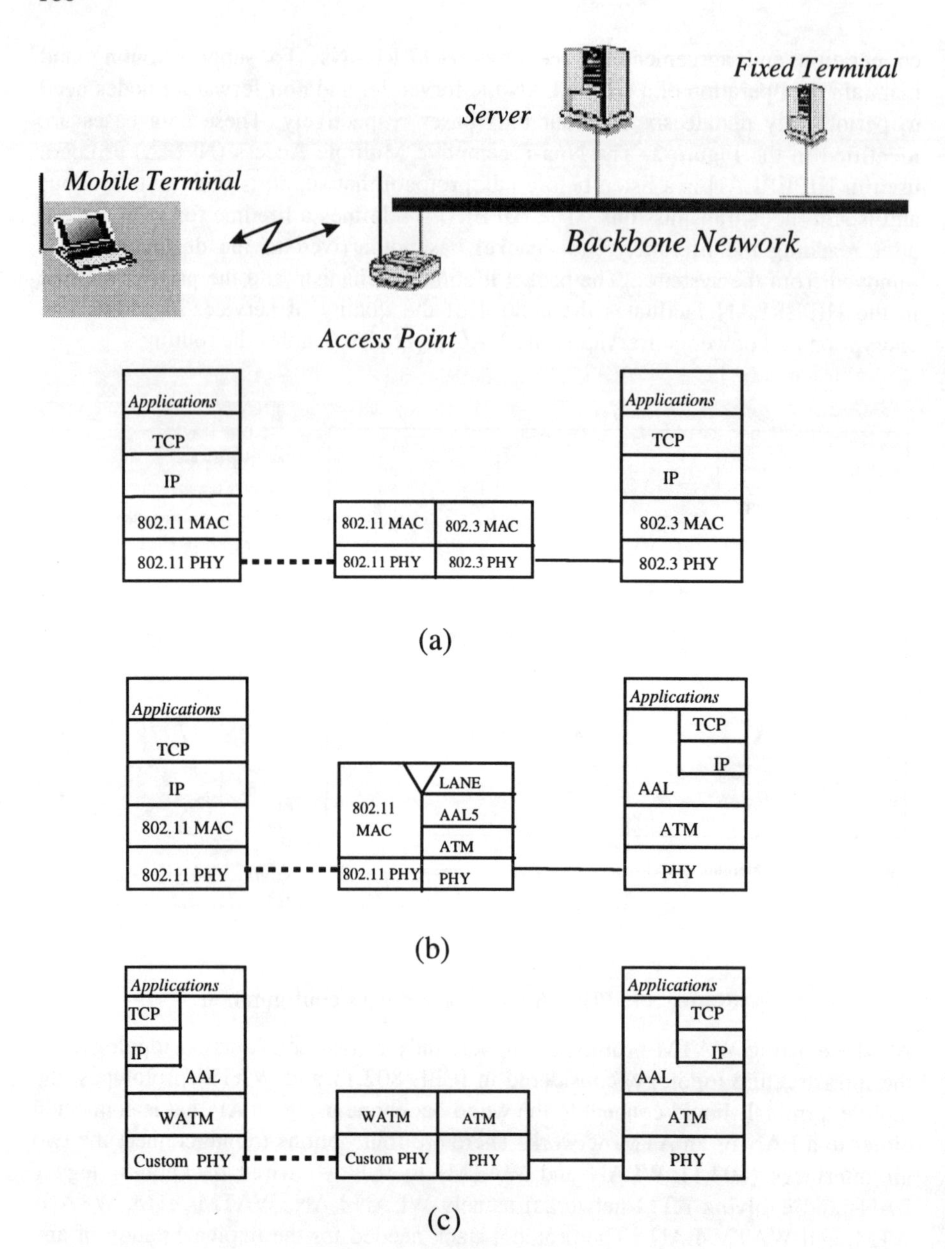


**Figure 4** WLAN connection to the backbone.

Table 2 compares the technical aspects of the IEEE 802.11 and HIPERLAN I with the three major WATM prototypes [AGR96, AYA96, ENG95, RAY97, WAN96]. The IEEE 802.11 is a mature standard with available commercial chip sets. HIPERLAN I is a standard with no commitment for product development. The WATM is under development. There are significant challenges for employing ATM in a mobile/wireless link that has to be resolved before this technology identifies itself as a legitimate standard. ATM was essentially designed for high bandwidth, highly reliable and static optical fiber channels. Wireless channels are inherently unreliable and time varying with limited available bandwidth. ATM is a connection oriented transmission scheme operating based on an initial negotiation with the network for a virtual channel with a specific QoS. In a mobile environment management of the virtual channel and the QoS is not simple because the route has to be continually modified as the terminals move during the lifetime of a connection. The BAHAMA project at Bell Labs tries to emulate a connectionless ATM network to overcome some of these drawbacks [AYA96].

## 1.7 CONCLUSIONS

The concept of WLAN was introduced around 1980. Since 1985 numerous companies have tried spread spectrum, infrared and traditional wideband radio technologies for variety of WLAN applications such as indoor wire-replacement, inter-LAN bridges, ad-hoc networks, campus area networks, and nomadic access to computing networks. Today, we are emerging to a new and exciting era for the wireless LANs. After a decade of self realization for the WLAN industry, the markets for WLAN and inter-LAN bridges are finding their way in the Health Care, Manufacturing, Finance, and Education environments. The sale of the WLAN products that started in 1992 have passed $200 million per year and it is expected to surpass a billion dollar early in the next millinum. The IEEE 802.11 standard for WLANs, operating in the unlicensed ISM bands, is emerging as a mature standard presenting a well defined technology that is adopted by the manufacturers and accepted by the users. Chipsets are developed for the 802.11 that makes it available for further software creativity to develop new applications to expand the market. ETSI's RES-10 has defined another alternative technology, HIPERLAN 1, that is more focused toward the ad-hoc networking applications and supports higher data rates up to 20Mbps. Research in finding a wireless ATM solution soars like an epidemic engaging numerous companies in examining the suitability of yet another alternative technology for the implementation of wireless LANs. The continual demand of the wireless LAN industry for more unlicensed bands in useful spectrums, initiated by WINForum, resulted in the release of 20 MHz unlicensed band around 2GHz for asynchronous and isochronous applications and the U-NII (formerly SUPERNet) won 350MHz of unlicensed bands in 5GHz earlier this year. On the other hand, UMTS is considering connectionless packet switched networks as class D bearer services and under ACTS the MEDIAN, WAND, SAMBA, and AWACS projects are all addressing WLAN services for the third generation wireless networks.

**Table 2** Comparison of WLAN standard technologies

| | | Frequency Band | Modulation Technic | Access Method | Topologies | MAC Services | QoS | Availibility |
|---|---|---|---|---|---|---|---|---|
| WIRELESS LAN | 802.11 | **Spread Spectrum Direct Sequense:** 2.4-2.4835 GHz, **Frequency Hopping:** 2.4-2.4835 GHz; **Diffused Infrared:** 850-950 nanometer | **Spread Spectrum Direct Sequense:** DBPSK, DQPSK; **Spread Spectrum Frequency Hopping:** 2GFSK, 4GFSK BT=0.5; **Diffused Infrared:** 16 and 4 PPM | Basic CSMA/CA, RTS/CTS, PCF with polling list, 20 frames | Ad-hoc, Infra-structure | Authentication, Encryption, Power conservation, Time bounded services | No explicit support for QoS, but includes infrastructure topology and priority scheme in PCF that are useful for quality assurance. | Technial standard finalized. Final administrative approval under progress. Products (e.g DEC Roamabout) and chipsets (e.g. Harris PRISM and Raytheon RAYLINK) are available. |
| | HIPERLAN | **5.15-5.30 GHz** | **Low bit rate**: FSK; **High bit rate**:GMSK (BT=0.3) | Non-Preemptive Multiple Access (NPMA), 10 PDU | Ad-hoc | Encryption, Power conservation, Routing and forwarding, Time bounded services | Advanced user priority scheme and packet lifetime mechanism to support QoS | Standard is finalized. No product in the market. Two prototypes: HIPERION, fully standards compliant, and LAURA, not fully compliant [Wil96]. |
| WIRELESS ATM | MII Bahama | 900 MHZ (Proposed 5 GHz U-NII Bands) | OFDM or GMSK with LMS or RLS Equalization | Distributed Queue Reservation Updated Multiple Access (DQRUMA): Reservation and Piggybacking | Infrastructure, ad hoc base station placement (optional) | ?? | Base station responsible for checking and guaranteeing QoS, connections with or without QoS guarantees possible. | Prototype at Bell labs in Lucent Technologies |
| | NEC | 2.4 GHz ISM Bands | $\pi/4$ - QPSK with decision feedback equalization | TDMA/TDD with Slotted ALOHA | Infrastructure based | Scheduling, multiplexing and demultiplexing of VCs | ABR, UBR, VBR and CBR slots are available but QoS support is not finalized | Prototype at NEC USA's C&C Research Laboratories, Princeton, NJ. |
| | Magic WAND | 5.2 GHz | 16 Channel OFDM | Reservation, Slotted ALOHA: Mobile Access Scheme based on Contention and Reservation (MASCARA) | Infrastructure Based | Scheduling, radio resource management and under further study | Worst case QoS estimate (cell delay or cell loss) to be used for determining the connection | Prototyping under the European ACTS AC085 project |

**References**

[ACT96] ACTS Homepage: `http://www.infowin.org/ACTS96` (*see Mobile Domain D4).*

[AGH96] H. Aghvami, "Wireless Data in Europe", *The Second IEEE Workshop on Wireless LANs,* Worcester Polytechnic Institute, October 24-25, 1996

[AGR96] P.Agrawal et al., "SWAN: A Mobile Multimedia Wireless Network", *IEEE Personal Communications Magazine,* pp.18-33 April 1996.

[AHM96] H. Ahmadi, A. Krishna, R. O. LaMaire, " Design Issues in Wireless LANs," *Journal on High Speed Networks (JHSN),* vol. 5, no. 1, 1996.

[AWA96] AWA System homepage: `http://www. netlab. tas. ntt. jp/ ACTS /AWA /awa .html` *at NTT.*

[AYA96] E. Ayanoglu et al., "Mobile Information Infrastructure", *Bell Labs Technical Journal,* pp.143-163, Autumn 1996.

[DEB95] J. DeBelina, "The Wireless Data Market", IEEE Communications Society, New York Chapter, 89th Seminar Proceedings, May 18, 1995.

[ENG95] K.Y. Eng et al., "BAHAMA: A Broadband Ad-Hoc Wireless ATM Local Area Network", *Proceedings of IEEE ICC 1995,* pp. 1216-1223, 1995.

[ETA95] ETSI TC-RES, "Radio Equipment and Systems (RES); High Performance Radio Local Area Network (HIPERLAN); Architecture for Time Bounded Services (TBS)," ETSI, France, October 1995.

[ETF96] ETSI TC-RES, "Radio Equipment and Systems (RES); High Performance Radio Local Area Network (HIPERLAN); Type 1; Functional Specification," ETSI, France, June 1996.

[ETS93] ETSI TC-RES, "Radio Equipment and Systems (RES); High Performance Radio Local Area Network (HIPERLAN); Services and Facilities," ETSI, France, February 1993.

[ETS94] ETSI TC-RES, "Radio Equipment and Systems (RES); High Performance Radio Local Area Network (HIPERLAN); System Definition," ETSI, France, July 1994.

[ETT96] ETSI TC-RES, "Radio Equipment and Systems (RES); High Performance Radio Local Area Network (HIPERLAN); Type 1; Functional Specification, Technical Correction" ETSI, France, December 1996.

[FCC96] "Notice of Proposed Rulemaking: In the matter of Amendment of the Commission's Rules to provide for Unlicensed NII/SUPERNet Operations in the 5 GHz Frequency Range", FCC ET Docket No. 96-102, FCC 96-193, May 1996.

[FCC97] "Report and Order: In the matter of Amendment of the Commission's Rules to provide for Unlicensed NII devices in the 5 GHz Frequency Range", FCC ET Docket No. 96-102, FCC 97-5, January 1997.

[FER80] P. Ferert, "Application of Spread Spectrum Radio to Wireless Terminal Communications," *Proc. of NTC'80*, Houston, Texas, Dec. 1980.

[GFE79] R.F. Gfeller and U. Bapst, "Wireless In-House Data Comunications via Diffused Infrared Radiations," *IEEE Proceedings*, Vol. 67, 1474-1486, 1979.

[HUD94] Hudson, Heather, "Universal Service in the Information Age", *Telecommunications Policy*, Vol.18, No. 8, pp.658-667, November 1994.

[IEE96] IEEE p802.11D3, Wireless LAN Medium Access Control (MAC) and Physical Layer (PHY) Specifications, Piscataway, NJ: IEEE Standards Dept., January 1996.

[LAM96] R. O. LaMaire, A. Krishna, P. Bhagawat, " Wireless LANs and Mobile Networking: Standards and Future Directions," *IEEE Communication Magazine*, August 1996.

[MAR85] M.J. Marcus, "Recent U.S. Regulatory Decision on Civil Use of Spread Spectrum," *Proc. Of the IEEE Globecom'85*, New Orleans, LA, Dec. 1985.

[MOR96] Masaharu Mori, "Wireless LANs in Japan", *The Second IEEE Workshop on Wireless LANs,* Worcester Polytechnic Institute, October 24-25, 1996

[MPT96] Japan Ministry of Post and Telecommunications: `http://www .mpt .go .jp /index-e.html` Press Releases July, October and November 1996.

[PAH85] K. Pahlavan, "Wireless Communications for Office Information Networks", *IEEE Comm. Soc. Mag.*, Sep. 1985.

[PAH95a] K. Pahlavan and A. Levesque, *Wireless Information Networks*, New York: John Wiley and Sons, 1995.

[PAH95b] K. Pahlavan, T. H. Probert, and M. E. Chase, "Trends in Local Wireless Networks", *Invited Paper, IEEE Comm. Soc. Mag.*, March 1995.

[RAY97] D. Raychaudhuri et al., "WATMNet: A Prototype Wireless ATM System for Multimedia Personal Communication", IEEE JSAC, Vol.15, No.1, pp.83-95, January 1997.

[STE94] David G. Steer, "Coexistence and Access Etiquette in the United States Unlicensed PCS Band", IEEE Personal Communications, pp. 36-43, Fourth Quarter, 1994.

[SWA96] R.S Swain, "RACE UMTS vision", Brussels, 1996.

[TEL96] Telecommunications Act of 1996, Pub. LA. No. 104-104, 110 Stat. 56 (1996) from the FCC available at http://www. fcc. gov/telecom. html.

[WAN96] WAND Annual Project Review Report AC085/NMP/MR/I/035/1.0, February 1996.

[WEI97] J. Weinmiller, M. Schlaeger, A. Festag, A. Wolisz, "Performance Study of Access Control in Wireless LANs - IEEE 802.11 DFWMAC and ETSI RES10 in Mobile Networks and Applications", Baltzer Science Publishers, July 1997.

[WIL95] T. Wilkinson, T.G.C. Phipps, S. K. Barton, "A Report on HIPERLAN Standardization," International Journal on Wireless Information Networks, vol. 2, April 1995.

[WOZ96] D. Woznicki, "Wireless LANs in Education, Health, and Finance", *The Second IEEE Workshop on Wireless LANs,* Worcester Polytechnic Institute, October 24-25, 1996

# 2 OPTIMAL JOINT HANDOFF AND CODE ASSIGNMENT IN CDMA CELLULAR NETWORKS

Evaggelos Geraniotis
Yu-Wen Chang

**Abstract:** In this paper, we formulate the joint the handoff and code assignment problem in a CDMA cellular network as a reward/cost optimization problem. The probabilistic properties of the signals and of the traffic in the cells involved in the handoff are used to define a Markov decision process. The optimal policy for a hard handoff or soft handoff decision and channel (CDMA code) assignment is obtained by minimizing a cost function consisting of the weighted sum of the switching cost and the blocking rates of traffic in the two cells subject to a hard constraint on the bit error rate or the outage probability of the signal received by the mobile. A value-iteration algorithm is applied to derive the optimal policy. The performances of the optimal policies for joint hard or soft handoff and code assignment are evaluated and compared to that of the direct admission/threshold handoff policies. It is established that The optimal soft handoff policy results in the smallest expected number of handoffs compared to the other policies but in larger average blocking rate than the hard handoff policy. Both the optimal hard and soft handoff policies outperform by a significant margin the direct-admission/fixed-threshold policies.

## 2.1 INTRODUCTION

One major issue in cellular radio system is the problem of handoffs. A mobile leaving one cell and entering a neighboring cell must transfer its call to the new cell. Traditional handoff schemes are threshold policies. The signal strengths from the current and candidate base stations are measured, and the difference between these two measurements is compared with a fixed threshold to determine the handoff action [1]-[2]. Because of the statistical fluctuations in signal strength, the threshold must be large enough to reduce the number of bouncing between two cells. However, increasing the threshold may result in higher probability of forced termination because the signal from the current base station may have been too weak before the threshold criterion is met.

An improved technique of the threshold policy was proposed in [3] where the number of unnecessary handoffs is reduced by limiting the handoffs to only those mobiles whose signal levels, in addition to satisfying the threshold criterion, also falls off below a certain threshold. In these models, no optimization regarding the number of handoffs or the call quality is achieved.

In [4]-[5] the phenomenon of handoffs is formulated as a reward/cost optimization problem. The received signal is treated as a stochastic process with an associated reward while the handoff is associated with a switching penalty. Necessary and sufficient conditions for handoff switching were derived. Since formulae for the transition probability of power measurements were not derived, no optimal policies were proposed. Only simulation results of suboptimal policies were presented.

In [6] an optimal strategy is proposed to minimize a cost function consisting of number of handoff switches and call quality, but traffics in the cells are not considered when deriving the optimal policy. However, this optimized policy is not applicable to CDMA cellular networks for which the channels (CDMA codes) used for handoff are sharing the same bandwidth with the traffic channels and thus the quality (BER) of the signal received by the mobile is affected by the total traffic (number of active users and CDMA codes assigned) of both base stations.

For channel assignment in CDMA cellular networks, an optimal admission policy was derived in our paper [7], which minimizes the long-term blocking rate of newly arrived calls. A Markov decision process formulation was used to obtain optimal code allocation policies. However, no handoff issues were addressed in that paper.

A handoff between two cells will change the number of active users in each cell, and the traffic conditions in the cells will also affect the handoff decision (for example, handoff can not be made if the new cell has no capacity left). Furthermore, the call quality depends on the other user interference in a CDMA network. Therefore, the handoff and channel assignment problems should be considered together to obtain an optimal policy, which will take into account the switching cost, the blocking rate and the call quality in the CDMA cellular system. This formulation and the resulting optimization of joint handoff and code assignment policies based on Markov decision processes are the contribution of this paper.

Finally, a distinct advantage of CDMA cellular networks is the soft handoff feature. The mobile remains connected to both base stations during the transition period of switching from one base station to another. It has been shown that CDMA soft handoff can provide larger cell coverage than hard handoff [8]. But no optimization or performance evaluation of policies for hard or soft handoff was carried out.

In this paper, we do compare the blocking rates of the traffic of two cells and the number of handoffs for both the optimal hard and soft handoff schemes used in conjunction with CDMA code assignment and compare them to each other and to traditional fixed-threshold/direct-admission policies.

The paper is organized as follows. In Section 2.2, the system models are described. The optimal handoff and channel assignment policy is derived in Section 2.3. In Section 2.4 performance of the optimal and traditional policies are analyzed. Numerical results and performance comparisons are presented in Section 2.5 while the conclusions of this paper are presented in Section 2.6.

## 2.2 SYSTEM MODEL

Our system model consists of two adjacent cells, cell 0 and cell 1, with a mobile moving inside these two cells (refer to Figure 2.1). Handoff is performed between two base stations. For the hard handoff scheme, only one base station can be connected to the mobile at any time instance, but for the soft handoff, diversity is achieved by connecting both base stations to the mobile. Background traffic in each cell is also considered with available channels assigned to new arrived calls. The traffic of each user is modeled as a two-state discrete-time Markov chain with transition probabilities $p_{01}$ and $p_{10}$ as shown in Figure 2.2. The population in each cell is denoted by $N_0$ and $N_1$.

The message of each active user is packetized with the same length, and time is divided into slots of duration equal to the transmission of one packet. The signal strength from the mobile for handoff decision is measured periodically at these regular time instants. Handoff decision and channel assignments are made after each measurement is made. The measured signal power is represented by the following.

$$P_t^i = U_1 - U_2 \log(d) + u(d) + 20 \log R_t^i \ \text{dB} \tag{2.1}$$

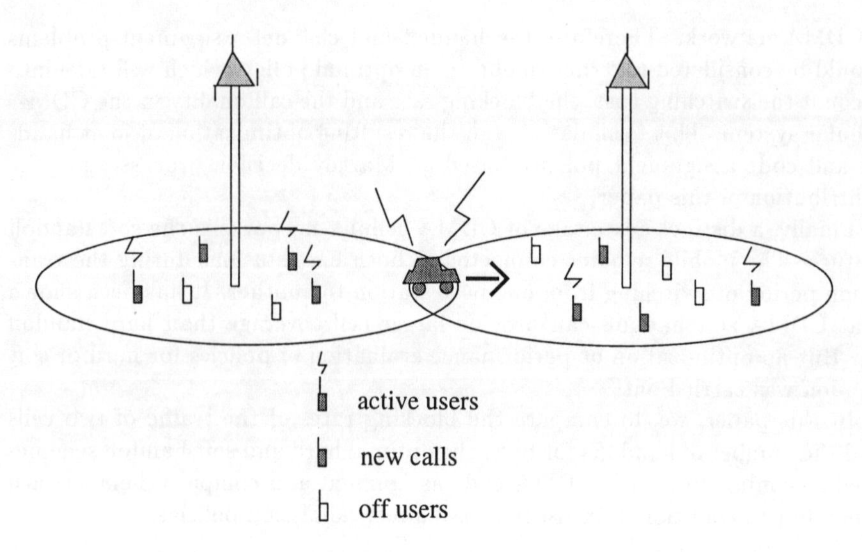


**Figure 2.1** System Model

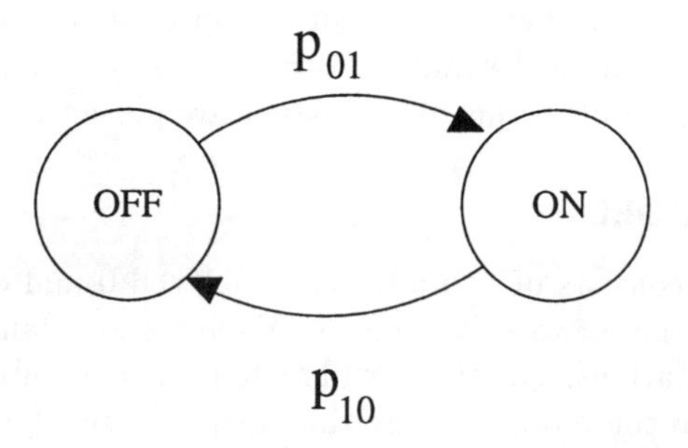


**Figure 2.2** Model for Voice Users

where $P_t^i$ is the signal level(dB) received at a distance $d$ from base station $i$ at time $t$, $U_1$ is the transmitting power, $U_2$ is the path loss exponent, $R_t^i$ is Rayleigh fast fading, and $u(d)$ represents the log-normal shadowing effects, which is a zero-mean stationary Gaussian process with an exponential correlation function

$$E[u(d_1)u(d_2)] = \sigma_i^2 \exp(-|d_1 - d_2|/D) \tag{2.2}$$

The mobile moves through the regions covered by two cells with finite possible positions at each measurement instant, for example, these positions can be a street grid in an urban area. Vehicle location tracking ability is assumed in the system so that the information of the distances between the moving mobile and two base stations are available when making handoff decision and channel assignment. The movement of the mobile is modeled by a transition matrix with each entry representing the probability of moving from one position to another position.

Handoff decision has higher priority than the channel assignment of background traffic, i.e., available channels are assigned to newly arrived calls only after the handoff decision is made and the channel is used by the mobile which requested the handoff. Although it can be extended to multiple handoff requests by increasing the state space of the system, we only consider the case of single handoff request in this paper. The strategy used in this paper can also applied to a multiple cell system.

## 2.3 OPTIMAL HANDOFF AND CHANNEL ASSIGNMENT

### *2.3.1 Markov Decision Process Formulation*

The handoff and channel assignment problem can be modeled by a MDP. Let the state space of the CDMA cellular network be

$$\mathbf{Z}_{t,hard} = (B_t, X_t, P_t^0, P_t^1, i_t^0, i_t^1, l_t^0, l_t^1) \tag{2.3}$$

$$\mathbf{Z}_{t,soft} = (B_t^0, B_t^1, X_t, P_t^0, P_t^1, i_t^0, i_t^1, l_t^0, l_t^1) \tag{2.4}$$

where $B_t$ is the current base station, $B_t^i$ is 1 if the mobile is connected with base $i$, 0 if not connected. $X_t$ is the mobile position, $P_t^0$ and $P_t^1$ are the power measurements from base station 0 and 1, $i_t^0$, $i_t^1$ are the number of active calls in each cell, and $l_t^0$, $l_t^1$ are the number of new arrivals in each cell.

The action corresponding to the state $\mathbf{Z}_t$ is given by

$$\mathbf{A}_{t,hard} = (H_t, a_t^0, a_t^1) \tag{2.5}$$

$$\mathbf{A}_{t,soft} = (H_t^0, H_t^1, a_t^0, a_t^1) \tag{2.6}$$

where $H_t$ is the handoff decision with value 0 or 1 representing the assignment of base station 0 or 1, $H_t^i$ is 1 if a connection is established between the mobile and base station $i$, 0 if disconnected, and $a_t^0$, $a_t^1$ are the number of accepted new arrivals, which come from the $l_t^0$ and $l_t^1$.

For the direct/threshold policy,

$$\begin{aligned} a_t^0 &= \min(l_t^0, K_0 - i_t^0) \\ a_t^1 &= \min(l_t^1, K_1 - i_t^1) \end{aligned} \tag{2.7}$$

$$H_t = \begin{cases} 0, & \text{if } \; B_t = 0, P_t^1 - P_t^0 \le \tau \; \text{ or } \; B_t = 1, P_t^0 - P_t^1 > \tau \\ 1, & \text{if } \; B_t = 0, P_t^1 - P_t^0 > \tau \; \text{ or } \; B_t = 1, P_t^0 - P_t^1 \le \tau \end{cases} \tag{2.8}$$

where $K_i$ is the capacity of cell $i$, and $\tau$ is the threshold for handoff.

The transition probabilities that at the next epoch $t+1$ the system will be in state $\mathbf{Z}_{t+1}$, if action $\mathbf{A}_t$ is chosen at the present state $\mathbf{Z}_t$, are the following:

$$
\begin{aligned}
& P_{\mathbf{Z}_{t+1,hard}|\mathbf{Z}_{t,hard}}(\mathbf{A}_{t,hard}) \\
= \; & Pr\,(P^0_{t+1}, P^1_{t+1}, X_{t+1}, i^0_{t+1}, i^1_{t+1}, l^0_{t+1}, l^1_{t+1}, B_{t+1} \\
& |P^0_t, P^1_t, X_t, i^0_t, i^1_t, l^0_t, l^1_t, B_t, a^0_t, a^1_t, H_t) \\
= \; & Pr(P^0_{t+1}, P^1_{t+1}, X_{t+1}|P^0_t, P^1_t, X_t) \times \mathbf{1}(B_{t+1} = H_t) \\
& \cdot Pr(i^0_{t+1}, i^1_{t+1}, l^0_{t+1}, l^1_{t+1}|i^0_t, i^1_t, l^0_t, l^1_t, a^0_t, a^1_t, H_t, B_t) \qquad (2.9)
\end{aligned}
$$

$$
\begin{aligned}
& P_{\mathbf{Z}_{t+1,soft}|\mathbf{Z}_{t,soft}}(\mathbf{A}_{t,soft}) \\
= \; & Pr(P^0_{t+1}, P^1_{t+1}, X_{t+1}, i^0_{t+1}, i^1_{t+1}, l^0_{t+1}, l^1_{t+1}, B^0_{t+1}, B^1_{t+1} \\
& |P^0_t, P^1_t, X_t, i^0_t, i^1_t, l^0_t, l^1_t, B^0_t, B^1_t, a^0_t, a^1_t, H^0_t, H^1_t) \\
= \; & Pr(P^0_{t+1}, P^1_{t+1}, X_{t+1}|P^0_t, P^1_t, X_t) \times \mathbf{1}(B^0_{t+1} = H^0_t, B^1_{t+1} = H^1_t) \\
& \times Pr(i^0_{t+1}, i^1_{t+1}, l^0_{t+1}, l^1_{t+1}|i^0_t, i^1_t, l^0_t, l^1_t, a^0_t, a^1_t, H^0_t, H^1_t, B^0_t, B^1_t) \qquad (2.10)
\end{aligned}
$$

where $\mathbf{1}$ is the indicator function. Because the power measurements and the mobile position are independent of the traffic variation, the transition probability can be decomposed into product of the power/position transition probability and the traffic transition probability as shown above.

Since $P^i_t$ and $P^i_{t+1}$ are Gaussian distributed and the correlation function between them is also known from $X_t$ and $X_{t+1}$, the conditional probability of $P^0_{t+1}, P^1_{t+1}, X_{t+1}$ given $P^0_t, P^1_t, X_t$ can be represented by

$$
\begin{aligned}
& Pr(P^0_{t+1}, P^1_{t+1}, X_{t+1}|P^0_t, P^1_t, X_t) \\
= \; & Pr(X_{t+1}|X_t) \cdot Pr(P^0_{t+1}, P^1_{t+1}|P^0_t, P^1_t, X_{t+1}) \\
= \; & Pr(X_{t+1}|X_t) \cdot N([P^0_t + K_2 \log d^0_t] \exp(-\Delta d_t/D) \\
& \qquad -K_2 \log d^0_{t+1}, \sigma^2_0[1 - \exp(-2\Delta d_t/D)]) \\
& \cdot N([P^1_t + K_2 \log d^1_t] \exp(-\Delta d_t/D) \\
& \qquad -K_2 \log d^1_{t+1}, \sigma^2_1[1 - \exp(-2\Delta d_t/D)]) \qquad (2.11)
\end{aligned}
$$

where $d^0_t(d^1_t)$ is the distance between $X_t$ and base station 0(1), $d^0_{t+1}(d^1_{t+1})$ is the distance between $X_{t+1}$ and base station 0(1), and $\Delta d_t$ is the distance of movement between the current and next mobile positions.

The traffic transition probability depends on actions $\mathbf{A}_t$ and can be written as

$$
\begin{aligned}
& Pr(i^0_{t+1}, i^1_{t+1}, l^0_{t+1}, l^1_{t+1}|i^0_t, i^1_t, l^0_t, l^1_t, a^0_t, a^1_t, H_t, B_t) \\
& = b(i^0_t, i^0_t + a^0_t - i^0_{t+1} + B_t - H_t, p_{10}) \cdot b(i^1_t, i^1_t + a^1_t - i^1_{t+1} + H_t - B_t, p_{10}) \\
& \cdot b(N_0 - i^0_t - a^0_t, l^0_{t+1}, p_{01}) \cdot b(N_1 - i^1_t - a^1_t, l^1_{t+1}, p_{01}) \qquad (2.12)
\end{aligned}
$$

$$Pr(i_{t+1}^0, i_{t+1}^1, l_{t+1}^0, l_{t+1}^1 | i_t^0, i_t^1, l_t^0, l_t^1, a_t^0, a_t^1, H_t^0, H_t^1, B_t^0, B_t^1)$$
$$= b(i_t^0, i_t^0 + a_t^0 - i_{t+1}^0 + H_t^0 - B_t^0, p_{10}) \cdot b(i_t^1, i_t^1 + a_t^1 - i_{t+1}^1 + H_t^1 - B_t^1, p_{10})$$
$$\cdot b(N_0 - i_t^0 - a_t^0, l_{t+1}^0, p_{01}) \cdot b(N_1 - i_t^1 - a_t^1, l_{t+1}^1, p_{01}) \quad (2.13)$$

where $b(M, m, p)$ denotes the binomial distribution with parameters $M$ and $p$, where $0 \le p \le 1$.

$$b(M, m, p) = \binom{M}{m} p^m (1-p)^{M-m} \quad (2.14)$$

*2.3.2 Cost Function and Constrained Optimization Formulation*

Let $C_1$ be the set-up cost incurred if a new connection is established between the mobile and a base station, and $C_{-1}$ be the disconnection cost if a connection is an existing connection is removed at each decision epoch. The blocking cost for the background traffic is the number of newly arrived calls in two cells rejected by the control center, which is $l_t^0 - a_t^0 + l_t^1 - a_t^1$. The goal of the optimal handoff and channel assignment policy is to minimize the weighted sum of the switching cost and the blocking cost,

$$E[\sum_t w_1(C_{H_t^0 - B_t^0} + C_{H_t^1 - B_t^1}) + w_2(l_t^0 - a_t^0 + l_t^1 - a_t^1)] \quad (2.15)$$

under a bit error rate(BER) or outage probability constraint on the mobile call quality, where $C_0 = 0$. By adjusting the weighting factors $w_1$ and $w_2$, we can obtain different tradeoffs between switching and blocking costs. Two alternatives for the formulation are possible: one is minimizing the switching cost under BER/outage probability and blocking rate constraints, and the other is minimizing the weighted sum of switching, blocking and BER/outage probability costs.

It has been shown that the error probability in a CDMA cellular network can be approximated by [9]

$$P_e \approx Q(\sqrt{SNR}) = Q(\frac{AT}{[\eta_0 T + \frac{A^2}{3}\sigma^2 T_c^2 L(K_0 - 1 + \sum_{i=1}^{K_1} \beta_i^2)]^{1/2}}) \quad (2.16)$$

where $A$ is the signal amplitude with fading, $T$ is the bit duration, $L$ is the processing gain, $T_c$ is the chip duration, $\sigma$ is the parameter of Rayleigh fading, $K_i$ is the number of users in cell $i$, and $\beta_i$ is the ratio of the distances between the mobile and two base stations. Therefore, to ensure the call quality of a mobile, the other user interference has to be limited by setting constraints on the number of active users in every cell.

Because a constraint, say $\alpha_1$, on the BER is equivalent to the constraint $SNR > [Q^{-1}(\alpha_1)]^2$, we can use the outage probability [10], which is defined below, as an alternative constraint.

$$\begin{aligned} P_{out} &= 1 - Pr(SNR > b) \\ &= Pr[R_0^2 > b \sum_{k=1}^{K_0+K_1-1} R_k^2 e^{\frac{v_k - v_0}{10}} (\frac{r_k}{r_0})^{-U_2/10}] \\ &= \int_{-\infty}^{\infty} \frac{dv_0}{\sqrt{2\pi}\sigma} e^{-\frac{v_0^2}{2\sigma^2}} [I(v_0, r_0)]^{K_0+K_1-1} \end{aligned} \tag{2.17}$$

where

$$I(v_0, r_0) = \int_{-\infty}^{\infty} \int_0^1 \frac{e^{-\frac{v^2}{2\sigma^2}} h(r) dr dv}{\sqrt{2\pi}\sigma(1 + b e^{v - v_0} (\frac{r}{r_0})^{-U_2/10})}, \tag{2.18}$$

$r_k$'s are the distances between mobiles in two cells and the base station, $h(r)$ is the distribution of mobile positions(here we assume uniformly distributed in a cell, which is a linear function of the distance to the base station), $v$ is Gaussian with mean zero and variance $\sigma^2$, and $K_i$ is number of users in cell $i$. Therefore, for each state with a distance $r_0$, the outage probability constraint can be transformed into a constraint on the number of users in each cell.

### *2.3.3 Dynamic Programming Formulation and Optimal Policy*

Since the cost function defined by (11) can be expressed as

$$E[\sum_{t=0}^{\infty} (1 - p_{10})^t (w_1 (C_{H_t^0 - B_t^0} + C_{H_t^1 - B_t^1}) + w_2 (l_t^0 - a_t^0 + l_t^1 - a_t^1))], \tag{2.19}$$

the optimization problem is an infinite horizon problem with discounted cost, and standard dynamic programming techniques can be applied to solve the problem. Because of the large state space, the **value-iteration algorithm** is used to obtain the optimal policy as below.

<u>*step 0*</u> : Choose $V_0(\mathbf{Z}_t)$ such that $0 \le V_0(\mathbf{Z}_t) \le \min_{\mathbf{A}_t}\{C(\mathbf{Z}_t, \mathbf{A}_t)\}$, for all $\mathbf{Z}_t$. Let $n := 1$.

<u>*step 1*</u> : Compute the recursive function of $V_n(\mathbf{Z}_t)$ for all $\mathbf{Z}_t$, from

$$V_n(\mathbf{Z}_t) = \min_{\mathbf{A}_t}\{C(\mathbf{Z}_t, \mathbf{A}_t) + \sum_{\mathbf{Z}_{t+1}} P_{\mathbf{Z}_{t+1}|\mathbf{Z}_t}(\mathbf{A}_t) \cdot V_{n-1}(\mathbf{Z}_{t+1})\} \tag{2.20}$$

and determine $\varphi(n)$ as a stationary policy [sequence of actions that minimize the RHS of the above equation].

<u>*step 2*</u> : Compute the bounds

$$l_n = \min_{\mathbf{Z}_t}\{V_n(\mathbf{Z}_t) - V_{n-1}(\mathbf{Z}_t)\} \tag{2.21}$$

$$L_n = \max_{\mathbf{Z}_t}\{V_n(\mathbf{Z}_t) - V_{n-1}(\mathbf{Z}_t)\} \tag{2.22}$$

The algorithm terminates and outputs policy $\varphi(n)$, when $0 \leq (L_n - l_n) \leq \varepsilon l_n$, where $\varepsilon$ is a prespecified bound on the relative error (accuracy). Otherwise, go to *step 3*.
<u>*step 3*</u> : $n := n + 1$ and go to *step 1*.

From the above algorithm, we obtain an optimal policy $\varphi$ for determining the handoff and channel assignment. This policy describes the optimal action $\mathbf{A}_t$ corresponding to the state $\mathbf{Z}_t$.

## 2.4 PERFORMANCE ANALYSIS

In this section, we analyze the performance of the mobile handoff and the traffics in two cells under the optimal and direct threshold policies. The performance measures include the expected number of handoffs during a call of the mobile, and the blocking rate of calls in both cells.

To analyze the performance, we have to obtain $Pr(\mathbf{Z}_t)$ first.

$$\begin{aligned} Pr(\mathbf{Z}_t) &= Pr(P_t^0, P_t^1, X_t, i_t^0, i_t^1, l_t^0, l_t^1, B_t) \\ &= Pr(l_t^0, l_t^1 | i_t^0, i_t^1) \cdot Pr(i_t^0, i_t^1, B_t) \cdot Pr(P_t^0, P_t^1, X_t) \end{aligned} \tag{2.23}$$

where

$$Pr(l_t^0, l_t^1 | i_t^0, i_t^1) = b(N_0 - i_t^0, l_t^0, p_{01}) \cdot b(N_1 - i_t^1, l_t^1, p_{01}) \tag{2.24}$$

and $Pr(i_t^0, i_t^1, B_t)$ and $Pr(P_t^0, P_t^1, X_t)$ are obtained by solving the following sets of linear equations:

$$\begin{cases} Pr(i_{t+1}^0, i_{t+1}^1, B_{t+1}) = \sum_{i_t^0, i_t^1, B_t} Pr(i_t^0, i_t^1, B_t) \cdot Pr(i_{t+1}^0, i_{t+1}^1, B_{t+1} | i_t^0, i_t^1, B_t) \\ \sum_{i_t^0, i_t^1, B_t} Pr(i_t^0, i_t^1, B_t) = 1 \end{cases}$$

$$\begin{cases} Pr(P_{t+1}^0, P_{t+1}^1, X_{t+1}) = \sum_{P_t^0, P_t^1, X_t} Pr(P_t^0, P_t^1, X_t) \\ \qquad\qquad\qquad \cdot Pr(P_{t+1}^0, P_{t+1}^1, X_{t+1} | P_t^0, P_t^1, X_t) \\ \sum_{P_t^0, P_t^1, X_t} Pr(P_t^0, P_t^1, X_t) = 1 \end{cases}$$

where

$$Pr(i_{t+1}^0, i_{t+1}^1, B_{t+1} | i_t^0, i_t^1, B_t) =$$

$$\sum_{l_t^0, l_t^1, P_t^0, P_t^1, X_t} Pr(P_t^0, P_t^1, X_t)\mathbf{1}(B_{t+1} = H_t) \cdot b(N_0 - i_t^0, l_t^0, p_{01})$$
$$\cdot b(N_1 - i_t^1, l_t^1, p_{01}) \cdot b(i_t^0, i_t^0 + a_t^0 - i_{t+1}^0 + B_t - H_t, p_{10})$$
$$\cdot b(i_t^1, i_t^1 + a_t^1 - i_{t+1}^1 + B_t - H_t, p_{10}) \quad (2.25)$$

A similar derivation can be obtained for the soft handoff case.

The expected number of hard handoffs is given by

$$\begin{aligned} N_{HO} &= E[\sum_{t=0}^{T} \mathbf{1}(H_t \neq B_t)] \\ &= E[\sum_{t=0}^{\infty} (1 - p_{10})^t \mathbf{1}(H_t \neq B_t)] \\ &= \sum_{t=0}^{\infty} (1 - p_{10})^t [Pr(H_t = 1, B_t = 0) + Pr(H_t = 0, B_t = 1)] \\ &= \sum_{t=0}^{\infty} (1 - p_{10})^t \sum_{H_t=1, B_t=0 \text{ or } H_t=0, B_t=1} Pr(\mathbf{Z_t}) \\ &= \sum_{H_t=1, B_t=0 \text{ or } H_t=0, B_t=1} Pr(\mathbf{Z_t})/p_{10} \end{aligned} \quad (2.26)$$

The expected numbers of set-ups and disconnections for soft handoff are

$$N_{setup} = E[\sum_{t=0}^{T} \mathbf{1}(H_t^0 - B_t^0 = 1 \text{ or } H_t^1 - B_t^1 = 1)] \quad (2.27)$$

$$N_{discon} = E[\sum_{t=0}^{T} \mathbf{1}(H_t^0 - B_t^0 = -1 \text{ or } H_t^1 - B_t^1 = -1)] \quad (2.28)$$

The blocking rate of background traffics is given by

$$P_B = \frac{\sum_{i_t^0, i_t^1, l_t^0, l_t^1} (l_t^0 - a_t^0 + l_t^1 - a_t^1) Pr(i_t^0, i_t^1, l_t^0, l_t^1)}{\sum_{i_t^0, i_t^1, l_t^0, l_t^1} (l_t^0 + l_t^1) Pr(i_t^0, i_t^1, l_t^0, l_t^1)} \quad (2.29)$$

## 2.5 NUMERICAL RESULTS

For the numerical results, we select a model of two base stations with distance 1000m. The mobile moves along the path connecting the two stations, with a random walk moving pattern. The offered traffic load in a cell is defined as

$$G_i = N_i \cdot \frac{p_{01}}{p_{01} + p_{10}} \quad (2.30)$$

The total population $N_i$ in each cell is 20, and $p_{10} = 0.05$. For the parameter values in eqs. (1) and (2), we use $K_1 = 0$, $K_2 = 30$, $\sigma_i = 6$ and $D = 20$.

In Figures 2.3 to 2.5 we show the performance measures versus offered traffic load. Figure 2.3 is the expected cost defined versus traffic load for the range

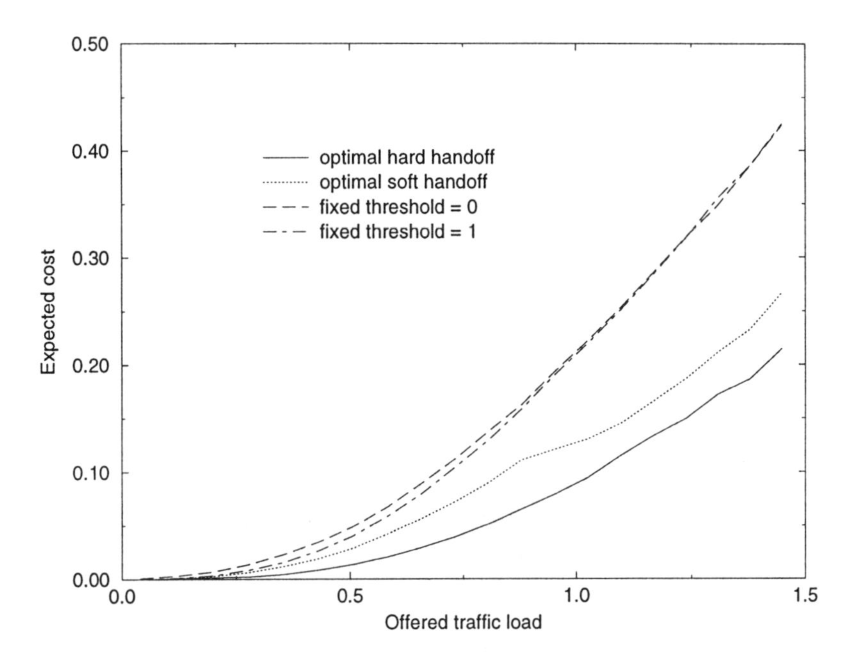


**Figure 2.3** Expected Cost vs Offered Traffic Load

from 0.04 to 1.45. The switching cost used is 0.001 and weighting factors are 1.

As expected the costs of both the optimal hard and soft handoff policies are substantially better than the direct/threshold policies, especially when the traffic load is heavy. The relation between the two costs (switching and blocking) of the soft and hard handoffs depends on the weighting factors $w_1$ and $w_2$. If more weighting is placed on $w_1$ ($w_1 > w_2$), then the switching cost is dominant and soft handoff will perform better. If $w_2 = w_1$ or $w_2$ is weighted more ($w_2 > w_1$) then the blocking cost dominates and the hard handoff will have less cost as shown in Figure 2.3. The costs of two direct/threshold policies with different thresholds are very close. In all cases, the cost increases as the load increases.

In Figure 2.4 we show the blocking probabilities versus offered traffic load. For all the range of traffic load, the optimal policies perform better than the

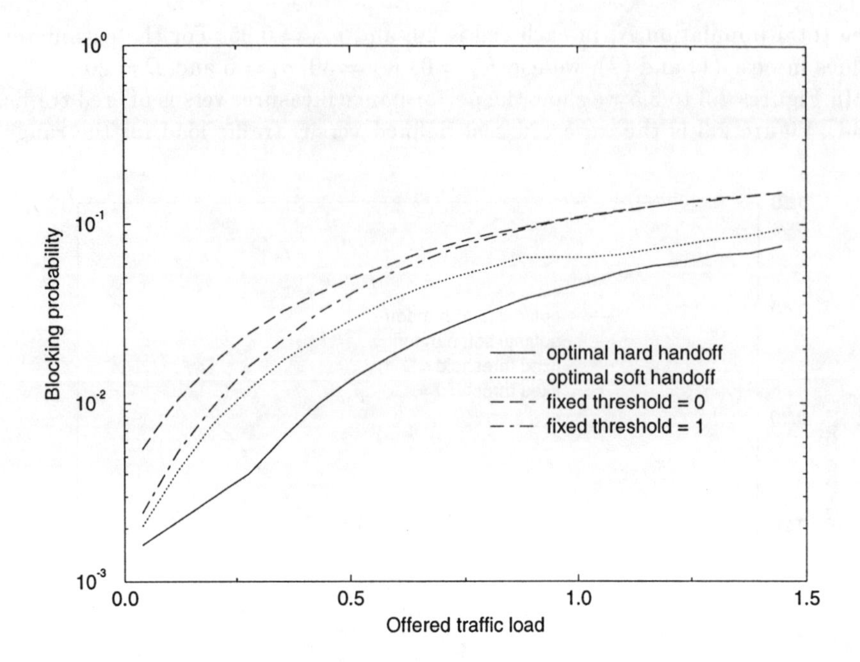


**Figure 2.4** Blocking Probability vs Offered Traffic Load

other two policies, and the blocking probabilities increase as the load increases. The optimal hard handoff policy has a better blocking probability than soft handoff because in the soft handoff more than one channel may be used by the mobile requesting handoff.

In Figure 2.5 the expected number of handoffs versus the traffic load is shown. Under low to medium load of traffic, the optimal hard handoff policy requires less handoffs than the direct/threshold policy, but for the soft handoff, the optimal policy always requires less number of handoff, and is less sensitive to the traffic load than the hard handoff. For the direct/threshold policy, the traffic load has no influence on the number of handoffs; whereas under the optimal policy heavier traffic tends to make the mobile handoff more frequently.

In Figures 2.6 and 2.7, we show the performance of the optimal hard handoff policy versus the switching cost $C$ for the range from $10^{-9}$ to $10^{-4}$ with weighting factors equal to 1 and traffic load equal to 0.2. Figure 2.6 shows the expected number of handoffs versus the switching cost. As expected when the switching cost increases, the handoff is discouraged, so the number of handoffs decreases. In Figure 2.7, the blocking probability is insensitive to the variation

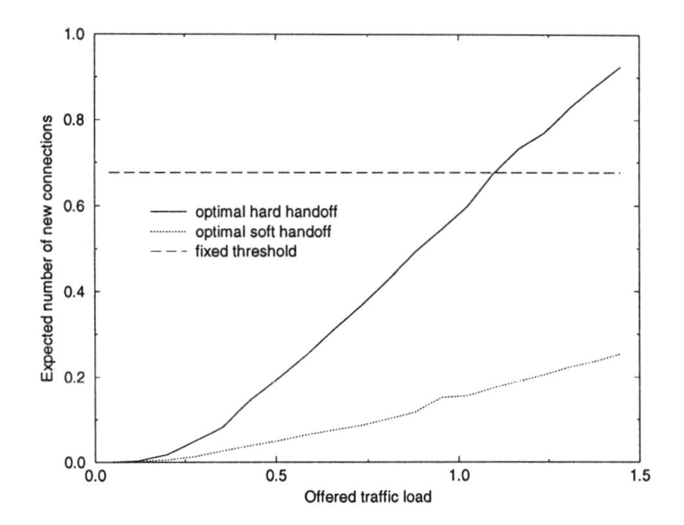


**Figure 2.5** Expected Number of Handoffs vs Offered Traffic Load

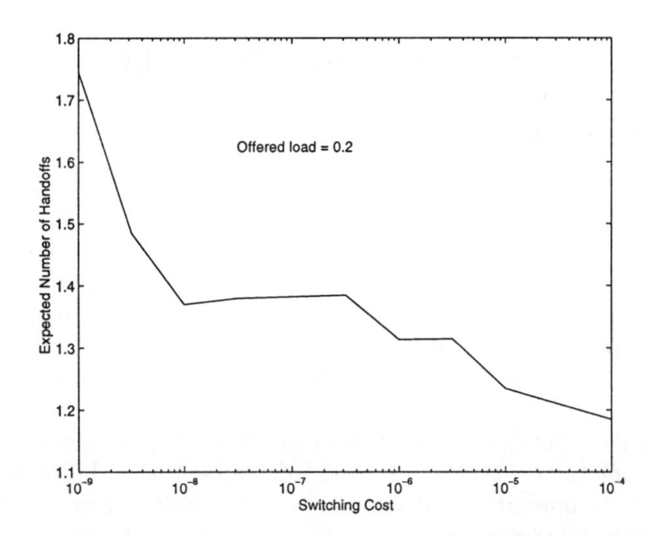


**Figure 2.6** Expected Number of Handoffs vs Switching Cost

of switching cost except that the blocking rate becomes slightly higher for a very low switching cost.

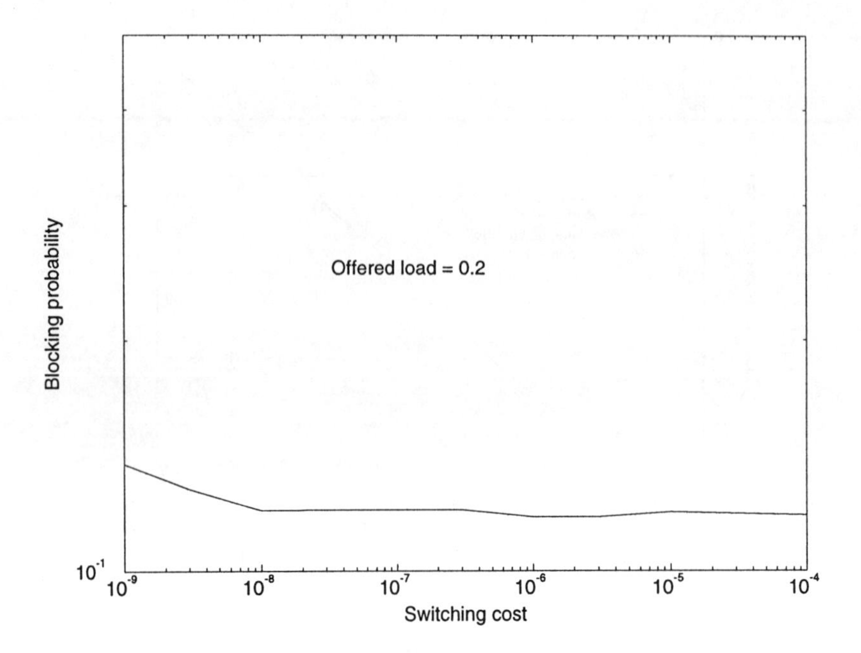


**Figure 2.7** Blocking Probability vs Switching Cost

## 2.6 CONCLUSIONS

In this paper, we have derived and analyzed the performance of optimal schemes for hard and soft handoff decisions and code assignment in a CDMA cellular network. After using a Markov decision process to model and formulate the joint handoff/code-assignment problem, the weighted sum of handoff switching cost and the blocking cost are minimized by using a value-iteration algorithm under a constraint of call quality (outage probability). The optimal scheme are shown to outperform the direct admission/threshold handoff policies especially for heavy traffic loads. Comparison of the performances of optimal soft and hard handoff policies shows that when the switching cost is considered more significant than the blocking cost, soft handoff must be preferred over hard handoff; the opposite trend is true when the blocking cost is judges more significant than the switching cost.

## References

[1] A. Murase, I. Symington, and E. Green, "Handover criterion for macro and microcellular systems," *Proc. 41st IEEE Vehicle Technology Conference*, pp. 524-530, May 1991.

[2] R. Vijayan and J.M. Holtzman, "A model for analyzing handoff algorithms," *IEEE Transactions on Vehicle Technology*, VT-42, pp. 351-356, August 1993.

[3] P.S. Kumar and J.M. Holtzman, "Analysis of handoff algorithms using both bit error rate (BER) and relative signal strength," *Proc. 3rd IEEE Int. Conf. on Universal Personal Communications*, pp. 1-5, 1994.

[4] M. Asawa and W.E. Stark, "A framework for optimal scheduling of handoffs in wireless networks," *Globecom*, pp. 1669-1673, 1994.

[5] M. Asawa and W.E. Stark, "Optimal scheduling of soft Handoffs in DS/CDMA communication systems," *Infocom'95*, pp. 105-112, 1995.

[6] R. Rezaiifar, A.M. Makowski, and S. Kumar, "Stochastic control of handoffs in cellular networks," *Proc. of VTC '95.*

[7] W.-B. Yang and E. Geraniotis, "Admission policies for integrated voice and data traffic in CDMA packet radio networks," *IEEE JSAC*, May 1994.

[8] A.J. Viterbi, A.M. Viterbi, K.S. Gilhousen, and E. Zehavi, "Soft handoff extends CDMA cell coverage and increase reverse link capacity," *IEEE JSAC*, pp.1281-1288, October 1994.

[9] L.B. Milstein, T.S. Rappaport, and R. Barghouti, "Performance evaluation for cellular CDMA," *IEEE JSAC*, pp.680-688, May 1992.

[10] M. Zorzi and R.R. Rao, "Capture and retransmission control in mobile radio," *IEEE JSAC*, pp. 1289-1298, October 1994.

# 3 WIRELESS ATM TECHNOLOGY: PRESENT STATUS AND FUTURE DIRECTIONS

Dipankar Raychaudhuri

**Abstract**

The concept of "wireless ATM", first proposed in 1992, is now being actively considered as a potential framework for next-generation wireless communication networks capable of supporting integrated, quality-of-service (QoS) based multimedia services. In this review paper, we outline the technological rationale for wireless ATM, present a system-level architecture, and discuss key design issues for both mobile ATM switching infrastructure and radio access subsystems. A brief summary of current research and technology development status (including an outline of NEC's WATMnet prototype) is given. The paper concludes with a view of future directions for wireless ATM technology and associated mobile multimedia applications.

## 3.1 INTRODUCTION

Broadband wireless network technologies such as wireless ATM are motivated by the increasing importance of portable computing/telecommunications applications in both business and consumer markets. The rapid penetration of cellular phones and laptop PC's during the previous decade is proof that users place a significant value on portability as a key feature which enables tighter integration of such technologies with their daily lives. In the last few years, first-generation multimedia capabilities (such as CD-ROM drives and MPEG decoders) have become available on portable PC's, reflecting the increasingly mainstream role of multimedia in computer applications. As multimedia features continue their inevitable migration to portable devices such a laptop PC's and PDA's, wireless extensions to broadband networks will be required to support user requirements. Such broadband wireless services could first start in the private local area network scenario, gradually moving to microcellular public PCS systems if the technology proves feasible for general consumer use. This future broadband wireless network concept is outlined schematically in Fig. 1. The basic idea is to provide a seamless wired + wireless networking environment with uniform protocols and applications across both mobile and fixed devices.

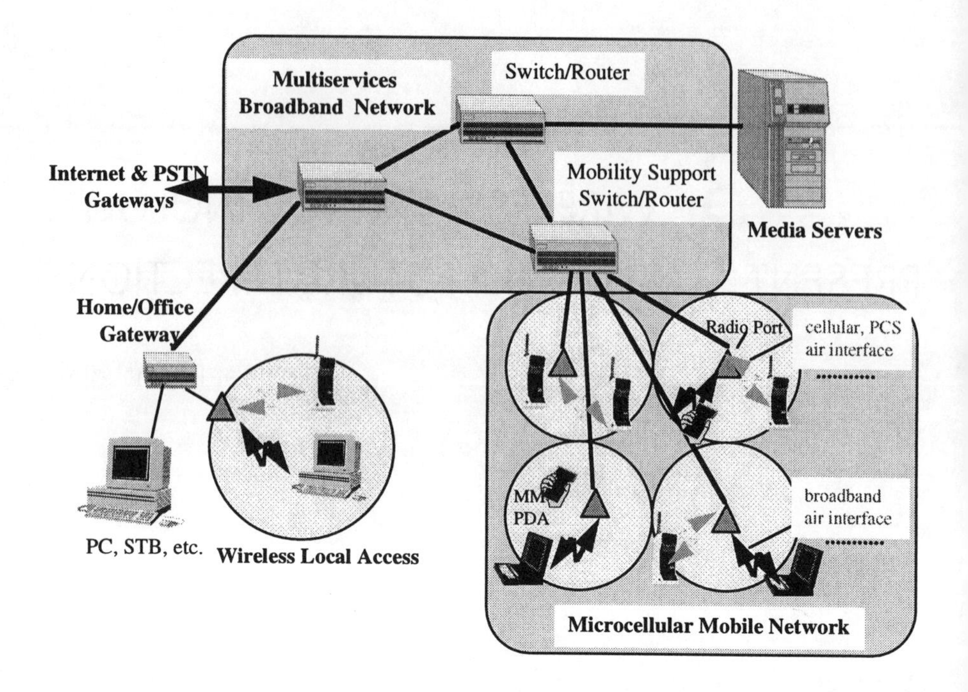


**Fig. 1** Seamless wired + wireless broadband network scenario

Fundamental network architecture and radio technology innovations are required to realize the above vision for a seamless wired + wireless broadband wireless network. At the architectural level, wireless systems need to migrate from the current model of single-application networks with custom protocol stacks towards generic integrated service networks with "plug-in" software and hardware components to support radio access and mobility. In terms of radio technology, it will be necessary to re-examine basic assumptions about frequency band, cell size, modem complexity, etc. to achieve service speeds of the order of 10-100 Mbps at reasonable reliability and spectrum utilization levels. On the network side, mobility which has traditionally been treated as an external function, must be recognized as a first-class feature to be considered during all phases of protocol and hardware design. Integrated mobility support within standard network protocols such as IP or ATM will facilitate a move towards the use of generic network API's and common applications on both mobile and fixed terminals. Finally, mobile multimedia services will require a networking framework which supports real-time streams with quality-of-service (QoS) in both wired and wireless segments of the system.

Wireless ATM, first proposed in [1,2], is a specific broadband wireless solution which substantially meets the architectural and performance goals outlined above. Work on wireless ATM has been motivated by the increasing acceptance of ATM switching technology as a basis for broadband networks which support integrated services with QoS control. The 53 byte ATM cell turns out to be quite reasonable for use as the basic transport unit over high bit-rate radio channels, taking into account both error control and medium access requirements. ATM signaling protocols (e.g. Q.2931) for connection establishment and QoS control also provide a suitable basis for mobility extensions such as handoff and location management. Early research results in this area [2-7] indicated that it is indeed possible to use standard ATM protocols to support seamless wired + wireless networking via incorporation of new wireless specific protocol sublayers (e.g. medium access control and data link control) into the ATM data plane, together with a limited number of mobility extensions to existing ATM control protocol layers. Subsequent proof-of-concept prototyping activities (e.g. NEC's WATMnet [8], ORL Cambridge's Radio ATM [9], Bell Labs' Bahama [10] & SWAN [11], NTT's AWA [12]) further confirmed the viability of wireless ATM technology for delivery of multimedia services to portable devices.

Wireless ATM technology is currently migrating from research stage to standardization and early commercialization. The ATM Forum started a wireless ATM (WATM) working group in June1996, with the objective of producing wireless and mobile extensions to the existing ATM UNI4.0 specification by the end of 1998. Significant R&D and trial product development efforts are now underway in various parts of the world, so that the technology should begin to reach the market over the next 2-3 years. The recent allocation of the so-called 5 GHz U-NII band by the U.S. Federal Communications Commission [13] together with earlier ETSI rulings on Hiperlan [14], represent a significant breakthrough at the regulatory level that should result in increased commercial activity in this area. We visualize the emergence of public and private microcellular wireless networks which provide a variety of broadband mobile services including high-speed Internet access and audio/video delivery. The availability of such a broadband wireless networks by the end of this decade should stimulate the evolution of a new generation of high-performance mobile computing devices in both vertical and horizontal markets worldwide. It is noted here that although wireless ATM is typically associated with mobile multimedia services, selected components of the technology can be applied to a variety of other application scenarios including PCS/cellular infrastructure, fixed/residential wireless access and microwave infrastructure links.

In following sections of this paper, we present an overview of wireless ATM system architecture, design of key subsystems, technology development status and standardization. The paper concludes with a view of future directions in the area of wireless broadband technology.

## 3.2 WIRELESS ATM SYSTEM

### *3.2.1 System Concept:*

In this section, we present a wireless ATM system model applicable to the basic microcellular/mobile broadband access scenario outlined in the previous section. It is noted here that WATM technology may also be applied to more complex scenarios (e.g. tactical networks) involving mobile switching platforms or ad-hoc mobile terminals [14,10], but these are beyond the scope of this paper.

The major hardware/software components which constitute a basic wireless ATM are shown in Fig. 2. From the figure, it can be seen that a wireless ATM system typically consists of three major components: (1) ATM switches with standard UNI/NNI capabilities together with additional mobility support software; (2) ATM "base stations" or "radio ports" also with mobility-enhanced UNI/NNI software and radio interface capabilities*; and (3) wireless ATM terminal with a

---

* Note that there are three terms used in this context: ATM base station, ATM radio port and WATM access point. In this paper, these terms are used with the following interpretation: base station: ATM

WATM radio network interface card (NIC) and mobility & radio enhanced UNI software. Thus, there are two new hardware components, ATM base station (which can be viewed as a small mobility-enhanced switch with both radio and fiber ports) and WATM NIC, to be developed for a wireless ATM system. New software components include the mobile ATM protocol extensions for switches and base stations, as well as the WATM UNI driver needed to support mobility and radio features on the user side.

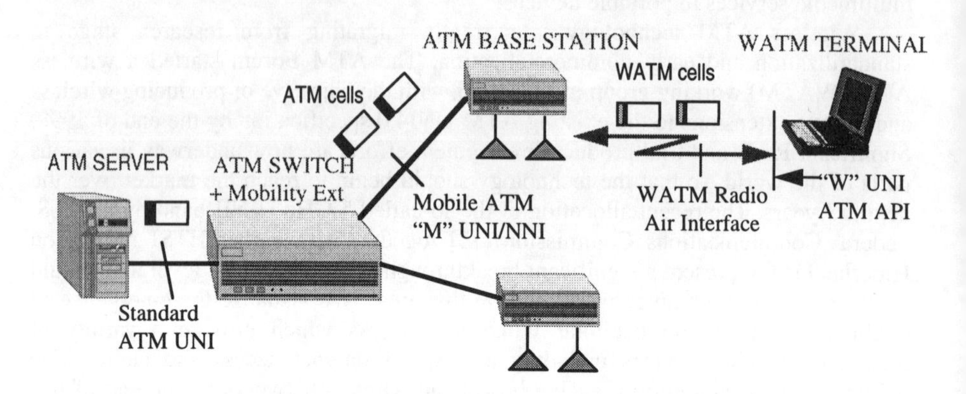


**Fig. 2** Wireless ATM Network Components

The system shown involves two new protocol interfaces: (a) the "W" UNI between mobile/wireless user terminal and ATM base station; and (b) the "M" UNI/NNI interface between mobility-capable ATM network devices including switches and base stations. Both these interfaces are required to support end-to-end ATM services at a mobile terminal such as that shown in Fig. 2. In particular, the WATM terminal sets up a connection using standard ATM signaling (UNI) capabilities to communicate with the ATM base station and network switches. All data transmitted by the WATM terminal is segmented into ATM cells with an additional radio link level header specified within the "W" interface. Migration of the WATM terminal (i.e. handoff and location management) is handled via switch-to-switch (NNI) signaling protocol extensions specified in the "M" interface.

It is noted here that subsets of the general WATM capability may also be useful for important application scenarios. For example, the "M" UNI/NNI specification can be used to construct a generic infrastructure for existing PCS, cellular and wireless data systems in addition to end-to-end WATM services, as shown in Fig. 3(a). In this configuration, ATM base stations convert wireless access protocols such as GSM, IS-136, CDMA or IEEE802.11 to a common ATM format which supports network level mobility requirements in a generic way. Such an ATM-based mobile infrastructure provides important service integration and cost/performance advantages over existing mobile networks, while facilitating smooth migration to broadband WATM services [16]. Alternatively, the WATM radio air interface specified in the "W" UNI can be used for wireless broadband access from stationary ATM devices in offices or homes, as illustrated in Fig. 3(b). This type of WATM access is expected to be an useful alternative to wired broadband access methods

---

switching (UNI/NNI) + radio interface(s); radio port: ATM interface (UNI) + radio interface(s); and access point: WATM radio air interface only.

such as SONET fiber or ADSL/VDSL, offering advantages of rapid deployment and statistical multiplexing of switch port resources.

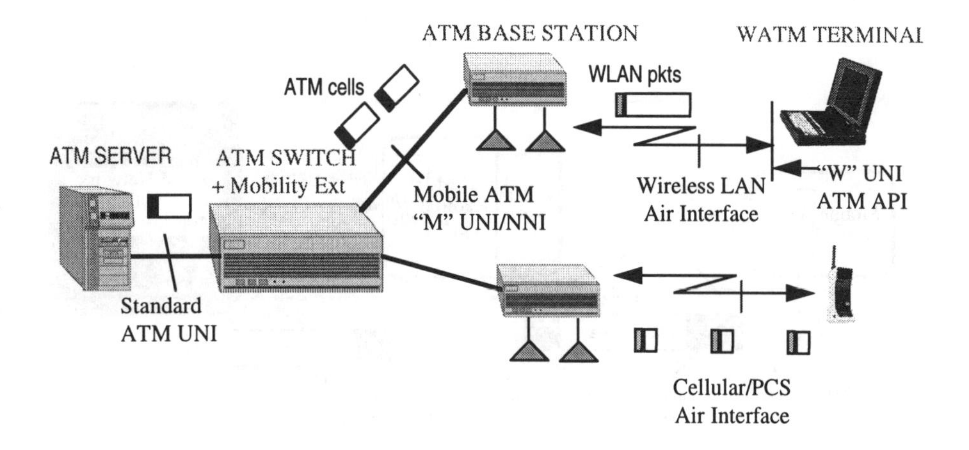


**Fig. 3(a)** Mobile ATM network used as infrastructure for legacy wireless services

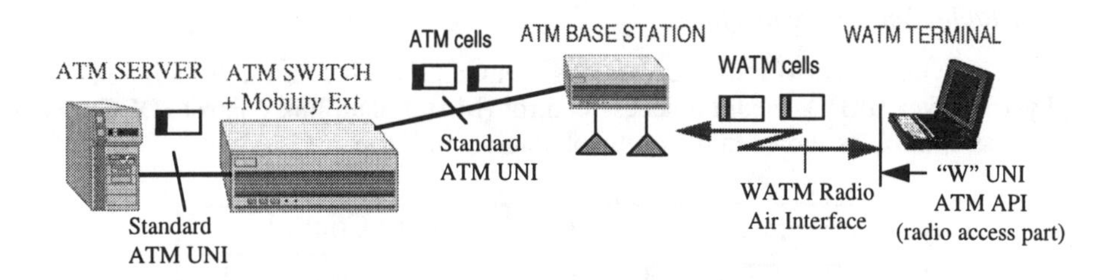


**Fig. 3 (b)** Wireless ATM radio air interface used for fixed broadband access services.

*3.2.2 Protocol Architecture:*

Wireless ATM protocol architecture is based on integration of radio access and mobility features as "first class" capabilities within the standard ATM protocol stack. The overall approach is summarized in Fig. 4, which shows that the ATM cell with 48 byte payload serves as the basis for both radio access and mobile network protocols, with minimal processing at the radio access segment to fixed network boundary [2]. The idea is to fully integrate new wireless PHY, medium access control (MAC), data link control (DLC), wireless control and mobility signaling/management functions into the ATM protocol stack as shown in Fig. 5. This means that normal ATM network layer and control plane services such as E.164 or IP-over-ATM addressing, VC multiplexing, cell prioritization (CLP), congestion/QoS control, Q.2931 signaling for call establishment, etc. will continue to be used for mobile services. As mentioned earlier, some extensions to ATM control plane protocols such as signaling and routing will be required to support terminal mobility related functions such as location management, VC handoff, etc.

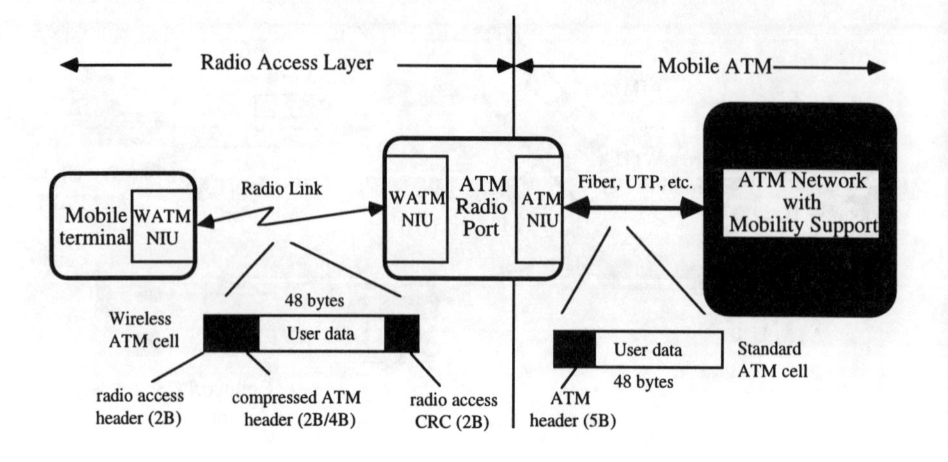


**Fig. 4** Wireless ATM protocol architecture summary

## 3.3 WATM RADIO ACCESS

### *3.3.1 Radio Access Architecture:*

The wireless ATM radio access subsystem consists of four major components: physical layer (PHY), medium access control (MAC), data link control (DLC) and wireless control, as shown in the protocol diagram in Fig. 5.

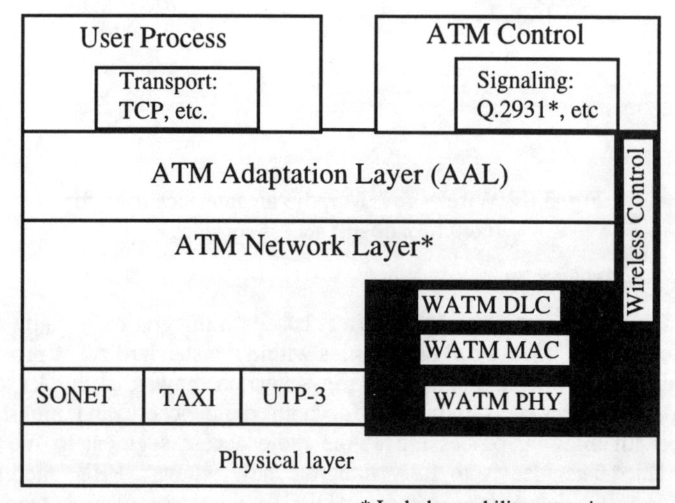


**Fig. 5** Wireless ATM protocol stack

For service data, the DLC sublayer serves as the interface to the existing ATM network layer, with a separate DLC process being invoked for every ATM virtual circuit in use. For control messages, the wireless control sublayer interfaces with the ATM control plane protocols (i.e. Q.2931 signaling, etc.). Both service data and control information must be multiplexed and scheduled on the shared wireless channel through the MAC layer, as shown in Fig. 6. The MAC layer in turn interfaces directly with the radio physical layer, which is designed to support bursts of one or more ATM cells which are received/transmitted by a mobile's MAC process.

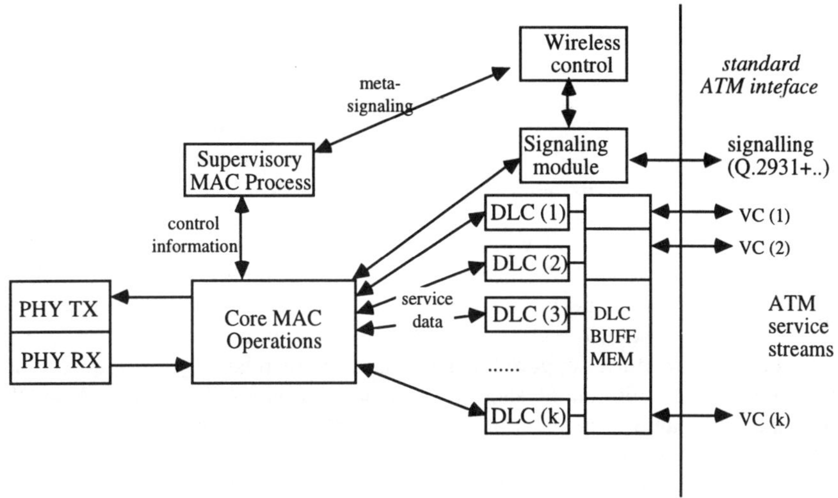


**Fig. 6.** WATM radio access protocol components and interface to ATM layer

Fig. 7 shows example formats for wireless ATM data cells and related wireless control packets used for DLC acknowledgments and radio link management. While the details may vary from one implementation to another, the main concept is to build on facilities currently provided by ATM keeping additional syntax to a minimum. In this case, the wireless header contains a cell sequence number for data link layer error recovery and control fields needed for supervisory MAC functions. ACK and wireless control messages (implemented as 8 byte short packets in this case) shown are needed to support wireless link level functions (such as MAC allocation, DLC retransmission, user authentication, handoff control, etc.) between the mobile terminal and the radio port.

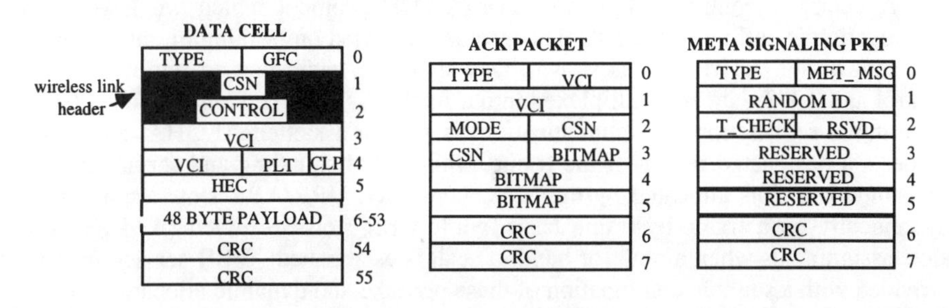


**Fig. 7** Typical wireless ATM cell and control packet formats

*3.3.2 Physical Layer:*

Wireless ATM requires a high-speed radio modem capable of providing reasonably reliable transmission in microcell and picocell environments with cell radius in the range of 100-500 m. Although such systems may operate in various frequency bands depending on national and international regulatory policies, WATM is often associated with the new 5 Ghz Unlicensed NII band in the US or the "Hiperlan" band in Europe. Higher frequency systems operating at either 20/30 Ghz or 60 Ghz may also be viable in the future, at least for certain non-mobile scenarios. Since the 5 Ghz U-NII band in the US is a technology neutral unlicensed band, WATM would be required to coexist with other radio access methods via suitable spectrum etiquette procedures to be specified in the future. In this context, a basic issue is that of designing a new technology neutral spectrum etiquette method which is compatible with stream-oriented MAC protocols associated with multimedia services. The situation is somewhat less complicated in Europe, where a single standard (including channelization, etiquette and MAC) based on WATM is planned for Hiperlan Type 2 services.

Typical target bit-rates for wireless ATM PHY are in the region of 25 Mbps, comparable to the 25 Mbps UTP specification adopted as a PHY option by the ATM Forum. The 25 Mbps value is based on a goal of per-VC service bit-rates in the range of 2-5 Mbps sustained and 5-10 Mbps peak. In addition to operating at a high bit-rate, the modem must support burst operation with relatively short preambles consistent with transmission of short control packets and ATM cells. Also, the modem should be able to compensate for r.m.s. delay spreads in the range of 200-500 depending on the service scenario and microcell size. Candidate modulation methods include equalized QPSK/GMSK [17], equalized QAM [18], multicarrier modulation such as OFDM [19] and wideband CDMA [20]. Based on R&D results available so far, equalized 25 Mbps QPSK/GMSK and OFDM appear to be the more prominent candidates for WATM PHY.

*3.3.3 Medium Access Control:*

A medium access control (MAC) layer is required to support shared use of the radio channel by multiple WATM users. The MAC layer must provide support for standard ATM services including UBR, ABR, VBR and CBR traffic classes, with associated quality-of-service (QoS) controls. A key factor in selection of the MAC protocol will be the ability to support these ATM traffic classes at reasonable QoS levels while maintaining a reasonably high radio channel efficiency. Specific techniques which have been considered for the WATM MAC layer include PRMA extensions [21], dynamic TDMA/TDD [2,22] and CDMA [23].

A centrally controlled dynamic TDMA/TDD protocol which we have been investigating is outlined in Fig. 8. This protocol is based on framing of channel time into control and ATM slots as shown. Downlink (i.e. radio port to remote station) control and ATM data are multiplexed into a single TDM burst. Uplink control (such as terminal registration and authentication) is sent in a slotted ALOHA contention mode in a designated region of the frame, while ATM signaling and service data is transmitted in slots allocated by the MAC controller. UBR/ABR slots are assigned dynamically on a frame-by-frame basis, while CBR slots are given fixed periodic slots assignments when a new (or handoff) call is established. VBR service may be provided with a suitable combination of these periodic and dynamic allocation modes [24]. A protocol of this type has been shown to provide reasonable performance at throughput levels up to about 0.6-0.7 after accounting for control and physical layer overheads.

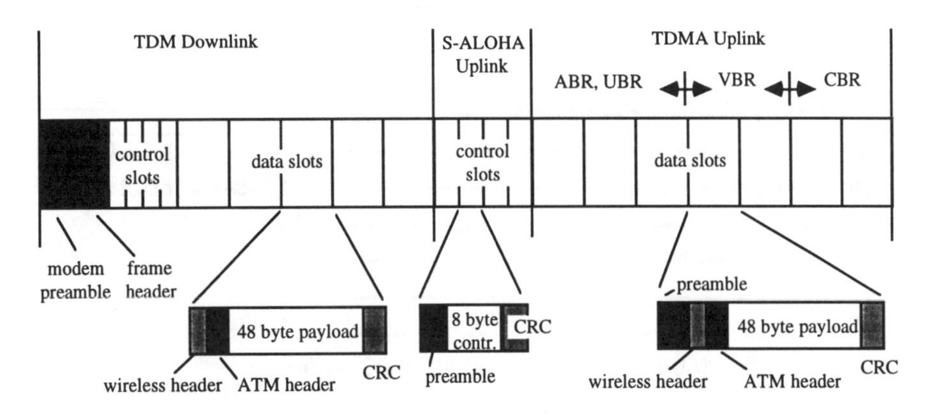


**Fig. 8** Dynamic TDMA/TDD protocol for wireless ATM access

*3.3.4 Data Link Control:*

A data link control (DLC) layer is necessary to mitigate the effect of radio channel errors before cells are released to the ATM network layer. Since end-to-end ATM performance is sensitive to cell loss, adequate error control procedures are a requirement for the WATM radio access segment. Available options include error detection/retransmission protocols and forward error correction methods (tailored for each ATM service classes). A unique consideration for wireless ATM is the requirement for relatively low delay jitter for QoS-based services such as VBR and CBR, indicating a possible need for new time-constrained retransmission control procedures.

In [25], we have proposed and evaluated a possible set of DLC techniques for wireless ATM services, applicable to both packet UBR/ABR and stream VBR/CBR. For ABR, this DLC method follows traditional SREJ ARQ procedures on a burst-by-burst basis, without time limits for completion. For CBR and VBR, the DLC operates within a finite buffering interval that is specified by the application during VC set-up. In this case since CBR or VBR allocation is periodic, additional UBR allocations are made at the MAC layer to support retransmitted cells. Recent experimental results on the WATMnet prototype [26] confirm that suitable DLC provides a significant increase in end-to-end TCP throughput at the WATM terminal, as shown in Fig. 9. Favorable results have also been obtained for CBR services, demonstrating that a substantial reduction in cell loss rate can be achieved with moderate buffering delay (~10 ms) within the DLC.

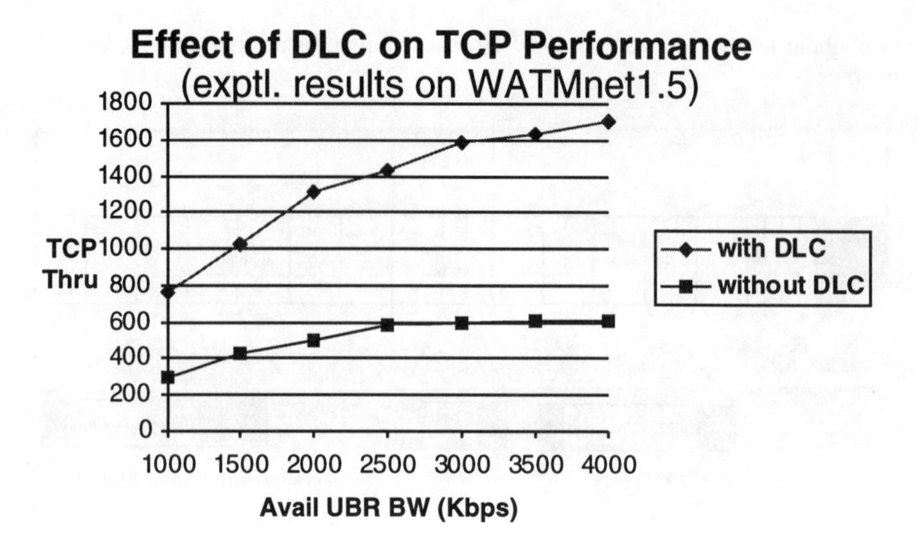


**Fig. 9** Example of measured DLC performance with TCP/UBR service on WATMnet prototype

## 3.4 MOBILE ATM NETWORK

*3.4.1 Mobile ATM Architecture:*

The term "mobile ATM" refers to a network consisting of ATM switches and base stations executing "M" UNI/NNI protocols capable of supporting mobility functions such as location management, handoff control and mobile QoS/routing [27,28]. As discussed earlier, the mobile ATM network can be used to support both broadband WATM access as well as legacy wireless systems in an integrated manner. In a typical implementation, various radio cards can be plugged into generic ATM base stations which convert applicable wireless protocols to the common mobile ATM protocol. In contrast to conventional circuit switched mobile networks (which typically have external attachments such as MSC, BSC, HLR and VLR for mobility support), location management and handoff control functions are fully integrated into the ATM switching software. This architecture is consistent with the concept of integrating mobility as a first-class network capability (as in mobile IP specified by the IETF), and avoids the need to partition address space and designate user terminals as mobile or static in an a-priori manner. Such a distinction may no longer be practical to administer as users begin to view portable devices as their primary computing or communication platform(s).

Fig. 10 schematically illustrates the handoff and location management functions carried out by a mobile ATM network. Each mobile terminal served by the network is associated with a "home" base station or switch which provides it with a permanent (home) ATM address. If the mobile terminal is an ATM device, this is the only address needed by the network to identify the user. For non-ATM terminals, the base station performs a mapping from the wireless device's telephone number or IP address to a home ATM address. All mobility functions are then handled within the ATM network framework via suitable association of a changing "foreign address" with the static home address. Further details on location management and handoff within such a mobile ATM framework are given in the subsections below.

*3.4.2 Location Management:*

The integrated ATM approach to mobile addressing outlined above provides a straightforward approach for realizing location management required to deal with terminal mobility [29]. As discussed, the home switch in mobile ATM maintains a pointer from the permanent home address to the current foreign address of the mobile via updates received from terminals as they register at new base stations. Connections to a mobile terminal are then established with a simple extension to the standard Q.2931 signaling procedure specified in existing ATM specifications. In particular, the standard CONNECT message is first used with the home address of the mobile as the routing destination. If the mobile is at home, the CONNECT message is processed by the home switch in the usual way, resulting in an immediate connection if resources are available. If the mobile is away, a standard RELEASE message is returned towards the source, with a new information element (IE) containing the foreign address of the mobile. The source (or any mobility-enabled switch on the returned path) can then establish a connection to the mobile by sending a second CONNECT message with the foreign address as routing ID. This connection procedure with integrated location management is illustrated in Fig. 10(a). It is remarked here that external location servers similar to those used in cellular/PCS systems have also been proposed for mobile ATM [30]. These two location management methods can coexist in a single network; for example, the integrated approach may be used within a single ATM domain, with external servers for inter-domain roaming.

*3.4.3 Handoff Control:*

Handoff is required to dynamically support active connections during the migration of mobile terminals from one radio access point to another. This feature is critical for mobile multimedia services delivered to a WATM terminal, as well as for efficient support of current PCS/cellular systems on an ATM infrastructure. Note that handoff is also beneficial for data users (especially those with higher volume and stream-oriented applications). Mobile IP does not currently provide dynamic handoff capability due to its connectionless framework; in contrast, mobile ATM can provide link-level handoff support for IP services running over its connection-oriented infrastructure. It is remarked here that handoff capabilities in ATM may also be beneficial for certain non-mobile scenario such as switching to hot-standby servers, call forwarding, etc.

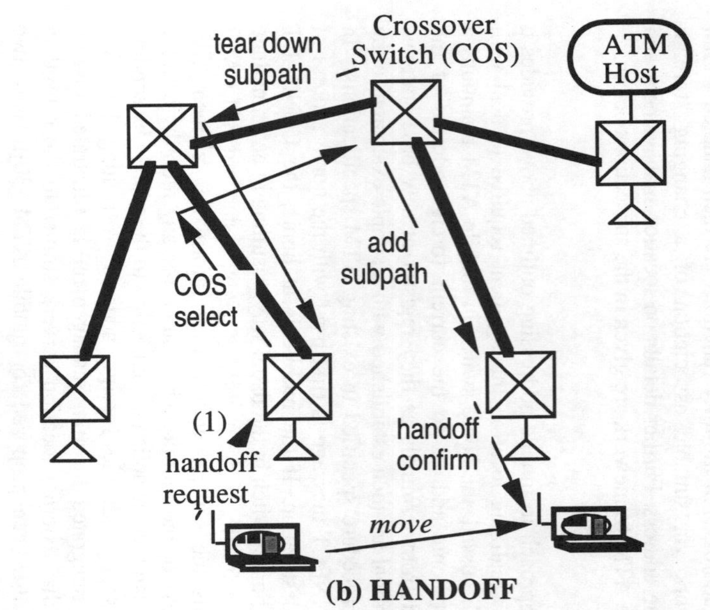


**Fig. 10** Handoff and location management in a mobile ATM network

In general, the handoff process involves migrating multiple VC's associated with a terminal from one ATM base station to another, while maintaining routing optimality and quality-of-service to the extent possible. As illustrated in Fig. 10(b), a typical mobile ATM handoff typically procedure consists of the following components [31,32]: (1) terminal initiates handoff by sending a suitable signaling message to its current base station (this case is called "forward handoff); (2) current base station initiates network procedure to move connection from current base station to specified new base station; (3) network switches and base stations collectively identify a "crossover" switch (COS) from which to reroute each VC; (4) the network routes a subpath from the COS to the new base station; (5) the cell stream is switched to the new base station, using applicable resequencing/recovery mechanisms (if any) to minimize cell loss; (6) the subpath from COS to old base station is torn down; (7) the terminal changes its radio connection to the new base station and confirms end-to-end handoff. Note that while the above handoff procedure is representative, a number other handoff control sequences are possible. For example, if the terminal loses wireless connectivity abruptly, a different "backward handoff" procedure would be initiated from the new base station.

Selected extensions to ATM signaling protocols (e.g. Q.2931) together with appropriate wireless control syntax have been proposed to support mobile ATM handoff [33,34]. New signaling messages such as handoff_request, handoff_confirm, handoff_join, etc. have been proposed, and related protocols verified via prototype implementation. At NEC's C&C Research Lab in Princeton, we have implemented handoff signaling modifications on experimental ATM radio ports and 2.4 Gbps ATM switches (such as Fore ASX-100 and NEC Model 5) and obtained reasonable latency and cell loss results [35]. Work on optimizing handoff performance (via switch/base station hardware or protocol improvements for zero cell loss, etc.) is currently in progress.

*3.4.4 Routing and QoS Control:*

Mobile ATM requires extensions to existing routing procedures (such as PNNI) to deal with route changes and optimizations associated with handoff. In general, a handoff event may result in a significant change in the optimal route of each active virtual circuit associated with the mobile terminal. Typically, this implies that a portion of the VC's route in the vicinity of the mobile terminal must be re-established according to suitable cost/performance criteria. One approach involves identification of an optimal "crossover" switch [36] at each handoff event. Alternatively, simple path extension techniques may be used at the time of a handoff, to be followed with periodic loop removal procedures, etc. In a microcellular system, path rerouting is generally preferable to path extension, but the complexity/performance trade-offs need to be evaluated further.

Routing in mobile ATM is closely related to quality-of-service (QoS) control for maintaining selected service parameters through the duration of a mobile connection. QoS/traffic control in mobile ATM must deal with several important effects unique to this scenario, i.e. the presence of a shared MAC layer in the radio link, traffic outages due to time-varying link impairments, and spatial movement of VC resources (in both infrastructure and access network segments) during terminal migration. A key question in this area is to determine the degree to which statistical service guarantees can be provided in such systems, and whether additional source declaration parameters (such as some measure of the degree of mobility) would be useful for this paradigm. Given the difficulty of providing firm statistical guarantees in this environment, we believe that the capability of the network and terminal to renegotiate "soft" QoS parameters during the course of a connection [37] is a key requirement for robust operation. Of course, QoS renegotiation also implies the need for a suitable QoS API and scalable multimedia applications at the terminal software level [38].

## 3.5 STATUS OF WATM TECHNOLOGY

In this section, we provide a brief outline of WATM related R&D and technology development status together with related standards activities. Note that the items covered here are reflect the authors' own experience and views, and are not intended to provide exhaustive coverage of ongoing WATM activities worldwide.

### *3.5.1 Radio Access Components:*

The radio access subsystem basically consists of three major technology components: (1) RF module; (2) modem LSI; and (3) WATM SAR/MAC LSI. The cost/performance of WATM systems as a whole is critically dependent on successful development of these modules at a high level of integration. GaAs MMIC's are currently being developed for the 5 Ghz U-NII and Hiperlan bands, and should become available as generic components during 1998. Modem LSI development depends to some extent on completion of emerging standards for WATM PHY. Several companies are currently developing either equalized QPSK/GMSK or OFDM modems for trial use at bit-rates varying between 10 and 100 Mbps, depending on the frequency band of use. However, there is still some uncertainty about the cost/complexity of the modem LSI due to the immaturity of current designs for burst-mode WATM. The WATM SAR/MAC device is also dependent on standards, although several companies are currently working on trial use hardware for TDMA/TDD class protocols [8-11]. No major technical barriers to LSI implementation of the WATM MAC/SAR are foreseen.

The above radio access components would typically be integrated with an embedded processor to produce a compact WATM network interface card (NIC) which can connect to existing portable computer interfaces (e.g. PC card). On the network side, the same WATM NIC serves as an "access point" which can plug into a mobility-capable ATM switch to form a base station. As discussed in [39], a high-performance WATM NIC is intrinsically more complex than a conventional ATM interface since it has to support both ATM layer functions (such as queue management, QoS control and SAR) and WATM MAC/DLC functions. At wireless link speeds of 25 Mbps or higher, hardware support is generally needed for MAC, DLC and SAR functions. Major WATM-specific hardware components of the NIC are the TDMA/TDD frame formatter and the cell format/memory pointer processor module as shown in Fig. 11. Higher level functions (e.g. supervisory MAC, radio channel control, etc.) are carried out in the embedded processor (typically a low power RISC such as the NEC V.851), which can also incorporate some of the ATM layer functionality if desired. Early commercial implementations of WATM NIC cards for 5 Ghz mobile services should become available in the 1998-99 time-frame.

Related standards activities include the Hiperlan Type 2 Wireless ATM specification being developed by ETSI BRAN (Broadband Radio Access Network, formerly known as RES10), intended for use in the 5.15-5.25 Ghz band. The ATM Forum's WATM working group (whose scope includes WATM radio access for the U.S. U-NII band) has adopted a strategy of coordinating radio related activities with ETSI BRAN. Both groups intend to produce a specification for the WATM radio access layer around the end of 1998. Several other countries are currently considering allocation of 5 Ghz spectrum for unlicensed broadband access similar to European Hiperlan or U.S. U-NII. This raises the possibility of a single world standard for WATM radio access in the unlicensed 5 GHz band.

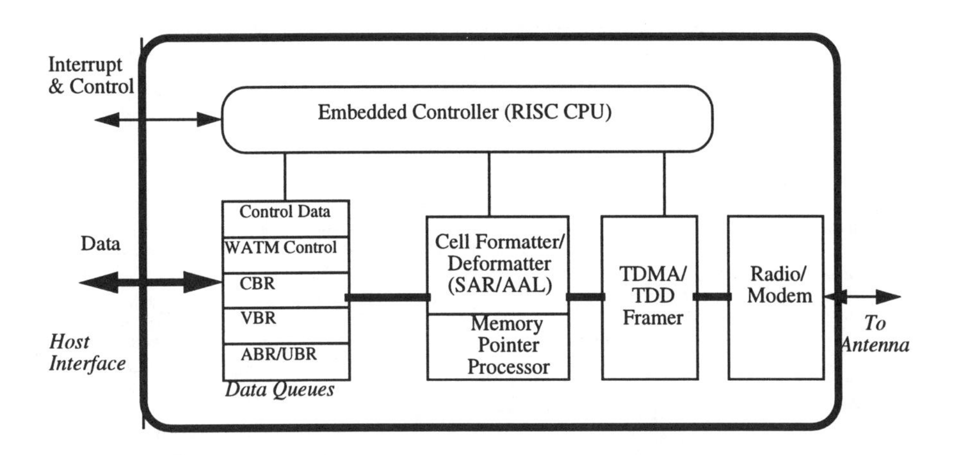


**Fig. 11** Hardware architecture of WATM Network Interface Card (NIC)

*3.5.2 Mobile Network Components:*

The major technology components of a mobile ATM network are: (1) ATM switch with mobility ("M" NNI) software enhancements; (2) ATM base station or radio port, also with mobility software ("M" UNI/NNI). Thus, the primary development items for mobile ATM are the "M" UNI/NNI software and the ATM base station/radio port hardware. For non-ATM wireless access, gateways from mobile ATM to existing systems such as GSM, IS-136 or CDMA would also be required.

The ATM base station or radio port hardware can readily be developed by leveraging available ATM edge switch, multiplexor or interface card technology depending on the desired functionality, cost and size. Compact, low cost radio ports without switching functionality appropriate for early WATM deployments can be developed by modifying the WATM NIC discussed in Sec. 5.2 to provide an ATM/SONET output that connects to a switch. A more general ATM radio port capable of terminating both WATM and non-ATM wireless services could be developed by augmenting an ATM interface (with embedded processor for "M" UNI signaling) to accept radio cards as plug-ins. Mobile ATM software development is subject to evolution in the standards process, and requires non-trivial changes in signaling/PNNI software to implement location management and handoff. In addition, switches running mobile ATM protocols will generally require an upgrade of processor performance in order to keep up with the significantly higher rate of signaling generated by mobiles.

Several R&D projects aimed at validating and refining proposed handoff and/or location management methods in ATM are currently in progress. We have developed a mobile ATM network prototype (shown schematically in Fig. 12) consisting of Pentium PC-based ATM radio ports and 2.4 Gbps NEC ATOMIS5 switches with external Pentium PC controllers. The system runs a preliminary version of the proposed "M" UNI/NNI ATM protocols, which have been implemented as extensions to available Forum 3.0 Ver UNI and NNI software. The prototype network has been used to demonstrate integrated location management and handoff support for both WATM terminal and non-ATM wireless devices (such as a laptop PC running IP over a wireless LAN). Work on protocol refinements and performance improvement (e.g. zero cell loss handoff, location caching, etc.) is currently in progress.

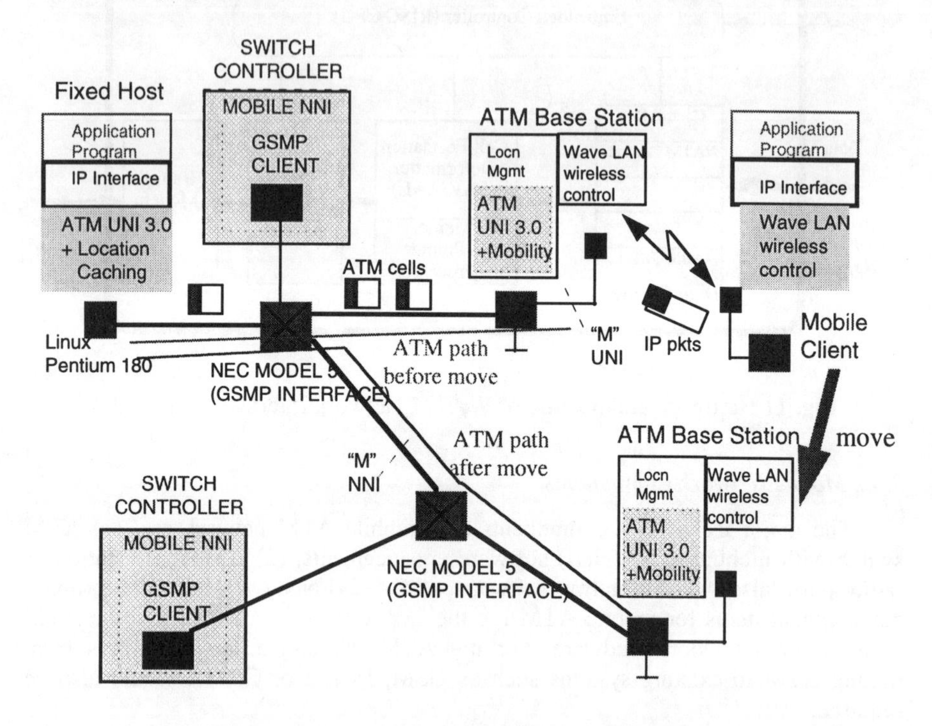


**Fig. 12** Outline of NEC's Mobile ATM network prototype

Related standards work on mobile ATM protocols is being done by the ATM Forum's WATM working group, which aims to complete an initial "M" UNI/NNI specification by mid-1998. This specification is intended to be in the form of optional enhancements to existing signaling, PNNI, etc. which operate in a backward compatible manner with current ATM switches. The working group has adopted a baseline architecture and is currently working on specific protocols for handoff and location management. Future work items on mobile ATM include IP support, cellular/PCS interworking, QoS and routing issues.

*3.5.3 System Prototypes:*

A number of system-level prototypes for wireless ATM have been developed by R&D labs in different parts of the world. These include NEC C&C Research Lab's "WATMnet" [8] Olivetti/Oracle Research Lab's "Radio ATM" [9], Bell Laboratories "Bahama" [10] & SWAN [4], and NTT's "AWA" [11]. Ongoing European technology demonstration projects in this area include ACTS Magic WAND [40] and MBS [41]. Although each of these prototypes has somewhat different technical characteristics, the overall result of these R&D activities has been to demonstrate that wireless ATM technology is viable and can effectively provide multimedia services to portable devices. Of course, the cost/performance of the technology can be proven only after at least the first stage of product development and trial deployment is completed over the next 2-3 years.

NEC's WATMnet prototype [8] (which first became operational in July 1995) was developed as a proof-of-concept demonstration of a complete wireless ATM system, which could also be used for experimental refinement of key subsystems such as MAC, DLC and mobility protocols. As shown in Fig. 13, the prototype consists of several laptop computers (NEC Versa-M) with high-speed TDMA/TDD radio interfaces, ATM radio ports, and mobility-enhanced ATM switches. The radio cards used in the first-generation WATMnet prototype operate at peak bit-rates up to 8 Mbps, using low-power 2.4 Ghz ISM-band modems. The system is capable of supporting standard ATM services at bit-rates up to ~5 Mbps, quality-of-service (QoS) control, and handoff of active connections from one microcell base station to another. The prototype has been used to demonstrate familiar TCP/IP based Internet applications (such as high-speed web access from a wireless laptop PC) as well as native ATM applications with media streaming and QoS control (such as MPEG video retrieval).

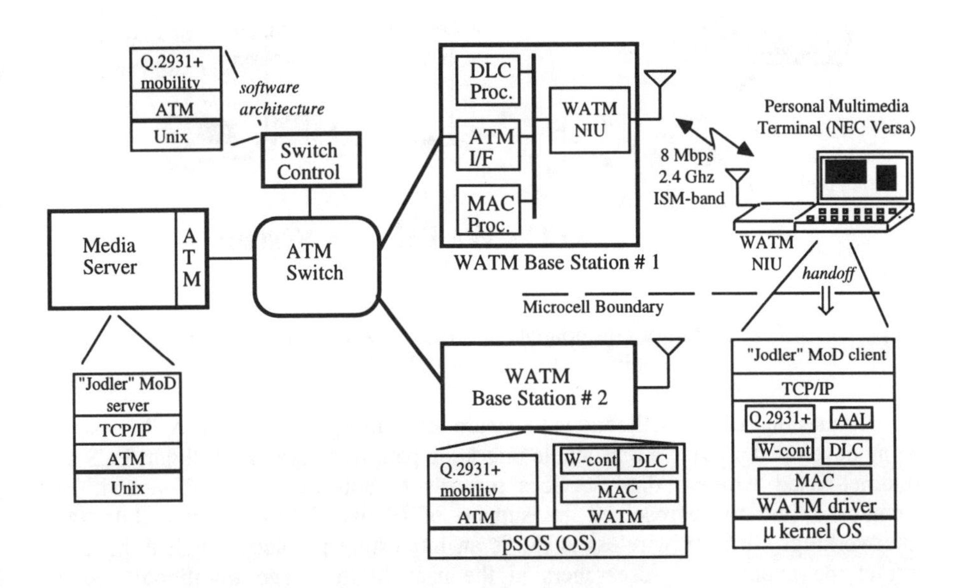


**Fig. 13** Hardware/software architecture of the WATMnet prototype

## 3.6 FUTURE DIRECTIONS

As discussed in previous sections, wireless ATM technology is steadily migrating from R&D stage to standardization, service trials and early products. In view of the growing significance of both multimedia and portable computing in various aspects of everyday life, there should be a large potential market for such broadband wireless network technologies if they can be offered at the right cost/performance points. Thus, the immediate challenge for WATM is to develop an open world standard that facilitates production of high-volume, generic LSI and software modules needed to build a cost-effective system. Significant technology development efforts on radio, modem and MAC/SAR LSI designs will be needed to achieve this objective. In parallel, it will be necessary to start consumer level field trials to demonstrate the use of WATM technology, and evaluate its effectiveness in different vertical and horizontal market scenarios. Fig. 14 shows the major components of a field-trial capable system ("WATMnet 2.0") we are currently developing for use in both public and private microcellular scenarios. This system will include compact 5 Ghz/25 Mbps WATM NIC's, ATM radio ports supporting both ATM and non-ATM wireless access, and mobile ATM switches (based on the future ATM Forum "M" specification).

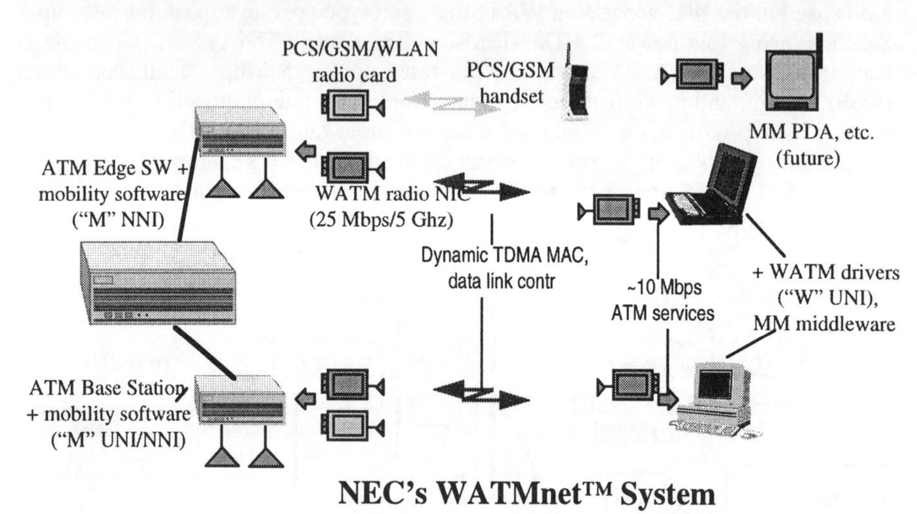


**Fig. 14** Major components of NEC's WATMnet2.0 system being developed for field trial use

The second major challenge is to ensure that wireless and mobile ATM systems being developed provide a reasonable migration path from existing cellular, PCS and Internet-based wireless data services towards broadband. In this context, key technical issues to be resolved are support of IP over WATM and cellular/PCS interworking. IP over wireless ATM is an interesting problem which is likely to attract the attention of researchers in the near future. The traditional way of supporting IP over ATM (called MPOA) can readily be extended to fit within the mobile ATM framework outlined in this paper. However, for IP-centric systems, it is also possible to use WATM hardware as a link level substrate for mobile IP services with QoS control, etc. in the same way as ATM hardware is being used as a core for gigabit IP routers. In particular, we are currently in the process of implementing a proof-of-concept demonstration of a specific IP switching technique [42,43] running on a seamless wired + wireless ATM network. In this system, IP protocols are used on a hop-by-hop basis at routers implemented in the control processor of each ATM switch or base station. IP flows are subsequently mapped on to the ATM and WATM hardware using appropriate flow<->VC and QoS mapping rules. Another alternative for a predominantly IP network would be to use only the

WATM NIC hardware to provide a QoS-capable wireless access link which interfaces directly with level 3 IP routers.

If broadband wireless technologies such as wireless ATM successfully enter the mainstream, this will have an important impact on portable terminal hardware and software. A high-bandwidth wireless connection will enable compact portable devices to retrieve multimedia streams from the network with low latency, thus enabling a variety of new applications. Since the concept of WATM is to provide a uniform API at both fixed and portable terminals, we look forward to a gradual blurring of the distinction between services available on stationary and mobile terminals. Further work will be required on mobility middleware [38,44,45] to deal with issues such as application scaling, adaptive QoS control and wireless link disconnections while using mainstream transport protocols and API's at portable devices. In terms of terminal hardware [46,47], the availability of a WATM connection would tend to stimulate development of a new generation of multimedia-oriented laptops, PDA's or PIA's (personal information appliances) with high-resolution displays and hardware support for media streaming, video decoding, etc. Such devices would be ideal for web-based media browsing, as well as for new location-aware mobile applications. Portable digital cameras now entering the mass market represent a different class of mobile multimedia devices with high-bandwidth uplink requirements which can also be effectively supported by wireless ATM.

## 3.7 CONCLUDING REMARKS

This paper has presented a review of wireless ATM technology, covering system concepts, protocol architecture, subsystem technologies, R&D status and future directions. Overall, we believe that broadband wireless techologies such as WATM will play an important role in the future network infrastructure. The specific wireless/mobile ATM approach described in this paper is steadily moving from R&D stage to standardization and early product, and thus has the promise of becoming a commercial reality over the next few years. Of course, a significant amount of technology development and standards work still needs to be done on key WATM subsystems such as modem/MAC, mobile network software (including IP support) and portable multimedia terminals. Availability of wireless ATM products at the right cost/performance point should stimulate future development of a new generation of mobile multimedia terminals and applications.

### References

[1] D. Raychaudhuri and N. Wilson, "Multimedia Personal Communication Networks: System Design Issues", 3rd WINLAB Workshop on 3rd Gen. Wireless Information Networks, April 1992, pp. 259-288. (also in: "Wireless Communications", Eds. J.M. Holtzman & D.J. Goodman, Kluwer Academic Pub., 1993, pp. 289-304).

[2] D. Raychaudhuri and N. Wilson, "ATM Based Transport Architecture for Multiservices Wireless Personal Communication Network", IEEE J. Selected Areas in Comm., Oct. 1994, pp. 1401-1414.

[3] D. Raychaudhuri, "Wireless ATM: An Enabling Technology for Multimedia Personal Communication", ACM/Baltzer J. Wireless Networks, 1996, Vol. 2, pp. 163-171.

[4] A. Acampora, "Wireless ATM: A Perspective on Issues and Prospects", IEEE Personal Comm. Mag., Aug. 1996, pp. 8-17.

[5] E. Ayanoglu, K. Eng and M.J. Karol, "Wireless ATM: Limits, Challenges and Proposals", IEEE Personal Comm. Mag., Aug. 1996, pp. 18-35.

[6] D. Raychaudhuri, "Wireless ATM Networks: Architecture, System Design & Prototyping", IEEE Personal Comm. Mag., Aug. 1996, pp. 42-49.

[7] Special Issue on Wireless ATM, IEEE J. Selected Areas in Comm., Jan 1997.

[8] D. Raychaudhuri, et al, "WATMnet: A Prototype Wireless ATM System for Multimedia Personal Communication", IEEE J. Selected Areas in Comm. , Jan 97, pp. 83-95.

[9] J. Porter & A. Hopper, "An Overview of the ORL Wireless ATM System", IEEE ATM Workshop, Wash. D.C., Sept. 30-Oct 1, 1995.

[10] K. Y. Eng et. al, "BAHAMA: A Broadband Ad-Hoc Wireless ATM Local Area Network", Proc. ICC' 95. pp. 1216-1123.

[11] E. Hyden, et. al, "SWAN: An Indoor Wireless ATM Network", Proc. ICUPC'95, Tokyo, Nov. 1995.

[12] M Umehira, et. al, "An ATM Wireless Access System for Tetherless Multimedia Services", Proc. ICUPC' 95, Tokyo, Nov. 1995.

[13] U.S. Federal Communications Communication, "Operation of Unlicensed NII Devices in the 5 Ghz Range", ET Docket 96-102, Jan 1997.

[14] ETSI-RES10, "High Performance Radio Local Area Network (HIPERLAN)", Draft Standard, Sophia Antipolis, France, 1995.

[15] L. Martinez, P. Sholander and L. Tolendino, "Effects of Mobile ATM Switches on PNNI Peer Group Operation", ATM Forum 97-0315, April 1997.

[16] D. Raychaudhuri and Y. Furuya, "ATM-based Wireless Personal Communication System with Migration to Broadband Services", Americas Telecom '96 Technology Forum, Rio De Janeiro, Brazil, May 1996.

[17] J. Tellado-Mouerelo, E. Wesel and J. Cioffi, "Adaptive DFE for GMSK in Indoor Radio Channels", IEEE J. Selected Areas in Comm., April 1996, pp. 492-501.

[18] R. Valenzuela, "Performance of Quadrature Amplitude Modulation for Indoor Radio Communications," IEEE Trans. on Commun., vol. COM-35, no. 11, November 1987, pp. 1236-38.

[19] L. Cimini, "Analysis and Simulation of a Digital Mobile Channel Using Orthogonal Frequency Division Multiplexing", IEEE Trans. on Comm., July 1985, pp. 665-675.

[20] J. T. Taylor and J. K. Omura, "Spread Spectrum Technology: A Solution to the Personal Communications Services Frequency Allocation Dilemma", IEEE Comm. Mag., Vol 29, No. 2, Feb. 1991, pp. 48-51.

[21] S. Nanda, D.J. Goodman and U. Timor, "Performance of PRMA: a packet voice protocol for cellular systems", IEEE Trans. Veh. Tech, Vol. VT-40, 1991, pp. 584-598.

[22] G. Falk, et. al., "Integration of voice and data in the wideband packet satellite network," IEEE J. Selected Areas in Comm., vol. SAC-1, no. 6, Dec. 1983, pp. 1076-1083.

[23] N. Wilson, R. Ganesh, K. Joseph and D. Raychaudhuri, "Packet CDMA vs. Dynamic TDMA for Access Control in an Integrated Voice/Data PCN", IEEE J. Selected Areas in Comm. , Aug. 1993, pp. 870-884.

[24] S.K. Biswas, D.J. Reininger and D. Raychaudhuri, "Bandwidth Allocation for VBR Video in Wireless ATM Networks", Proc. ICC'97, Montreal, CA, June 1997.

[25] H. Xie, R. Yuan and D. Raychaudhuri, "Data Link Control Protocols for Wireless ATM Access Channels", Proc. ICUPC'95, Tokyo, Nov. 1995.

[26] P. Narasimhan, et al, "Design and Performance of Radio Access Protocols in WATMnet, a Prototype Wireless ATM Network", Proc. Winlab. Workshop, April 1997.

[27] R. Yuan, S.K. Biswas, L.J. French, J. Li and D. Raychaudhuri, "A Signaling and Control Architecture for Mobility Support in Wireless ATM Networks", J. Mobile Networks and Applications, 1996, Vol. 1, pp. 287-298.

[28] A. Acharya, S.K. Biswas, L.J. French, J. Li, D. Raychaudhuri, "Handoff and Location Management in Mobile ATM Networks", Proc. 3rd. Intl. Mobile Multimedia Comm. (MoMuC-3) Workshop, Princeton, NJ, Sept 25-27, 1996.

[29] A. Acharya, J. Li and D. Raychaudhuri, "Primitives for Location Management and Handoff in Mobile ATM Networks", ATM Forum/96-1121/WATM, Aug 1996.

[30] G. Bautz and M. Johnsson, "Proposal for Location Management in WATM", ATM Forum 96-1516, Dec 1996.

[31] B. Rajagopalan, A. Acharya and J. Li, "Signaling and Connection Rerouting for Handoff Control Management" ATM Forum 97-0338,April 1997.

[32] H. Mitts, et al, "Microcellular Handover for WATM Release 1.0: Proposal for Scope and Terms of Reference", ATM Forum 97-0226, April 1997.

[33] J. Li, A. Acharya and D. Raychaudhuri, "Signaling Syntax for Handoff Control in Mobile ATM", ATM Forum 97-..., Feb. 1997.

[34] B. Akyol and D. Cox, "Signaling Alternatives in a Wireless ATM Network", IEEE J. Selected Areas in Comm., Jan 97, pp. 35-49.

[35] J. Li and R. Yuan "Handoff Control in Wireless ATM: An Experimental Study", Proc. ICUPC'96, Boston, MA, Sept. 1996.

[36] C.K. Toh, "Crossover Switch Discovery in Wireless ATM LAN's" to appear in ACM Mobile Networks and Nomadic Applications Journal, 1(4), 1996.

[37] D. Reinginer, D. Raychaudhuri & J. Hui, "Dynamic Bandwidth Allocation for VBR Video in ATM Networks", IEEE J. Selected Areas in Commun., August 1996, pp. 1076-1086.

[38] M. Ott, G. Michelitsch, D. Reininger and G. Welling, "QoS Aware Browsing in Distributed Multimedia Systems", Proc. IDMS'97 (Interactive Distributed Multimedia Systems & Telecommunications Services).

[39] C. Johnston, "A Network Interface Card for Wireless ATM Networks", Proc. PIMRC'96, Oct. 1996, Taiwan.

[40] P. Nikos, et al, "MAGIC WAND: Wireless ATM MAC Overall Description", Report 3D1, Dec. 1996.

[41] B. Walke, D. Petrass, and D. Plassmann, "Wireless ATM: Air Interface and Networks Protocols of the Mobile Broadband System", IEEE Personal Comm. Magazine, Aug. 1996, pp. 50-56.

[42] P. Newman, T. Lyon and G. Minshall, "Flow Labelled IP: A Connectionless Approach to ATM", Proc. IEEE Infocom 96, pp. 1251-60.

[43] A. Acharya, R. Dighe and F. Ansari, "IP Switching over Fast ATM Cell Transport (IPSOFACTO)", IEEE Broadband '97 Workshop, Tucson, AZ, Jan 1997.

[44] M. Ranganathan, A. Acharya, S, Sharma and J. Saltz, "Network-Aware Mobile Programs", Proc. USENIX Conference, Jan 1997.

[45] A. Campbell, C. Aurrecoechea and L. Hauw, "A Review of QoS Architectures:, ACM Multimedia Systems Journal, 1996.

[46] S. Sheng, A. Chandrashekharan, R. Broderson, "A Portable Multimedia Terminal", IEEE Communications Magazine, Dec. 1992, pp. 64-75.

[47] C. Chien, et al, "Design Experience with an Integrated for Wireless Multimedia Computing", Proc. Mobile Multimedia Comm. Workshop (MoMuC-3), Princeton, NJ, Sept. 1996.

# 4 A SYSTEM FOR WIRELESS DATA SERVICES

Krishan K. Sabnani
Thomas Y.C. Woo
Thomas F. La Porta

**Abstract**

Wireless data services are poised for tremendous growth, accelerated by both the Internet and the rapid build-out of the necessary infrastructure. To support this growth, it is critical to design applications, protocols and system architecture that are adapted for a wireless operating environment. In this paper, we identify the key issues and challenges in implementing systems for providing wireless data services, and present design approaches and principles. As a concrete example, we describe our design and implementation experience with WDS, a prototype we have built for providing wireless data services to cellular phones over cellular SMS channels.

## 4.1 INTRODUCTION

The growth of wireless data has been slow. In particular, most applications of wireless data have only been in vertical markets, such as field services, dispatch, and telemetry. In the horizontal consumer market, wireless data has not been successful, standing in sharp contrast to the popularity of mobile voice service.

The problems hindering the growth of wireless data are manifold. Broadly speaking, they fall into two categories: transport and software. The former refers to the lack of ubiquitous and standardized infrastructures for transporting wireless data, while the latter concerns the inadequate understanding of the necessary protocol and system architecture required to deliver wireless data services.

Fortunately, this is poised to change. A number of factors are now in place to greatly accelerate the growth of wireless data. These include:

- **Internet** - The explosion in wireline data demand can be attributed in large part to the growth of the Internet, which provides rich connectivity and hosts enormous content.

  By the same token, Internet will be the key driver for wireless data. The added value of wireless data is the ability to connect to the Internet from anywhere at any time.

- **New Spectrum and Air Interfaces** - More spectrum are being opened up and dedicated for wireless data. For example, in the U.S., the government recently auctioned off a band of frequencies for Narrowband PCS services [8], and approved a new unlicensed National Information Infrastructure band for campus-area wireless Internet access [5].
  For air interfaces, the new digital cellular and PCS standards have provisions for wireless data. For example, both the new U.S. TDMA (i.e., IS-136 [19]and CDMA (i.e., IS-95 [18]) standards specify various types of data capabilities, ranging from Short Messaging Service (SMS), to circuit, and even packet data (e.g., IS-657 [17]). Other data standards such as CDPD [4] and Motorola's ReFlex [13] are also being deployed.
- **Infrastructure** - Ubiquity is critical for wireless data services. To achieve it, however, can incur a prohibitively high cost of building out the necessary infrastructure. This has long been a dilemma for wireless data.
  The new air interface standards (e.g., CDPD, IS-136, IS-95) overcome this problem by overlaying or piggybacking the data channels on top of the mobile voice infrastructure. This not only eliminates the need to construct a separate infrastructure for wireless data, thus significantly reducing the cost, but it also allows wireless data to inherit the coverage of mobile voice, thus greatly enhancing its ubiquity.
- **End Devices** - Wireless data end devices have traditionally been specialized and expensive. Technological advances have made available general purpose end devices at reasonable cost. Devices such as Personal Digital Assistants (PDAs), palmtops, and laptops are becoming commonplace.
  Also, *smartphones*, i.e., cellular/PCS phones with integrated software applications that take advantage of the data capabilities, are beginning to appear.

With the above factors converging, the transport problem for wireless data should be eased. The limiting factor then becomes software. The unique characteristics and constraints of a wireless operating environment imply that designs that have been used in wireline applications may not apply. In particular, key issues concerning the following must be reconsidered.

**Applications.** Because of lower bandwidth, higher latency, and less capable end devices, not all applications can perform in a reasonable manner in a wireless environment. Thus, a first challenge in providing wireless data services is to identify a suite of applications that collectively provide some useful and coherent functions under a specific set of wireless constraints.

Also, application semantics may need to be adapted for wireless operation. For example, an asynchronous semantics may be preferred over a synchronous semantics for high latency wireless channels.

**Protocol Design.** Wireless links exhibit significantly different characteristics from wireline links. For optimal performance, protocols must be explicitly designed to accommodate or match these characteristics.

In addition, diverse characteristics may exist even among wireless links that are simultaneously available. For example, in GSM [14], data channels come in at least 4 different flavors: SMS, single-slot circuit, multi-slot (or high-speed) circuit, and packet (GPRS).

A protocol should try to make use of the different channels for its different operations (e.g., notification on SMS, bulk data transfer on circuit), and dynamically adapt itself as availability of channels changes.

**System Architecture.** The diverse variety of wireless access technologies and end devices is a rule, rather than an exception. A wireless system architecture must be designed to be flexible and extensible. In particular, it should not presuppose a specific combination of access methods and end devices.

A wireless system architecture must address additional concerns that are not present in a wireline system. Two key ones are mobility and disconnection. Mobility implies that the system needs to track and locate subscribers. Disconnection implies that the system must be prepared to handle data on behalf of disconnected subscribers.

In this paper, we present the design and implementation of a system for providing a selected set of wireless data services. The focus of this paper is in identifying the key issues and challenges, and providing design approaches and principles, at both protocol and system levels.

As will become clear, the recurring theme in this paper is how to provide a maximal and interesting set of wireless data services while keeping the complexity of the end device and bandwidth usage (especially in the uplink direction) consistent with the wireless operating constraints.

For concreteness, we have embodied our design and techniques in the construction of a set of servers for wireless data, which we called collectively a *wireless data server* (WDS). A prototype of WDS has been implemented and is operational at Bell Laboratories.

The balance of this paper is organized as follows. In Section 2, we motivate the key issues and challenges in building wireless data servers. We also present a basic architecture for such servers. In Section 3, we describe the actual WDS prototype we have built. We explain the services it provides, present the details of its system architecture, and provide some system performance measurements. In Section 4, we conclude.

## 4.2 WDS - OVERVIEW

The most challenging environment in which to offer data services is outdoors via wireless access. This environment also provides the most potential in terms of user base and services. To meet this potential, a system must accommodate multiple types of end devices, access protocols, transport networks, and services.

End devices will vary depending on applications. For example, many users may wish to extend their desktop to a mobile environment. In these cases, the system must support end devices, most likely laptop computers, and applications, that are normally accessed from workstations via high-speed wired connections. On the other hand, some users may subscribe to services tailored to the wireless mobile environment, such as real-time navigational services [1]. These may require special software that executes on existing portable computing devices, or special hardware to assist and display the navigational application. It is also expected that current popular wireless services, namely, voice and paging services, will continue to grow. In these cases, the end devices will be cellular phones and pagers.

Depending on the application requirements, different media access protocols will be used. For example, for real-time streaming applications, such as voice, admission control will likely be performed at connection establishment, and some static media sharing protocol that provides guaranteed resources, such as time division multiple access, will be used. For non-real-time applications, contention protocols may be used to support connectionless services. Other applications may require hybrid solutions.

Users will access data services via mobile networks over several types of transport networks. The most straightforward access is via a dedicated connection through the telephone network using a cellular modem. This type of connection is commonly used for fax service. Connectionless packet networks, such as the Internet and Intranets, will be used for direct access to services such as Web browsing and email Other services may be offered over other types of packet networks, such as by using special protocols running on the common channel signaling networks supported by Signaling System Number 7 transport networks [12].

In the following subsections, we review the constraints of the operating environment presented by outdoor wireless networks, discuss the impact on services, and present the objectives and basic architecture of our system.

### *4.2.1 Constraints*

Systems that offer data services via wireless mobile networks are constrained by both the end devices and communication networks.

End devices are primarily constrained by the fact that they must be portable. For this reason, they are made as simple and small as possible to support their targeted applications. The more general purpose a mobile end device is designed to be, the more complex it becomes. The resources that are most limited in end devices are its input/output capabilities, processing power, memory, and battery capacity.

To extend a desktop to the mobile environment, an end device requires a keyboard and reasonable size display to be useful. This interface is available on laptop computers. Unfortunately, for many users, this type of end device is too large to be easily portable. For this reason, other input/output models have received attention, including touch screens and pen-based interfaces with handwriting recognition. Devices can also be designed to support specific services. Examples include pagers which have small displays and a limited number of buttons for input, and simple wireless tablets which are typically used in vertical applications such as delivery services, rental car returns, and inventory systems. These devices provide only the input/output capabilities required to support their applications.

Similar observations may be made about the processing and memory resources, and power consumption of an end device. Again, the more general purpose the device, the more processing power and memory it will require since it will likely run a general purpose operating system and be required to execute complex applications. As a result, general purpose devices consume more power and have shorter battery lifetimes. Special purpose devices can use simpler operating systems, and be customized in terms of hardware, algorithms, and protocols, to be more efficient. For example, pagers, due to their limited functions, are simple devices with long battery lifetimes. The battery lifetime is extended because of their limited displays, simple processing, and use of algorithms that allow pagers to be in a dormant state most of their operating lifetime, only becoming active periodically to receive messages [10].

Wireless access to networks that support mobility also presents many system constraints in terms of system bandwidth, channel quality, and availability. Outdoor wireless links typically provide 3-20 Kbps of usable bandwidth to mobile devices, compared to 10-100 Mbps available on wireline links. This reduction in bandwidth must be addressed at the application layer, network algorithms, and protocol design.

In addition to the limited bandwidth available, characteristics of the system may make the distribution of bandwidth asymmetrical, i.e., there may be significantly (2-10 times) more bandwidth available in the downlink than in the uplink. This is caused by two main factors. First, network-based transmitters, such as base stations, may transmit with much higher power than a portable mobile device. Therefore, to maintain an acceptable bit error rate, mobile devices may have to transmit at a lower rate than network-based transmitters.

Second, a base station is a centralized source of traffic, and may therefore efficiently schedule downlink traffic for transmission. Multiplexing between base stations is typically performed using a static technique, such as assigning different frequencies to different base stations. On the uplink, however, depending on the access protocol being used, mobile stations may use various Media Access Control (MAC) layer protocols for network access. Because of channel characteristics and propagation delay, typical MAC protocols such as CSMA/CD do not work well. As a result, distributed MAC layer protocols requiring real-time feedback from the network are often used. These protocols add delay and overhead, thus reducing the effective bandwidth, in the uplink direction. If resources can be dedicated on a per mobile basis on the uplink, this second cause of bandwidth asymmetry is not a factor.

The quality of wireless links varies, and is often harsher than wireline links. Bit error rates can be as high as $10^{-2}$ for sustained periods of time. Mobility also must be accounted for when providing data services. As users move, the quality of their access links may change. Also, users must be located in order for data to be delivered to their devices. This requires update and location tracking algorithms that may affect routing algorithms. The combination of mobility and poor channel characteristics may cause users to be disconnected from the network. Applications must be designed to react to a disconnected period, perhaps will the aid of network elements.

In summary, systems that provide wireless data services are constrained by:

1. End devices - limited input/output, processing and memory resources, and battery capacity; and
2. Network resources - limited and asymmetric bandwidth, high latency and error rates, disconnectivity, and mobility.

Application, algorithm, protocol, and system design must take these constraints into consideration as discussed in the next subsection.

### *4.2.2 Impact on Services*

The constraints imposed by end devices and networks obviously affect which types of applications can be supported and what quality of performance will be experienced. In some cases, it is possible to support applications without changing user behavior or semantics. In many others, it is required to change the semantics of interaction between application end points, and perhaps even users, to provide applications that are usable in a natural way. Two representative examples are given below:

1. Many applications are designed to be interactive. For example, database queries for information retrieval applications, such as getting a stock price or sports score, typically return a result in a matter of seconds while the user waits for the response. In a system with limited bandwidth, large delays, and periods of disconnectivity, real-time interaction may be transformed into a non-real-time request-response application, in which a user may place a request to which the response will be delivered at a later time. In this way, a synchronous application is transformed into an asynchronous one.
2. Many applications use blocking semantics for resource access, such as opening a file. As with interactive applications, these semantics may be changed to non-blocking to overcome large delays and periods of network disconnectivity. In this way, a system may continue to do useful work while waiting for a resource or response from a remote application.

Besides changing the semantics of interaction, the design of certain system elements and architecture will depend on the application type being supported. Real-time applications will require different algorithms than non-real-time applications. Likewise, differences in streaming vs. non-streaming (or transaction-types) of applications may lead to different system designs. For real-time applications, performance considerations are of the utmost importance. Real-time streaming applications cannot tolerate the introduction of processing delays or non-optimal routing; real-time transaction applications must receive low latency performance.

When designing new systems, including applications specifically for wireless environments, different constraints can be accommodated in different places of the system. An example is the messaging application which is described in Section 3. This is a non-real-time transaction application that was designed specifically for the wireless environment. As a result application layer protocols are designed to accommodate asymmetric bandwidth.

When attempting to provide wireless services without modifying applications that were designed to execute in a wireline environment, system constraints can only be addressed in protocol and algorithm design. An example is the Web access service described in Section 3. To provide this service, new elements were added to the network to overcome device and network constraints without requiring changes to Web servers.

### *4.2.3 Objectives*

Our main objective is to build a system that supports a wide range of data applications via a wireless network. This requires defining an architecture and network elements that may be inserted to provide reasonable service depending on the applications supported and the current operating environment. This includes various protocols and network elements that may be used to supplement simple end devices, or may be excluded when more sophisticated end devices are being used.

In addition, we provide multiple network interfaces so that users may access services residing on multiple types of networks, such as the Internet, private data networks, or telephone networks. We also support multiple air interfaces so that network access may be made by a variety of devices. Obviously, one design goal is to limit the impact of the different protocols on the core system elements.

The method we use to achieve these objectives will become clear as we discuss our basic architecture.

### *4.2.4 Basic Architecture*

In this subsection, we give a high-level description of the key architectural components of WDS. We defer the details to Section 3 when we describe the specific prototype we have built.

Figure 1 shows the basic architecture of WDS. On the left side is a generic end device, and on the right side is the network. The architecture follows that of a standard client-server approach, client on the end device side and server in the network side, except with two slight variations.

First, the client is simple and small in size, both relative to the server as well as in absolute terms. This is to accommodate for the asymmetry in processing power, battery capacity, and memory limitation between the end device and the network. The sophistication and capabilities of the end device are always going to lag that of the network. This type of client is called a "thin" client.

Second, to assist and supplement the simple end device, each end device has a corresponding *user agent* [9, 15, 16] resident inside the network to act on its behalf. A user agent provides proxy functionalities for an end device. Using user agents, the problem of wireless client-server communication can be logically broken up into two subproblems, namely, client-user agent and user agent-server communications. The former can be tackled by specialized protocols optimized for wireless access, while the latter is standard wireline communication. Therefore, to be accurate, WDS follows a "thin" client-user agent-server architecture.

The internal structure of the "thin" client also deserves special attention. At the lowest level is an operating system that provides basic management of system resources, which include, among others, the user input (e.g., keypad, pen input) and output (e.g., display) interfaces, and wireless channels. The specific combination of wireless channels available is dependent on the air interface used.

The applications are shielded from the operating system by a *middleware* layer, which serves two key purposes. First, it provides a uniform and platform-independent abstraction of the available system resources to the applications, thus allowing applications to be portable across different end devices. Second, it encapsulates the "thin" client-user agent communication protocol. This protocol should ideally be *asymmetric*, that is, protocol data units sent in the uplink direction are fewer and much smaller than those sent in the downlink direction.

On the network side, WDS makes use of *service gateways* to provide some of its value-added services. A service gateway is typically used to bridge specific wireline services for wireless access. For example, both email and Web access are provided using service gateways.

A service gateway is an application-level gateway. It understands the semantics of the service it is providing, and is specially adapted for wireless access. A service gateway preserves the "essence" of the bridged wireline service, while making it usable in a constrained wireless environment.

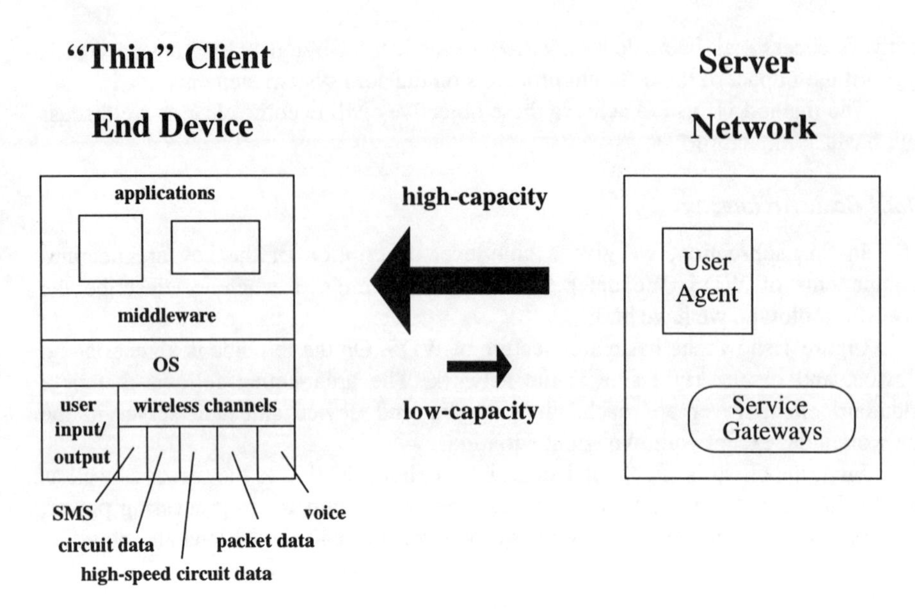


**Figure 1** Basic Architecture

## 4.3 WDS - CURRENT PROTOTYPE

In a practical deployment of WDS, its general design as described in the last section must be mapped to match specific operating constraints. In this section, we describe an instantiation of WDS we have implemented under the new digital cellular standard IS-136. Specifically, our prototype uses IS-136 SMS for wireless access, and prototype IS-136 phones as end devices.[1] IS-136 is a TDMA cellular air interface that supports two-way messaging or SMS over designated control channels.

Our initial focus on a cellular environment is deliberate. Cellular phones are by far the most popular wireless consumer device and because of their limited user input/output interfaces, they represent one of the most constrained end devices, hence a challenge, for WDS.

In the following subsections, we first describe the set of services available in our current prototype. Then we present its design and architecture, and finish with a discussion of its performance characteristics.

### *4.3.1 Services*

The choice of SMS for access and cellular phones as end devices directly impact the services that can be provided. Certain services, e.g., Web browsing, would be too awkward to be interesting. The high latency of SMS also suggests an asynchronous service model would be more appropriate.

The initial set of services we have chosen to support are two-way paging+ and selected Internet services. We elaborate on them below. Two-way paging+ provides wireless two-way messaging functions. Like its wireline counterpart, it is asynchronous in nature. Two-way paging+ also serves as a message transport layer

[1] Most IS-136 phones do not currently support mobile-originated user messages. We had to modify the existing phone software to port our "thin" client software onto the phones

upon which other value-added (e.g., Internet) services are built. This layered approach allows the value-added service to automatically inherit the asynchronous nature of messaging.

An earlier system that provides only two-way paging+ functionalities and from which WDS evolves was described in [20].

#### *4.3.1.1 Two-Way Paging+*

The basic service offered by WDS is two-way paging+. Basic two-way paging extends traditional one-way paging with message origination (new messages as well as replies) capability and reliable delivery (i.e., messages are acknowledged). WDS's two-way paging+ enhances basic two-way paging with the following features: (1) message query, the ability to query and retrieve status about messages sent; (2) multicast, the ability to address and deliver a message to multiple recipients with only one uplink message; (3) transactions, the ability of the system to actively tracks and correlate replies to their requests; (4) heterogeneity, the ability of the system to deliver a message in a variety of media formats (e.g., page, email, phone, fax) as proposed by the sender or specified in the recipient's profile.

Aside from the above, WDS's two-way paging+ is distinguished from basic two-way paging by the class of messages it transports. In most existing two-way paging services, e.g., Skytel [11], only fixed pre-canned or completely free-form messages can be sent. WDS makes use of a novel class of messages, called *flexible messages*, that is specially designed and optimized for two-way paging. In simple terms, flexible messages are message templates that contain variable components that can be dynamically customized on demand. It encompasses both fixed pre-canned and free-form messages, and is expressive for most practical applications, especially server-based ones. We describe flexible messages in greater details in Section 3.2.2.

#### *4.3.1.2 Internet Services*

WDS currently provides two Internet services, namely email and a restricted form of Web access. We provide a brief overview of them here, the details of their design and implementation is presented in Sections 3.2.2 and 3.2.4.

WDS email follows a client-server approach. Incoming mail are maintained on a mail server in the network, and the phones act as remote clients. The phones do not maintain a local mailbox, instead they cache mail messages from the server; this eliminates the problem of synchronization. This approach to email is similar in spirit to what has been proposed in IMAP4 [6].

WDS defines and uses a remote mail access protocol that is optimized for wireless access. The protocol supports the following features: (1) Asynchronous notification: when new mail arrives, notification is sent asynchronously to the phone. The notifications could be batched or conditional (e.g., sender-based or periodic). (2) Server-based filtering: filter/action pairs could be sent to the mail server to trigger mail processing on the server. (3) Selective retrieval: each piece of mail and its components are individually retrievable, thus allowing a user to receive and pay for exactly what he or she needs. (4) Incremental downloading: objects are always downloaded to the phone incrementally. This applies mostly to email bodies and components. All the above features are designed to put most of the processing load on the server and to minimize the amount of data transferred over the wireless link.

For Web access, the constraints of SMS and cellular phones preclude general browsing. Instead, WDS offers the ability to retrieve specific information from the Web. This function is provided using Internet service gateways. To retrieve data from the Web, a user selects and originates the request for a particular Web page from the phone. This request terminates at a Internet gateway, which translates it into an actual HTTP [3] request and forwards it on to the designated Web server. Upon receiving the response from the Web server, the Internet gateway filters out the irrelevant materials, formats the desired information for display on the phone, and returns it to the phone.

The Internet gateway contains a generic filtering engine. The specific filtering steps required for processing a specific Web page is pre-defined in *filter scripts*, which are written in a high-level language.

The key advantage of WDS's gateway approach is that it is transparent to the Web servers. Specifically, the Web servers as well as their information contents need not be modified or reformatted. Its main limitation is that a separate gateway filter script must be written to access each distinct Web pages, though this is somewhat alleviated by the ease of writing filter scripts, and by allowing filters to be installed by individual subscribers in their own user agents.

### *4.3.2 System Architecture*

Figure 2 shows the WDS system architecture. For ease of exposition, we break the system down into three logical parts: the Messaging Server - providing two-way paging+ functionalities, the Internet Gateways - providing email and Web access, and the User Agents - network-resident proxies for end devices. We describe them in greater details in the subsections below.

A key distinction of WDS from existing systems is its distributed and modular nature. Specifically, in WDS, system operations are separated into distinct functions that are implemented in individual servers, which communicate by message passing. In this way, a provider of wireless data services who would like to implement only a subset of WDS services or desires an incremental rollout of WDS can deploy only the components required for their desired level of service.

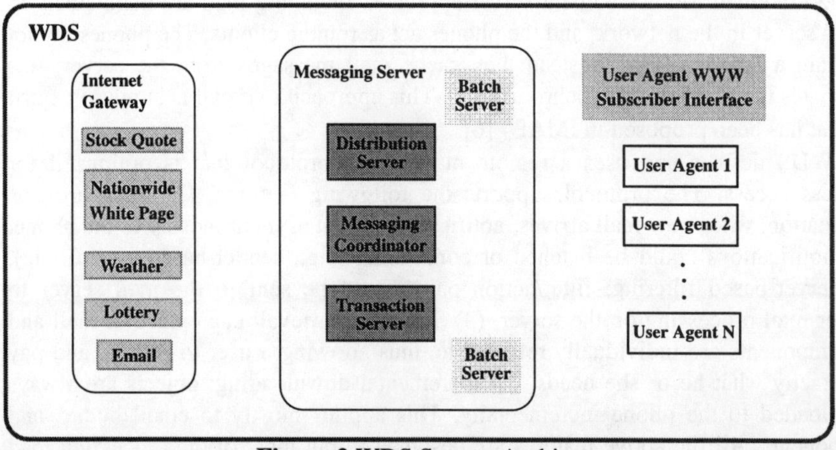


**Figure 2** WDS System Architecture

*4.3.2.1 Messaging Server*

The messaging server part is composed of three core servers: a messaging coordinator, a transaction server and a distribution server. The batch servers perform peripheral functions, and serve mainly as relays to the base stations. They forward messages to and receive messages from base stations. On the downlink direction, batch servers batch messages to support sleep mode operation of end devices.

The messaging coordinator, as its name suggests, coordinates the activities of the other servers. It receives originating messages (from the batch servers), coordinates with other servers (including the user agents) to determine the location and format in which the message should be delivered, invokes value added services (e.g., multicast, transactions), and finally routes the messages to the distribution server which can deliver them.

The transaction server is responsible for tracking transactions, which are request/response pairs. This involves correlating message requests, their replies and acknowledgments. The transaction server supports several transaction types (one-to-one for point-to-point messaging and one-to-many for multicast messaging), reports the status of transactions when requested, and closes transactions when complete.

The distribution server is responsible for routing messages to their final destinations in the proper format. As such, the distribution server supports the delivery of messages to multiple destinations (multicasting), and the delivery of messages to destinations in various formats (page, email, etc.).

To support multicast message delivery, the distribution server determines, from the location information provided to it by the user agents, how to most efficiently distribute messages. For example, if several destination end devices are currently in the same cluster, the distribution server will forward one copy of the message to the cluster along with a list of recipient addresses. If recipients are in different clusters, the distribution server will forward one copy of the message to each cluster.

To support heterogeneous message format delivery, the distribution server invokes different procedures to deliver messages in different formats. For example, if a message is to be delivered by telephone, the distribution server must first forward the message to a text-to-speech convertor, and then call the destination phone number to play the converted message. Convertors for email, fax, etc., are also available.

*4.3.2.2 User Agents*

User agents are introduced in the design of WDS for two key purposes: (1) to mitigate air interface and end device constraints; and (2) to enhance personalizability.

Regarding (1), a user agent allows reduced uplink bandwidth usage by supporting the transmission of coded messages on the uplink. This accounts for the asymmetric nature of wireless links. The compact coding is made possible by the use of flexible messages (see below). Second, a user agent serves as a repository for the information about its end device, e.g., its on/off and message buffer status, and thus can act as the termination point for wireline signaling protocols. This keeps end device processing and state simple, and reduces the amount of signaling over the air interface. Third, the user agent eases mobility management by maintaining and providing location information to the rest of the system. Fourth, a user agent, being network-resident, is always online, unlike its end device that can be powered off or go out of range. It allows the system to gracefully handle disconnected users. Finally, by performing intelligent processing in the user agent instead of on the end device, we account for the asymmetric nature of processing power between the network and the end device.

Regarding (2), a user agent can perform customized messaging functions, e.g., selective message forwarding and message screening. It does this by maintaining a profile for the subscriber. WDS user agents are also programmable, i.e., a user can potentially install new programs into their own user agents, thus making it a true personal server. This adds another dimension to the functionalities that may be provided.

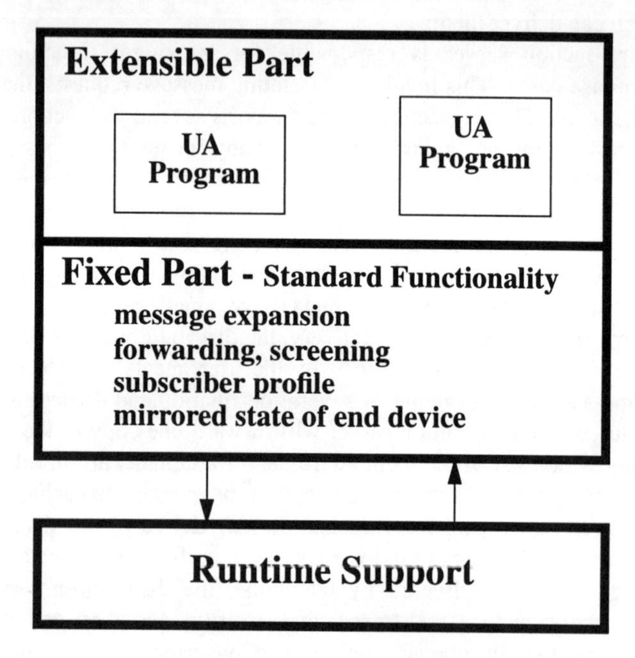


**Figure 3** User Agent Structure

In our design, a user agent has a fixed and an extensible part (see Figure 3). The fixed part implements the basic messaging functionalities discussed above. These functionalities are generic for all user agents. For its operation, this part mimics the context of the end device (e.g., the address table, the message table) and keeps information about ongoing message delivery. The extensible part, as the name suggests, can be programmed to perform specific tasks as desired by a subscriber. Some examples include maintaining a personal calendar, retrieving specific information from a Web page, etc.

The operation of the fixed part is straightforward; the extensible part, on the other hand, is much more complex. Strictly speaking, the extensible part specifies only a framework in which additional functions can be added to a user agent in the form of *UA programs*. A UA program is a collection of program blocks, each of which contains codes to handle messages of a specific pattern. The framework follows an event-driven model; it contains a kernel that pattern-matches incoming messages and dispatches them to the appropriate program blocks in UA programs.

A UA program can be used to provide third-party value-added functions to a message flow or it can serve as the destination end point of a message flow itself. An example of the former use is a UA program that monitors and oversees stock transactions for a subscriber. Specifically, it examines all order placement requests from a subscriber and ensures that sufficient funds are available before forwarding them on to their destination server.

An example of the latter use is message status query, where query messages sent uplink are addressed to and processed by user agents, possibly with the help of other servers in the system. It is in this regard that the extensible part bears special significance in relationship to signaling. Specifically, by addressing a message to the user agent itself, the extensible part could be used to implement direct user agent signaling functions. This turns an end device into a remote control for its user agent, and a user agent into a personal server for a subscriber.

The precise specification of UA program syntax and semantics is beyond the scope of this paper.

### *Flexible Messages*

In WDS, user messages are expressed in a novel class of messages, called *flexible messages*, which has been specially designed and optimized for use in wireless two-way messaging. The flexible messages are supported through cooperation between the user agents and end devices.

Flexible messages are a form of "active" messages. One way to understand them is as small programs that are interpreted when they are sent or replied to. Like programs, flexible messages are constructed from a number of well-defined basic building blocks. Each basic building block provides a distinct form of dynamic customization. They can be combined or mutually nested to achieve maximum flexibility.

Each basic building block has associated with it a unique processing logic and input convention. They have been designed such that their implementation is well suited to a device with limited input capabilities, and their combinations can meet the demand required for most personal and server-based applications. As we will see, the processing logic, as well as the input interface, required to process these building blocks are relatively simple.

**Table 1** Basic Building Blocks for Flexible Messages

<table>
<tr><th>Component</th><th>Use</th><th>Syntax in MDL</th><th>Example</th></tr>
<tr><td>Plain text</td><td>regular text</td><td>text</td><td>Care for lunch?</td></tr>
<tr><td>Rich text</td><td>highlight message parts</td><td>\< rich text specifier >{...}</td><td>Care for lunch \emphasize{now}?</td></tr>
<tr><td>Optional component</td><td>include or exclude message parts</td><td>\optional{...}</td><td>Care for lunch \optional{soon}?</td></tr>
<tr><td>User-defined selection</td><td>specifies a list of choices</td><td>\choice{...|...|...}</td><td>Care for \choice{lunch\dinner}?</td></tr>
<tr><td>Pre-defined variable</td><td>system defined selections</td><td>specific for each variable</td><td>Care for lunch at \ time?</td></tr>
<tr><td>If conditional</td><td>two-way branch</td><td>\if{expr}...\else ... \ fi</td><td>\if{$r==10}high\else low\fi</td></tr>
<tr><td>Case conditional</td><td>pattern branch</td><td>\case{expr}...\or...\or...\esac</td><td>\case{$c} OK \ or Probably Not \ esac</td></tr>
<tr><td>Reply component</td><td>delineate replies</td><td>\reply{...}</td><td>\reply{\choice{Yes | No}}</td></tr>
</table>

Table 1 gives a summary of the major building blocks. We omit a full description for brevity. Instead, we explain using two examples (see Figure 4).

```
Message M1:
Care for \ choice {lunch | dinner}?
\case{$c}
    \choice{McDonald's | Taco Bell}
\or
    \choice{Olivia Garden | Red Lobster}
\esac

Message M2:
Care for\choice{lunch | dinner}\optional{soon}?
\reply{
    \choice{Sure | I am busy}
    \case{$c}
        at\time
    \or
    \esac
}
```

**Figure 4** Flexible Message Examples

Consider the message *M*1 in Figure 4. It is composed of a plain text "Care for", followed by a user-defined selection "\ choice {lunch | dinner}, and then a case-conditional. The branch condition "$c" denotes the last choice selected; it evaluates to a value between 0 and $n$-1, where $n$ is the number of choices in the last selection. In the case "lunch" is selected, the first branch of the case will be taken, i.e., the user will be asked to select between the choices "McDonald's" and "Taco Bell." The case when "dinner" is selected is similar.

Consider message *M*2. A *reply component* is used to embed the possible replies along with the request. This reduces the reply step to that of processing the reply component, thus eliminating the need for free-form input. The embedded reply in this case is constructed with a choice component together with a case-conditional. The meaning of the message should be clear, we omit its explanation here.

All user messages are expressed in flexible message format. A flexible message is processed by an end device when it is being sent or replied to. The former is referred to as the *send-side* processing while the latter *receive-side* processing. Send-side processing is performed before a new message originates from a messaging device. It allows a sender to customize a request. Receive-side processing is performed when forming a reply. It lets a recipient answer each component of a request.

The actions taken in send-side and receive-side processing are similar. In both, each basic building block has associated with it an evaluation procedure that, when invoked, returns a value. For example, the evaluation of a plain text returns itself, while the evaluation of a user-defined selection returns one of the choices. The evaluation of a building block may trigger the evaluation of its component building blocks. In other words, evaluation of building blocks proceed from the outermost level in a recursive fashion.

As an example, the evaluation of message *M*2 in Figure 4 proceeds as follows: In the send-side processing, the request part of the message, i.e., "Care for\choice{lunch | dinner}\optional{soon}? is evaluated. The evaluation proceeds sequentially from the first component, the plain text block "Care for", then the next component, the user-defined selection "\choice{lunch | dinner}", then the next component, the optional component "\optional{soon}", and finally to the last component, the plain text "?". The values from each of the evaluation are then combined to form the final request.

This final request is transferred on the uplink in a so called *transfer encoding*, which highly compresses the message and relies on the user agent to decode and expand the message.

As described earlier, a user agent mirrors the state of its end device. In particular, they both store the same set of addresses and messages. Thus, to send a message uplink, all that needs to be transferred is: (1) a list of address aliases; (2) reference to the desired message; and (3) a modifier that encodes the customizations to be applied to the message. This is an application of the standard *differencing* technique, where the difference (encoded by the modifier in this case) from a base (the reference in this case) is sent instead of the value itself. The technique is most useful when the size of the difference is small compared to the true value; this is almost always the case in WDS.

Essentially, the modifier encodes the sequence of actions that needs to be taken (by the user agent) to recover the desired message from the stored message. The generation of the modifier is performed in an incremental fashion during receive-side processing.

A study of typical messages sent using WDS reveals that the average payload size of transfer encoded messages in only 3% of the unencoded message size. When mapped to the IS-136 SMS [19], this results in a bandwidth savings of approximately 65% [15].

The receive-side processing is similar, except that it acts only on the reply component and the components contained therein.

*4.3.2.3 Email Service Gateway*

As mentioned earlier, WDS email uses a client-server approach. The phone hosts a WDS email client, which serves as a remote mail reader, and the network hosts the mail server and the associated mailboxes.

The key protocol commands in the WDS email protocol is shown in Figure 5. They are best explained by describing the basic operation of WDS email. As new mail messages arrive, they are deposited in the server mailboxes. The phone client is informed about the new mail messages via a NOTIFY message, which contains only status information about the mailboxes. To retrieve mail, a two-step process is followed: headers are requested first using GETHEADER, then selected bodies are requested using GETBODY based on user choices. To be accurate, the email bodies themselves are downloaded incrementally in 256 byte chunks, this allows a user to read part of a message without transferring the whole message. This two-step procedure allows a user to selectively retrieve only the mail messages he or she desires. In fact, the GETHEADER request itself may optionally contain a keyword pattern that further reduce the headers downloaded to those that match the keyword. In addition, the headers are first simplified before downloading because the standard RFC822 [7] format is too verbose.

We make several observations about WDS email protocol design here. First, most commands include a `uid` as a parameter. A `uid` is a unique identifier for a message. Each new incoming mail message is assigned a distinct `uid`, which serves as a handle for the message for all client-server communications. A `uid` allows each side to refer to a message without exchanging the actual message. Second, the protocol supports a number of server-based operations, e.g., FORWARDMSG, FAXMSG, to reduce bandwidth usage. On wireline, these operations are typically carried out by the mail client itself. That is, a mail message being forwarded is first downloaded to the mail client, and then resent as a new message by the client. Third, an explicit delete request via DELETEMSG is required to purge a mail message from the server mailbox. This is because a WDS email client does not maintain its own mailbox. Rather, it only caches mail from the server.

**Client Requests:**

| | |
|---|---|
| GETHEADER: `[keyword]` | retrieve headers of unread mail, or mail that matches keyword `keyword` if present |
| GETBODY `uid` | retrieve body of message with UID `uid` |
| DELETEMSG `uid1 uid2` | delete messages with UID `uid1`, `uid2`, etc. |
| REPLYMSG `uid {1\|A}` "message body" | reply to the message with UID `uid`, 1 for reply to sender only, A for reply to sender as well as all recipients, "message body" is the reply text |
| FORWARDMSG `uid user1 user2...` | forward the message with UID `uid` to addresses `user1`, `user2`, etc. |
| FAXMSG `uid faxno1 faxno2...` | send the message with UID `uid` as FAX to phone numbers at `faxno1`, `faxno2`, etc. |
| SENDMSG `user1 user2...` "message body" | send a new message to addresses `user1`, `user2`, etc. with "message body" as message text |

**Server Responses**

| | |
|---|---|
| NOTIFY `n` | asynchronous notification, `n` is the number of new messages received |
| HEADER `uid1` "header1" texttuid2 "header2" ... | downloading headers for message with UID `uid1`, `uid2`, etc. |
| BODY `uid` "message body" | downloading body for message with UID `uid` |

**Figure 5** Protocol Commands for WDS Email

#### *4.3.2.4 Web Service Gateway*

The Web service gateway is essentially a protocol convertor. It converts between the "thin" client-gateway protocol, i.e., the WDS messaging protocol, and the gateway-Web server protocol, i.e., the standard HTTP protocol. As part of the protocol conversion function, it filters out irrelevant materials from the Web response, thus cutting down significantly on the bandwidth usage.

The Web service gateway has a filtering engine that "understands" HTML [2], the encoding format of Web documents. The operation of the filtering engine is driven by a filter script. The processing proceeds as follows. To access a particular Web page, a user originates an uplink message in the same manner as normal two-way paging+. This message terminates at the Web service gateway, which forms an actual HTTP request based on the parameters contained in the uplink message, and sends it to the designated Web server. Abstractly speaking, the uplink message is a lightweight equivalent of the actual HTTP request. When the Web service gateway receives a Web response, it is first parsed and stored in an internal representation. In this internal representation, each HTML object (e.g., headings, tables, rows) is tagged with an unique identifier. The filter script is then interpreted to extract out the desired components. In simple terms, the filter script is a specification of what are the objects of interest. All the materials not addressed in the filter script are discarded.

An example filter script is given in Figure 6. It is the actual filter script we use for extracting stock quote information from quote.yahoo.com. Terms such as <^tab3.tr2.td1> are unique identifiers for specific HTML objects in the Web page. For example, <^tab3.tr2.td1> refers to the 1st data entry (<td1>) in the 2nd row (<tr2>) of the 3rd table (<^tab3>) on the page. Both absolute (denoted by "^") and relative (denoted by ".") addressing are supported. The scripting language also supports looping, as shown in the term <^tab5.tr1-*>, which loops through all the rows in the 5th table, "tr1-*" specifies a range. A full specification of all the available scripting constructs is outside the scope of this paper.

### *4.3.3 An Example Service Scenario*

To tie all the pieces together, we outline here a basic service scenario of WDS. This scenario illustrates the operation of two-way paging+, which is also the foundation
for other services.

Consider the scenario whereby a sender *S* sends a message to a receiver *R*, who receives the message and replies back to *S* (see Figure7).

```
#  Filter Script Yahoo Stock Quotes Services
#  URL: http://quote.yahoo.com/quotes
#
#  <@label>: set access point at "label"
#  <^label>: access -- absolute
#  <.label>: access -- relative to the access point


#  extract the current price, etc
<^tab3.tr2.td1> " " <^tab3.tr2.td3> " " <^tab3.tr2.td4.para1.line1>
    " " <^tab3.tr2.td5.para1.line1> " " <^tab3.tr2.td6.para1.line1> <br>


# headline news
<^tab5.tr1-*> {
    "* " <.td5.para1.line1> <br>
}
```

**Figure 6** An Example Filter Script

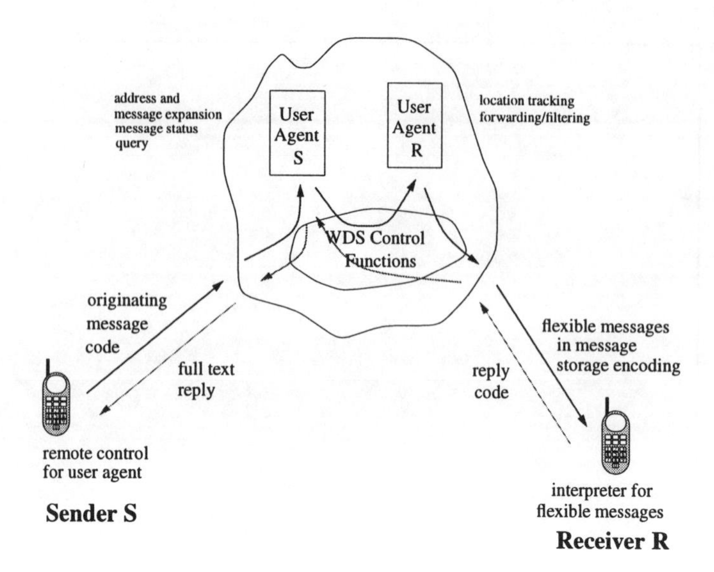


**Figure 7** basic Protocol Flow

1. **Message Origination** - To send a message, *S* first selects it from a set of pre-stored flexible messages. This initiates the send-side processing, which results in a transfer-coded uplink message.
2. **Sender User Agent Processing** - The uplink message is first routed to the user agent of *S*, which reconstructs the full text of the message from the transfer coding and its message and address tables.
3. **Receiver User Agent Processing** - The message is next routed to the user agent of *R*, which provides the location of *R* and determines how the message should be forwarded to *R* based on the profile of *R* it maintains.
4. **Message Reply** - Finally, the full message is delivered to *R*. The reply from *R* traces similar steps as the origination except it bypasses the user agent of *R* as the necessary information has been cached in the core part of WDS.

Both email and Web accesses follow the same basic scenario, except that the message is routed to the proper service gateway as opposed to a subscriber.

*4.3.4 System Performance*

A WDS prototype has been implemented and is operational at Bell Laboratories. The WDS servers are implemented using C++ as Unix processes on the SUN Solaris 2.X platform. A graphical protocol visualization tool has also been developed to provide a real-time view of all message exchanges between the servers.

Our end device is a prototype IS-136 cellular telephone handset. The wireless access protocol used is IS-136 SMS. We have also built a software simulated phone to work with the prototype.

The IS-136 phone uses a 8-bit Motorola microcontroller as its CPU, has 256K ROM (for code) and 8K EEPROM (for message and address tables, send and receive buffers, etc.) memory, and is controlled by an embedded message-passing operating system. Its user interface consists of a 3-line LCD with 12 characters per line, 10 scrollable soft keys and a standard keypad.

The prototype implements the three data services we describe earlier: two-way paging+, modified Web access, and email. For Web access, a number of filters have been written, including nationwide White Pages, stock quotes, weather and lottery information.

In the following, we evaluate our prototype in terms of two quantitative measures: code size of the "thin" client on the end device, and the amount of bandwidth used on the uplink to retrieve information.

#### *4.3.4.1 "Thin" Client*

One goal of our system design is to simplify and reduce the amount of code that is resident on the mobile device. The motivation is to reduce the memory and processing requirements on the end device which is limited in these areas as well as battery capacity. We simplify the client code by placing intelligence in the user agents and Internet gateways which are resident in the network.

The main software components added to the phone to support our applications are the user interface for flexible messaging, the message decoding/encoding functions, and memory management. Note that by using the two-way paging+ application and interface as the basis for all of the supported applications, application-specific code is minimized. Instead, when any application requests information from the network, such as email headers, email bodies, stock quotes, etc., the request is formed as a message and sent to the network using the same processing and protocols as in two-way paging+. Likewise, the response is received and displayed as a regular message in two-way paging+.

The total size of the code added to the phone, excluding the lower-layer messaging protocols which are part of the IS-136 standard, is 14 Kbytes. Approximately 65% of this code was for the user interface, 25% for the message encoding/decoding, and 10% for memory management. Each of the three IS-136 messaging protocol messages required an additional 1K of code. By contrast, a full Web browser is approximately 3 MB in size. In WDS, this code is resident on a network server. This demonstrates the successful distribution of application software between the network and the mobile device to drastically reduce the end device requirements.

#### *4.3.4.2 Bandwidth Usage*

A second system design goal is to reduce the amount of bandwidth required to provide services to the mobile device. This goal has two components: first, because the amount of bandwidth available over the air interface is often limited, the overall amount of information exchanged between the mobile device and network should be limited; and second, because the end device has limited battery capacity, and because bandwidth in the uplink direction is often lower than in the downlink, the uplink bandwidth usage should be particularly curtailed.

Below, we compare the bandwidth usage for the WDS Web access applications with the bandwidth usage for regular Web browsing on workstations in a wireline environment. We call this second system the standard system. The WDS system uses the protocol and network elements described above. The standard system uses a Netscape Navigator running on a Sun Workstation. We also present measurements of the WDS email application.

We compare the Web access applications of the systems with two measures: the amount of application data that is exchanged, and the number of messages that are exchanged. The first measure is an indication of the effectiveness of the gateway filters of WDS. The second measure is an indication of the differences between the protocols used by the systems. Our results show a dramatic decrease in bandwidth usage and the number of messages exchanged. Of course, the services provided by the systems are drastically different in that WDS allows a user to request only a specific piece of information while the standard system allows full Web browsing and email access.

*Modified Web Access*

Table 2 compares the amount of application data and number of messages exchanged in the uplink (from a workstation) and downlink (from a server) for the WDS (standard) system. A comparison of the absolute numbers may be misleading as the services provided are different. However, the comparison does highlight some interesting points.

**Table 2** Comparison of Web Access Overhead

| | WDS | | | | | | | |
|---|---|---|---|---|---|---|---|---|
| | Payload (bytes) | | Number of Messages | | Payload (bytes) | | Number of Messages | |
| Appica tion | Uplink | Downlink | Uplink | Downlink | Uplink | Downlink | Uplink | Downlink |
| Stock Quotes | 2.5 | 36 | 3 | 3 | 849 | 9,869 | 19 | 21 |
| White Pages | 12 | 80 | 3 | 3 | 2,387 | 15,957 | 47 | 48 |

The downlink payload differences are due to the filtering performed in WDS. The stock quote application, from yahoo.quote.com, included the download of 5 GIF images to the workstation in the standard system. The White Pages application, from www.switchboard.com, included the download of 13 GIF images to the workstation in the standard system. These images are filtered in WDS, therefore significantly lowering the downlink load.

The uplink payload differences are due to the fact that in WDS the actual fetching of the information from the Web is performed inside the network, and not from the mobile device. For that reason, the mobile device makes only a single short request for information, while the repeated HTTP requests are made from inside the network. Conversely, in the standard system, the White Page access incurred 8 HTTP request and the stock quote access incurred 3 HTTP requests.

The differences in the number of messages is due primarily to protocol differences. Because the WDS system completes all transactions with a single request-response from the mobile device, there are very few messages exchanged with the end device. There are some additional messages that are part of the IS-136 procedures which are reflected in Table 2. The standard system uses HTTP over TCP/IP, and therefore many acknowledgments and TCP connection-related messages are exchanged.

The conclusion from these results is that it is possible, using the filtering techniques and messaging semantics, to provide limited access to Web data using a small amount of communication resources. Of course, the service is limited with respect to full Web browsing, but a Web browsing application would require 200-300 times the communication resources.

*Email*

As described in Section 3, the semantics of the WDS email application differs significantly from standard email. In WDS, the email application has three main steps. First, a mobile device is asynchronously notified that there is email present. Second, the mobile can request a listing of email headers. Finally, the mobile can request the email body in 256 character segments. This technique allows the user to download a minimal amount of information (headers only) before making decisions on how to use their communication resources.

In our implementation, the notification that email has arrived requires 5 bytes of application data to be sent to the mobile device; no application data is sent in the uplink. To retrieve the list of headers requires a 22 byte payload to be sent uplink from the mobile device. The amount of data sent in the downlink depends on the number of headers in the mailbox. For a particular listing of 10 headers, 480 bytes of information were sent in the downlink. Finally, to request the email body, a 44 byte payload is sent uplink from the mobile station. Up to 256 bytes is sent on the downlink, depending on the size of the message being retrieved.

This illustrates how this application allows a user to judiciously use their bandwidth and retrieve only information they require.

## 4.4 CONCLUSION

As we have described, WDS is not a single server, but a distributed group of servers that collectively provides a tightly integrated set of wireless data services.

WDS is unique in many ways. From the service perspective, it offers a maximally useful set of services given the constraints of SMS and cellular phones. The use of flexible messages is particularly novel. From the design perspective, WDS embodies many techniques for mitigating the limitations of wireless channels. For example, its asymmetric protocol design matches the asymmetric nature of wireless links. Separately, user agents mirror the state of an end device and allow processing to be off loaded from an end device.

WDS is also extensible. It is independent of the physical layer air interface, and thus can easily support new kinds of wireless access protocols and end devices. New services can also be easily added with the use of service gateways.

To conclude, we offer two lessons we learn in the design of WDS. First, the usual service model for a particular service may have to be adapted for wireless. For example, both the email and Web access services in WDS follow an asynchronous model, as opposed to the usual synchronous model they use on the wireline. Second, application level awareness is crucial for achieving good performance under wireless. In other words, optimization at the transport level or lower alone is not sufficient. For example, WDS email provides selective message retrieval. This capability can only be provided at the application level as it requires understanding of service semantics and application data units.

## References


[1] E. Ayanoglu, K.H. Chen, S. Dar, N.H. Gehani, W.D. Roome, and K.K. Sabnani. COLUMBUS: Location-based services. In *Proceedings of AT&T Workshop on Wireless Communications and Mobile Computing,* Holmdel, New Jersey, October 3-4 1994.

[2] T. Berners-Lee and D. Connolly. *RFC 1866: Hypertext Markup Language-2.0.* Internet Network Information Center, November 1995.

[3] T. Berners-Lee, R. Fielding, and H. Nielsen. *RFC 1945: Hypertext Transfer Protocol-HTTP/I.O.* Internet Network Information Center, May 1996.

[4] CDPD Forum. *Cellular Digital Packet Data System Specification-Release 1.1*, January 19 1995.

[5] S. Cisler. Securing radio spectrum for wireless internet access. In *Proceedings of INET,* Montreal, Canada, June 24-28 1996.

[6] M. Crispin. RFC 2060: *Internet Message Access Protocol—Version 4 Rev 1.* University of Washington, December 1996.

[7] D. Crocker. *RFC 822: Standard for the format of ARPA Internet text messages*, August 13 1982.

[8] Federal Communications Commission. *Narrowband PCS Fact Sheet, 1997.* http://www.fcc.gov/wtb/nbfctsh.html.

[9] M.T. Le, F. Brughart, S. Seshan, and J. Rabaey. InfoNet: The networking infrastructure of InfoPad. In *Proceedings of Compcon,* pages 163-168, San Francisco, California, March 5-9 1995.

[10] B. Mangione-Smith. Low power communications protocols: paging and beyond. In *Proceedings of IEEE Symposium on Low Power Electronics,* pages 8-11, San Jose, California, October 9- 11 1995.

[11] Mobile Telecommunication Technologies Corporation. *Skytel 2-Way Technology Backgrounder, 1995.* Available from http://www.skytel.com/products/st2way.html.

[12] A.R. Modarressi and R.A. Skoog. Signaling system no. 7: A tutorial. *IEEE Communications Magazine,* 28(7):19-35, July 1990.

[13] Motorola Advanced Messaging Group. *The FLEX Story - FLEX Technology,* 1995. Available from http://www.mot.com/MIMS/MSPG/AMG/white papers/index.html.

[14] M. Mouly and M.B. Pautet. *The GSM System for Mobile Communications.* 1992.

[15] T.F. La Porta, R. Ramjee, T.Y.C. Woo, and K.K. Sabnani. Experiences with network-based user agents for mobile applications. *ACM/Baltzer Mobile Networks and Nomadic Applications.* accepted for publication.

[16] R. Ramjee, T.F. La Porta, and M. Veeraraghavan. The use of network-based migrating user agents for personal communications services. *IEEE Personal Communications Magazine,* 2(6), December 1995.

[17] Telecommunications Industry Association. TIA/EIA IS-657 Packet Data Service Options for Wideband Spread Spectrum Systems.

[18] Telecommunications Industry Association. *TIA/EIA IS-95 Mobile Station - Dual-Mode Wideband Spread Spectrum Cellular System.*

[19] Telecommunications Industry Association. *TIA/EIA IS-136 800 MHz TDMA Cellular-Radio Interface-Mobile Station -Base Station Compatability-Digital Control Channel*, December 1994.

[20] T.Y.C. Woo, T.F. La Porta, and K.K. Sabnani. Pigeon: A wireless two-way messaging system. In *Proceedings of IEEE Symposium on Personal, Indoor and Mobile Radio Communications,* pages 693-697, Taipei, Taiwan, ROC, October 15-1X 1996.

# 5 WIRELESS PACKET AND WIRELESS ATM SYSTEMS

Ender Ayanoglu

**Abstract**

The need for wireless packet communication systems is increasing in the local area and for access networks, as data and integrated services networking are becoming more widely available. Addressing this need poses significant problems in lower layers of the communications networking hierarchy. This paper addresses these issues and points to the technical problems that need to be solved in order to implement efficient wireless packet systems.

## 5.1 INTRODUCTION

During the last few years, the popularity of mobile cellular communications, the Internet, and laptop computing has resulted in development efforts to transmit data packets over the air, both for local area networks and for residential access. This development has the potential to enable efficient mobile computing, as well as to provide residential high speed data access capability without building new infrastructure. Although wireless packet radio technology was developed for the defense sector about a decade ago, the transmission efficiencies involved need to be significantly improved for a public service based on this technology. There exists an IEEE standard, 802.11, for wireless local area networks that operate at 2 Mbps. Another standard, HIPERLAN, was developed in ETSI for wireless local area networks operating at 20 Mbps (however, no prototype was built and there are no known product plans). Yet another standardization effort is ongoing at the ATM Forum and ETSI to develop ''wireless ATM.'' Cellular voice networks are beginning to support limited forms of data services. In addition, recently, a service offering for Internet access based on a proprietary transmission scheme has become popular in some cities and on some university campuses in the U.S. In spite of all this activity, indicative of the demand for wide-scale wireless packet transmission, it is widely recognized in the engineering community that technology development for packet data transmission over the air is not straightforward. Efficient transmission across unknown and time-varying channels in the burst mode requires significant complexity. Further, what is really needed today is the support of a variety of transmission formats: data, voice, and video, or integrated services. Efforts for networking standards to provide quality-of-service support for integrated services over a variety of networks are under way at various standardization bodies. The issue is more complicated for the wireless channel, mainly because of its highly noisy nature. This paper is a survey of the issues in wireless packet networking. The emphasis is on wireless ATM networking, although the general conclusions to be drawn are applicable to general wireless packet networks as well.

## 5.2 NETWORKING ISSUES

There are two networking alternatives that need to be taken into consideration in developing wireless packet systems: IP and ATM. IP encompasses a general suite of networking protocols that were developed for Internet data networking. IP packets have variable size, and can be from about 40 bytes to 65 kilobytes in size. The IP suite has transport layer protocols such as TCP or UDP that are based on variable-size packets of their own. IP is now being extended beyond data networking to incorporate support for real-time services. To this end, so far, two protocols have been developed: RTP and RSVP. RTP provides mechanisms for timestamping and control to different streams of data in the Internet. RSVP is a protocol for reservation of resources in the network by the receiver of an application. However, at this time, neither of these protocols can assure end-to-end real-time delivery. In-time delivery requires the support of lower layers that have control over resources in switches and routers, which is currently not available. ATM was developed with the purpose of providing high-speed integrated services. Its basic transmission unit is a 53-byte packet, known as an ATM cell. The ATM cell carries a 5-byte header and a 48-byte payload. The header has the address, type, and some flow control information. ATM cells were chosen to have fized size in order to simplify switching. Various high-speed ATM switch chips are becoming commercially available. These chips can be employed to build ATM switching systems. Because of their wide availability, they are also being used as the core of switching hardware for IP packets. At this point, it is unknown whether IP or ATM will be the dominant protocol in the future of data networking or integrated services networking. On the one hand, IP-based protocols are widely deployed and are an inevitable part of Internet services. On the other hand, switching favors ATM. Although it is possible to run IP protocols over ATM, this multiple layering of protocols is inefficient. Nevertheless, the convenience of supporting existing data networking software with widely available and scalable switching technology will likely overweigh the disadvantage due to multiple layering, and some form of IP-over-ATM is attractive for data and may prevail for integrated services networking.

There are three proposals for internet working using IP over ATM. One is developed by the Internet Engineering Task Force and the other two are developed by the ATM Forum. The first, known as IP over ATM (or Classical IP over ATM) was developed by the Internet Engineering Task Force. It employs a translation between IP and ATM addresses via an ATM address resolution protocol server in an IP network. An end station queries the ATM address resolution protocol server for an ATM address to the end station it desires to connect to. After receipt of this address, the end station itself uses ATM AAL 5 cell format to transmit to the destination. This solution is restricted to single IP subnetworks. It cannot serve other protocols, e.g., IPX. Most importantly, it cannot take advantage of ATM's quality-of-service advantages. The second proposal is known as LAN Emulation. It is similar to IP over ATM, except it emulates the Media Access Control sublayer of the Open Systems Interconnection model. Any network layer protocol that works with the OSI model (IP, IPX, NetBIOS, and DECnet) can work with that solution to travel over an ATM network without modification. However, this solution cannot take advantage of the quality-of-service advantages of ATM either. To overcome this disadvantage, the ATM Forum developed the Multiprotocol over ATM solution. This solution takes advantage of LAN emulation and builds upon it. Unlike the previous two solutions, it can route between different IP subnetworks and can make use of the quality-of-service advantages of ATM.

The impact of IP versus ATM in wireless packet networking mainly concerns the support of fixed- or variable-size packets. Handling fixed-size packets is simpler for a wireless system as well, however, the short size of ATM cells can result in a wireless system that has a high overhead due to synchronization and equalization. On the other hand, IP packets can also be very short (just a few bytes with PPP header compression), and therefore inefficiencies due to IP packet transmission over the air can be significant. Both multiple-access and quality-of-service considerations favor short fixed-size data transmission units over the air. Thus, it can be argued that ATM is better suited to implementing a wireless packet system.

ATM switching technology was originally designed for wide-area high-speed integrated services networking by the telecommunications industry. Various wide-area service providers have announced plans to introduce ATM switching to their backbone networks within the next decade for wide-scale data and integrated services networking. After its introduction as a wide-area networking technology, ATM was adopted for local area networking by the computer industry as an answer to high-speed multimedia networking needs. However, the development of switched ethernets at 100 Mbps and 1 Gbps slowed the introduction of ATM to the local area. With the introduction of new technologies such as ADSL and VDSL, or HFC and FTTC, which employ ATM for switching to bring high-speed networking to the home via the existing telephone or cable plant, ATM is likely to become a networking technology for access networking before is adoption in the wide area.

## 5.3 PHYSICAL LAYER ISSUES

The physical layer deals with the actual transmission of data over the physical medium by means of a radio or an optical transmitter/receiver pair. At this time, radios that operate in a burst and multiaccess mode at 20+ Mbps are not commercially available. However, many research projects are ongoing to build such radios. In this effort, the main challenge is overcoming multipath reception from stationary and moving objects. Multipath reception results in a space- and time-varying dispersive channel. This section will list a number of issues, together with main options, their fundamental limits, pertinent implementation issues, and conclusions reached with the available data. A more detailed survey can be found in [1].

*5.3.1 Infrared vs. Radio*

In general, communication systems based on infrared transmission can be less expensive to build than those based on radio transmission. The main reason for this is that infrared receivers are based on detection of the amplitude or position of optical signals, not the frequency or the phase. Therefore, infrared systems can be built by simple power detection. Since a significant part of the cost of a radio is in frequency conversion or quadrature demodulation, this results in an inexpensive solution. Infrared frequencies are not regulated, and therefore licensing is not an issue. Not having to go through the expensive licensing process is an added advantage to infrared, and so is the privacy advantages by keeping transmissions within a room with infrared.

It is possible to diffuse an infrared beam by means of a lens. Then, a system similar to radio in terms of an indoor coverage area can be obtained. The problem is, this system suffers from technical problems similar to those of radio, especially in terms of multipath and synchronization. In addition, providing multiple access with an infrared system is difficult. These systems also have shorter range than radio of the same rate. One place where such systems can find use is when electromagnetic interference is a problem. For example, in hospitals electromagnetic interference with sensitive medical equipment cannot be risked.

By using direct (or aimed) infrared, multipath problems can be avoided, and therefore, very high transmission rates are possible. There are several systems that have been demonstrated that operate at more than 100 Mbps. Such a system requires line-of-sight transmission and a high degree of accuracy in pointing the infrared beam. It is not always possible to obtain a line-of-sight system, especially in populated indoor and outdoor environments. Pointing the beam is nontrivial, and there may be a need for frequent realignment. Highly directed beams make multiaccess even more difficult. For outdoor applications, extreme intensity of solar radiation is a serious problem for infrared systems.

In summary, for high-speed indoor or outdoor wireless packet networks, radio is the preferred solution since it is not restricted to line-of-sight, it does not require pointing, and multiaccess is simpler to attain. Infrared is an option for high-speed networks with pointed links that essentially operate as a collection of point-to-point communication links, or for solving electromagnetic interference problems. At low speeds, diffused wireless LANs restricted to a single room are possible.

*5.3.2 Circuit-Switched vs. Packet-Switched Operation*

Although we are interested in building a wireless transmission system that will deliver a packet-switched system to the user, the transmission of packets over the air can be carried out either in circuit-switched or a packet-switched mode.

Circuit-switched transmission is attractive for modem design since the link is continuously active, data is always available, and timing and carrier recovery and channel equalization functions can be performed continuously. This mode of operation simplifies the significant operational problems present in the burst modem operation. For circuit-switched data, or continuous bit rate operation, this alternative is the natural solution. On the other hand, in a packet scenario, it provides an option to carry out packet segmentation and reassembly at either the mobile or the base station. Multiple access for this system requires a separate signaling channel dedicated to call setup and tear-down. For this reason, this solution is not well-suited to individual packet transmission.

Packet switching is the natural operation mode for variable bit rate operation. The potential need to transmit single packets requires large overhead due to equalization, carrier and timing recovery. The design of the radio to accommodate the use of data from previous packet transmissions to reduce this overhead is possible, but complicates the design. The need to accommodate different qualities of service makes packet-switched operation significantly more complex.

In summary, due to the presence of variable bit rate services, a pure circuit-switched system is not desirable for wireless packet transmission. The most desirable system is a packet-switched one with quality-of-service guarantees. The operational difficulties of providing any quality-of-service need makes a hybrid system a final alternative, by trading off a potentially more complicated implementation.

### *5.3.3 Operating Frequency*

A very important issue to consider when building a high-speed wireless packet system is the frequency of operation. There exist physical as well as regulatory limitations. The physical limitations are due to device limitations: Silicon elements do not operate beyond about 10 GHz, Gallium Arsenide (GaAs) operates but it is expensive. Another problem is propagation: due to the standard propagation formula, attenuation increases with the square of the operating frequency. This restricts the range. For less than 10 GHz, FCC allocations exist. The issue in that case is which band to use, and whether to operate in a licensed or an unlicensed band.

### *5.3.4 Licensed vs. Unlicensed Bands*

In the United States, the FCC has allocated certain spectral bands for uses potentially open for applications of wireless packet transmission. Some of these bands are designated unlicensed: the band can be used by certain equipment as long as it adheres to certain operational procedures. Two general sets of these bands are known as ISM (Industrial, Scientific, Medical) and PCS (Personal Communications Services). The remainder of the bands are licensed, operation in those bands is contingent upon licensing agreements with the FCC, which is difficult to obtain.

In 1985, three ISM bands were opened by the FCC to communications applications with the condition that they use spread spectrum techniques or very low power. The first, 915 MHz band is only available in North America, is highly crowded, and is expected to get even more crowded. Many existing users are non-spread-spectrum applications, which is a problem for spread spectrum communications. The second, the 2.4 GHz band is lightly loaded, but interference from microwave ovens is a problem. The third, the 5.8 GHz band is lightly loaded, there exists some radar interference, but has limited coverage (80\% with directional antenna) and is only available in North America. The requirement of spread spectrum makes ISM bands undesirable for Wireless ATM at rates of about 25 Mbps since the needed chip rates are prohibitively high.

Also, the FCC opened a band at 1.9 GHz for unlicensed PCS operations. This band is from 1850 MHz to 1990 MHz. The part between 1910 and 1920 MHz is allocated for unlicensed asynchronous or packet-switched applications, and that between 1920 and 1930 MHz is allocated for unlicensed synchronous or circuit-switched applications; the rest of the PCS band is allocated for licensed applications. In the U.S., use of this band requires the 'etiquette" (listen before transmit; if there are others, do not transmit). There is not enough bandwidth available in this band to transmit 25 Mbps. Recently, in response to a request by an industry organization known as WINForum, the FCC opened 300 MHz of bandwidth for unlicensed operation in a new band at 5 GHz. This is the most promising band in terms of wireless packet applications.

The licensed bands require FCC approval, which is a long and difficult process. The performance will be a premium in licensed bands since the band is not shared with other users. As such, the licensed band operation is preferable for outdoor applications, in particular for residential broadband delivery.

For experimental purposes, it is desirable to start with a system that operates in the unlicensed band, and carry the design over to the licensed band for premium applications. The decision has significant impact on the proposed system.

### *5.3.5 Spread Spectrum*

There are higher capacity arguments in the literature for using spread spectrum systems for telephony. Two types of spread spectrum techniques exist: direct sequence and frequency hopping. For a 10-Mbps system, with a direct sequence spread spectrum technique and processing gains of 10-100, chip rates of 100 Mbps to 1 Gbps are needed, which are difficult to achieve, and the system is therefore expensive. In addition, the near-far problem requires power control, which results in added complexity. Frequency hopping spread spectrum is implementable, and it helps solve the multipath problem; however, frequency hopping is difficult for data applications because of synchronization problems. Although it is true that there are currently several wireless LANs that employ spread spectrum techniques, they were designed to operate in the ISM bands, and therefore were required to use spread spectrum.

In summary, for high-speed wireless packet systems, spread spectrum is difficult at high bit rates and is not really needed. For low bit rate access applications, direct sequence spread spectrum may have a place due to potential capacity increase, and additionally, for voice applications, to provide soft handoffs. The capacity calculations in direct sequence spread spectrum are controversial, and depend on the basic assumptions. This subject is an open research topic. In general, it is not expected that high-speed wireless packet systems will employ spread spectrum due to the difficulty of achieving high bit rates.

### *5.3.6 Modulation*

Linear modulation techniques known as BPSK, QPSK, DPSK, and QAM are possible for the radio. Due to the high level of noise and the difficulty of equalization, multi-level constellations are difficult to achieve. The disadvantages of such a system are that a significant equalization effort is needed, and amplification is difficult. The amplification problem can be solved by constant amplitude modulation schemes, in particular by MSK and GMSK. A very desirable property of MSK and

GMSK is that although they have constant amplitude, they can be implemented using a quadrature-type architecture (similar to linear modulation techniques).
Another modulation scheme, known as Orthogonal Frequency Division Multiplexing (OFDM), has several desirable properties: it simplifies equalization considerably, it has graceful performance degradation, and because of the absence of equalization, lower complexity.

In summary, a radio that operates at about 25 Mbps using QPSK or GMSK is feasible. The OFDM alternative is attractive, but requires further research before wide-scale adoption. At this point, technologies based on TCM remain out of question for bit rates as high as 10 Mbps.

### *5.3.7 Equalization*

The Least Mean Square (LMS) algorithm is the most commonly used equalization algorithm because of its simplicity and stability. Its main disadvantage is its relatively slow convergence. For QPSK, LMS converges in about 25-250 bytes.

A faster equalization technique is known as Recursive Least Squares (RLS). There exist various versions of RLS with somewhat different complexity (computation and coding complexity) and convergence tradeoffs. RLS is more difficult to implement than LMS, but it converges in fewer number of symbols. In indoor propagation simulations, about 25 bytes have been observed to be sufficient for RLS to converge. In addition to increased complexity requirements, RLS has known stability problems. This makes LMS the first choice in implementing equalization. RLS for wireless applications is still a research topic.

### *5.3.8 Carrier and Timing Recovery*

Carrier and timing recovery add delay to processing, and therefore reduce the efficiency of a burst modem. Carrier recovery takes 5-10 μ at 2.4 and 5.2 GHz. At 20 Mbps, this is equal to about 12-25 bytes. Differential techniques reduce the need for accurate carriers; differential GMSK is attractive and is an open research topic. Timing recovery is less significant than carrier recovery; it can be accomplished by two-pass algorithms at the expense of additional delay.
Carrier and timing recovery add overhead in a burst modem. This overhead is less than needed for equalization, but it is still large. The reduction of this overhead is an open research topic.

### *5.3.9 Channel Coding*

There is no consensus in the literature on the feasibility of channel coding for wireless channels, especially for indoor applications. The main issue is the bursty nature of the wireless channel errors. The bursts are very long, and typically break the interleaving depths used for conventional physical layer coding systems. Large interleaving depths are more difficult to implement, and increase the end-to-end delay. The wireless channel, especially the indoor wireless channel, can be characterized as being bimodal: either it has no errors, or it has many errors. Under these circumstances, conventional physical layer error correction will not help. However, system performance improves by incorporating error correction in unconventional ways. In addition, physical layer coding has the potential to improve the range of the system, and may be a viable technique for the multi-rate system, with a variable coding rate.

### *5.3.10 Multi-Rate System*

A multi-rate system is one where high bit rates are employed at high signal-to-noise ratios, and when the signal-to-noise ratio is low, the system is switched to a low bit rate. Such a system is not difficult to implement provided that a highly reliable feedback channel exists. This system provides a communication link at low signal-to-noise ratios, which is a highly desirable feature.

### *5.3.11 Multiple Antennas*

Multiple antennas improve performance. Even using multiple antennas and choosing the one with the best signal-to-interference ratio provides a definite performance improvement. More complicated techniques, such as adaptive antenna arrays will improve the performance substantially.

### *5.3.12 Transmission Capacity*

In general, a radio that can approach a bandwidth efficiency of 1-2 bps/Hz with a single nondirectional antenna is considered feasible, while with multiple antennas, this figure increases linearly with the number of antennas.

## 5.4 DATA LINK LAYER ISSUES

### *5.4.1 Encapsulation*

Encapsulation is a technique used for transporting protocol data units of a protocol within those of another. In its simplest form, the protocol data units of the former protocol are placed within the headers and the trailers of the new protocol, which are stripped off at the end where the latter protocol is terminated. The advantage of this technique is the transparency it provides; the disadvantages are the added overhead and the encapsulation and decapsulation delays. Encapsulation and decapsulation delays can be minimized by using cut-through techniques (switch right after reading the header), and the added overhead can be minimized by reducing the unnecessary overhead to a minimum, for example, by using header compression.

### *5.4.2 Header Compression*

One of the issues in Wireless ATM is the 10% header overhead inherent in ATM. This overhead, if present in every ATM cell, causes a large degree of inefficiency, which is not actually needed. In media ATM is designed for, bandwidth is abundant, and the 10% overhead is not an issue. In the wireless medium, however, this is not tolerable. On the other hand, most of the time, the header a mobile host utilizes is fixed, or is one of a few alternatives. Thus, the information content of a header can easily be represented with a fewer number of bits than the 40 bits used in conventional ATM. There are several obvious possibilities for header compression; one is described in [1].

### *5.4.3 ARQ vs. FEC*

The best error control technique is a combination of ARQ and FEC. Although several hybrid ARQ/FEC techniques have been reported in the literature, particular characteristics of the wireless channel and the availability of the reverse and the forward channels under the multiaccess protocol change many of the boundary conditions of the problem, and new research results are needed. In general, data applications that are not delay sensitive can be transmitted with ARQ, while real-time applications such as video and audioare better transmitted with FEC.

### *5.4.4 Quality-of-Service Issues*

As stated above, it is difficult to satisfy quality-of-service requirements for delay-sensitive applications by using pure ARQ. In the absence of channel errors, quality-of-service issues can be taken care of by the multiaccess layer. When channel errors are present, however, the multiaccess layer has to work in conjunction with the data link layer to satisfy varying requirements. Satisfaction of all possible quality-of-service requirements under wireless channel noise in a multiaccess system is a tall order. A more realistic proposal is to divide the quality-of-service requirements into a few classes, with zero probability of loss at one end, and as small delay and delay variation as possible at the other. The former is served best with ARQ whereas the latter with FEC.

### *5.4.5 An Example Protocol*

An example of a reliable data link layer protocol to transport ATM cells over a wireless point-to-point link which ensures that the cells are transported reliably by combining a sliding window transport mechanism with selective repeat ARQ and FEC, which minimizes ATM header overhead over the air by means of header compression, which provides per-cell FEC whose size can be changed adaptively, and which provides parity cells for recovery from errors that cannot be corrected using the per-cell FEC field is described in [1].

## 5.5 MULTIACCESS LAYER ISSUES

Many MAC protocols have been proposed and studied during the past few decades. The protocols allow shared, wireless access by multiple, mobile users. In a wireless packet (ATM) network that supports an integrated mix of multimedia traffic, the MAC protocol needs to be designed such that mobiles share the limited communications bandwidth in an efficient manner: maximizing the utilization of the frequency spectrum and minimizing the delay experienced by mobiles. Also, to provide acceptable end-to-end ATM performance, it is important to define an efficient MAC protocol that can satisfy quality-of-service parameters such as cell delay variation and cell loss rate, and support various services such as constant-, variable-, available-, and unspecified-bit rate.

An efficient demand-assignment channel access protocol, called Distributed-Queueing Request Update Multiple Access (DQRUMA) is described in [1]. This protocol is for a cellular or microcellular wireless network and for fixed-length packet transmission. On the uplink, the protocol has a time-slotted system in which a Request-Access (*RA*) Channel and a Packet-Transmission (*Xmt*) Channel are formed on a slot-by-slot basis. The base station receives transmission requests (by listening to the uplink *RA* channel) from the mobiles and it updates the appropriate entries in a Request Table. It schedules uplink transmissions and transmits the schedule to the mobiles on the downlink channel. The base station also schedules and transmits the responses to mobile transmissions on the downlink channel. The mobiles transmit on the uplink *Xmt* channel based on the schedule from the base station. Along with their packet transmissions, mobiles can also piggyback (conflict-free) requests for additional transmission slots.

## 5.6 SUMMARY AND CONCLUSIONS

It is believed that ATM has inherent advantages in providing QoS guarantees and is a strong candidate for the transport of broadband integrated communications. High-speed ATM switching fabrics will soon be available in increasing capacities and decreasing prices. As ATM begins to become widely available, accessing ATM networks will become important. Among various access technologies, wireless access to ATM networks may satisfy unique needs quickly. However, wireless ATM has strong requirements from the lower layers of the communication hierarchy. In this paper we provided an overview of the need for wireless ATM and surveyed the requirements placed on the physical, data-link, and multiaccess layers by wireless ATM. We conclude that although wireless ATM is feasible, further research and development is needed to implement *efficient* wireless ATM systems.

## References

[1]. E. Ayanoglu, K. Y. Eng, M. J. Karol, "Wireless ATM: Limits, Challenges, and Proposals," *IEEE Personal Communications Magazine*, Vol. 3, pp. 18--34, August 1996.

# 6 OVERVIEW OF WIDEBAND CDMA

Donald L. Schilling


## Abstract

The need for wired-line quality voice and high speed data up to 2Mb/s is driving the international community to a third generation standard based on Wideband-CDMA (WCDMA), known as FPLMTS. FPLMTS is earmarked for the year 2000. In addition, the need for privacy and mobile fax or Internet hookups, as well as wired line voice quality, is driving cordless cellular telephones to third generation technology. Another application for W-CDMA technology is the wireless local loop ("WLL"). The WLL permits a long distance carrier to bypass the local service provider and thereby reduce consumer costs while increasing profits. However, for this to be a successful venture it is required that the wireless technology yield the same quality as the wired technology. None of the first or second generation technologies can meet that objective. Another application of WLL is providing wireless telecommunication service to customers throughout the world who, today, do not have such wired service. However, even in this case of "catch-up," the local telephone service providers demand the highest quality. The demand for high quality can be met today only through the use of third generation Wideband-CDMA systems. These and number of other applications of Wideband CDMA will be reviewed in the paper.


## 6.1 INTRODUCTION

Over the last twenty years a revolution in telecommunications technology has occurred. Cellular communications using analog FM Technology represented the first step in this revolution. In this system, the user wanting to place a telephone call asked the base station for a radio channel. Communication then took place. However, FM is an inefficient technology, since adjacent cells were forced to use different frequency bands to avoid interference. Indeed the spectrum efficiency of an analog FM cellular system is extremely poor, since only 15% of the allocated spectrum is used in each cell.

As the number of total users grew to significant numbers, access problems arose which resulted in second generation cellular, called Time Division Multiple Access ("TDMA"). The United States standard TDMA system was called IS54 and the

European standard, used throughout most of the world, was called GSM. Both of these systems are TDMA, and hence, digital systems. In these systems the user not only is given a frequency channel over which to transmit but also a time interval during which to burst his communication. In an IS54 system, three users share the same frequency channel by transmitting during time slots one, two or three. In a GSM system, used primarily outside the United States, eight users share a wider frequency band. GSM is a superior system to IS54; however, the system is still relatively inefficient.

Another second generation technology called narrowband-CDMA ("N-CDMA") received considerable attention in the early 1990's, but commercial service was delayed until recently due to performance problems and availability. As a result, GSM has become the de facto second generation world standard for cellular communications.

Code Division Multiple Access ("CDMA"), allows each user, in each cell, to transmit on the same frequency channel and at the same time. In order to separate messages, the data to be sent is given a digital address, in much the same way as we put an address on a letter before mailing, to insure that it reaches the desired party. In the case of CDMA, this address insures privacy and someone receiving a CDMA letter can read it only if he possesses the correct address. The procedure is similar to encryption. However, CDMA, a technology used by the military for over 40 years to avoid detection and interference, requires a wide bandwidth. Indeed, the wider the better. Thus, it is expected that N-CDMA, will provide an efficiency no greater than that of either of the two TDMA systems which have been in commercial operation since 1990.

Even though the TDMA technologies are becoming increasingly popular, the demand for wireless access is increasing at a pace which exceeds the available increase in capacity. As a result, governments throughout the world authorized the use of five to seven additional frequency bands to provide additional service. These systems are called Personal Communications Services (PCS) systems, and use the frequency bands near 2GHz, in contrast to cellular systems which use the frequency bands near 900MHz. Although commercial PCS operation has just begun in the United States, such systems are very popular in Europe and Asia. The technology being used is primarily GSM TDMA.

Both cellular and PCS systems concentrate mostly on satisfying the customer's need for voice communications and the first and second generation technologies are not able to transmit high speed data, such as ISDN or high quality video, without major changes in their design.

The need for wired-line quality voice and high speed data up to 2Mb/s is driving the international community to a third generation standard based on Wideband-CDMA (W-CDMA), known as FPLMTS. FPLMTS is earmarked for the year 2000. In addition, the need for privacy and mobile fax or Internet hookups, as well as wired line voice quality, is driving cordless cellular telephones to third generation technology.

Another application for W-CDMA technology is the wireless local loop ("WLL"). The WLL permits a long distance carrier to bypass the local service provider and thereby reduce consumer costs while increasing profits. However, for this to be a successful venture it is required that the wireless technology yield the same quality as the wired technology. None of the first or second generation technologies can meet that objective. Another application of WLL is providing wireless telecommunication service to customers throughout the world who, today, do not have such wired service. However, even in this case of "catch-up," the local telephone service providers demand the highest quality. The demand for high quality can be met today only through the use of third generation Wideband-CDMA systems.

In addition, W-CDMA can be used for packet switching in High Performance LAN (HIPERLAN) systems. These systems are characterized by the transmission of short packets of 10 Mb/s or more data. W-CDMA can accomplish such communication while providing frequency diversity and time diversity (RAKE) to protect against multipath interference.

## 6.2 TYPICAL W-CDMA SPECIFICATIONS

- Bandwidth: Up to 20Mhz
- Data Rates: ISDN-based, 64; 144; 384; 2,048 kb/s. Also 10Mb/s for HIPERLAN applications.
- Frame Length: 5-10 ms
- Pilot Code or Header: Most designs employ a header at the start of each frame. The header provides a reference for the bit stream following it. In addition, the use of the header permits acquisition of each frame and therefore is in consonance with a packet being transmitted.
- PN Sequences: Multirate codes, where the basic component is a Gold Code, are typically chosen.
- FEC Coding: Concatenated codes: Rate 1/3 or 1/2 convolutional codes with RS coding of rate 7/8 or 9/10. 10 ms interleaving is employed.
- Time Diversity (RAKE): RAKE is used in all systems. Correlator demodulators use either 1 search finger for selection diversity or 3-4 search fingers for maximal ratio combining. MF demodulators can, in theory, combine as many multipaths as there chips in a bit. However, typically the 4 largest multipaths are combined, using maximal ratio combining.
- Adaptive Power Control: Adaptive power control is used with minimum step size of 0.5dB-1dB. Power control is adjusted every 0.5-1 ms, depending on the system.
- Bit Error Rates: Designs vary from $10^{-3}$ to $10^{-5}$ for voice, $10^{-6}$ for data and $10^{-7}$ for video communications.
- Tolerable Doppler: 300Hz to 500Hz Doppler shifts are expected.
- Interference Cancellation: **A GOOD IDEA!**

## 6.3 IS THERE A CUSTOMER FOR W-CDMA?

The first W-CDMA system standardized was my B-CDMA system, standardized by the TIA and ITU. The leaders in this standardization effort were IDC and OKI. Unfortunately, when the effort was completed, interested service providers were already looking at newer designs.

There are three major service providers currently looking for W-CDMA systems that meet their designs. NTT-Docomo issued an RFP in October 1996 for a W-CDMA system. Several Japanese companies, Ericsson, Motorola, etc., are currently building this equipment. Daecom in Korea has also issued an RFP and Korean companies are currently building this equipment. In the USA, AT&T, the worlds largest service provider, has specified its requirements which includes HIPERLAN operation. The AT&T design is somewhat unique as it requires rapid acquisition of a header, which is necessary for packet reception. Thus, the AT&T system uses the GBT MF design.

Each of these service providers expects to begin commercial service before the year 2000.

## 6.4 CONCLUSION

Services, such as wired-line quality voice, ISDN, compressed video and multimedia service, all require the transmission of high data rates and their reception with a low probability of error. Only W-CDMA can accomplish this goal. It is, therefore, my opinion that wireless communication during the first part of the 21st century will be performed using W-CDMA.

# 7 IMPULSE RADIO

Robert A. Scholtz
Moe Z. Win

**Abstract**

Impulse radio, a form of ultra-wide band signaling, has properties that make it a viable candidate for short range communications in dense multipath environments. This paper describes the characteristics of impulse radio, gives analytical estimates of its multiple access capability, and presents propagation test results are their implications for the design of the radio receiver.

## 7.1. A RATIONALE FOR IMPULSE RADIO

Typically when a baseband pulse of duration $T_m$ seconds is transformed to the frequency domain, its energy is viewed as spanning a frequency band from d.c. up to roughly $2/T_m$ Hz. When this pulse is applied to an appropriately designed antenna, the pulse propagates with distortion. The antennas behave as a filters, and even in free space a differentiation of the pulse occurs as the wave radiates. Impulse radio communicates with pulses of very short duration. typically on the order of a nanosecond thereby spreading the energy of the radio signal very thinly from near d.c. to a few gigahertz.

Impulse radios, operating over the highly populated frequency range below a few gigahertz, must contend with a variety of interfering signals, and also must insure that they do not interfere with narrowband radio systems operating in dedicated bands. These requirements necessitate the use of spread-spectrum techniques. A simple means for spreading the spectrum of these ultra-wideband low-duty-cycle pulse trains is time hopping, with data modulation accomplished by additional pulse position modulation at the rate of many pulses per data bit. As a simple numerical example, suppose that an impulse radio operates over a 2 GHz bandwidth with a pulse rate of $10^6$ pulses per second at an average transmitted power level of 1 milliwatt and a data rate of $10^4$ bits per second with binary modulation. This radio then has a transmitted power spectral density of less than a microwatt per megahertz with a processing gain well over 50 dB.

Potentially there must be a real payoff in the use of impulse radio to undertake the difficult problem of coexistence with a myriad of other radio systems. For the bandwidth occupied, impulse radio operates in the lowest possible frequency range, and hence has the best chance of penetrating materials with the least attenuation. This capability, when combined with multipath resolution down to a nanosecond in differential path delay (equivalently down to a differential path length of one foot) make impulse techniques a viable technology for high quality fully mobile indoor radio systems. Lack of significant multipath fading may considerably reduce fading margins in link budgets and allow low transmission power operation. With its significant bandwidth, an impulse radio based multiple access system may accommodate many users, even in multipath environments. Monolithic implementation appears possible, implying that an impulse radio may be manufactured inexpensively.

The same qualities that make this radio attractive also provide the design challenges. Regulatory considerations over a such a wide band will limit the radiated power, ultra-fine time resolution will increase sync acquisition times and may require additional correlators to capture adequate signal energy, full mobility will exacerbate power control needs in multiple-access networks, etc. Impulse radios have been implemented and demonstrated single-user links up to at least 150 kbps, and hence the basic principles of operation have been validated. In the following review of impulse radio, we will survey our work on this novel technology.

## 7.2 MULTIPLE ACCESS TECHNIQUES FOR IMPULSE RADIO

### *7.2.1. Time-Hopping Format Using Impulses*

A typical time-hopping format employed by an impulse radio in which the $k^{\underline{\text{th}}}$ transmitter's output signal $s_{\text{tr}}^{(k)}\left(t^{(k)}\right)$ is given by

$$s_{\text{tr}}^{(k)}\left(t^{(k)}\right)=\sum_{j=-\infty}^{\infty} w_{tr}(t^{(k)}-jT_f-c_j^{(k)}T_c-\delta d_{\lfloor j/N_s \rfloor}^{(k)}, \tag{1}$$

where $t^{(k)}$ is the $k^{\underline{th}}$ transmitter's clock time, and $w_{tr}(t)$ represents the transmitted pulse waveform, referred to as a *monocycle,* that nominally begins at time zero on the transmitter's clock.

The frame *time* or *pulse repetition time* $T_f$ typically may be a hundred to a thousand times the monocycle width, resulting a signal with a very low duty cycle. To eliminate catastrophic collisions in multiple accessing, each user (indexed by *k*) is assigned a distinctive time-shift pattern $\left\{c_j^{(k)}\right\}$, called a *time-hopping sequence,* which provides an additional time shift to each monocycle in the pulse train. The $j^{\underline{\text{th}}}$ monocycle undergoes an additional shift of $\left\{c_j^{(k)}\right\}T_c$ seconds, where $T_c$ is the duration of addressable time delay bin. The elements $c_j^{(k)}$ of the sequence are chosen from a finite set {0,1, . . ., $N_h$-1}, and hence hop-time shifts from 0 to $N_hT_c$ are possible. The addressable time-hopping duration is strictly less than the frame time since a short time interval is required to read the output of a monocycle correlator and to reset the correlator.

For performance prediction purposes, the data sequence $\left\{d_j^{(k)}\right\}_{j=-\infty}^{\infty}$ is a modeled as a widesense stationary random process composed of equally likely binary symbols. A pulse position data modulation is considered here in which it is assumed that the data stream is balanced so that the clock tracking loop S-curve can maintain a stable tracking point. With more complicated schemes, pulse shift balance can be achieved in each symbol time. The parameter $\delta$ is a modulation factor that can can be chosen to optimize performance. If $\delta > T_{\mathrm{m}}$, then the transmitted signals representing 0 and 1 are orthogonal.

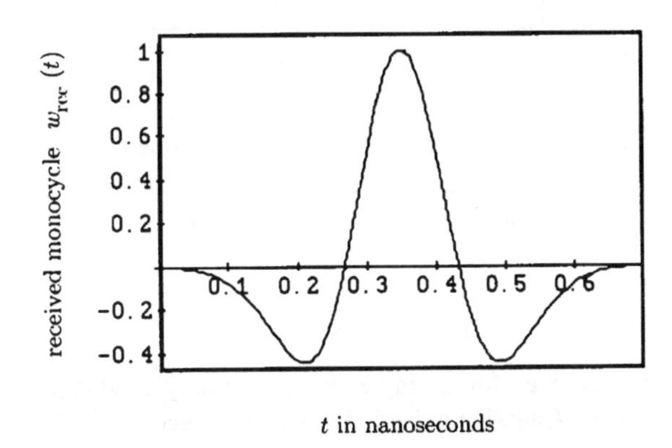


**Fig. 1**. A typical received monocycle $w_{\mathrm{rec}}(t)$ at the output of the antenna subsystem as a function of time in nanoseconds.

### *7.2.2 The Multiple Access Channel*

When $N_{\mathrm{u}}$ users are active in the multiple-access system, the composite received signal $r(t)$ at the output of this receiver's antenna is modeled as

$$r(t) = \sum_{k=1}^{N_{\mathrm{u}}} A_k s_{\mathrm{rec}}^{(k)}(t-\tau_k) + n(t) \,, \tag{2}$$

in which $A_k$ models the attenuation over the propagation path of the signal, $s_{\mathrm{rec}}^{(k)}\left(t-\tau_k\right)$, received from the $k^{\text{th}}$ transmitter. The random variable $\tau_k$ represents the time asynchronism between the clock of transmitter $k$ and the receiver, and $n(t)$ represents other non-monocycle interference (e.g., receiver noise) present at the correlator input.

The number of transmitters $N_{\mathrm{u}}$ on the air and the signal amplitudes $A_k$ are assumed to be constant during the data symbol interval. The propagation of the signals from each transmitter to the receiver is assumed to be ideal, each signal undergoing only a constant attenuation and delay. The antenna/propagation system modifies the shape of the transmitted monocycle $w_{\mathrm{tr}}(t)$ to $w_{\mathrm{rec}}(t)$ at its output. An idealized received monocycle shape $w_{\mathrm{rec}}(t)$ for a free-space channel model is shown in Fig. 1. This channel model ignores multipath, dispersive effects, etc.

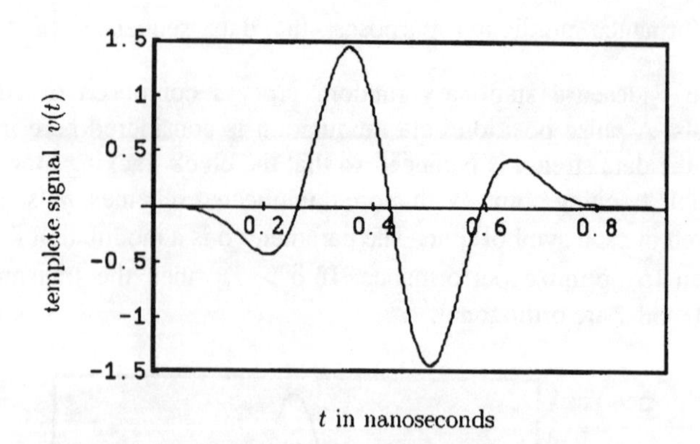


**Fig. 2.** The template signal $v(t)$ with the modulation parameter $\delta$ chosen to be 0.156 ns. Since the template is a difference of two pulses shifted by $\delta$, the non-zero extent of the template signal is approximately $\delta$ plus the monocycle width, i.e., about 0.86 ns.

### *7.2.3. Impulse Radio Receiver Signal Processing*

The optimum receiver for a single bit of a binary modulated impulse radio signal in additive white Gaussian noise is a correlation receiver [1],[2],[3], which can be reduced to

$$\text{"decide } d_0^{(1)} = 0\text{"} \quad \Leftrightarrow \quad \underbrace{\sum_{j=0}^{N_s-1} \overbrace{\int_{\tau_1+jT_f}^{\tau_1+(j+1)T_f} r(t)v(t-\tau_1 jT_f - c_j^{(1)}T_c)dt}^{\text{pulse correlator output} \,\hat{=}\, \alpha_j}}_{\text{test statistic} \,\hat{=}\, \alpha} > 0, \tag{3}$$

where the *correlation template signal* is $v(t) \hat{=} w_{rec}(t) - w_{rec}(t - \delta)$.

The optimal detection in a multi-user environment, with knowledge of all time-hopping sequences, leads to complex receiver designs [4], [5]. However, if the number of users is large and no such multi-user detector is feasible, then it is reasonable to approximate the combined effect of the other users' dehopped interfering signals as a Gaussian random process [1], [6]. Hence the single-link reception algorithm (3) is used here as a theoretically tractable receiver model, amenable as well to practical implementations.

The test statistic $\alpha$ in (3) consists of summing the $N_s$ correlations $\alpha_j$ of the correlator's template signal $v(t)$ at various time shifts with the received signal $r(t)$. The signal processing corresponding to this decision rule in (3) is shown in Fig. 3. A graph of the template signal is shown in Fig. 2 using the typical received waveform given in Fig. 1.

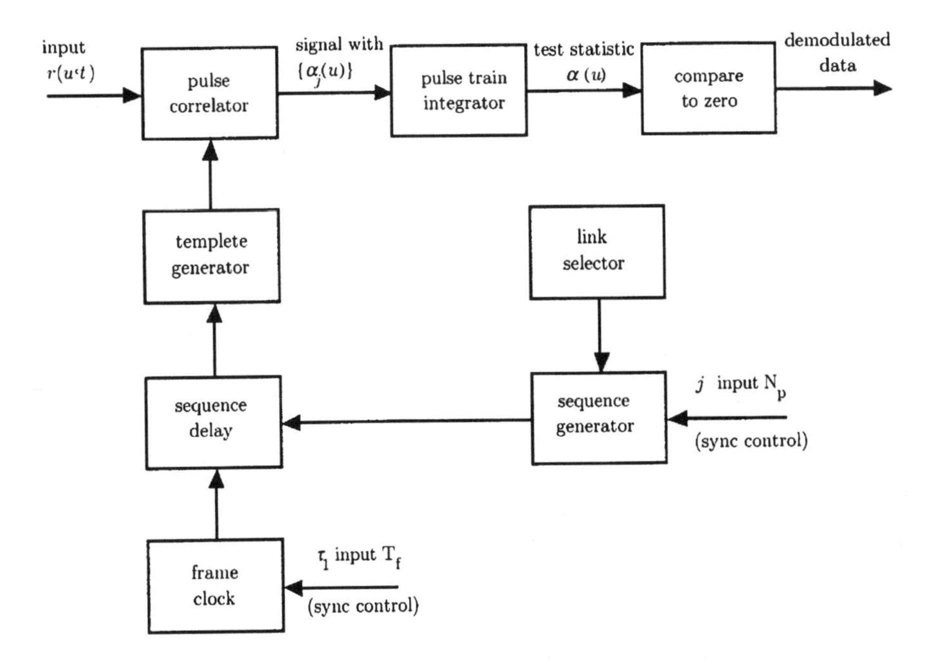


**Fig. 3.** Receiver block diagram for the reception of the first user's signal.

*7.2.4 Multiple Access Performance*

Using the approach of [7], the average output signal-to-noise ratio of the impulse radio is calculated in [1] for randomly selected time-hopping sequences as a function of the number of active users $N_u$ as

$$SNR_{out}(N_u) = \frac{(N_s A_1 m_p)^2}{\sigma_{rec}^2 + N_s \sigma_a^2 \sum_{k=2}^{N_u} A_k^2} \tag{4}$$

Here $\sigma_{rec}^2$ is the variance of the receiver noise component at the pulse train integrator output. The monocycle waveform-dependent parameters $m_p$ and $\sigma_a^2$ in (4) are given by

$$m_p = \int_{-\infty}^{\infty} w_{rec}(x-\delta)v(x)dx, \qquad \text{and}$$

$$\sigma_a^2 = T_f^{-1} \int_{-\infty}^{\infty} \left[ \int_{-\infty}^{\infty} w_{rec}(x-s)v(x)dx \right]^2 ds,$$

respectively.
The $SNR_{out}$ $(N_u)$ of the impulse radio can be rewritten as

$$SNR_{out}(N_u) = \left\{ SNR_{out}^{-1}(1) + M \sum_{k=2}^{N_u} \left( \frac{A_k}{A_1} \right)^2 \right\}^{-1}, \tag{5}$$

where the parameter $M$ is given by

$$M^{-1} \triangleq \frac{N_s m_p^2}{\sigma_a^2} \tag{6}$$

Let's suppose that a specified signal-to-noise ratio $SNR_{spec}$ must be maintained for the link to satisfy a performance specification. If this specification is to be met when $N_u$ - 1 other users are active, then it follows that $SNR_{out}(1)$ in (5) represents the required equivalent single link signal-to-noise ratio (ignoring multiple access noise) such that $SNR_{out}(N_u)= SNR_{spec}$. Therefore the ratio of $SNR_{out}(1)$ to $SNR_{out}(N_u)= SNR_{spec}$ represents the fractional increase in every transmitter's power that is required to maintain its signal-to-noise ratio at a level $SNR_{spec}$ in its receiver in the presence of multiple-access interference caused by $N_u$ -1 other users. We define the fractional increase in required power (in units of dB) as $\Delta P \triangleq 10\log_{10}\left\{SNR_{out}(1) / SNR_{spec}\right\}$.

Under the assumption of perfect power control, the number of users that the multiple access impulse radio system can support on an aggregate additive white Gaussian noise channel for a given data rate is shown in [8] to be

$$N_u(\Delta P) = \left\lfloor M^{-1} SNR_{spec}^{-1}\left\{1 - 10^{-(\Delta P/10)}\right\}\right\rfloor + 1 \tag{7}$$

which is a monotonically increasing function of $\Delta P$. Therefore

$$\begin{aligned} N_u(\Delta P) &\leq \lim_{\Delta P \to \infty} N_u(\Delta P) \\ &= \left\lfloor M^{-1} SNR_{spec}^{-1} \right\rfloor + 1 \triangleq N_{max}. \end{aligned} \tag{8}$$

Hence the number of users at a specified bit-error rate (BER) based on $SNR_{spec}$ cannot be larger than $N_{max}$, no matter how large the power of each user's signal is. In other words, when the number of active users is more than $N_{max}$, then the receiver can not maintain the specified level of performance regardless of the additional available power. Similar results for direct sequence code division multiple-access systems can be found in [7].

### *7.2.5 A Performance Evaluation Example*

The performance of the impulse radio receiver in a multiple-access environment is evaluated using a specific example. The duration of a single symbol used in this example is $T_s = N_s T_f$. For a fixed frame (pulse repetition) time $T_f$, the *symbol rate* $R_s$ determines the number $N_s$ of monocycles that are modulated by a given binary symbol via the equation $R_s = \frac{1}{T_s} = \frac{1}{N_s T_f}\,\text{sec}^{-1}$.

The modulation parameter $\delta$ in (1), which affects the shape of the template signal $v(t)$, affects performance only through $m_p$ and $\sigma_a^2$ implicitly, and can be adjusted to maximize $SNR_{out}(N_u)$ under various conditions. When the receiver noise dominates the multiple-access noise, e.g., when there is only one user or when there is a strong external interferer, then it can be shown that the optimum choice of modulation parameter is the one that maximizes $|m_p|$. On the other hand, when the receiver noise is negligible and $SNR_{out}(1)$ is nearly infinite, then the optimum choice of δ, suggested by (4), is the one that maximizes $|m_p|/\sigma_a$. For the monocycle waveform of Fig. 1 which we will use in this example, these considerations imply that δ should be chosen as either 0.144 ns or 0.156 ns, and little is lost in choosing either of these values. Choosing δ = 0.156 ns and $T_f$ = 100 ns, then $m_p$ = -0.1746, $\sigma_a^2 = 0.006045$, and the unitless constant that is required for calculating $M^{-1}$ in (6) is $m_p^2/\sigma_a^2 \approx 504$. With a data rate $R_s$ = 19.2 kbps, the quantity $M^{-1}$ is calculated to be $2.63 \times 10^5$.

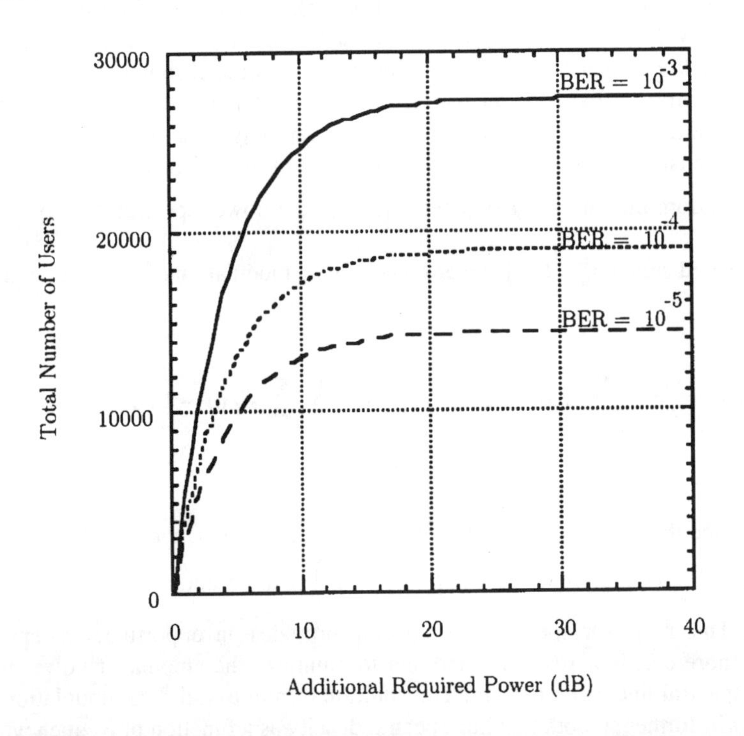


**Fig. 4**. Total number of users versus additional required power (dB) for the impulse radio example. Ideal power control is assumed at the receiver. Three different bit error rate performance levels with the data rate set at 19.2 Kbps are considered.

The number of users versus additional required power $\Delta P$ for multiple access operation with ideal power control is plotted for typical BERs in Fig. 4 for this example. To maintain BER of $10^{-3}$, $10^{-4}$, and $10^{-5}$ in a communications system with no error control coding, $SNR_{spec}$ must be 12.8 dB, 14.4 dB, and 15.6 dB respectively. Note that the number of users increases rapidly as $\Delta P$ increases from 0 to 10 dB. However, this improvement becomes gradual as $\Delta P$ increases from 10 to 20 dB. Beyond this point, only negligible improvement can be made as $\Delta P$ increases and $N_u$

approaches $N_{max}$. In practice, impulse radios are expected to operate in regions where the increase in the number of users as a function of $\Delta P$ is rapid. The values of $N_{max}$ is calculated to be 27488, 19017, and 14426 for BERs of $10^{-3}$, $10^{-4}$, and $10^{-5}$ respectively, and these are the asymptotic values on the curves in Fig. 4. It is worth noting that if a direct sequence CDMA system with roughly the same bandwidth were analyzed, one would find comparable numbers of users in the same communication environment.

### *7.2.6. Comments on Sequence Design*

The above performance evaluation is based on average results for randomly selected timehopping sequence designs. In reality some sort of pseudonoise generator must provide transmitter and receiver with, previously agreed upon, time-hopping sequences for each communication link. Techniques for providing sets of sequences with good Hamming correlation are well known [9] and may be adapted to the time-hopping application to provide quasi-orthogonal signaling schemes.

The ability of the receiver to reject narrowband interference and the ability of the transmitter to avoid interfering with other radio systems depends on the power spectral density of the time-hopped monocycle pulse trains. For a given periodic pseudorandom time-hopping sequence $\left\{c_j^{(k)}\right\}$, the power spectral density $S_{tr}(f)$ of a time-hopped signal $s_{tr}^{(k)}\left(t^{(k)}\right)$ in a absence of data modulation of can be computed as

$$S_{tr}(f) = \frac{|W(f)|^2}{T_p^2} \underbrace{\left| \sum_{n=0}^{N_p-1} \exp\left\{-j2\pi f(nT_f + c_n T_c\right\} \right|^2}_{\triangleq C(f)} \sum_{k=-\infty}^{\infty} \delta_D(f - k/T_p). \tag{9}$$

Notice that the delta functions which compose the line spectral density are now separated by the reciprocal of one period $\left(\frac{1}{T_p}\right)$ of the pseudorandomly time-hopped signal. This narrower spectral line spacing provides an opportunity to spread the power more evenly across the band and to minimize the amount of power that any single spectral line can represent. The addition of non-trivial data modulation on the signal will further smooth this line spectral density as a function of frequency.

The envelope of the lines in the spectral density has two frequency-dependent factors, namely $|W(f)|^2$ and $C(f)$, the latter being time-hopping sequence dependent. Note that when $T_f$ is a multiple of $T_c$, $C(f)$ is periodic in $f$ with period $1/T_c$, so attempts to influence one portion of the frequency spectrum by sequence design will have an affect on another portion of the spectrum. There may be an opportunity to make $C(f)$ better than approximately flat as a function of frequency, e.g., make $C(f) \approx 1/|W(f)|^2$ over a specified interval.

There may be some lines in the power spectral density which cannot be reduced by clever design of time-hopping sequence. For example, suppose that $T_f/T_C = m'/n'$, where $m'$ and $n'$ are relatively prime integers. Then $C(f)= N_p^2$ for all frequencies $f$ that are integer multiples of $n'/T_c$, and lines exist in $S_{tr}(f)$ at these frequencies. The heights of these spectral lines are independent of the time-hopping sequence and can only be influenced by the energy spectrum $|W(f)|^2$ of the monocycle waveform.

## 7.3 DESIGN FOR INDOOR MULTIPATH

Analysis for the aggregate additive white Gaussian noise channel is a reasonable way to obtain an initial understanding of how a communication link works, but does not accurately predict performance in a complicated propagation environment. We will now describe ultra-wideband propagation measurements of indoor communication channels, and their use in impulse radio design.

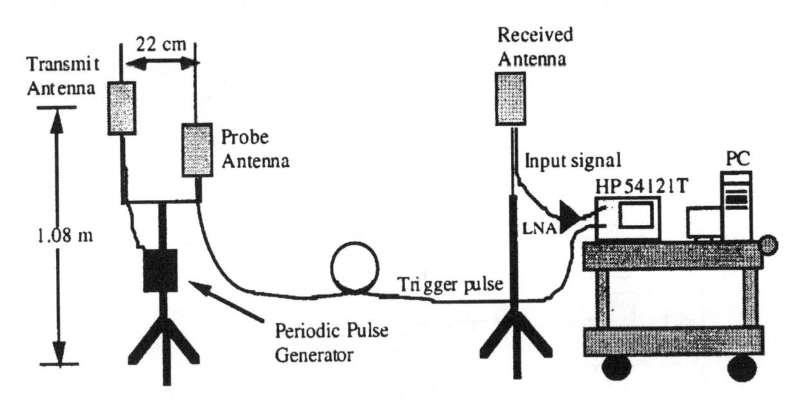


**Fig. 5**. A block diagram of the measurement apparatus.

### *7.3.1 A. Propagation Measurement Technique*

The measurement technique employed here is to probe the channel periodically with a sub-nanosecond pulse and to record the response of the channel with a digital sampling oscilloscope (DSO). Path resolution is possible down to about 1 ns of differential delay, corresponding to about one foot differential path length, without special processing. The repetition rate of the pulses is $2 \times 10^6$ pulses per second, implying that multipath spreads up to 0.5 ns can be observed unambiguously.

A diagram of the measurement apparatus is shown in Fig. 5. One of the three ultra-wideband antennas is set in close proximity to the transmit antenna to supply a trigger signal to the DSO over a long fixed length coaxial cable. Therefore, all recorded multipath profiles have the same absolute delay reference, and time delay measurements of the signals arriving to the receiving antenna via different propagation paths can be made. During each of the multipath profile measurements, both the transmitter and receiver are kept stationary.

### *7.3.2 Measurement Results*

Propagation measurements were made on one floor of a modern laboratory/office building having the floor plan shown in Fig. 6. Each of the rooms is labeled alphanumerically. Walls around offices are framed with metal studs and covered with plaster board. The wall around the laboratory is made from acoustically silenced heavy cement block. There are steel core support pillars throughout the building, notably along the outside wall and two within the laboratory itself. The shield room's walls and door are metallic. The transmitter is kept stationary in the central location of the building near a computer server in a laboratory denoted by F. The transmit antenna is located 165 cm from the floor and 105 cm from the ceiling.

Figure 7 shows the transmitted pulses measured by the receive antenna, located 1 m away from the transmit antenna at the same height. Measurements were made while the vertically polarized receive antenna is rotated about its vertical axis in 45° steps (see Fig. 7). The antenna, which is flat and roughly the size of a playing card, displays nearly circularly symmetric patterns about its vertical axis. The building layout diagram of Fig. 6 indicates that the closest object to the measurement apparatus is the south wall of laboratory F, which is at least 1 meter away. The signal arriving at the receiving antenna, except for the line-of-sight (LOS) signal, must travel a millimum distance of 3 meters. The initial multipaths come from the floor and ceiling, 5.2 ns

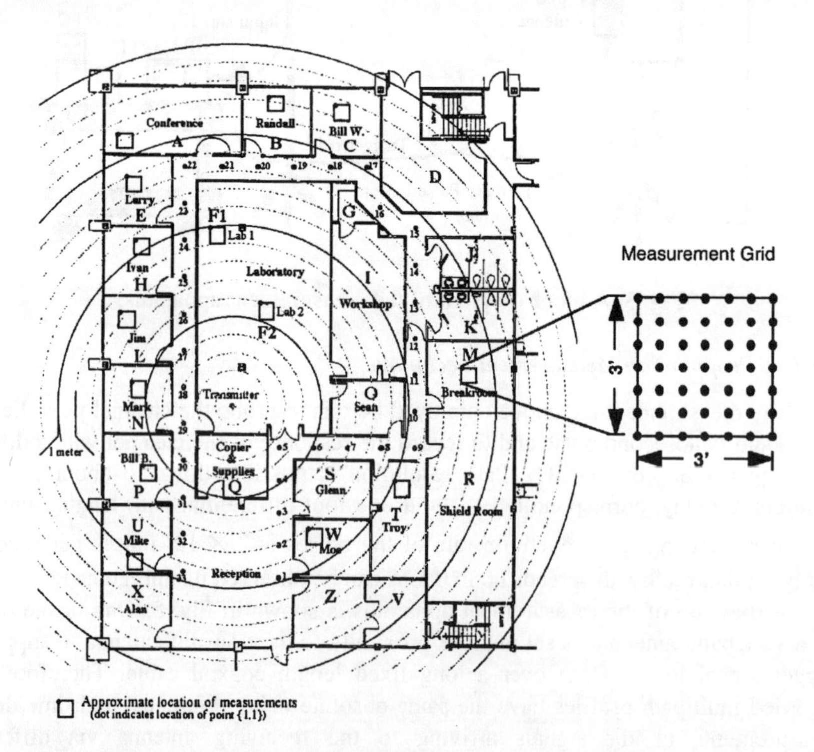


**Fig. 6**. The layout diagram of a typical modern office building where the propagation measurement experiment was performed. The concentric circles are centered on the transmit antenna and are a spaced at 1 meter intervals.

and 4.1 ns after the LOS signal respectively, and hence the first 10 ns of the recorded waveforms in Fig. 7 represent clean pulse arriving via the direct LOS path and not corrupted by multipath components.

Multipath profiles are measured at various locations (see Fig. 6) in 14 rooms and hallways through out the building. In each room, 300 ns long response measurements are made at 49 different locations over a 3 feet by 3 feet square grid with 6 inch spacing between measurement points. One side of the grid is always parallel to north wall of the room. The receiving antenna is located 120 cm from the floor and 150 cm from the ceiling.

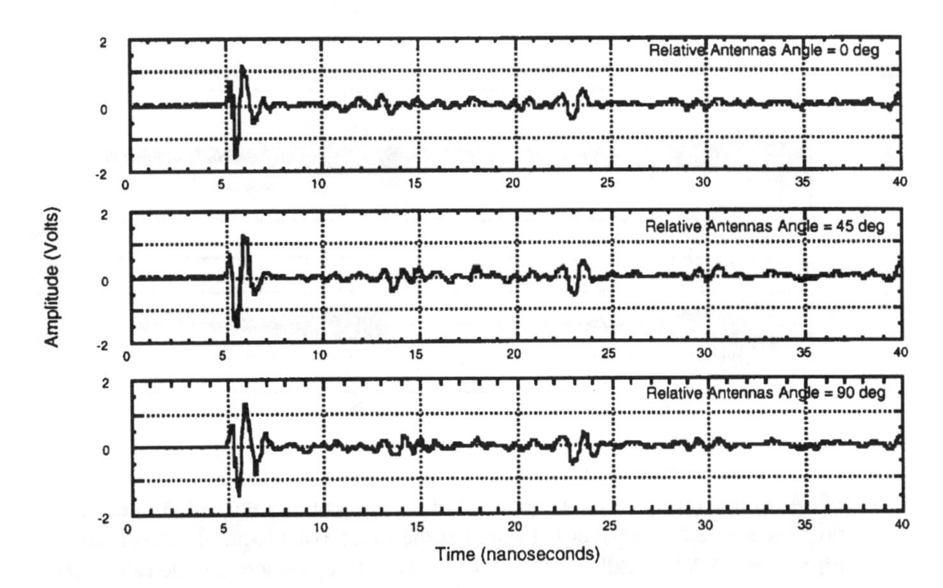


**Fig. 7.** Transmitted pulses measured by the receiving antenna located 1 m away from the transmit antenna with the same height. Measurements were made while the vertically polarized receiving antenna is rotated about its axis, where 0° refers to the case where the transmit and the receiving antennas are facing at each other.

Profiles measured over one microsecond in offices U, W, and M are shown in Fig. 8. The approximate distances between the transmitter and the locations of these measurements are 10, 8, and 13 meters respectively. Figure 8 also shows that the response to the first probing pulse has decayed almost completely in roughly 200 ns, and has disappeared before the response to the next pulse arrives at the antenna. The multipath profiles recorded in the offices W and M have a substantially lower noise floor than those recorded in office U. This is explained, with the help of Fig. 6, by observing that Office U is situated at the edge of the building with a large glass window and is subject to more external interference, while Offices W and M are situated roughly in the middle of the building. Furthermore, offices W and M are near to room R which is shielded from electromagnetic radiation. Interference from radio stations, television stations, cellular and paging towers, and other external electromagnetic interference (EMI) sources are attenuated by the shielded walls and multiple layers of other regular walls. In general, an increased noise floor is observed for all the measurements made in offices located at edges of the building with large glass windows.

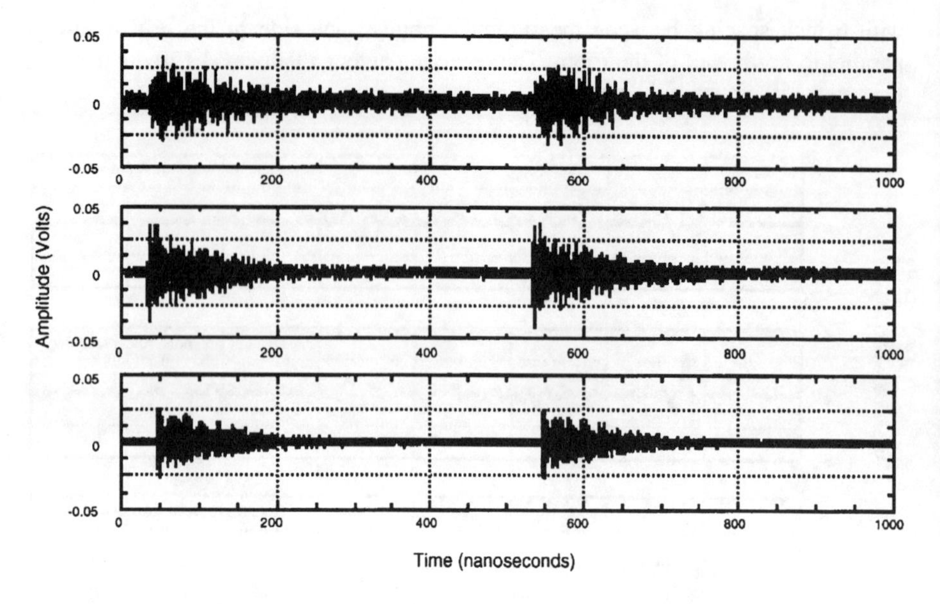


**Fig. 8.** Average multipath measurements of 32 sequentially measured multipath profiles where the receiver is located at the same exact locations in offices U (upper trace), W (middle trace), and M (lower trace) where the measurement grids are 10, 8, and 13 meters away from the transmitter respectively.

Figure 9 shows the averaged multipath profiles measured in office P and H at three different aligned positions one foot apart in the measurement grid. The positions of the receiving antenna in rooms P and H are located 6 and 10 meters away from the transmitter respectively, representing typical UWB signal transmission for the "high SNR," and "low SNR" environments. Since there are walls and reflectors along the line of sight, the direct path response cannot be isolated from multipath. Notice that the leading edge of the direct path response suggests that the location of the receiver for the lower trace is closer to the transmitter than that of the upper trace.

To understand the effect of office doors, two multipath profiles were recorded at the same location in office B, one with the office door open and the other with the office door closed. No noticeable difference between these two measurements was observed. The effect of the large computer monitor was also considered. When the receiving antenna was placed near a large computer monitor in office C, a slight increase in noise floor was observed when the computer monitor was turned on.

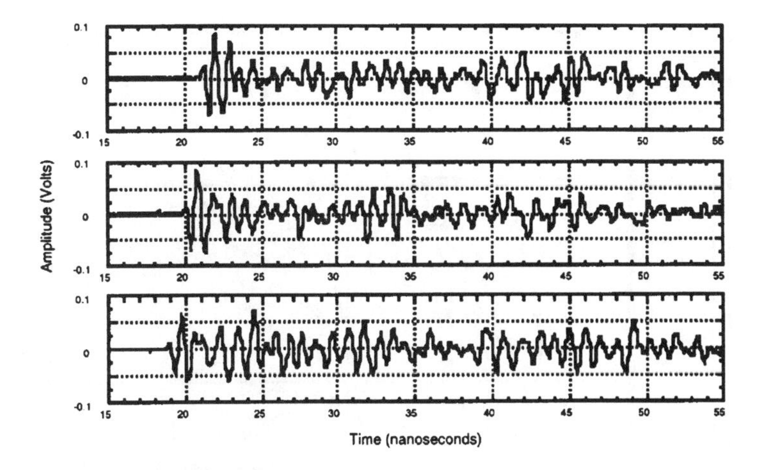


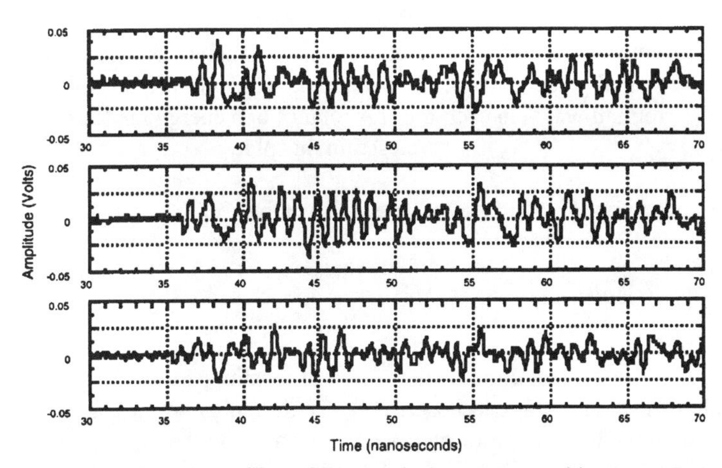


**Fig. 9.** Averaged multipath profiles of 50 ns window measured in rooms P (top three traces) and H (bottom three traces) along the horizontal cross section of the grid at three different aligned positions of one foot apart. The positions of the receiving antenna in rooms P and H are located 6 and 10 meters away from the transmitter respectively, representing typical UWB signal transmission for the "high SNR," and "low SNR" environments.

### *7.3.3 Robustness in Multipath*

Robustness of the impulse radio signal to multipath can be assessed by measuring the received energy in various locations of the building relative to the received energy at a reference point. Mathematically, the signal quality at measurement grid location $(i, j)$ can be defined as

$$Q_{i,j}=10\log_{10}E_{i,j}-10\log_{10}E_{ref} \quad [dB]. \tag{10}$$

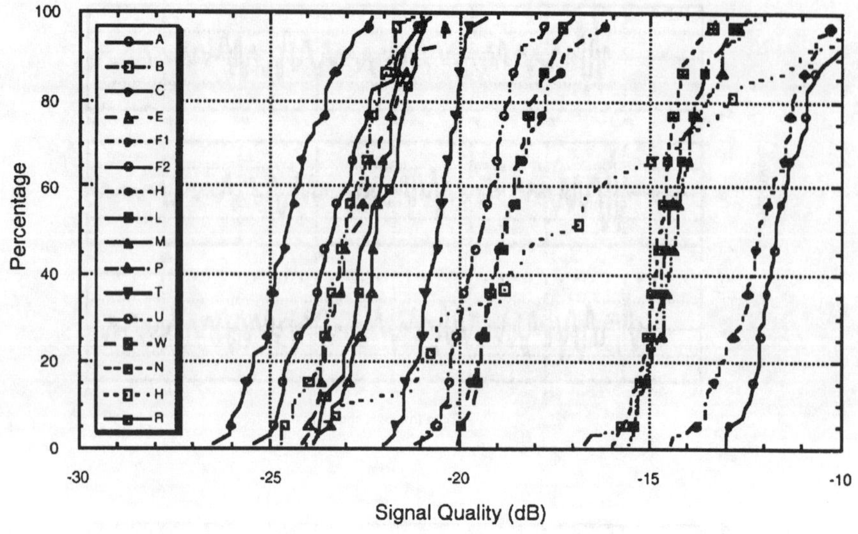


**Fig. 10.** The cumulative distribution function of the received energy collected locally over 49 spatial points (except 21 spatial points for room R, and 34 spatial points for hall ways) in each room.A total of 686 energy measurements are used in this plot.

The received energy $E_{i,j}$ at a location $(i, j)$ is given by

$$E_{i,j} = \int_0^T \left| r_{i,j}(t) \right|^2 dt, \tag{11}$$

where $r_{i,j}(t)$ is the measured multipath profile at location $(i,j)$ in the grid and $T$ is the observation time. The reference energy $E_{ref}$ is chosen to be the energy in the LOS path measured by the receiver located 1 meter away from the transmitter.

The signal quality $Q_{i,j}$ is calculated for the measurements made at 686 different locations (14 different rooms with 49 locations/room, 21 locations in the shield room, and 34 locations in the hall ways).

Table I shows the estimates of the mean and the variance of the signal quality in each room based on the samples taken in that area. The cumulative distribution function of the signal quality for measurements made in these locations are shown in Fig. 10. This data indicates that the signal energy per received multipath waveform varies by at most 5 dB as the receiving position varies over the measurement grid within a room. This is considerably less than the fading margin in narrowband systems, and indicates the potential of impulse radio and ultra-wideband radios in general for robust indoor operation at low transmitted power levels.

**Table I** Signal Quality Statistics

| Office | distance (meters) | Minimum (dB) | Maximum (dB) | $\hat{\mu}$ (dB) | Median (dB) | $\hat{\sigma}$ (dB) | # of Samples |
|---|---|---|---|---|---|---|---|
| F2 | 5.5 | -12.9970 | -9.64586 | -11.5241 | -11.6813 | 0.8161 | 49 |
| N | 5.5 | -16.0060 | -13.2949 | -14.7260 | -14.7690 | 0.5892 | 49 |
| P | 6.0 | -15.5253 | -12.2185 | -14.2373 | -14.2820 | 0.8091 | 49 |
| L | 8.0 | -16.6966 | -12.4310 | -14.4500 | -14.5538 | 0.8342 | 49 |
| W | 8.5 | -20.0157 | -17.0351 | -18.7358 | -18.7425 | 0.7622 | 49 |
| Fl | 9.5 | -14.4064 | -9.79770 | -12.0986 | -12.1407 | 1.0563 | 49 |
| H | 10.0 | -21.0415 | -16.1628 | -18.7141 | -18.8142 | 1.1240 | 49 |
| U | 10.0 | -21.1719 | -17.6232 | -19.4275 | -19.4092 | 0.8024 | 49 |
| T | 10.5 | -21.9113 | -19.2986 | -20.6100 | -20.5419 | 0.5960 | 49 |
| R | 10.5 | -23.7221 | -20.8867 | -22.2675 | -22.3851 | 0.8686 | 21 |
| M | 13.5 | -23.8258 | -20.9277 | -22.2568 | -22.2064 | 0.6439 | 49 |
| E | 13.5 | -24.1454 | -20.2000 | -22.5973 | -22.7824 | 1.0332 | 49 |
| A | 16.0 | -25.4171 | -20.7822 | -23.2826 | -23.3541 | 1.1512 | 49 |
| B | 17.0 | -24.7191 | -21.2006 | -22.9837 | -22.9987 | 0.8860 | 49 |
| C | 17.5 | -26.4448 | -22.3120 | -24.4842 | -24.5777 | 1.0028 | 49 |
| Hallways | | -23.8342 | -6.72469 | -16.9317 | -17.3286 | 4.5289 | 34 |

*7.3.4 Infinite Rake Receiver*

The ultimate goal of a Rake receiver is to construct correlators or filters that are matched to the set of symbol waveforms that it must process. If the propagation measurement process is carried out by sounding the channel with the monocycle $w_{tr}(t)$ from which an impulse radio constructs its time-hopping signal, then the measurements can be used directly to estimate the performance of the receiver and to carry out various aspects of the receiver design.[1]

We use the term *infinite Rake (IRake)* to describe a receiver with unlimited resources (correlators) and infinitely fast adaptability, so that it can, in principle, construct matched filters or correlators arbitrarily well.

The performance of any ideal synchronous receiver operating over a single-link additive white Gaussian noise channel depends on the autocorrelation matrix of the signal set. Assuming that the multipath spread plus the maximum time-hop delay is less than the pulse repetition time, no intersymbol or interpulse interference is present and the performance of a perfectly synchronized impulse radio in such multipath can be predicted from the autocorrelation function $R_{\text{IRake}}(\zeta)$ of an accurately measured multipath profile $s(t)$ given by

$$R_{\text{IRake}}(\zeta) \approx \int_0^T s(t)s(t-\zeta)dt. \tag{12}$$

[1] [1]Note that the measurements reported in Section 7.3 were made with pulse generator having wider pulse width than the pulse waveform model used in Fig. 1 and Section 7.2.

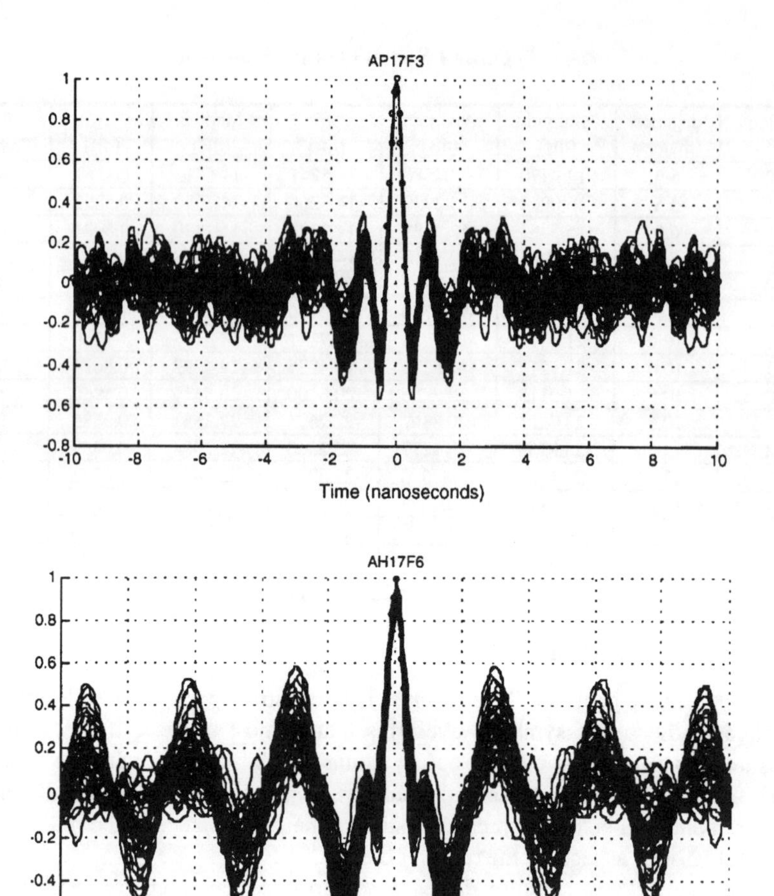


**Fig. 11**. Joint normalized plot of 49 different $R_{\text{IRake}}$ ($\zeta$) functions obtained in rooms P (upper plot) and H (lower plot).

Examples of these correlation functions for the set of measurements from rooms P and H are shown in Fig. 11. As one would expect, when normalized to equal energy these correlation functions look approximately the same for values of the shift parameter $\zeta$ less than a pulse width (< 1 ns). However these correlation functions vary considerably for larger values of $\zeta$ because the set of differential path delays varies for different points on the measurement grid. For large values of $\zeta$, these plots may look quite different for different rooms.

For the monocycle used to sound rooms and produce the curves of Fig. 11, a reasonable choice of the PPM data shift parameter $\delta$ is the location of the first aggregate minimum next to the peak in these curves, roughly 0.5 ns. Notice that the BER performance prediction will vary somewhat from position to position within the measurement grid because the correlation properties of the multipath profile will vary. This technique of using measured multipath profiles to evaluate signal designs has been used to compare different possible 4-ary PPM designs [10].

### *7.3.5 Selective Rake in a Multipath Environment*

The infinite Rake receiver is not realizable. In fact, as the resolution of a radio system becomes finer, performance approaching that of an infinite Rake receiver becomes more difficult to achieve. We can use propagation measurements made with a monocycle pulser to predict the performance of an impulse radio system, and that technique is described here for a single impulse radio link.

A selective Rake receiver is one that diversity combines the outputs of a fixed number of correlators, each correlator locked to the signal coming over a particular propagation path. The performance of a selective Rake impulse radio receiver with coherent diversity combining of $L$ correlator outputs, each correlator matched to an ideal reference template $w_{rec}(t)$ for detecting binary PPM modulation, is shown in Fig. 12. For simplicity in this analysis, we have assumed that clocks and hop times are chosen so that no path is subject to interpulse interference. The technique used in constructing these curves was to determine from the multipath profile $s(t)$ how much received signal energy could be captured by a correlator as a function of its reference delay time, and to chose the reference delay times of the $L$ strongest measurements as the settings for selective Rake reception. The BER curves were then constructed using standard techniques.

The real payoff in selective Rake systems is the fact that system synchronization and communication can be maintained when a propagation path is broken, provided that other selected paths are not interrupted at the same time. Obviously the careful analysis of the likelihood of loss of system lock in a selective Rake system is a complex geometry-dependent problem.

## 7.4. A CLOSING COMMENT

The potential of impulse radio to solve difficult indoor mobile communication problems is apparent because of its fine multipath resolution capability. As with most systems that push the capabilities of current technology, we believe that eventually (and quite possibly soon) impulse radio will become a practical solution to these problems.

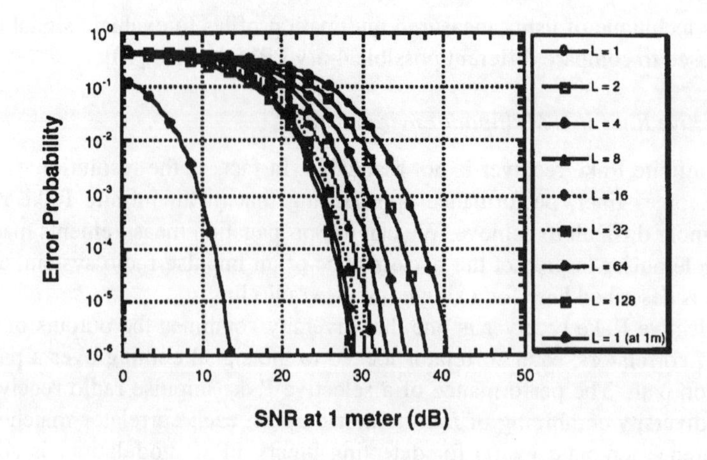


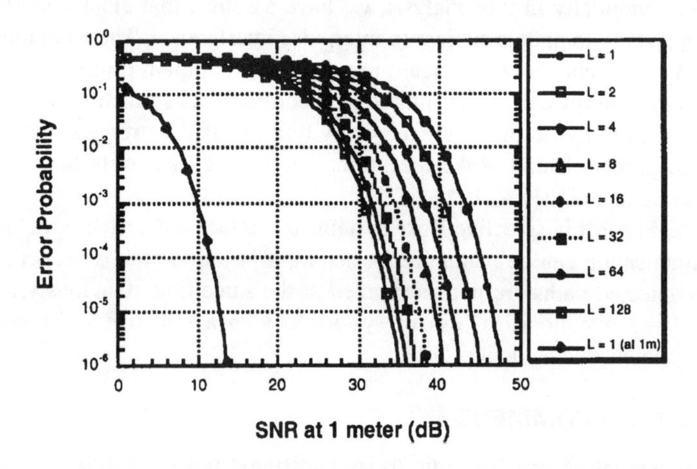


**Fig. 12.** Performance of Selective Rake receiver in rooms P (upper plot) and H (lower plot). These performance curves are averaged over the 49 measurements made in each room.

## References

[1] R. A. Scholtz, "Multiple access with time-hopping impulse modulation," in *Proc. MILCOM,* Oct. 1993.

[2] J. M. Wozencraft and I. M. Jacobs, *Principles of Communication Engineering.* London: John Wiley & Sons, Inc., first ed., 1965.

[3] M. K. Simon, S. M. Hinedi, and W. C. Lindsey, *Digital Communication Technigues: Signal Design and Detection.* Englewood Cliffs, New Jersey 07632: Prentice Hall, first ed., 1995.

[4] H. V. Poor, "Signal processing for wideband communications," *IEEE Information Society Newsletter,* June 1992.

[5] S. Verdu, "Recent progress in multiuser detection," in *Multiple Access Communications: Foundations for Emerging Technologies,* pp. 164-175, IEEE Press, 1993.

[6] M. Z. Win, R. A. Scholtz, and L. W. Fullerton, "Time-hopping SSMA techniques for impulse radio with an analog modulated data subcarrier," in *Proc. IEEE Fourth Int. Symp. on Spread Spectrum Techniques Applications,* Sept. 1996. Mainz, Germany.

[7] C. L. Weber, G. K. Huth, and B. H. Batson, "Performance considerations of code division multiple access systems," *IEEE Trans. on Vehicul. Technol.,* vol. VT-30, pp. 3-9, Feb. 1981.

[8] M. Z. Win and R. A. Scholtz, "Ultra-wide bandwidth (UWB) time-hopping spread-spectrum impulse radio for wireless multiple access communications," *IEEE Trans. Clommun.,* Jan. 1997. submitted.

[9] M. K. Simon, J. K. Omura, R. A. Scholtz, and B. K. Levitt, *Spread Spectrum Communications Handbook.* McGraw-Hill, Inc., revised ed., 1994.

[10] F. Ramirez-Mireles, M. Z. Win, and R. A. Scholtz, "Signal selection for the indoor wireless impulse radio channel," in *Proc. 47th Annual Int. Veh. Technol. Conf.,* May 1997.

# 8 OVERVIEW OF WIRELESS IN LOCAL LOOP

Ramjee Prasad
Chandan Kumar Chatterjee

**Abstract**

This paper presents a bird's eye view on the hot topic of present day in telecommunications viz., Wireless in Local Loop. After laying down the description of WILL and stating its advantages and disadvantages, it goes on to discuss the details of different types of architecture. In brief it also presents the different standards in Europe, Japan and USA where all the development work has been done so far. The paper further dwells upon a few products relating to each standard and finally ends with a brief discussion on the air interface multiple accessschemes. Mention or description of certain products in this paper should not, in any way, be construed by the reader as a recommendation for the product. These have been taken only as examples.

## 8.1. INTRODUCTION

### *8.1.1 Background*

Little did the author of the phrase "Where there is WILL there is a way" realize that the developers and researchers in the field of telecommunications would apply the phrase literally in the telecommunications field. As we know, for extending the telephone service to a user, conventionally, underground physical cables are laid to the user's premises. Such cable laying has its own problems in terms of time required, costs, logistics etc. At many a place it is well nigh impossible to lay the cable because of narrow and congested bye-lanes or solid rocky terrain. The wireless overcomes all such barriers and **Wireless in Local Loop (WILL)** is seen as a solution to all these problems and much more. Thus, where there is 'WILL', a way will be found to reach the user.

It is well known that telephony was invented in the year 1885, slowly it spread all over the world and even today the global demand has not been satiated. Contrary to it, the demand is increasing by the day. The spread has mainly been met with individual copper pair connections from the Central Office (C.O) to each user. This method has been found wanting in some aspects and research had been going on to find better and cheaper means. In this paper we have attempted to examine the growth of the alternative viz., wireless, and since wireless is used in the local loop as well as in mobile communication services we have done an overall examination of radio access to the communication network. Wireless in Local Loop (WILL) is also popularly known as Wireless Local Loop (WLL).

*8.1.2 Past, present and future*

Even though wireless was discovered in the year 1895 by G. Marconiyet it was only in the year 1921 that the first field tests in mobile communications were done in USA by the Detroit Police. In the year 1948 the first commercial fully automatic mobile telephone system was deployed in Richmond, USA. With the advancement of technology, the deployment of analog cellular systems spread the world over in the 1980's. The analog systems have been overshadowed by digital systems in the 1990's. A chronological list of events in the development path of wireless is given in Table 1 [1].

This paper is organized as follows. Section 8.2 first introduces the local loop and then discusses the challenges facing WILL. Advantages and disadvantages are also presented in Section 8.2. Architecture of WILL is presented in Section 8.3. Various standards and developments are discussed in Section 4 and 5, respectively. Section 6 presents the issues related to air interface multiple access systems. Security aspects, cost issues, market issues and extent of deployment are discussed in Section 8.7, 8.8, 8.9 and 8.10, respectively. Finally, our conclusions are given in Section 8.11.

**Table 1** Chronological List of events in development of wireless.

(a) Pioneer Era

| | |
|---|---|
| 1860's | James Clark Maxwell's EM waves postulates |
| 1880's | Proof of the existence of EM waves by Heinrich Rudolf Hertz |
| 1890's | First use of "wireless" and first patent of wireless communications by Gugliemo Marconi |
| 1905 | First transmission of speech and music via a wireless link by Reginald Fessenden |
| 1912 | Sinking of the Titanic highlights the importance of wireless communication on the seaways;<br>in the following years marine radio telegraphy becomes established |

(b) Pre Cellular Era

| | |
|---|---|
| 1921 | Detroit Police Department conducts field tests with mobile radio |
| 1933 | In USA four channels in the 30-40 MHz range |
| 1938 | In USA, ruled for regular services |
| 1940 | Wireless communication is stimulated by World War II |
| 1946 | First commercial mobile telephone system operated by the Bell system and deployed in St. Louis |
| 1948 | First commercial fully automatic mobile telephone system is deployed in Richmond, USA |
| 1950's | Microwave telephone and communication links are developed. |
| 1960's | Introduction of trunked radio systems with automatic channel allocation capabilities in USA |
| 1970's | Commercial mobile telephone system operated in many countries, e.g. 100 million moving vehicles on US highways, "B-Netz" in West Germany, etc. |

(c) Cellular Era

| | |
|---|---|
| 1980's | Deployment of analogue Cellular systems |
| 1990's | Digital cellular Deployment and dual mode operation of digital systems |
| 2000's | Future Public Land Mobile Telecommunication systems (FPLMTS)/ International Mobile Telecommunications-2000 (IMT-2000)/ Universal Mobile Telecommunication Systems (UMTS) will be deployed with multimedia services |
| 2010's | Wireless broadband communications will be available with B-ISDN and ATM networks |
| 2010's + | Radio over fibre (such as fibre-optic microcells) |

## 8.2. WHAT IS A LOCAL LOOP?

The term 'Local Loop' stands for the medium that connects the user premises equipment with the telephone switching equipment. There are two types of local loop namely, physical (fixed) local loopand wireless local loop. These have been discussed in the following sections.

### *8.2.1 Physical Local Loop*

Conventionally this has been made up of copper pairs leading from the C.O. to the user premises in the form of underground cables. To meet the requirements of specific localities depending upon the telephone density, points of flexibilityare introduced in the cable route. Such points are called 'Cabinets' and 'Pillars'. The cables emanating from the C.O.'s are typically of 1000 or 2000 pairs to the cabinets and of smaller sizes between the cabinets and pillars. At the user end there are distribution pointsgenerally catering to 10, 20 or 100 users. From this point the pairs are lead into the user premises and terminated finally on to the user equipment like telephone, fax etc. Figure 1 illustrates a physical pair local loop set up.

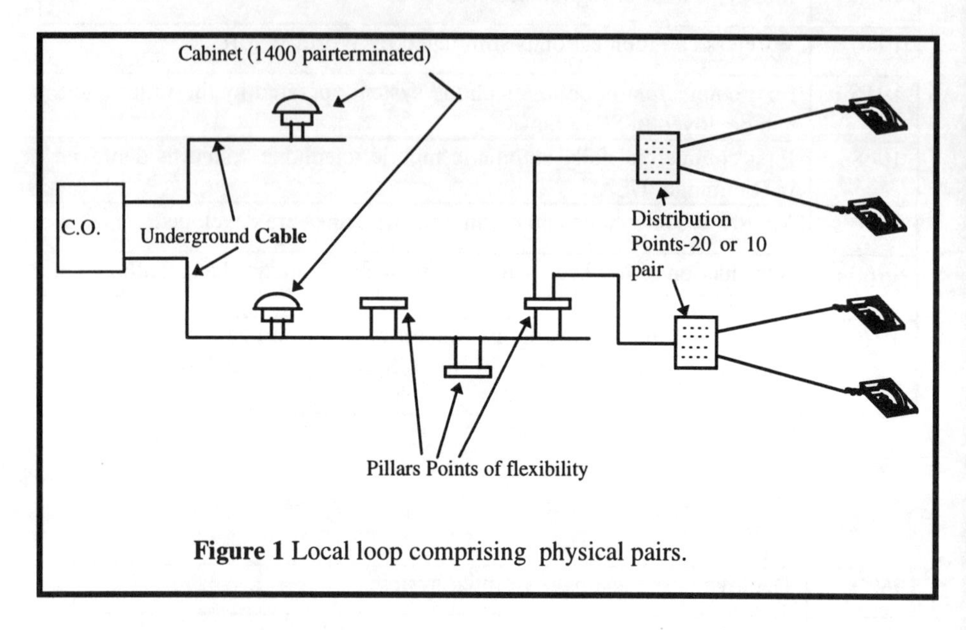


**Figure 1** Local loop comprising physical pairs.

*8.2.2 Wireless Local Loop*

The above stated physical wiring has certain drawbacks and substitute for the same has been found in application of the wireless technology. Thus the phrase Wireless in Local Loop (WILL) was born. With the improving and changing technologies, the application of wireless in the local loop has found acceptance and application to various extents in different countries. Figure 2 depicts a basic wireless local loop system. The wireless access to the telecommunication network is also termed as Radio Access and for the local loop it is being called "Fixed Radio Access". Thus the radio access can be for both modes of communication viz., mobile communication and simple wireless access for stationary services. Some of the developments have been strictly for fixed access whereas in other cases the mobile system at hand has been put to use for fixed access. The overlap of the two types will be clear when we recognize that after all a mobile phone can be put to fixed usage and certain amount of mobility already exists in cordless phones which are fixed phones.

*8.2.3 Challenges*

The challenges that WILL faces in being acceptable to the user are placed in two categories some of which are psychological and some technical. Since it replaces the copper cable for connecting the user with the local exchange the user may have apprehensions on reliability, secrecy and interference with wireless appliances like radio, TV etc. (man made noise) and other WILL users. On the other hand WILL will have to prove itself at least as good, if not better, than the services provided by the physical cable. It should be able to carry and deliver voice, data, state-of--art multimedia (with some add-ons), leased line connectivity, dial-up, Internet connectivity, high speed data and what have we, as efficiently as the age old fixed line does. Onus rests on the researchers and developers to prove that it is so. Only time will tell.

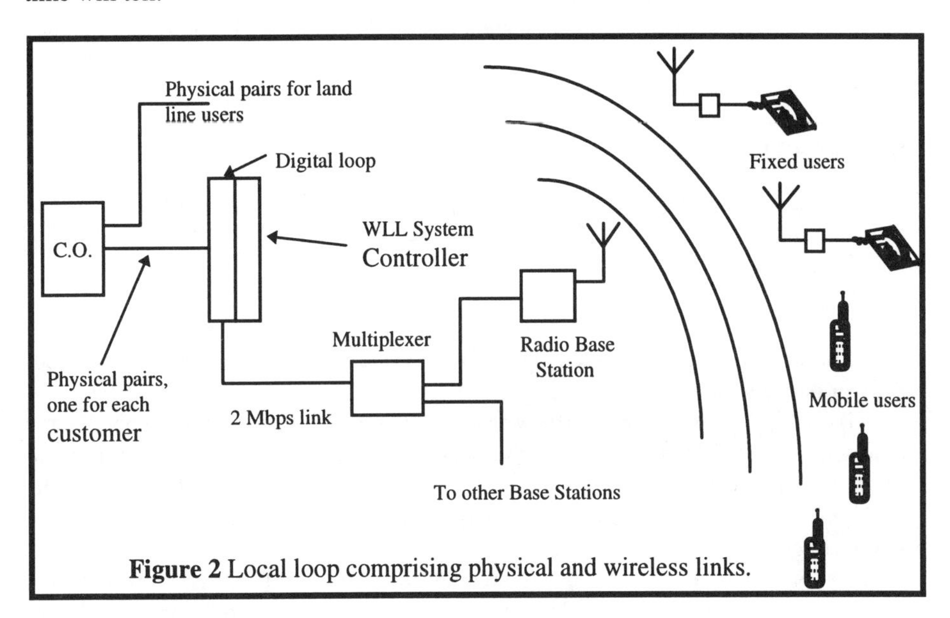


**Figure 2** Local loop comprising physical and wireless links.

### *8.2.4 Advantages of WILL*

The advantages of WILL technology are summarized as follows [2 ]:-

a) Deployment is fast as compared to the physical cable laying.
b) Low incremental cost for adding users once the base stations and interconnection with PSTN (Public Switched Telephone Network) has been established.
c) The higher cost is balanced by the early completion of work which results in early cash inflow.
d) Network maintenance costs are lower.
e) Less prone to failure due to external hazards like theft of copper cable, damage of cable due to third party digging etc.
f) Time taken to repair is much lesser since it is not necessary to carry out extensive testing for localization of fault and no permission from local authority is required to dig specific areas.

### *8.2.5 Disadvantages of WILL*

On the other hand the disadvantages can be stated as follows:-

a) In developing countries, where exists the potential market for WILL and where continuous supply of power may not be so certain, the base stations and user's equipment will need power supply locally and in the event of power failure the service to a user or a group of users will be lost.

b) The technology has still not stabilized and, as a result, the performance of present day wireless communication services is not of top quality with frequent dropping of calls, unsatisfactory levels of noise etc.

c) With obsolescence of the equipment installed, there being fast development in this area, replacement of the same at considerable expenses may have to be done by the service provider.

d) In the absence of sufficient technically skilled personnel at grass root levels, particularly in developing countries, the repair or replacement of base stations and users' units will cause problems.

## 8.3 ARCHITECTURE OF WILL

WILL can be categorized in three ways viz., a) physical categorization, b) usage categorization and c) technology based categorization. Given below are brief descriptions of the three categories.

### *8.3.1 Physical (implementation) categorization*

The categorization is based upon the way the WILL has been implemented-fully or partly wireless.

i) Partly physical and partly wireless :-In such an architecture generally the connection from C.O. to the user locality is physical viz., of copper or fiber. Beyond the street crossing it is wireless. Figure 2 shows such a system. This solves the problems associated with cable laying only partly but can still be effectively used in really congested areas.

ii) Fully wireless :- In this architecture the end to end link is wireless and is built with points of flexibility. These points of flexibility correspond to the cabinets and pillars in the fixed wire scenario. Figure 3 illustrates such a scenario.

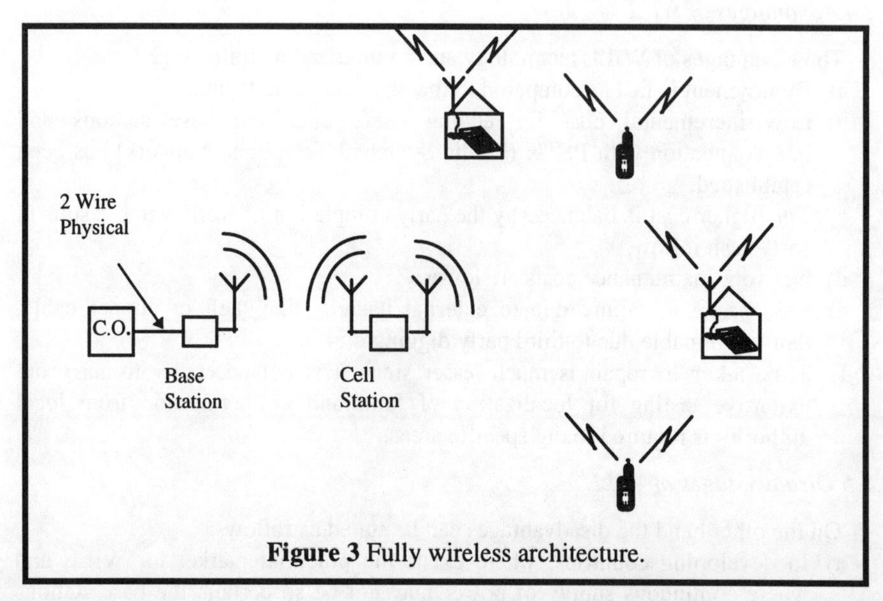


**Figure 3** Fully wireless architecture.

### *8.3.2 Configurational categorization [3]*

WILL can be categorized based upon configuration in the following ways :

i) Fixed Radio Access.

In this case the user terminal is fixed with absolutely no mobility. Each end user communicates with the Radio Network Controller directly or through Base Stations. Figure 4 shows such a configuration.

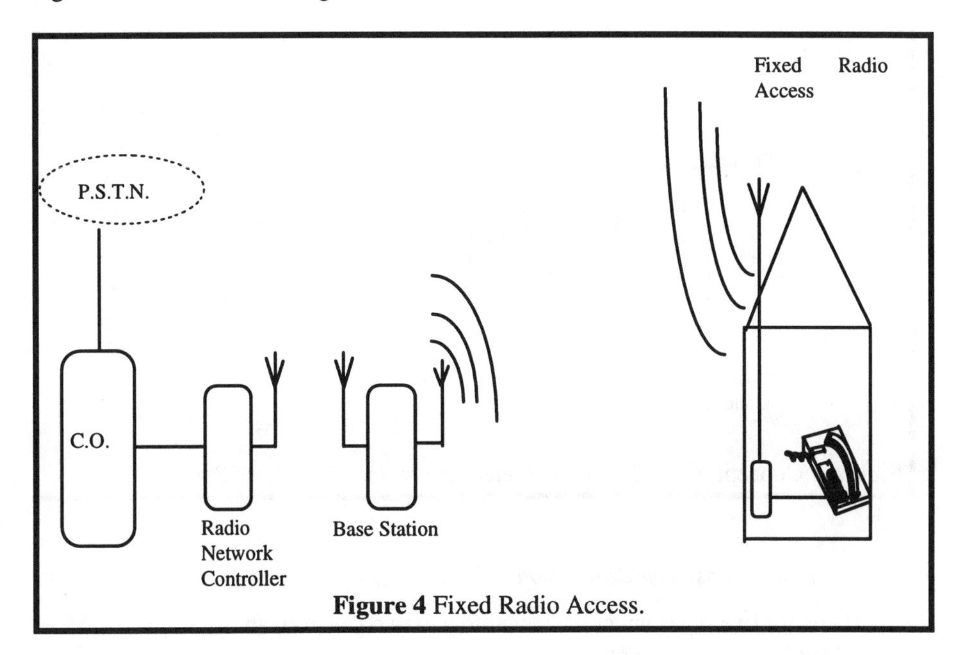


**Figure 4** Fixed Radio Access.

ii) Neighborhood Telephony

In this case the user is able to move around in the house or in the immediate neighborhood. He has a cordless phone unit giving him very limited mobilityHand over is possible within the Base Stations of the same Radio Network Controllerbut not beyond. Figure 5 shows such a configuration.

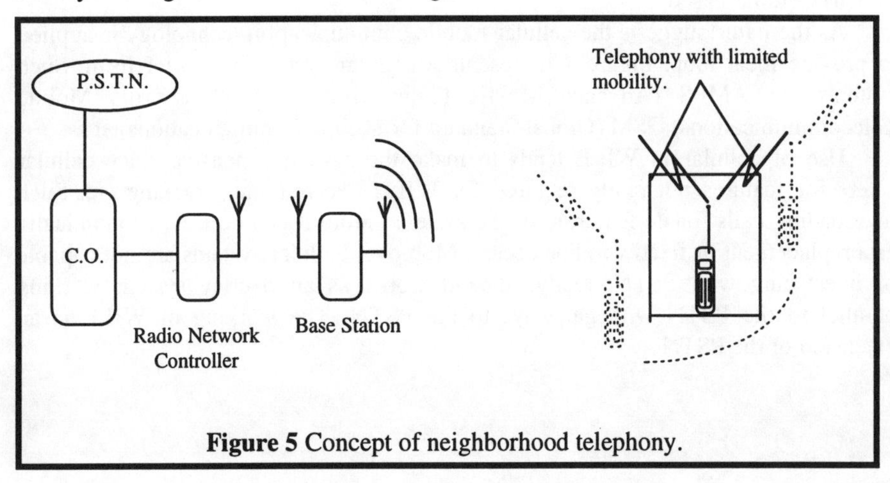


**Figure 5** Concept of neighborhood telephony.

ii) Neighborhood Telephony with Indoor Base Station.

In this case at the user end there is a Radio Private Automatic Branch Exchange (PABX) interacting with end user on one hand and Base Station on the other. In figure 6 such a system has been shown.

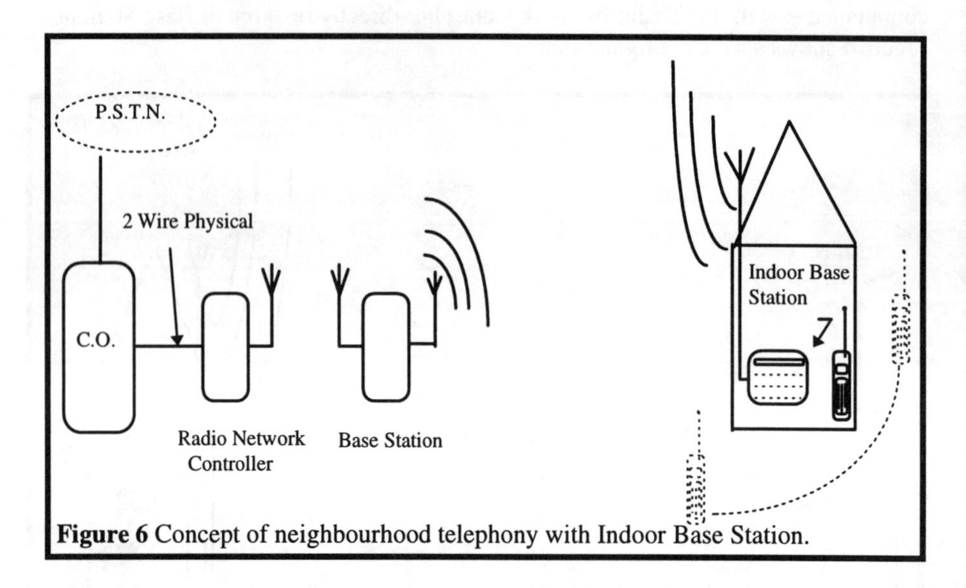


**Figure 6** Concept of neighbourhood telephony with Indoor Base Station.

*8.3.3 Technology based categorization [3]*

There are five types under the technology based categorization.

i) Cordless Telephony based

Such systems are Cordless Telephony-2 (CT2) DECT (Digital Enhanced Cordless Telephoneearlier the word 'European' was used in place of 'Enhanced'.) and PHS (Public Handyphone System) The last two are expanding speedily whereas the first one is losing ground fast.

ii) Cellular based

As the name suggests the cellular mobile communication technology is applied to provide local loop. Figure 7 represents a cellular system in a brief form. Such systems are AMPS (Advance Mobile Phone System) NMT (Nordic Mobile Telecommunications) GSM (Global Standard for Mobile Communications) etc.

Use of cellular in WILL tends to make the system expensive since cellular caters for features not really required for WILL like roaming, tracking etc. Such overloading calls for de-featuring of the system before it can be used economically as a replacement of fixed wireline access. Moreover cellular systems are not capable of interfacing with PSTNdirectly since it works as an overlay system working parallel to the PSTN with gateways to the PSTN. The working of WILL is an extension of the PSTN.

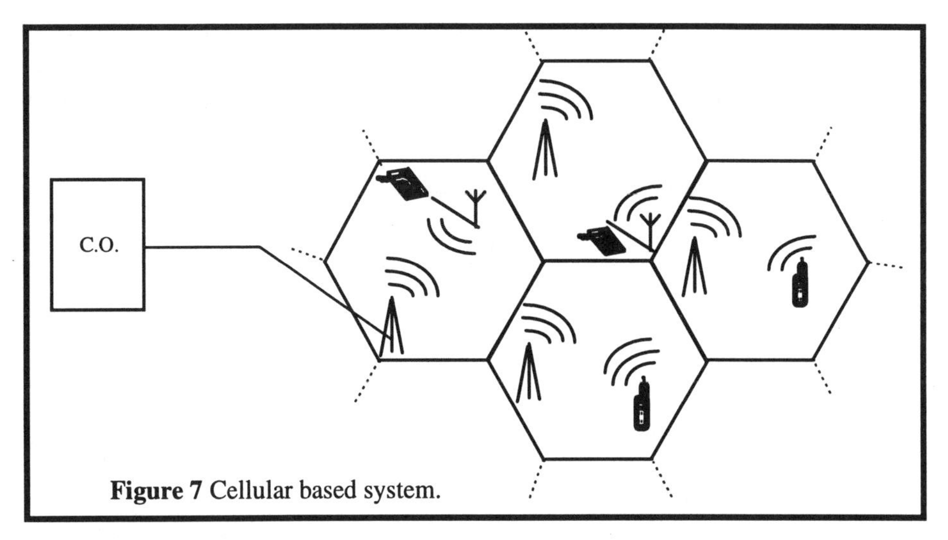


**Figure 7** Cellular based system.

iii) Point to Multi-Point Microwave (PMPM).:- In the microwave system again the end to end connectivity is non-physical but consists of microwave link. The PMPM systems can be further sub-divided into a) Rural Technolgies System (RTS) e.g., IRT 2000and b) Large Bandwidth Point to Multi-point (LBPMP) system e.g., Ericsson's ACTRAN. These are more suited for rural areas where large distances are to be covered. Since microwave has been in use for quite sometime and has proven to be capable of providing satisfactory levels of grade of service and good PSTN connectivity, its application in WILL can be presumed to be providing equally high standards in these respects. It is capable of even providing ISDN (Integrated Services Digital Network) services.

iv) Satellite based.

In this case there is a link from the exchange to a satellite and all the users have small dishes for linking their telephones with the satellite. Such user linkages could be at fixed points or could be mobile. Land mobile satellite system has been shown in figure 8. In January 1997 the first global personal satellite phone service, belonging to Comsat Personal Communications has gone commercial. The service is named Planet1 and provides communication from portable, note-book sized phones [4]. With mobility is related the size of antenna one can carry, resulting in the restriction on the types of services one can get.

v) Dedicated 2nd generation Fixed Wireless Access.

Products of various manufacturers fall under this category e.g., Nortel, Proximity1Airloop and product of Lucent Technologies [3].

## 8.4 STANDARDS

From time to time the need for the radio access has been felt and the development has proceeded in two streams viz., the cellular streamand the cordless stream. The former is the stream of mobile telephony and the latter the cordless telephone approach which is used for the purpose of mobility within the house. Developments have continued in USA, Europe and Japan and different standards have come up using both analog and digital techniques. In the following paragraphs various wireless access methods have been covered, fixed as well as mobile, since even systems for mobile communications are being deployed for fixed access, i.e., Wireless Local Loop. The forerunnners of the present day technologies were Advance Mobile Phone

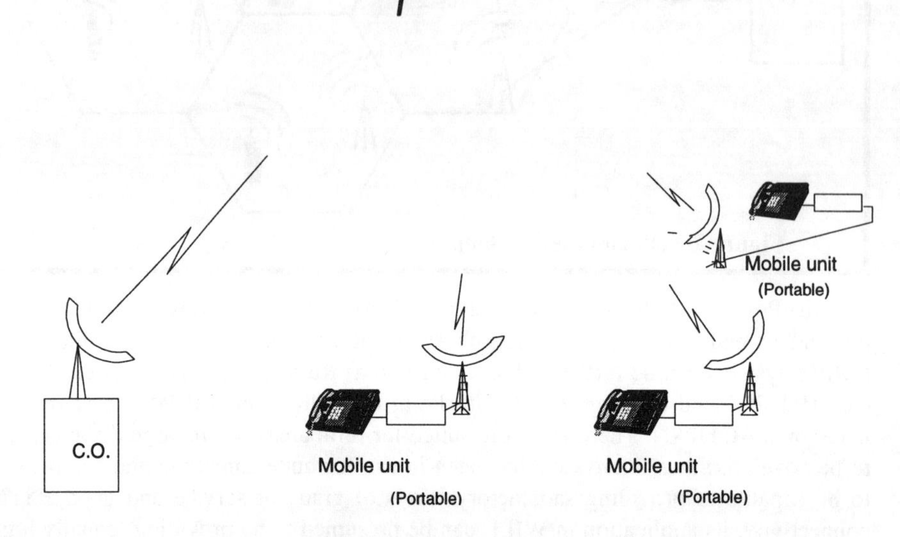


**Figure 8** Land mobile satellite

System (AMPS) of USA, Total Access Communication System (TACS) developed in UK etc. Presently a keen competition is on between different technologies supported by their proponents for market share.

*8.4.1 The American standards*

It is being estimated that about 50% of the market will go to the Code Division Multiple Access System (CDMA) systems and the balance will be distributed amongst others viz., TDMA-FDD ( TDMA- Frequency Division Duplex ), E-TDMA, DCS1900 (Digital Cordless System 1900) etc[5]. The more important TDMA based system is the Personal Access Communication System (PACS). The PACS Providers' Forum was founded in late 1995 to develop the PACS market in North America. PACS is based upon TDMA and Wireless Access Communications System (WACS). Members of the forum are Bellcore, GCIHughes Network Systems, NEC (Nippon Electric Company) America, Panasonic, PCSI(R) and Siemens Stromberg-Carlson. PACS was originally developed jointly by Motorola, Hughes and Bellcore. It is claimed by the developers of PACS that the hand off speeds of up to 70 mph are possible. It is, however, widely accepted that 30 to 40 mph is more realistic[6 ]. The main features of WACS are given in Table 2 [7].

**Table 2** Technical information on different standards [8 ]

| | WACS | DECT | DCS1800 | CT-2 | PHS |
|---|---|---|---|---|---|
| Frequency Band GHz | 1.85 to 2.2 | 1.88-1.9 | 1.71-1.785 1.805-1.88 (UK) | .8641-.8681 | 1.895 to 1.9185 |
| Access method | TDMA-FDD | TDMA-TDD | TDMA-FDD | TDMA-TDD | TDMA-TDD |
| No. of plexes per carrier | 8 | 12 | 8 | 1 | 4 |
| Carrier spacing KHz | 300 | 1728 | 200 | 100 | 300 |
| Channel bit rate Kbps | 400 | 1152 | 270.833 | 72 | 384 |
| Speech coding Kbps | 32 ADPCM | 32 ADPCM | 13 PRE-LTP | 32 ADPCM | 32 ADPCM |
| Cell Station Transmit Power mW | 800 peak | 250 peak | - | 10 peak | 160 peak |
| Portable transmit power mW | 200 peak | 250 peak | - | 10 peak | 80 peak |
| TDMA Frame msec | 2.5 | 10 | - | 2 | 5 |
| Bandwidth per traffic channel KHz | 75 | 144 | - | 100 | 75 |
| Appeared in | 1995 | 1992 | 1993 | 1989 | 1994 |

The latest development in radio access has been the induction of CDMA technique. The leader in CDMA technology has been Qualcomm. The Telecommunications Industry Association (TIA) of North America has standardized Qualcomm's CDMA technology as the Interim-Standard-95 (IS-95). Though the most significant developments in CDMA have been pioneered by companies of US origin, yet the specifications they have used are of UK and the first demo's also have been in UK [5 ]. AT&T Network Systems, now known as Lucent Technologies , has developed the Airloop, partly, on the specifications of Liberty Communications and DSC Communications has developed Airspan on BT's specifications. In October 1995 the first CDMA based commercial systems were commissioned in USA and Hongkong [5]. Variety of products and architectures have been developed to meet the requirements of the market.

*i) DIVA 2000 System*

This link is implemented with fully or partly wireless media and comprises DIVA 2000 Radio/Switch Controller, DIVA 2000 Modular Base Stations, DIVA 2000 User Unit and DIVA 2000 System Management Software. The connections between R/S Controller and the Modular Base Station and the R/S Controller and the local switch are established through twisted pair, by microwave links or by 30 channel PCM (Pulse Code Modulation) systems on physical pairs. The connection between user unit and Modular Base station is through radio. The system supports voice, group 3 fax and 2.4 Kbps data. Figure 9 depicts a DIVA 2000 configuration [8].

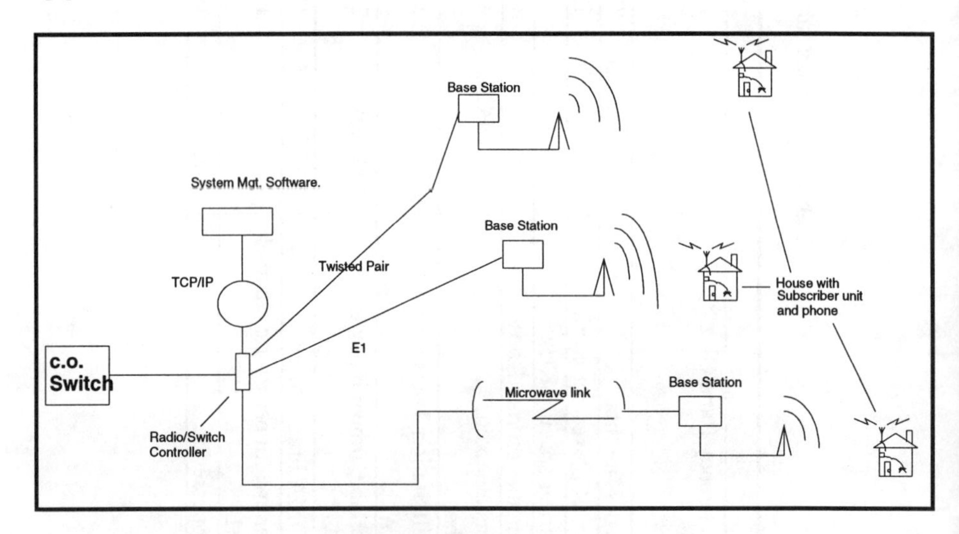


**Figure 9** DIVA 2000 System

*ii) AT&T's Airloop [5,9]*

AT&T has developed Airloop based upon Liberty Communications'(UK) standards. This system comprises 4 units viz., Wireless Line Transceiver (WLT), Network Interface Unit (NIU), User Transceiver Unit (STRU) and Operation and Maintenance Center (OMC). The WLT and OMC are co-located with the local exchange whereas NIU and STU are at the user premises. The WLT interfaces with the local exchange and the NIU. In turn the NIU interfaces between the user terminal and the air interface NIU. The product supports voice, leased line, metering for payphones, voiceband data upto 28.8 Kbps, G3 fax and ISDN services 64 Kbps B+D and 2B+D. Figure 10 shows a schematic of Airloop.

*8.4.2 European Standards*

In the earlier stages majority of the work in developing standards and products has been done in Europe. Also, as stated earlier, at least in 2 cases though the developing agency is from USA yet the development has taken place based upon European standards. The DECT, Digital Cordless System (DCS) 1800 and CT2 are the main standards developed in Europe. DCS 1800 is a derivative of the much talked about GSM. The main features of these are given below.

CT-2 is one of the earliest standards and is now on the decline with reports from Asia Pacific region regarding drastic fall in the number of users as well as the newer candidate countries for radio technology preferring other more advanced standards [10]. GSM is a digital system and was developed as a mobile communication system. It has spread world wide (178 Operators in 110 countries-[11]) but suffers from the handicap of voice quality though it is a step better than the analog systems. There are reports of development of full rate voicecoder, absence of which has been the reason for the poor voice quality. Efforts are on to make the voice quality of GSM near land line quality.

A field trial on DECT was carried out in 1994 in Sweden to assess the possibility of user reaction and replacing the copper wire with wireless. The results showed that DECT could be an efficient way of replacing the copper loop [12]. The users regarded the limited mobility as a value added service. The DECT comprises Central Control Fixed Part (CCFP) Radio Base Station (RBS) and the Portable Part (PP). The CCFP connects on one side with the main network like PSTN, X.25, ISDN etc. and on the other side to the RBS'. It does the mobility management (movement of users between RBS') and call related functions [13 ]. Several hundred RBS' may be parented to one CCFP. A standard RBS has the capacity to handle 12 simultaneous calls. On average it handles 5 Erlangs traffic with a blocking rate of less than .5%. In general an RBS provides an indoor range of 20 to 60 metres and up to 500 metres outdoor with omni-directional antennas. With directive antennas the range can be increased to several kilometers. The PP caters for the need of the main network with which ultimately the DECT is going to interwork [13]. DECT is shown in Figure 11.

Last year trials were under way in Sweden with the Dynamic Beam Switched method using the Time Division Multiple Access / Time Division Duplex (TDMA/TDD) at 10 MHz [14]. The radio channel capacity is determined by the number of TDM time slots. In the TimeSpace Radio (TSR) system there are 256 time slots, each of 64 Kbps, which gives a total of 16 Mbps duplex. Apparently this method has certain advantages like gain in link budget, absence of co-channel interference, allocation of capacity on demand to users, no frequency planning required, more efficient use of available spectrum and alternative network topographies possible. On the other hand there is a more complicated network management system [14]. For technical details of the standard the reader is referred to Table 2. Brief descriptions of some of the products are given below.

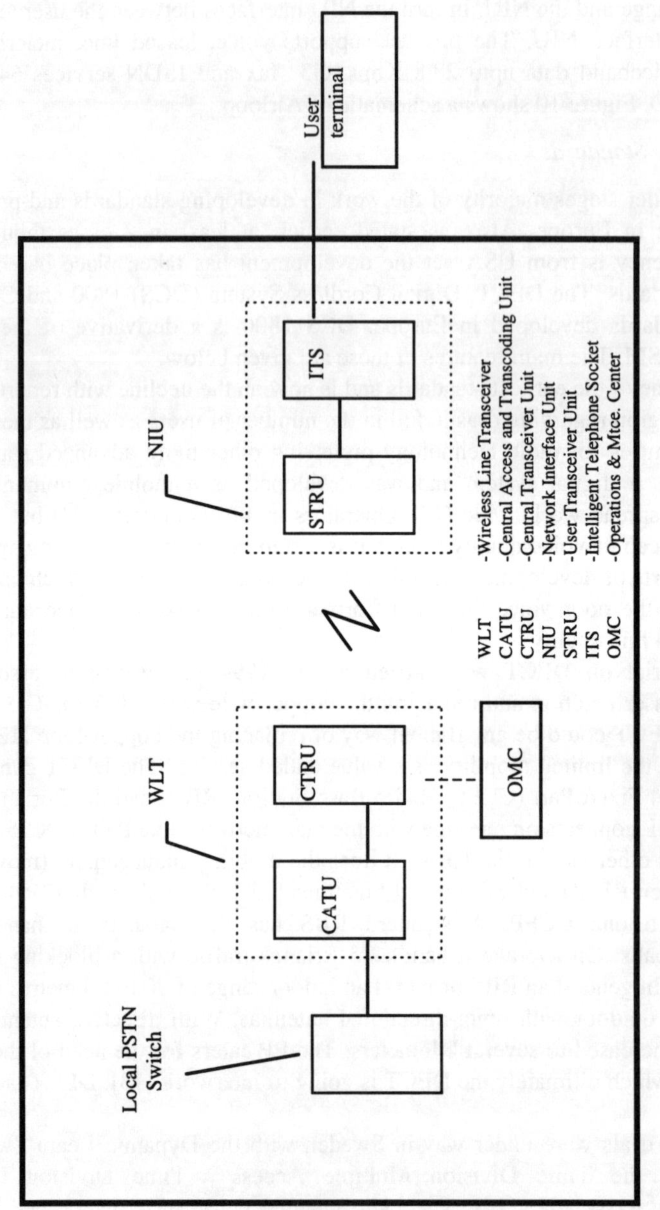


**Figure 10 AT&T's Airloop**

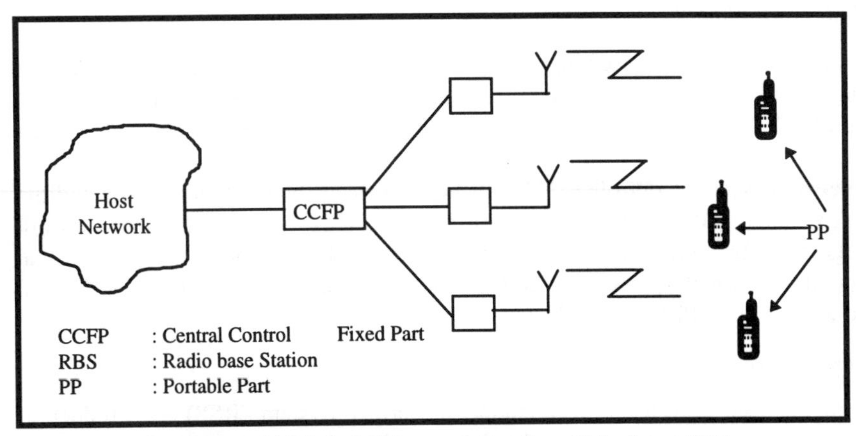


**Figure 11** Digital Enhanced Cordless Telephone [13]

*i) Siemens' DECTLink [15]*

This product is implemented as a fully or partly wireless system. It interfaces with the local exchange through Radio Distribution Unit(RDU) which supports a maximum of 480 users. The RDU connects with a maximum of 4 Radio Base Station Controllers (RBSC) through copper, fiber or microwave at 2 Mbps. Each RBSC in turn can support upto 4 cores, each core being able to handle 15 Radio Base Stations (RBS). The connection between RBSC and RBS can be radio access or wired access for ISDN. Finally each RBS can connect upto 12 Radio Network Terminating (RNT) units for single users. This last connection between RBS and RNT is necessarily onradio. Thus DECTLink connection from exchange to user premises can be a mix of copper, fiber, microwave and radio. Figure 12 represents a block schematic of DECTLink.

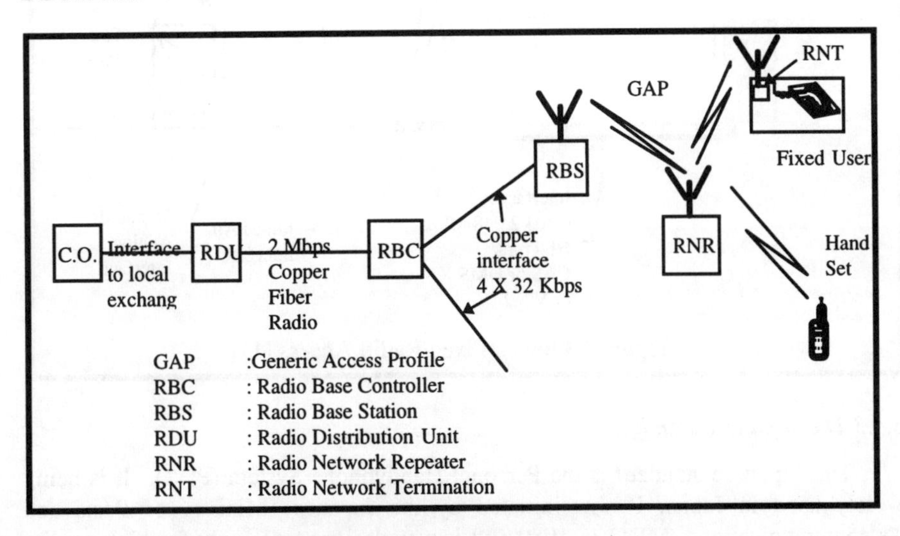


**Figure 12** Siemen's DECTLink [15]

*ii) Philips' IRT4000 [16]*

This is a system based upon SWING which is a DECT standard. As per definition the SWING system comprises 1) the handset, 2) the Cordless Terminal Adapter (CTA), 3) the Radio Fixed Part (RFP) which handles the channels, 4) the Radio Control Unit (RCU) which concentrates the users connected to the RFP's, 5) the Network Access Controller (NAC) which interfaces between the RCU's and the local exchange and 6) SWING Network Manager (SNM) which does the network management functions. The system covers both local and remote users. The local ones (within 5 Kms) access using DECT terminals whereas the remote ones (5 to 50 Kms) access using IRT Terminals.

*iii) Ionica Fixed Radio Access [17]*

At the user's premises Residential Subscriber System (RSS) , a product of Nortel, provides the connection to the Customer Premises Equipment (CPE) through a physical pair. The Base Station (BS), which has Ionica's proprietary interface and concentrates the users, links with the users' premises through a 3.5 GHz radio link. From the BS to the local exchange connectivity is provided through microwave or physical pair. Figure 13 shows the main components of Ionica Fixed Radio Access.

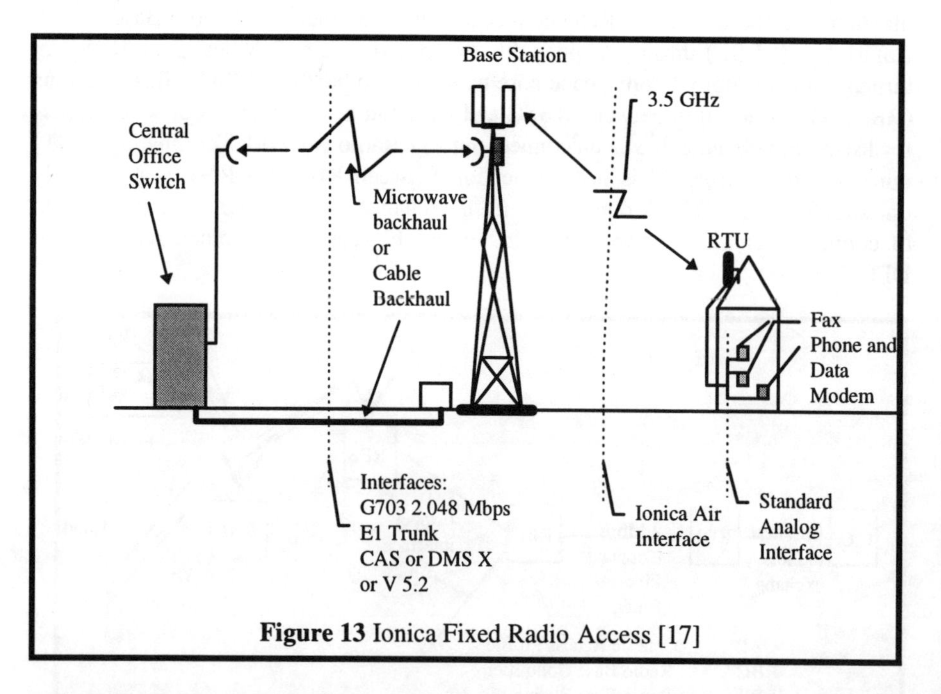


**Figure 13** Ionica Fixed Radio Access [17]

*8.4.3 The Japanese standard*

The Japanese standard is the Personal Handyphone System (PHS). It is being widely used in Japan. It was initiated by the Japanese Ministry of Posts and Telecommunications (MPT) in 1989 and it was designed not only for high quality voice but also for data, still pictures and video. The first field trials, however, for PHS based Personal Communication System (PCS) were conducted in the year 1992 and were completed in around October 1994. Some of the companies which participated in these trials were NTT (Nippon Telephone and Telegraph), DDI Pocket Telephone, JT/KDD and Teleway Japan. The trials confirmed that the speech quality and convenience of use were impressive and interconnection between ISDN and PHS could be achieved. The first PHS service was started only on 1st July 1995. PHS employs Time Division Multiplexing-Time Division Duplex (TDM-TDD). The voice coding is done by 32 Kbps Adaptive Differential Pulse Code Modulation (ADPCM) to provide toll quality speech. Though the system has been developed for mobile communications yet it is a candidate for use as WILL alone owing to its support to slow mobility [18]. Main features of the standard can be seen in Table 2.

As stated earlier it has been established that interconnection between PHS and ISDN is feasible. Data application has already been established by a demonstration given by NTT in Telecom 95. However the range being small, of the order of 300 to 500 meters, for good performance the number of base stations required is large. The speed of mobility can be held as one of the drawbacks of the system, being the speed of a slow moving car. For the use as WILL this speed is satisfactory since in such application the speed of a slow moving car is more than sufficient [7].

The characteristics of PHS have been its unique numbering plan, high speech quality and low-power phones. It follows a ten digit numbering plan which covers the code for PHS, service provider's code and finally the terminal unit number [18].

The more important parameters of PHS development have been ISDN, C7 as signaling backbone and X.25 signaling protocol [18].

With the trend of liberalization setting in globally, giving rise to fast expansion of the telephone network, and with the explosion of Internet and its entry into the developing world, there is a convergence of explosion in WILL market and requirement of 64 Kbps data support on the WILL. This leads to the necessity that PHS should very soon provide high speed data applications without which the fixed wire will steal the march over it. What we have seen till now is 4.8 Kbps and at the most 32 Kbps data applications.

PHS basically being a Mobile technology, application of the same to WILL fixed radio access use would tend to make it expensive, but with volumes and some extent of de-featuring the cost can come down.

*i) NEC's DCTS (Digital Cordless Telephone System) [19]*

NEC's (DCTS) is one of the important developments in the PHS technology. A DCTS network comprises Base Stations, Cell Stations(CS) and User Terminals(UT). The BS is connected with the local exchange on copper pair (analog or V5.2 interface). The number of users handled by a base station is 120 or 60 depending upon the type of interface used between BS and the local exchange. A CS is located nearly at the geographical center of the group of users it is handling. Each CS can handle upto a maximum of 128 users depending upon the traffic. There are 3 types of UT's. Fixed Terminal(FT), intended for stationary application at a place far from the CS and radio linkage with it through Yagi antenna, 2 wire FT, connected by two wires and is to work within the building and Portable Terminal(PT) which is designed for slow moving mobile usage but within a radius of a few hundred metres. The 2 wire FT supports G3 fax and voice band data transmission of 4.8 Kbps. The connectivity from UT to the CS is thus DCTS. On the other hand the connectivity between the CS and BS (or Repeater unit) is through Digital Radio Multiple Access Subscriber System (DRMASS). Figure 14 shows NEC's DCTS with DRMASS [22].

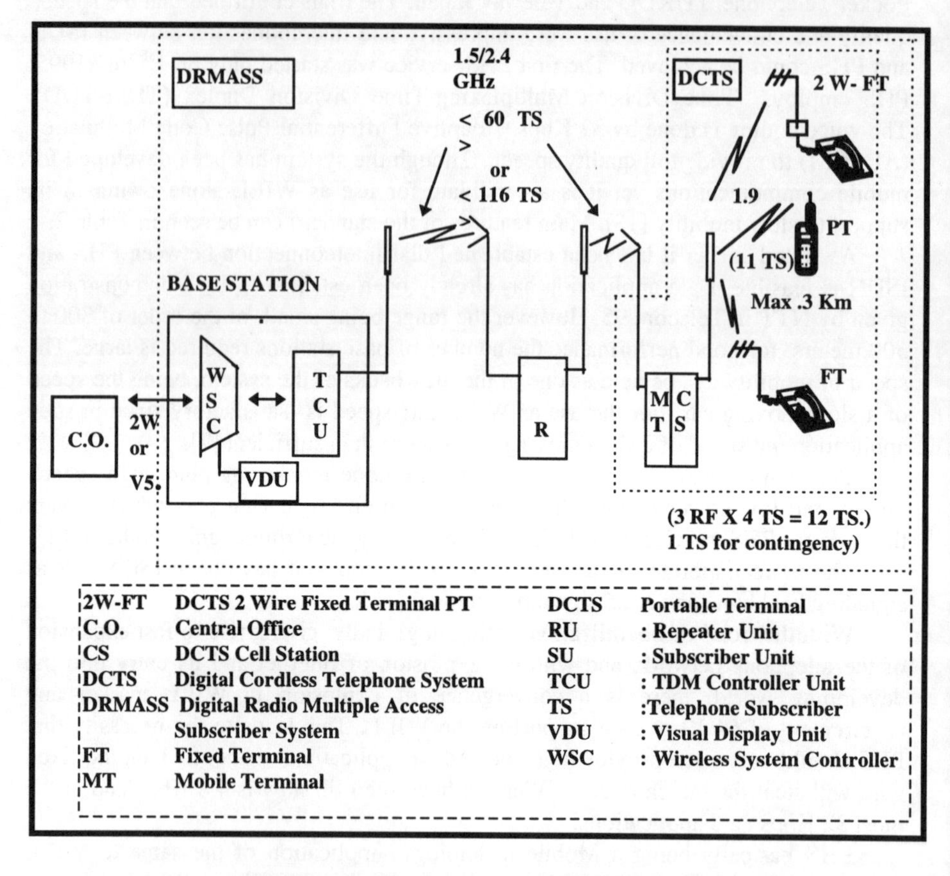


**Figure 14** NEC's DCTS with DRMASS [19]

## 8.5 DEVELOPMENT

Development of the third generation wireless system has been taking place all over the world known as FPLMTS (Future Public Land Mobile Telecommunication System) / IMT-2000 (International Mobile Telecommunication by 2000) In Europe the third generation system is known as UMTS (Universal Mobile Telecommunications System). This is a project of laying down standards and it is estimated that by the year end 1999 the basic standards would be laid down and advance features would be added by the end of 2001. Finally UMTS is targeted for operation by 2005. The main objectives of the UMTS are 1) to have user bit rates of up to 144 Kbps with wide mobility and coverage and upto 2 Mbps with local mobility and coverage, 2) services via handheld, portable, vehicular mounted, movable and fixed terminals, in all radio environments based on single radio technology, 3) high spectrum efficiency compared to the present systems, 4) speech and service quality at least as good as fixed line networks, 5) flexibility for introduction of new services and technical capabilities and 6) radio resource flexibility to multiple networks and traffic types within frequency bands. This is said to be the 3rd generation wireless system after the 1st generation of 1980's with analog systems and 2nd generation of digital systems of late 80's viz., GSM, Personal Digital Cellular (PDC), IS-54 (US TDMA) and IS-95 (US CDMA). In between there is a 2.5 generation which is the result of evolution of the 2nd generation to 100 Kbps data capability and packet switched data.[20].

## 8.6. AIR INTERFACE MULTIPLE ACCESS

A proper choice of air interface multiple access plays an important role in the efficient functioning of WILL. It is particularly important because a WILL system is to serve both the developing and developed markets, therefore a diverse range of services must be supported by the air interface [21]. CDMA seems to be one of the strongest candidates for the future wireless communication[22].

A short note here on CDMA will be appropriate. CDMA is the more recent access technology and as expected there are divergent views from the proponents of the technology and their competitors.

Spread spectrum communication was first used officially during the World War II for secured communication between the leaders of the allied forces and later on quite a few systems were developed in 1970's by the US Military. CDMA came as a spin off for commercial applications and the first such application was the Global Positioning Satellite System (GPSS). CDMA is not known to have been used for telephony applications until 1990. The CDMA could not develop earlier because of non-availability of high speed digital signal processing chips, which have become available now and as a result the technology has developed. CDMA has developed in the USA as a replacement for analog cellular systems and in the UK as a radio access in replacement of the physical local loop[5].

The following benefits have been claimed by the supporters of CDMA [5]:-

1. Efficient use of frequency spectrum.
2. Higher capacity than competing TDMA based systems.
3. Lower transmit power.
4. Ability to combat multipath.
5. Low processing delay.
6. Does not create interference in other systems such as TDMA.

The major gains of CDMA are claimed as simplified frequency planning and for mobile communications frequency re-use in adjacent cells[5].

CEO Qualcomm has defended CDMA giving the following statistics [23] :-

a) MCI has announced purchasing of 10 billion minutes of PCS from NextWave, a CDMA supporting carrier, to offer services to more than 110 million individuals in 63 markets.

b) Korea Mobile Telecom has added 210,000 CDMA customers from January to August 1996 and approximately 515,000 CDMA customers by the end of 1996.

c) Capacity gains of 10 to 20 times of AMPS will be achieved. In Hongkong, where the systems have not been loaded to capacity , the figures were 11 subscribers per MHz per base station for AMPS, 85 for GSM and 283 for CDMA.

It is important to note the capacities of different systems in use in terms of number of calls per cell possible. Without going into the mathematical derivations involved it may be stated that CDMA has a capacity of 5.4 times of GSM and 2.5

times of TDMA [21]. The frequency reuse factor is 1 for CDMA, 7 for TDMA and 3 for GSM. Comparison of TDMA, narrowband CDMA and wideband CDMA from the point of view of trunking efficiency, shows that the optimum solution for WILL is the wideband CDMA scheme [21]. In order to support high quality, high bit rate services on large number of basic telephony circuits, the use of a wideband CDMA system for WILL is very attractive. Conventional CDMA techniques as used for mobile cellular application are not well suited for WILL application since the bit error rate requirements demand the use of costly channel coding if a viable number of simultaneous users are to be supported.

## 8.7. SECURITY ASPECTS [25]

Security in wireless communication is one aspect with which the user has more than a cursory interest. Developers have put in efforts to provide a tough security. At the same time the opposite party has also been active to break the security code. In an article in the International Herald Tribune (dated March 21, 1997) Mr. John Markoff states that a team of computer-security experts has declared that they have cracked a key part of the electronic code meant for protecting calls made with the new digital generation of American cellular telephones. The article however goes on to dispel the fear about GSM technology and states that GSM is still safe since it has a tougher security. It only means that what is safe today may not be safe tomorrow.

It is interesting to note here that when the analog wireless communication system was developed no security was provided and even now the analog systems do not provide any security. With the advent of digital technology even though security was provided yet not much attention was given to privacy. Moreover sometime back under pressure from the US Government the industry accepted a less strong security code in USA. The reasons can be the following:-

1. Expenses.
2. Can't wait for the security system to be in place.
3. To meet US Government's objections.
4. To meet federal export standards in USA.

A very tough encryption could make it easy for criminals and terrorists to conspire.

The existing analog services are progressively getting converted to digital networks as the latter supports more features and is much less noisy. The problem is thus getting compounded.

With the present cracking of the code, the technical details of the code, being available on Internet, it is possible that people start manufacturing and selling small devices for listening in to conversations. Such listening in can be possible from 300 metres to several kilometers. As per the present US laws, manufacturing and selling such devices is not illegal but using them would be. This is not deterrent enough.

Setting the things right at this stage is going to be an enormous task. The software in each switch handling the cellular calls will have to be modified and in USA alone there are over 45 million cellular phone users. Each one of them will have to be retrofitted with the new secure version of the embedded software. There may be a possibility to make it optional on the cellular phone user whether he wants a new version of secure cell phone or continue with the old version running with it the risk of being eavesdropped. This may reduce the size of the problem.

## 8.8. COST ISSUES [26]

Figure 15 compares the lifetime costs for two technologies, the conventional underground copper cable and wireless. The figure is a simplified one to give an idea of the comparison. The wireless curve relates to GSM. The actual costs will depend upon many factors discussed later. It is seen that, for a user density of 250 to 300 per sq km, wireless systems have lower costs. But with advances in wireless technology wireless systems are becoming competitive for higher densities than 300.

The investment in WILL systems is dependent upon the network configuration being targeted and the services that are proposed to be provided The following can act as a list of the factors driving the investment figures [3]:-

a) Technology used.

Decision can be taken by the service provider/investor keeping in view the competitors in the target area. Though in some it has been noticed that the regulatory authorities have laid down criteria which limit the choice on part of the investor , e.g. if the local government lays down that a specific technology has to be used.

b) Services.

WILL systems generally offer a combination of services e.g., voice at 16, 32 or 64 Kbps. Higher the speed higher will be the cost.

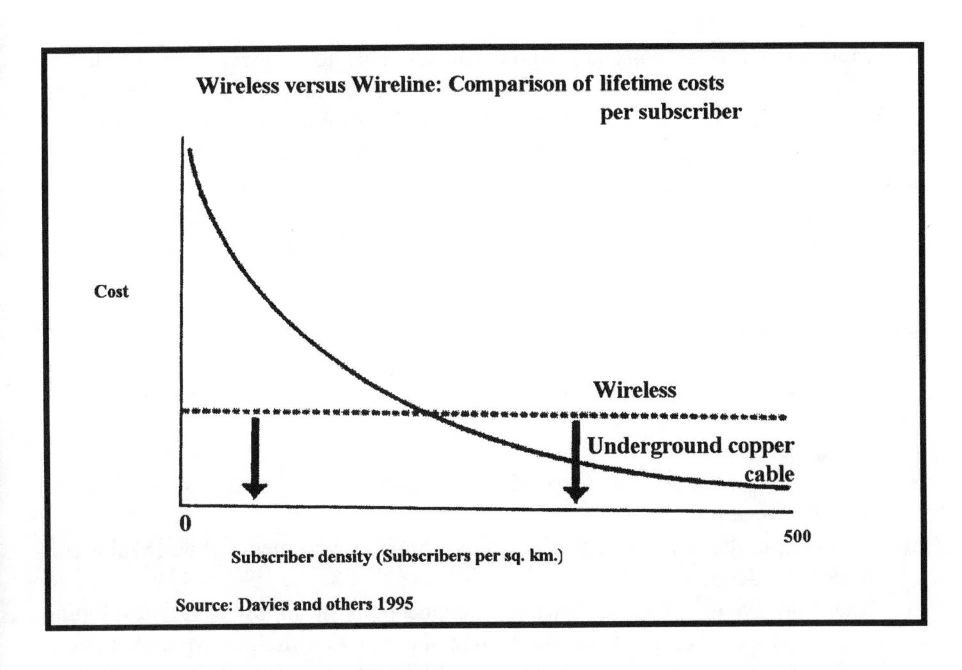


**Figure 15** Wireless versus wireline cost comparison [26]

c) Service quality.

It is natural to expect that the cost per line of WILL will vary with the quality of service provided, higher for better and lower for inferior quality. A trade off can be done here between the quality and cost.

d) Frequency spectrum.

The number of cell sites depends inversely upon the frequency bandwidth available. Lower the bandwidth higher is the number of cell sites required and vice versa. It is evident that the investment will vary inversely with the bandwidth available and is not directly under the control of the investor.

e) Traffic per user.

Higher the traffic generated higher capacity is required. Thus the expected traffic generated will control the cost.

f) User density.

Every WILL system will have a range. If the user density is low, more number of cell sites will be required since the users will be spread far and wide.

g) Existing infrastructure.

The existing infrastructure like underground cables and optical fibre, for connecting the Radio Base Stations and RNC's, may mean lesser investment. Where there is poor infrastructure microwave links may have to be established which will be fast but involve investment.

h) Switching technology.

Most of WILL systems are digital and are able to connect with the digital switches of the telephone network directly. However, in the case of the analog variety of WILL there is a requirement of multiplexers etc. which means additional investment. With the passing of analog WILL slowly into history this aspect is of course not very important.

The investment required would usually vary from US$ 900 to 1200 in the case of high density (urban) areas and 1000 to 1500 in rural areas. This would cover all the WILL equipment and installation cost.

## 8.9 MARKET ISSUES [2]

It can be reasonably stated that the need for telephone access arises from the following :-

a) Pent-up demand,
b) Economic pressures to expand a region or a nation's access to telecommunications and
c) The effect of the new found practice of almost all governments of de-regulation.

a) Pent-up demand.

There are waiting lists for telephone connections in almost every developing country. The reasons for these lists are seen as shortage of switching equipment, non-availability of trunk capacity and finally insufficient local loop access. All these factors either individually or severally result in waiting lists. Even where two are available the absence or delay in availability of the third puts a drag on the other two. Digital switching equipment occupying much lesser space as compared to the old electro-mechanical systems, the building requirement has reduced drastically and the time required for the building construction is minimal. Thus whereas switching equipment is available and also the trunking which is being met by microwave and in many cases optical fibre, the cable laying time is the same resulting in delays. The demand thus waits and it has been estimated that for every 100 applications on the waiting list, on an average, there area about 170 applications not registered, the total demand being 270 . The local loop is, thus, the culprit and there is a need to speed it up, either by conventional means or by newer methods.

b) Economic pressure.

Telephone service is one of the most vital aspects in the economic growth of a country. The per-capita telephone lines is an indicator of the health of the economy of a country. The value ranges from about 0.8 per 100 in India (not the lowest) to about 50 in USA and 85 in Monaco. (Figures are not the latest ones [2]). India alone was requiring about 10 million lines per year to keep pace with the world average. The global requirement of network access, needless to state, is mind boggling.

c) De-regulation.

Historically the telephone services belonged to the governments clubbed with Public Works Department, Postal, Transport etc. It was seen more as a social obligation than as an economic necessity. As the services have grown the demands have grown in terms of features, spread and inter-connectivity. The governmental agencies have been found not able to keep pace and a global phenomenon of de-regulation has started backed by USA and World Bank opinion. With de-regulation the services are expected to improve and prices move downwards. These aspects will within themselves attract more users. Also the new service providers will necessarily have to provide for their own access mechanism since the existing ones will only lead to the existing switching systems and not to the new ones. Thus de-regulation puts two fold pressure on the market for network access.

It is estimated that by the year 2005 radio in local loop access will be required in 25% of all new residential connections and 15% of all new business connections. Out of this 65% will be in Asia Pacific region whereas 16% in Eastern Europe, 15% in Latin America and 4% in Africa [3].

## 8.10 EXTENT OF DEPLOYMENT

Fixed radio access systems are already being deployed in over 50 countries. Recently the Indian Government has enjoined upon the prospective basic service providers to provide the local loop through WILL. In Germany since 1990 over 50000 lines [2] have been provided using Ericsson and Nokia systems. In Budapest an 8500 lines WILL has been installed [2]. In Hungary, Czech Republic and Russia over 300,000 lines are in the process of deployment [2]. The Spanish Government expects its WILL related users to grow to over 800,000 by the end of the current year [2]. The figures will keep rising with the confidence in WILL getting established.

In Japan already there were 3.56 million users in August 1996 using PHS system which gives radio access and the figure was expected to grow to 6 million by March 1997 . The number of base stations was 180000 in March 1996 and will be more than 400,000 by March 1997 [7]. It is interesting to note here that the MPT had estimated that by the year 2000 there would be 3 million users of PHS, 20 million by 2005 and 38 million by 2010 [18]. Now that the estimate for the year 2000 has already been exceeded it has to be seen how soon the years 2005 and 2010 arrive in terms of those estimates. PHS is also spreading to other countries. Australia, Hongkong, Indonesia, Singapore, Thailand have already decided to adopt it where as field trials have started in Malaysia. In Indonesia about 100,000 users' lines and equal amount of fixed telephone terminals are being installed for WILL application which will be operational in the current year [27]. The expansion of PHS in Japan has actually encouraged GTE , a leading US cellular operator, and some European groups to join in Astel Tokyo for providing PHS services in the Tokyo region [28]. In Mexico 1.5 million new users are proposed to be connected through wireless. WILL has already made entry into Columbia.

China, India, Indonesia, Philippines are having a hard look at the possibility of using WILL in a big way to clear their backlogs of telephone lines.

## 8.11 CONCLUSIONS

WILL holds a great promise towards fulfilling the requirement of a huge visible and pent up demand of telephone connections fast. With WILL becoming a commonplace item in the communication arena it will come under pressure to quench the thirst of the user for more and better services since more one gets the more one demands. On the other hand with advance of technology wireless is offering better services in terms of quality and capacity, improved and miniaturized chips resulting in smaller, lighter and lesser power consuming equipment, more advanced digital signal processing capabilities and more interference free working. The end user is, however, not bothered whether it is TDMA, FDMA or CDMA, American standard or European. What he will remain concerned about is that how much is he required to pay for the service and calls and what is the quality? To that extent whatever the proponents have to say in support of their products the user **will** remain the king.

## References

[1] Ramjee Prasad, "Overview of Wireless personal Communications: Microwave Perspectives".

[2] Alan Jacobson, "Wireless Local Loop Technology : Motivations and Alternatives"- Diva Communications on //www.diva.com/wpwll.html.

[3] R Hekmat, KPN Research, Netherlands, in personal discussions with the authors.

[4] Stuart Sharrock, "Start the Countdown on Satellite Roaming", International Herald tribune of 19/2/1997.

[5] Peter Bacon, Netcom Consultants (UK) Ltd., "CDMA-Myth or reality".

[6] Thomas Kaneshige, "The Answer to a Prayer?" in Communication International , Dec. 1995.

[7] Shujo Kato and Kazuhiko Seki, "Development of PCS in Japan", IEEE Fourth Symposium on Communication and Vehicular Technology in the Benelux, Gent, Belgium, October 1996.

[8] Diva Communication, "Features and benefits of the Diva-2000 Wireless Local Loop" at //www.diva.com/features.html.

[9] D R Pulley, "The use of wideband time controlled DS-CDMA for wireless loop", IEE Colloquium organized by Professional Group ES( Radiocommunication System), Savoy Place, London December 1995.

[10] Ian Channing, "Hanging up on CT2", Communication International, 1996.

[11] Peggy Salz-Trautman, "Have Phone Will Travel" in International Herald Tribune of 19/2/1997.

[12] Gustaf Ekberg, Stefan Fleron, Anne-Sofie Jolde of Telia AB, "Accomplished Field trial Using DECT in the local loop", IEEE Communication Magazine-1995.

[13] Bjorn Erik Eskedal, "The DECT System", Communication International, 1996.

[14] Douglas Poslethwaite, "Radio Days", Communication International, April 1996.

[15] Siemens' product brochures.

[16] Peter Kiddle, OBE, "Radio Access Architectures and Equipment Approaches", IEE Colloquium organized by Professional Group ES( Radiocommunication System), Savoy Place, London, December 1995.

[17] Jeffrey Searle, Colin Kellett, Keith Edwards, Richard Hankins, Jie Lin and R McArthur. "Ionica Fixed radio Access System", IEE Colloquium organized by Professional Group ES( Radiocommunication System), Savoy Place, London, December 1995.

[18] Stewart Hampton, "Handy Handset Hints", Communication International, October 1995.

[19] NEC's presentation documents on DRMASS and DCTS.

[20] R Prasad, "UMTS Schedule" from "Third Generation Wireless Mobile Communication", Workshop on Trends in Wireless Personal Communication : Solution for Tomorrow's Society, Delhi December 1996.

[21] D R Pulley and R L Davies, "The use and capacity of Wideband- Time Controlled DS-CDMA for Wireless Local Loop", International Wireless and Telecommunication Symposium, Shah Alam, malaysia, May 1997.

[22] Ramjee Prasad, "CDMA for Wireless Personal Communication Networks", Artech House, 1996.

[23] Dr Irwin Jacobs, CEO, Qualcomm's views in Communications International, November 1996.

[24] Vijay K Garg and E L Sneed, Lucent Technologies Inc., "Digital Wireless Local Loop System", IEEE Communications magazine, October 1996.

[25] John Markoff, "Cracked: Cell Phones' Security Code", International Herald Tribune of 21/3/97.

[26] Peter Smith, Senior Telecom Policy Specialist, Industry and Energy Deptt., World bank note on Internet site www.worldbank.org/html/fpd/notes/63/63Smith.html.

[27] News Release from NEC:

http://www.nec.co.jp/english/today/newsrel/9602/1401.html.

[28] Advanced Cordless Communications: http://www.telecoms.com/acc-02.htm.

# Part 3

# Antennas & Propagation

# 1 SPATIAL AND TEMPORAL COMMUNICATION THEORY USING SOFTWARE ANTENNAS FOR WIRELESS COMMUNICATIONS

Ryuji Kohno

**Abstract:** An adaptive array antenna or a kind of smart antenna is named *a software antenna* because it can programably form a desired beam pattern in flexible if an appropriate set of antenna weights is provided in software. It must be a typical tool for realizing a software radio. I consider it as an adaptive filter in space and time domains for radio communications, so that the communication theory can be generalized from a comventional time domain into both space and time domains. This paper introduces a spatial and temporal communication theory based on a software antenna, such as spatial and temporal channel modeling, equalization, optimum detection for a single user and multiuser for CDMA, precoding and joint optization of both a transmitter and a receiver. Such spatial and temporal processing promises drastic improvement of performance as a practical countermeasure for multipath fading in mobile radio communications.

## 1.1 INTRODUCTION

Recent research interests in a field of personal mobile radio communications have been moving to the third generation cellular systems for higher quality and variable speed of transmission for multimedia information[1], [2], [3]. For the demand in the third wireless personal communications, however, we have several problems which must be conquered. Especially, signal distortion is one of the serious problems; the signal distortion is classified as ISI (InterSymbol Interference) due to the signal delay by going through the multipath channel and CCI (Co-Channel Interference) due to the multiple access. There have already been many measures for combatting signal distortion. A traditional equalizer in a time domain is useful for short time delay signals. However, when the delay time is large, the complexity of the equalization system increases.

An Array antenna, on the other hand, is defined as a group of spatially distributed antennas. The output of the array antenna is obtained from combining properly each element antenna output. By this operation, it is possible to extract the desired signal from all received signals, even if the same frequency band is occupied by all signals. An array antenna can reduce the interference according to the arrival angles or direction of arrival (DOA). Even if the delay time is large, the system complexity does not increase because the array antenna can reduce the interference by using the antenna directivity. Thus, the combination of an array antenna and a traditional equalizer will be able to obtain good performance by compensating each other. That is why it is possible to increase the user capacity, which means the number of available users at one base station, by using an array antenna not only in a time domain but also in a space or an angular space domain. Therefore, spatial and temporal, i. e. two dimensional signal processing based on an array antenna will become a break-through technique for the third generation of wireless personal communications. This concept has been also successfully used for a long time in many engineering applications such as a radar and aerospace technology.

Much research for spatial and temporal signal processing using an adaptive array antenna has pursued in recent years[5]–[4]. Research of adaptive algorithm for deriving optimal antenna weights adaptively in a time domain such as LMS, CMA[23] etc, has been proposed from a viewpoint of extending techniques of a digital filter. On the other hand, there is also research based on DOA estimation from the viewpoint of spectral analysis in a space domain, such as DFT[6], MEM[7] and MUSIC[8]. Adaptive schemes of obtaining the optimal weights are classified into these two groups. *SDMA* that is space division multiple access, a new concept of access scheme, is comparable with FDMA, TDMA and CDMA, and can be combined with them for more user capacity. Its research interest is to investigate how much capacity is improved by using an array antenna. Moreover, since communications technology continues its rapid transition from analog to digital, the fundamental processes, i.e., modulation, equalization, demodulation, etc., have been implemented in software. This is referred to as *software radio architecture* [9][10]. In this sense, we call an adaptive array antenna as a *software antenna* as well. Thus, the analysis of radio communication systems are perfectly simulated on a computer. The design of a radio communication system which includes an air interface has to consider the combination of each fundamental process. Furthermore, hardware implementation of an adaptive array antenna has been recently reported to ensure performance improvement and to evaluate complexity of implementation. A typical style of software antenna is a digital beamformer which is implemented by combination of a phased array, down-converter, A/D converter and field programable arrays or digital signal processers[11],[12].

As the above-mentioned trend, the research area for an adaptive array antenna is expanding to many subjects of signal processing in wireless personal communications. However, there is no communication theory covering the entire subjects based on adaptive array antennas. Therefore, the author's group has been researching a spatial and temporal communication theory based on adaptive array antennas[13] –[14]. This paper briefly introduces an overview of spatial and temporal communication theory for design and analysis of wireless communication systems using adaptive array antennas from a viewpoint of extending traditional communication theory. I hope readers will have interest in capability of an adaptive array antenna by this paper and accelate its R & D.

## 1.2 ADAPTIVE ARRAY ANTENNA

An adaptive array antenna is an antenna that controls its own pattern by means of feedback control while the antenna operates. An adaptive tapped delay line (TDL) array antenna in Fig.1.1, which has TDL or digital filter in each antenna element, also control their own frequency response. The pattern of an array is easily controlled by adjusting the amplitude and phase of the signal from each element before combining the signals.

When the input signal to the TDL array antenna is $x(t)$, the antenna output is represented by

$$y(t) = \sum_{n=0}^{N-1} \sum_{m=0}^{M} x(t - mT_0) w_{n,m} \exp(-jn\varphi), \tag{1.1}$$

where $T_0$ is the delay between adjacent taps, $w_{n,m}$ is the $m$th complex tap coefficient of the $n$th, $N$ is the number of elements and $M$ is the number of taps. $\varphi$ is the phase difference between the received signal at adjacent antenna elements in a line array is given by

$$\varphi = \frac{2\pi L \sin\theta}{\lambda}, \tag{1.2}$$

where $\lambda$, $L$ and $\theta$ are the wave-length of an input signal, the distance between adjacent elements and the arrival angle of the received signal or direction of arrival(DOA), respectively. The antenna transfer function in both spatial and temporal domains is written by

$$H(\omega, \theta) = \sum_{m=0}^{M} \exp(-jm\omega T_0) \sum_{n=0}^{N-1} w_{n,m} \exp(-jn\varphi) \tag{1.3}$$

Equation (1.3) represents the antenna pattern when $\omega$ is a constant, while it represents the frequency response when $\theta$ is a constant.

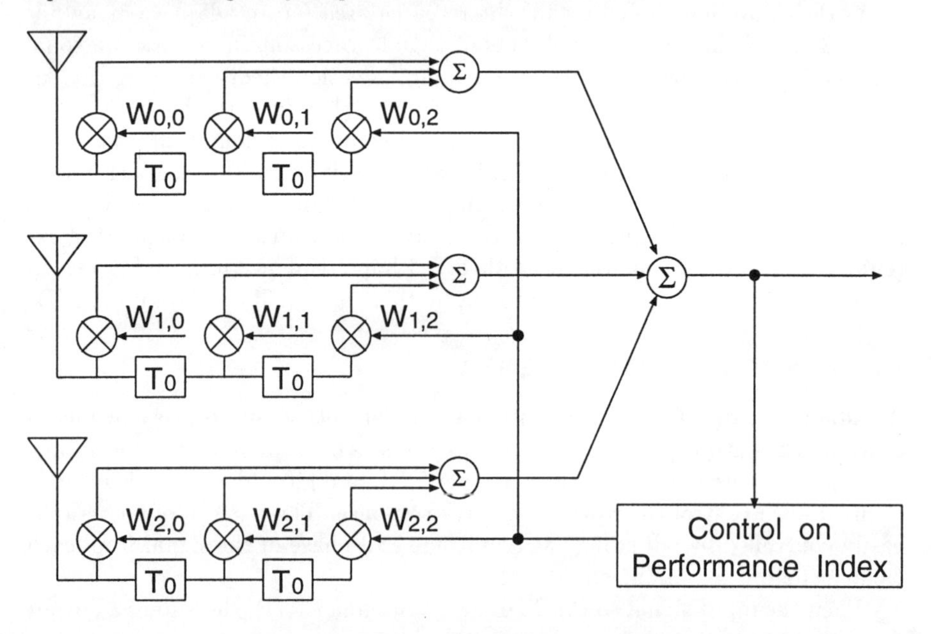


**Figure 1.1** An Adaptive TDL Array Antenna

Therefore, the adaptive TDL array antenna can be employed as a tool of signaling, equalization and detection in spatial and temporal domains.

## 1.3 SPATIAL AND TEMPORAL CHANNEL MODEL

A multipath fading channel such as a mobile radio channel is modeled in which a transmitted signal from one signal source arrives at the receiver with different angles and delays. The received signal is represented by using time $t$ and arrival angle $\theta$.

Each propagation path in a channel is defined by its impulse response at the arrival angle $\theta$ of the received signal. Thus, the channel can be represented by a spatial and temporal, two dimensional model like Fig.1.2. Fig.1.3 illustrates such a spatial and temporal channel model measured by a practical measurement system From this figure, it is noted that individual propagation paths defined by DOA's have different impulse responses.

Therefore, the impulse response of the $k$th path $h_k(t)$ with the arrival angle $\theta_k$, $(k = 1, 2, \cdots, K)$ is represented by

$$h_k(t) = \sum_{i=1}^{I_k} g_{k,i}\delta(t - \tau_{k,i}) \exp(j\psi_{k,i}), \tag{1.4}$$

where $g_{k,i}$,$\tau_{k,i}$ and $\psi_{k,i}$ denote path amplitude, path delay and path phase of the $i$th delayed signal through the $k$th path, respectively. $I_k$ is the number of delayed signals in the $k$th path. $\delta(t)$ is the Dirac delta function.

An equivalent complex baseband signal of received signal $R_n(t, \theta_k)$ in the $n$th antenna element is written by

$$\begin{aligned} R_n(t, \theta_k) &= \sum_{i=1}^{I_k} g_{k,i} S_b(t - \tau_{k,i}) \\ &\quad \times \exp(-j2\pi n L \frac{\sin\theta_k}{\lambda}) \exp(j\phi_{k,i}) \end{aligned} \tag{1.5}$$

where $S_b(t)$ is the complex baseband transmitted signal and $\phi_{k,i}$ is the net phase offset.

## 1.4 SPATIAL AND TEMPORAL EQUALIZATION

By using the above-mentioned spatial and temporal channel model, we can derive an extended Nyquist therem for a known channel. Moreover, for unknown or time-varying channels, various algorithms for updating antenna weights are discussed.

### *1.4.1 Spatial and Temporal Nyquest Theorem*

The Nyquist criterion in both spatial and temporal domains can be derived from Eqs.(1.1) and (1.5). The array output $y(t)$ can be replaced by $y(t, \Theta)$ because the antenna output depends on time $t$ and arrival angle set $\Theta = (\theta_1, \theta_2, \cdots, \theta_K)$.

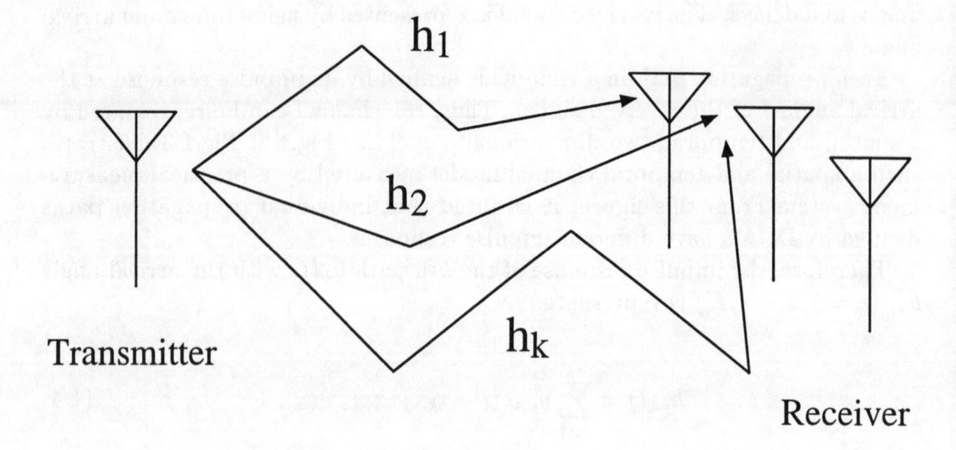


**Figure 1.2** Spatial and Temporal Multipath Channel Model

The sampled antenna output is rewritten as

$$y(lT_d; \Theta) = \sum_{k=1}^{K} p(t, \theta_k), \tag{1.6}$$

where $p(t, \theta_k)$ is defined as

$$\begin{aligned} p(t, \theta_k) \quad &= \quad \sum_{n=0}^{N-1} \sum_{m=0}^{M} \sum_{i=1}^{I_k} w_{n,m} g_{k,i} \exp(j\phi_{k,i}) \\ &\quad \exp(-j2\pi nL \frac{\sin\theta_k}{\lambda}) S_b(t - \tau_{k,i} - mT_0). \end{aligned} \tag{1.7}$$

Assumed that $\theta_1$ represents the desired arrival angle. If $p(t, \theta)$ equals to the symbol $s_l$ at $t = lT_d$ and $\theta = \theta_1$, and equals to zero elsewhere, ISI must be zero. This condition is named the generalized Nyquist criterion in both spatial and temporal domains. Then, the criterion is represented by

$$p(lT_d, \theta) = s_l \delta'(t - lT_d, \theta - \theta_1), \tag{1.8}$$

where $\delta'(t, \theta)$ is the two dimensional Dirac delta function and $s_l$ represents the transmitted symbol at $t = lT_d$. This includes the usual Nyquist criterion when $p(t, \theta_k)$ is a function of time only.

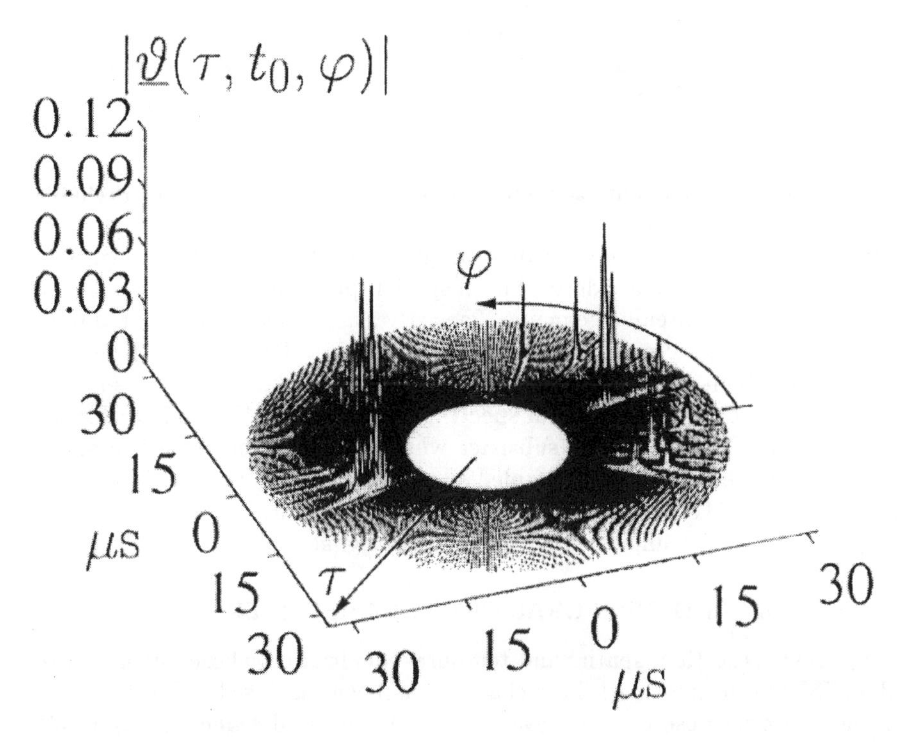


**Figure 1.3** A Real Measured Impulse Response According to DOA (from Ref[15])

### *1.4.2 Adaptive S & T Equalization for Reducing ISI*

Several criteria for spatial and temporal equalization such as ZF (Zero Forcing) and MSE (Mean Square Error), are available to update the weights and tap coefficients. The ZF criterion satisfies the generalized Nyquist criterion in case of noise free if there are infinite number of taps and elements. Since finite number of taps and elements is available in practical noisy multipath channels, there may be some equalization errors in adaptive equalization based on temporal updating algorithms. If the permissible equalization error is given, we can obtain several combinations of taps and elements which achieved the same equalization error. Therefore, the number of antenna elements can be reduced by increasing the number of taps when the difference in arrival angles is large.

A temporal updating algorithm of adaptive array antennas, such as LMS and Applebaum arrays, beamforms to track the desired signal and to suppress interference by nulls so as to maximize array output signal-to-noise ratio (SNR). Applebaum array is also useful when the arrival angle of the desired signal is known in advance. LMS array doesn't require any knowledge for the arrival angle of the desired signal. As long as reference signal correlated with the desired signal can be obtained, the array pattern will adaptively track the desired signal so as to maximize SNR. However, it is difficult to obtain a reliable reference signal in time-varying channels such as a mobile radio channel.

We have proposed and investigated an algorithm for controlling weights of antenna elements which are derived from spatial spectrum of spatially sampled signals by array of antenna elements. The arrival angles can be estimated from spatial frequency spectrum, which can be obtained by DFT or MEM for spatially sampled signals. The weight coefficients are updated by Wiener solution derived from the estimated spatial spectrum. Moreover, MUSIC algorithm estimates arrival angles in noise subspace which is defined by eigen vectors of covariance matrix of spatially sampled signals, while DFT and MEM do it in signal subspace. MUSIC has better estimation performance than MEM if noise subspace is larger for uncorrelated signals than signal subspace.

## 1.5 SPATIAL AND TEMPORAL OPTIMUM RECEIVER

In the previous section, spatial and temporal equalization whose purpose is to reduce ISI due to multipath in a channel has been discussed. Viterbi equalization whose purpose is to achieve maximum likelihood sequence estimation utilizing ISI can be also generalized in spatial and temporal domains if an array antenna is employed. The tandem structure of a matched filter(MF) and a maximum likelihood sequence estimator(MLSE) or Viterbi Detector(VD) has been considered to be an optimum receiver in the presence of ISI and AWGN. The optimum receiver is generalized into spatial and temporal domains in this section.

### *1.5.1 Spatial and Temporal Matched Filter*

First, the spatially and temporally whitened matched filter (ST-WMF) is derived using a TDL array antenna. The SNR at the TDL array antenna output at $t = K_0 D$ is represented using the delay operator $D$ as

$$\mathrm{SNR} = \frac{\sigma_s^2 \left| \mathbf{w}^T(D,\theta) \sum_{l=0}^{L-1} h_l(D) \mathbf{q}(\theta_l) \right|^2}{\sigma^2 \left| \mathbf{w}(D,\theta) \right|^2}, \tag{1.9}$$

where $h_l(D)$, $\mathbf{w}(D,\theta)$ and $\sigma_s^2$ denote the impulse response of th $l$th path, $N$-dimensional impulse response of the array at the arrival angle $\theta$, and the variance of the input sequence $x(D)$, respectively. From Schwarz's inequality, the optimal weight vector $\mathbf{w}(D,\theta)$ ($N$-dimensional) so as to maximize the SNR at the TDL array antenna output is given by the time inversion $h_l(D^{-1})D^{K_0}$ ($l = 0, 1, \cdots, L-1$) of the impulse response and the directivity information $\phi$ as

$$\mathbf{w}(D,\theta) = \sum_{l=0}^{L-1} h_l(D^{-1})D^{K_0}\phi, \tag{1.10}$$

where $K_0$ satisfied the duration time $max_{l=0,1,\cdots,L-1}\{K_l\} \leq K_0$, the time inversion of the impulse response denotes a temporal WMF (T-WMF), and a directivity information denotes a spatial WMF (S-WMF) represented by $\phi = [w_0, w_1, \cdots, w_{N-1}]^T$. A S-WMF for the $l$th path is realized by the complex conjugated operation of received phase difference $\exp(jn\pi \sin\theta_l)$ due to the arrival angle. Therefore, a S-WMF coefficient $w_n$ in the $n$th antenna element is

$$w_n = \sum_{l=0}^{L-1} \exp(jn\pi \sin\theta_l). \tag{1.11}$$

This is the generalization of the WMF in both spatial and temporal domains.

### *1.5.2 Spatial and Temporal Optimum Receiver*

Fig.1.4 shows a VD, which is connected to a ST-WMF built using a TDL array antenna. We call it a receiver with ST-WMF and VD or a spatial and temporal optimum receiver. As a special case, a receiver with a S-WMF & VD and that with a T-WMF & VD are included. The detection algorithm in the proposed receiver is described as follows. (i) Each antenna element receives signals. (ii)Received signals in each antenna element are filtered by a ST-WMF, which is matched to the transmission channel impulse response. (iii)The maximum likelihood sequence is estimated for the ST-WMF output. The symbol error probability $P(e)$ of the proposed optimal receiver in the spatial and temporal domains is bounded by

$$P(e) \leq \alpha Q(d_{min}/2\sigma), \tag{1.12}$$

where $d_{min}$ is the minimum Euclidean distance, and $\alpha$ is a small constant. $Q(\cdot)$ is the error function.

Since ISI is taken into account, the transmission rate $\mathcal{R}_{st}$ is derived as

$$\mathcal{R}_{st} = W \log \frac{\sigma_s^2 \sum_{k=0}^{2K-1} |g_k|^2 + \sigma^2}{\sigma^2}, \tag{1.13}$$

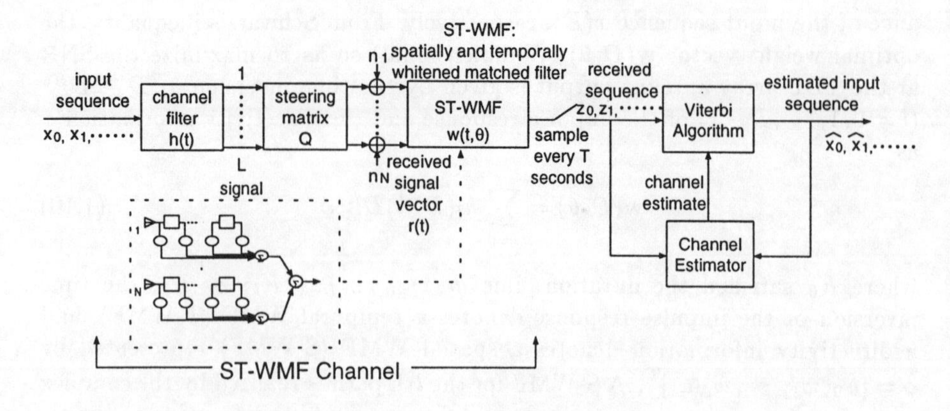


**Figure 1.4** Spatail and Temporal Optimum Receiver

where the input has variance $\sigma_s^2$. Fig.1.5 shows the BER and the transmission rate of the proposed ST-WMF and VD receiver in comparison with other receivers.

### *1.5.3 Spatial and Temporal Optimum Multiuser Receiver for CDMA*

A direct-sequence CDMA(DS/CDMA) mobile radio communication channel is modeled as a channel with both ISI due to multipath and co-channel interference(CCI) due to the correlation between spreading sequences of simultaneously accessing users. The optimum multiuser receiver for DS/CDMA detects every user's data in a sense of MLSE by utilizing CCI as redundant information which multiple accessing users share. By using an adaptive TDL array antenna, we derive such spatial and temporal optimum multiuser receiver that MLSE for every user's data can be achieved with both CCI and ISI in DS/CDMA.

The receiver has an extended structure in Fig.1.4 so that correlators for every user are located in front of the ST-WMF, the ST-WMF is modified to be multiple input/output structure with cross coupling and is followed by multiple VD. The detection algorithm in the proposed receiver is described as follows. (i)Each antenna element receives signals. (ii)Received signals in each element are filtered by each user's correlator. (iii)Each user's correlator output vector is filtered by each user's ST-WMF, which is matched to each user's channel impulse response. (iv)Each user's maximum likelihood sequence is estimated for each user's ST-WMF output, where the path metric is calculated taking into

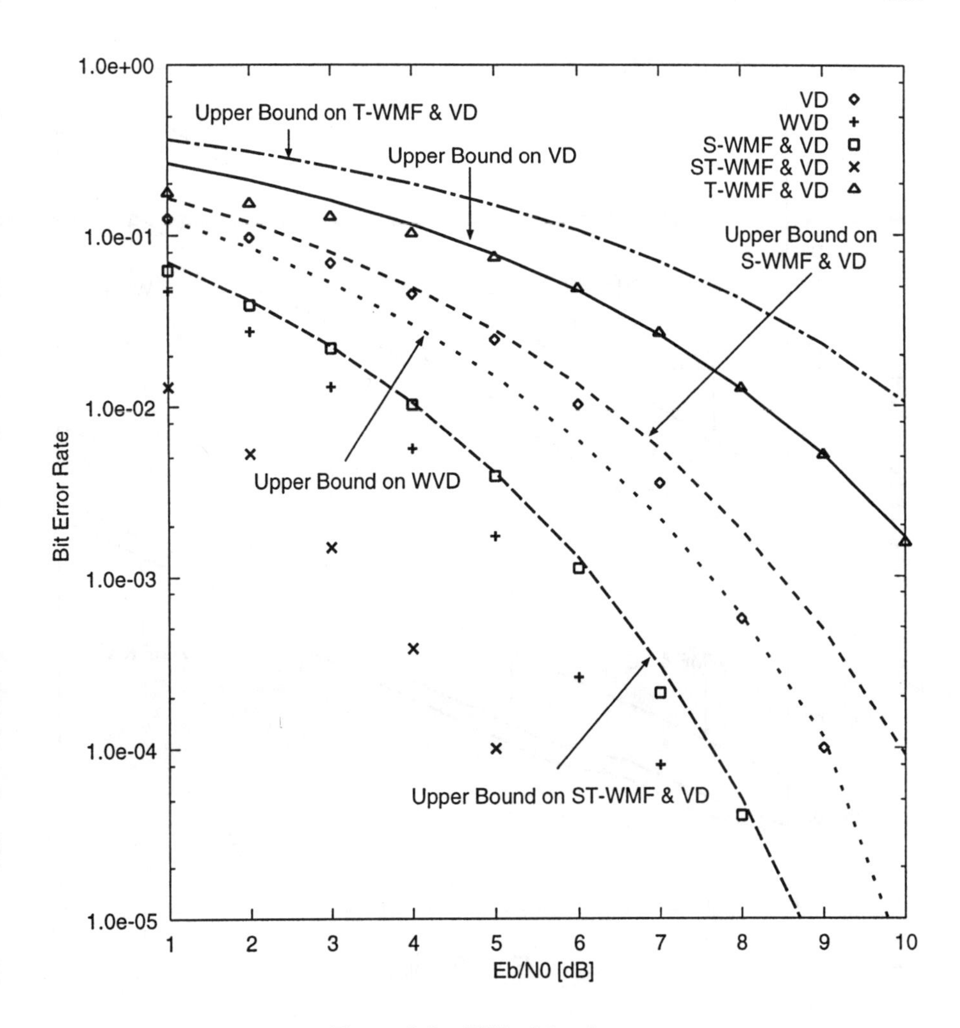


**Figure 1.5** BER of Receivers

consideration the influence of CCI. Fig.1.7 shows the BER of the ST optimum multiuser receiver in comparison with other receivers.

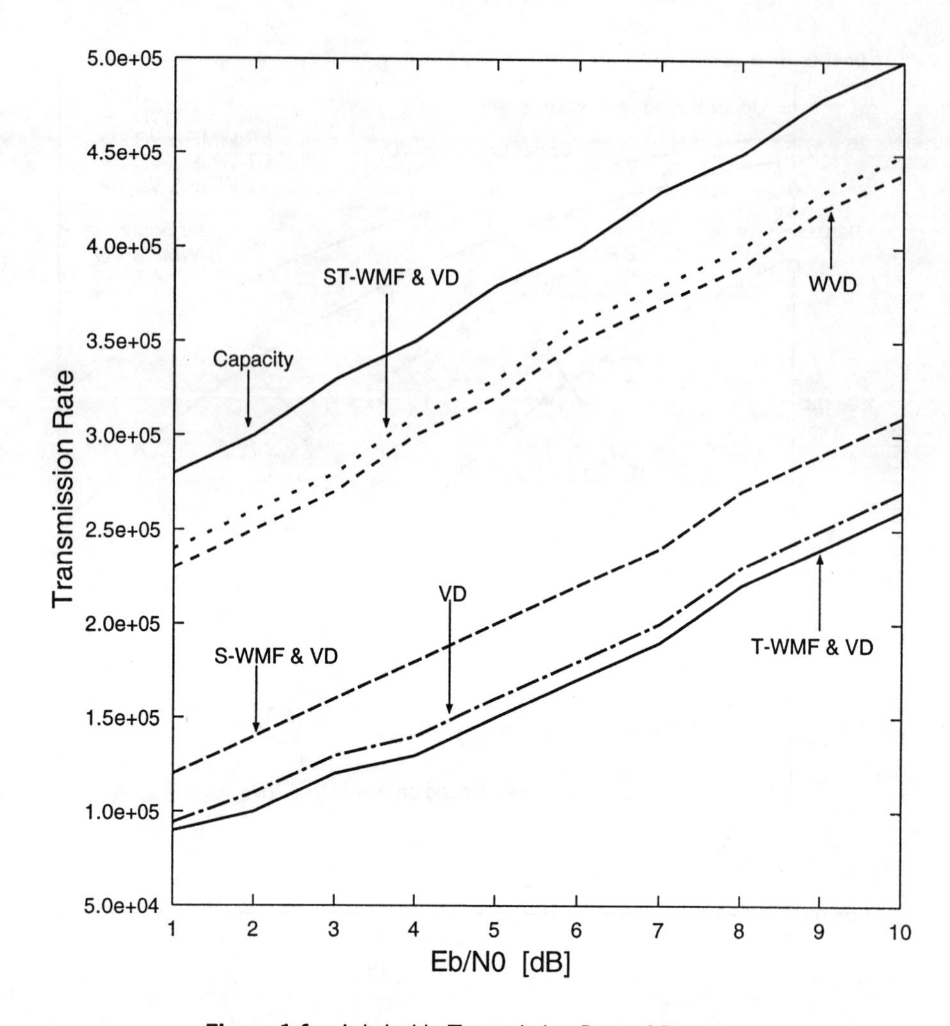


**Figure 1.6** Acheivable Transmission Rate of Receivers

## 1.6 SPATIAL AND TEMPORAL JOINT EQUALIZER IN TRANSMITTER AND RECEIVER

### *1.6.1 Signal Configuration for Transmitter and Receiver*

The input sequence is defined using delay operator $D$ ($D$-transform) as

$$x(D) = x_0 + x_1 D + x_2 D^2 + \cdots . \tag{1.14}$$

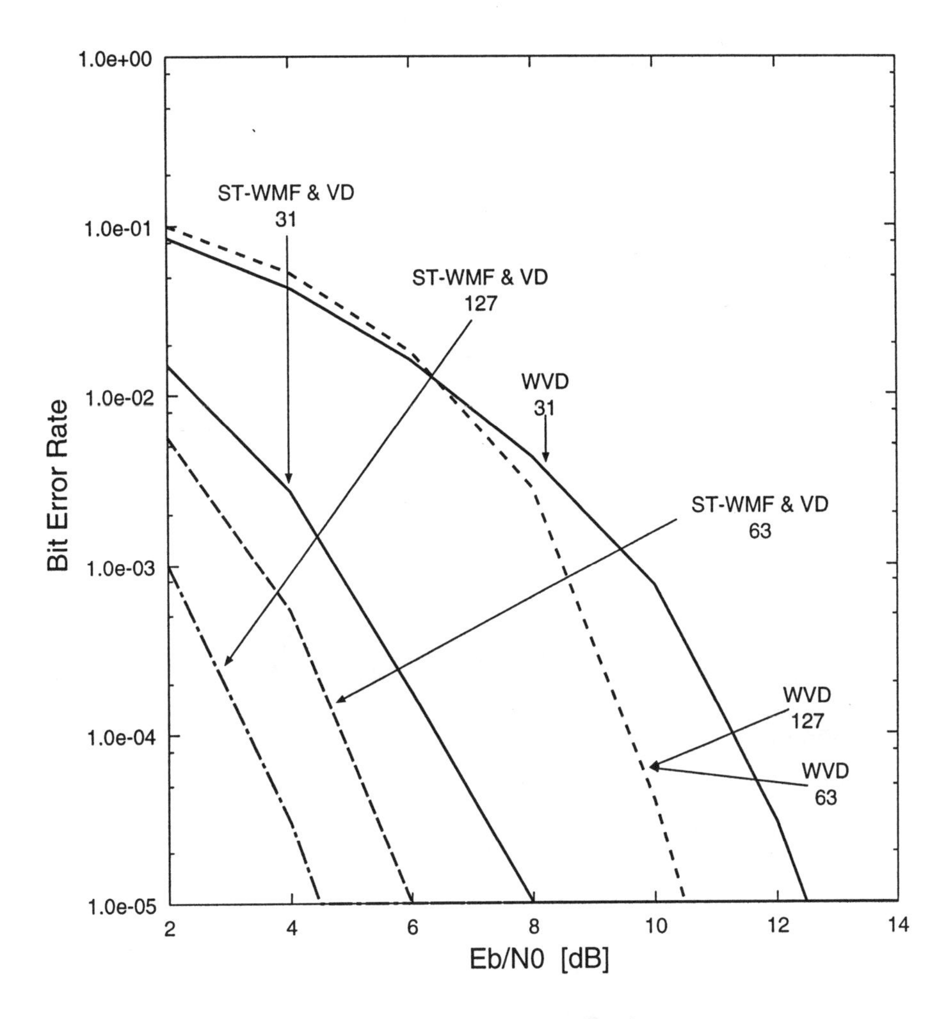


**Figure 1.7** BER of Multiuser Receivers

It is assumed that the signals can be transmitted for $K$ directions. Figure 1.9 shows the transmission array antenna. The number of antenna weight sets is $K$ and these antenna weight sets can be transmitted for $K$ direction. The transmission array antenna in temporal part are characterized by the vector of finite impulse response $W_{t,i}(D)$ ($i$=0,1,$\cdots$,$K$–1) ( $M$-dimensional ) of length $M_i$ symbol interval; i.e., $M_i$ is the smallest integer such that $W_{t,i}(t)$=0 for

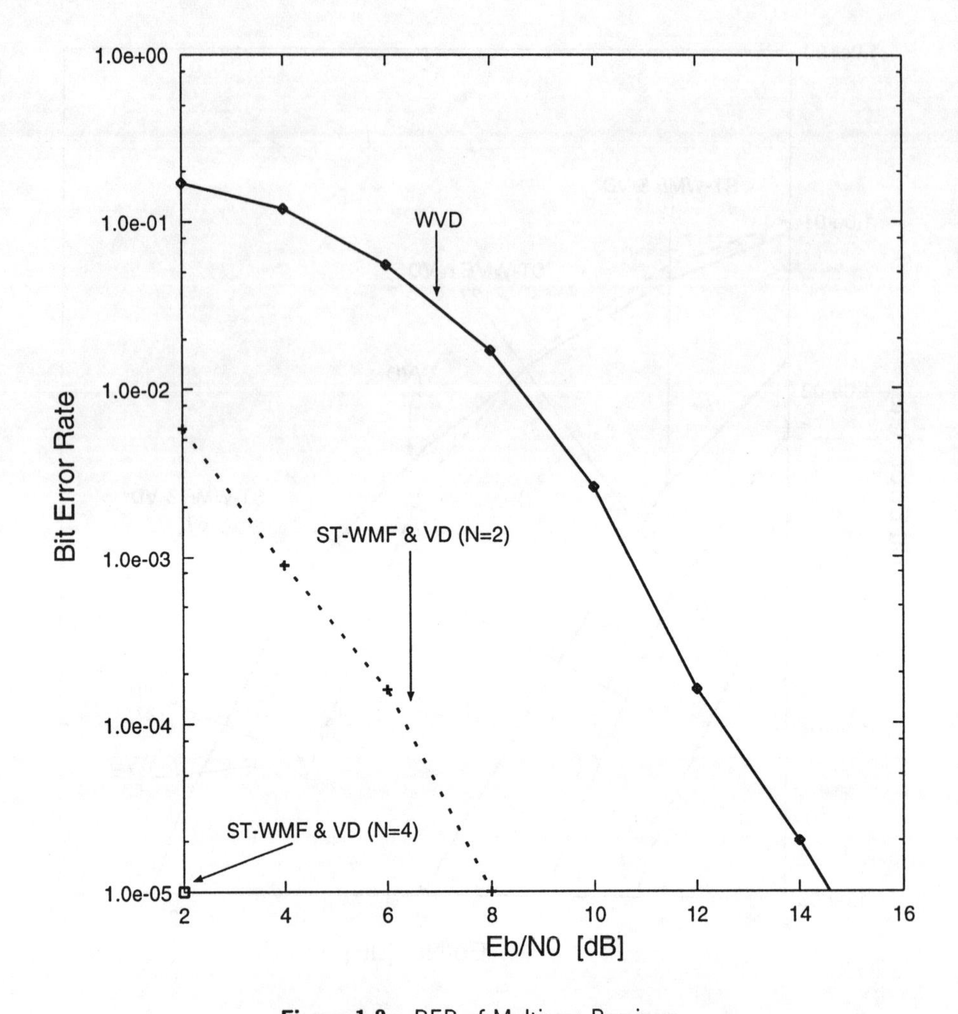


**Figure 1.8** BER of Multiuser Receivers

$t \geq M_i T$. Where $M$ donates the number of transmission antenna elements. The transmission vector is defined by

$$W_t(D) = \begin{bmatrix} W_{t,0}(D) & W_{t,1}(D) & \cdots & W_{t,K-1}(D) \end{bmatrix}^T \tag{1.15}$$

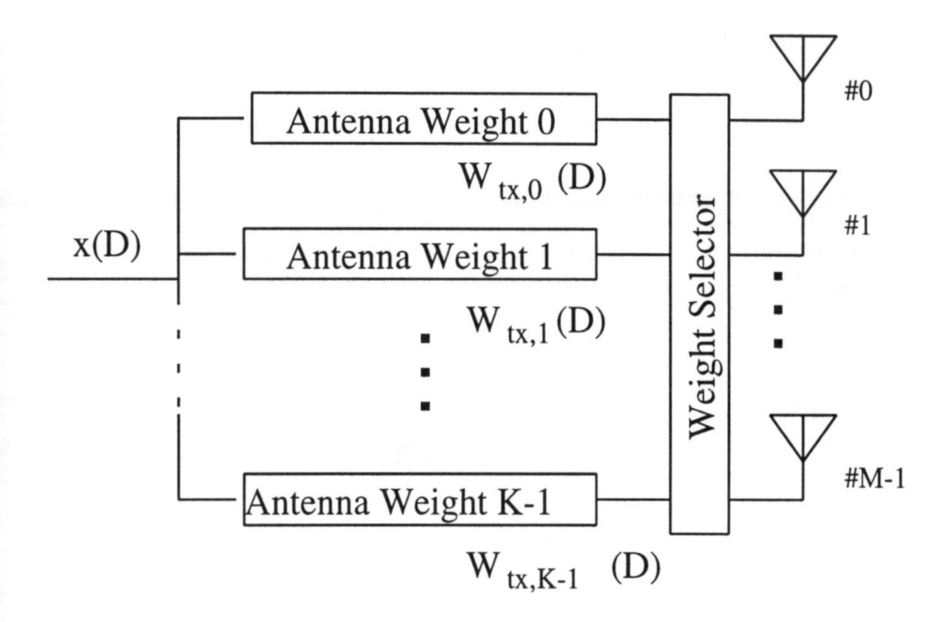


**Figure 1.9** Structure of Transmission Array Antenna

Then, this vector is changed to $K$-dimensional vector by transmission steering converter in spatial domain.

$$Q_t = \begin{bmatrix} q_t(\phi_0) & 0 & \cdots & 0 \\ 0 & q_t(\phi_1) & \cdots & 0 \\ \vdots & \vdots & \cdots & \vdots \\ 0 & 0 & \cdots & q_t(\phi_{K-1}) \end{bmatrix} \tag{1.16}$$

Let the steering vector for bearing $\phi$ be $q_t(\phi)$ ( $K$-dimensional ) given by

$$q_t(\phi) = \left[ \begin{array}{cccc} 1 & e^{-j\pi \sin\phi} & \cdots & e^{-j(M-1)\pi \sin\phi} \end{array} \right]. \tag{1.17}$$

The transmitted sequence is represented by the convolution of the input sequence $x(D)$ and the transmission array antenna weight $Q_t W_t(D)$ as

$$\begin{aligned} X(D) &= (Q_t W_t(D))^T x(D) \qquad (1.18) \\ &= \left[ \begin{array}{cccc} X_0(D) & X_1(D) & \cdots & X_{K-1}(D) \end{array} \right]^T. \end{aligned}$$

It is assumed that the channel for each arrival angle $\theta_l$ at the receiver array antenna is characterized by a vector

$$h_l(D) = [\ h_{l,0}(D) \quad h_{l,1}(D) \quad \cdots \quad h_{l,K-1}(D)\ ] \tag{1.19}$$

where $l$=0,1,⋯,$L-1$ and $L$ denotes the total number of paths. We defined the path as its past history due to bearing. The $l^{th}$ signal sequence $s_l(D)$ is represented by the transmitted sequence vector $X(D)$ and the $l^{th}$ channel vector $h_l(D)$ as

$$\begin{aligned} s_l(D) &= h_l(D)X(D) = \sum_{k=0}^{K-1} h_{l,k}(D)X_k(D) \\ &= \sum_{k=0}^{K-1} h_{l,k}(D)q_t(\phi_k)W_{t,k}(D)x(D). \end{aligned} \tag{1.20}$$

Therefore, the spatial and temporal channel change the signals from $K$ inputs to $L$ outputs.

Since a receiver using a TDL array antenna can sample signals in the spatial and temporal domains, each transmission path is represented not only by time but also by bearing or direction of arrival (DOA). If the receiver is a omnidirectional antenna, the signal sequence is simply composed of sequences through all paths as

$$s(D) = \sum_{l=0}^{L-1} s_l(D). \tag{1.21}$$

It is assumed that

1. A uniform linear array (ULA) is employed in the receiver.

2. All incoming signals are plane waves.

3. Bearing is defined as the angle from broadside.

Let the reception steering vector for bearing $\theta$ be $q_r(\theta)$ ($N-$dimensional) given by

$$q_r(\theta) = [\ 1 \quad e^{-j\pi \sin\theta} \quad \ldots \quad e^{-j(N-1)\pi \sin\theta}\ ], \tag{1.22}$$

where $N$ denotes the number of antenna elements.

When the signals are received by the array antenna which has $N$ antenna elements, the signal sequence vector $\mathbf{r}_l(D, \theta_l)$ through the $l^{th}$ path ($N$-dimensional) is given by $D$ and the bearing $\theta_l$ as

$$\mathbf{r}_l(D, \theta_l) = q_r(\theta_l)s_l(D), \tag{1.23}$$

where $\theta_l$ denotes the bearing of the $l^{th}$ path. Then the received signal sequence vector $\mathbf{r}(D,\theta)$ actually is composed of signal sequence vectors through each path and a white Gaussian noise vector of zero mean and variance $\sigma^2$ as

$$\mathbf{r}(D,\theta) = \sum_{l=0}^{L-1} \mathbf{r}_l(D,\theta_l) + \mathbf{n}_0. \tag{1.24}$$

In this section, it is assumed that the received signal via the first path is the desired one. Therefore, the received signal sequence vector $\mathbf{r}(D,\theta)$ is represented by the sum of the desired signal sequence vector $\mathbf{s}_0(D,\theta_0)$, the interfering signals sequence vectors $\mathbf{r}_l(D,\theta_l)$ $(l \neq 0)$ and the noise sequence vector $\mathbf{n}_0$ as

$$\mathbf{r}(D) = \mathbf{r}_0(D,\theta_0) + \sum_{l=1}^{L-1} \mathbf{r}_l(D,\theta_l) + \mathbf{n}_0. \tag{1.25}$$

Let the array impulse response vector be $W_r(D,\theta)$ ($N$−dimensional). The output sequence $z(D)$ is given by

$$z(D) = W_r^T(D,\theta)\mathbf{r}(D,\theta). \tag{1.26}$$

The channel model is represented by Fig. 1.10. The matrix $H$, $Q_r$ and vector

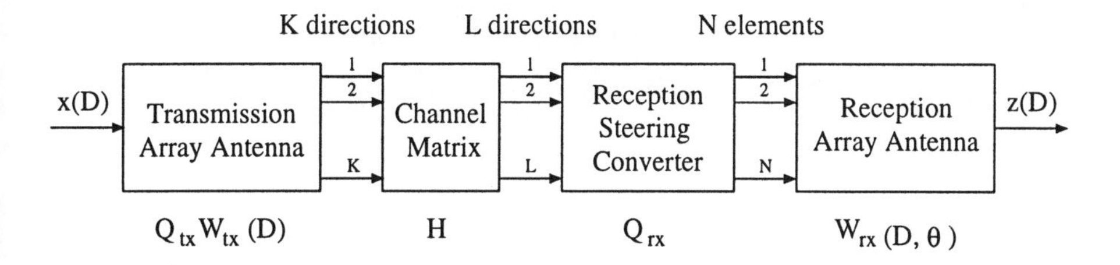


**Figure 1.10** Channel Including Transmitter and Receiver

$W_r$ are represented by

$$H = \begin{bmatrix} h_{0,0}(D) & h_{0,1}(D) & \cdots & h_{0,K-1}(D) \\ h_{1,0}(D) & h_{1,1}(D) & \cdots & h_{1,K-1}(D) \\ \vdots & \vdots & \vdots & \vdots \\ h_{L-1,0}(D) & h_{L-1,1}(D) & \cdots & h_{L-1,K-1}(D) \end{bmatrix}, \tag{1.27}$$

$$Q_r = \begin{bmatrix} q_r(\theta_0) & 0 & \cdots & 0 \\ 0 & q_r(\theta_1) & \cdots & 0 \\ \vdots & \vdots & \cdots & \vdots \\ 0 & 0 & \cdots & q_r(\theta_{L-1}) \end{bmatrix}, \tag{1.28}$$

$$W_r = [\; W_{r,0}(D) \quad W_{r,1}(D) \quad \cdots \quad W_{r,N-1}(D) \;]. \tag{1.29}$$

### *1.6.2 Derivation of the ST-WMF*

If the noise spectral density is large, the conventional Viterbi decoder (VD) may produce an estimate that is different from the input sequence due to the influence of noise. To improve the performance, a whitened matched filter (WMF) is considered. A usual WMF, i.e., temporally WMF (T-WMF), maximizes the SNR at the filter output by compensating for time difference or ISI in the received signals after they are merged in a tapped delay line (TDL) array antenna. However, there are also phase differences between the received signals at the input of the TDL array antenna. That is, a TDL array antenna can sample signals in both spatial and temporal domains. Therefore, in this section we design a spatially and temporally WMF (ST-WMF) using a TDL array antenna. The derivation of a ST-WMF is as follows. The SNR at the TDL array antenna output at $t = \lambda D$ is given by

$$\text{SNR} = \frac{\sigma_s^2 \left| W_r^T(D,\theta) \sum_{l=0}^{L-1} \mathbf{q}(\theta_l) g_l(D) \right|^2}{\sigma^2 \left| W_r(D,\theta) \right|^2} \tag{1.30}$$

$$g_l(D) = \sum_{k=0}^{K-1} h_{l,k}(D) \mathbf{q}_t(\phi_k) W_{t,k}(D), \tag{1.31}$$

where $\sigma_s^2$ denotes the variance of the input sequence $x(D)$. From Schwarz's inequality, the optimal weight vector $W_r(D,\theta)$ ($N$-dimensional) so as to maximize the SNR at the TDL array antenna output is given by the time inversion and conjugate $g_l^*(D^{-1})D^\lambda$ $(l = 0, 1, \cdots, L-1)$ of the impulse response and the directivity information $\mathbf{\Phi}$ as

$$w_r(D,\theta) = \sum_{l=0}^{L-1} g_l^*(D^{-1}) D^\lambda \mathbf{\Phi}, \tag{1.32}$$

where $\lambda$ satisfied the sum of the maximum delay of the channel and the delay of the transmission filter and * donates the complex conjugate. The time inversion and conjugate of the impulse response denotes a T-WMF, and a directivity information denotes a spatially WMF (S-WMF) represented by

$$\boldsymbol{\Phi} = [w_0, w_1, \cdots, w_{N-1}]^T. \tag{1.33}$$

A S-WMF for the $l^{th}$ path is realized by the complex conjugate operation of received phase difference $\exp(jn\pi \sin\theta_l)$ due to bearing. Therefore, a S-WMF coefficient $w_n$ in the $n^{th}$ antenna element is

$$w_n = \sum_{l=0}^{L-1} \exp(jn\pi \sin\theta_l). \tag{1.34}$$

This is the generalization of the WMF in both spatial and temporal domains. A T-WMF and a S-WMF are considered to be special kinds of the above mentioned ST-WMF.

The temporally optimal weight vector $W_T(D)$ is given by making the directivity information $w_n = 1$ $(n = 0, 1, \cdots, N-1)$ in Eq. (1.32). That is

$$W_T(D) = \sum_{l=0}^{L-1} g_l^*(D^{-1}) D^{\lambda}. \tag{1.35}$$

While, the spatially optimal weight vector $W_S(\theta)$ is consisted of only the directivity information $\boldsymbol{\Phi}$ so as to beamform to track the signal component due to ISI. That is

$$W_S(\theta) = \boldsymbol{\Phi}. \tag{1.36}$$

Using the above ST-WMF, we propose an optimal receiver in the spatial and temporal domains that uses a ST-WMF and a VD which performs MLSE on the TDL array antenna output. As a special case, a receiver with a S-WMF and VD and a receiver with a T-WMF and VD are included.

### *1.6.3 Joint Transmitter-Receiver System*

Figure 1.11 shows a MLSE, which is connected to a ST-WMF built using a TDL array antenna. The detection algorithm for received sequence is described as follows.

1. Each antenna element receives signals.
2. Received signals in each antenna element are filtered by a ST-WMF, which is matched to the transmission channel impulse response.
3. The maximum likelihood sequence is estimated for the ST-WMF output.

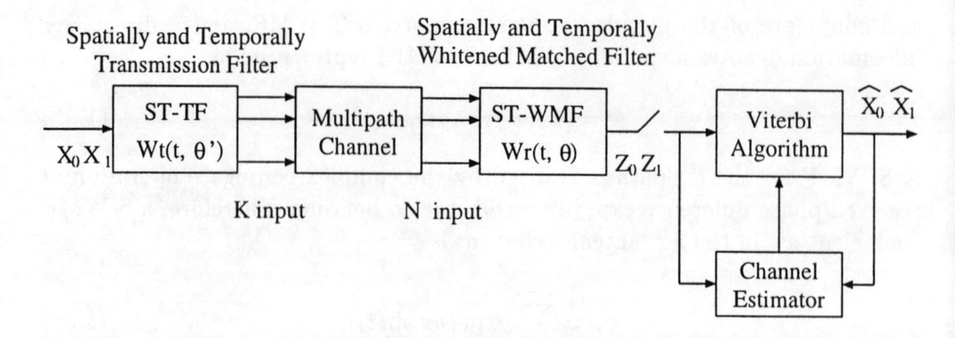


**Figure 1.11** MLSE Based on ST-WMF and VD

If AWGN is filtered by the WMF, the noise component at the WMF output becomes correlated. Therefore, the effect of the correlated noise must be removed by the noise whitening filter. Since the noise becomes correlated due to filtering by the WMF, the noise whitening filter is the inverse filter of the ST-WMF. Thus, the difference between the output sequence $z(D)$ and it's estimate $\hat{z}(D)$ is filtered by the noise whitening filter.

### *1.6.4 Upper Bound of Symbol Error Rate*

The output sequence $z(D)$ of a ST-WMF is represented by the sum of the signal component $y(D)$ and the whitened noise component $n(D)$ as

$$z(D) = y(D) + n(D), \tag{1.37}$$

where $y(D)$ is described with the $l^{th}$ signal sequence vector $\mathbf{s}_l(D, \theta_l)$ and the array impulse response vector $\mathbf{w}_r(D, \theta)$ as follows

$$\begin{aligned} y(D) &= \mathbf{w}_r^T(D, \theta) \sum_{l=0}^{L-1} \mathbf{r}_l(D, \theta_l) \\ &= f(D)x(D). \end{aligned} \tag{1.38}$$

The impulse response $f(D)$ of the joint ST-WMF channel which includes a transmitter, a channel and a receiver, is given by

$$\begin{aligned} f(D) &= \mathbf{w}_r^T(D) \sum_{l=0}^{L-1} \mathbf{q}(\theta_l) g_l(D) \\ &= f_0 + f_1 D + \cdots + f_{2\lambda-1} D^{2\lambda-1}, \end{aligned} \tag{1.39}$$

$2\lambda-1$ is the total memory of the joint channel including the transmitter and receiver. The squared minimum Euclidean distance $d^2_{min}$ for the joint ST-WMF output is defined by

$$d^2_{min} = ||f||^2. \tag{1.40}$$

The energy of the impulse response $f(D)$ is defined by

$$||f||^2 \equiv \sum_{i=0}^{2\lambda-1} f_i^2. \tag{1.41}$$

The symbol error probability $P(e)$ of the proposed optimal receiver in the spatial and temporal domains is bounded by

$$P(e) \leq \alpha Q(d_{min}/2\sigma), \tag{1.42}$$

where $\alpha$ is a small constant and $Q(\cdot)$ is the error function.

### *1.6.5 Acheivable Transmission Rate $\mathcal{R}_{st}$*

It is assumed that the receiver would ideally filter out all received signals and noise outside the band $(-W, W)$, and the noise samples are independent and have variance $\sigma^2$. The transmission rate $\mathcal{R}$ [bps] is given by

$$\mathcal{R}_{st} = W \log \frac{\sigma_s^2 \sum_{i=0}^{2\lambda-1} |f_i|^2 + \sigma^2}{\sigma^2}, \tag{1.43}$$

where the input has variance $\sigma_s^2$. This formular takes into account the ISI.

### *1.6.6 S & T Joint Multiuser Transmitter-Receiver System for CDMA*

The performance in a multi-user environment is decided by not only the noise but also the CCI. Therefore, the Viterbi Equalizer has to be revised to take into account the influence of CCI.

The multi-user equalization system is represented by Fig. 1.14 At first, the received signals at the antenna elements are processed by multiplying each user's PN sequence. Then each user's output at each antenna element is processed by a spatial filtering bank and divided into the signals whose arrival angles are different. These signals include the CCI component or cross-correlation of other users.

The Viterbi Equalizer in a multi-user environment takes into account the CCI component when the path metric is calculated. The $j^{th}$ user's path metric

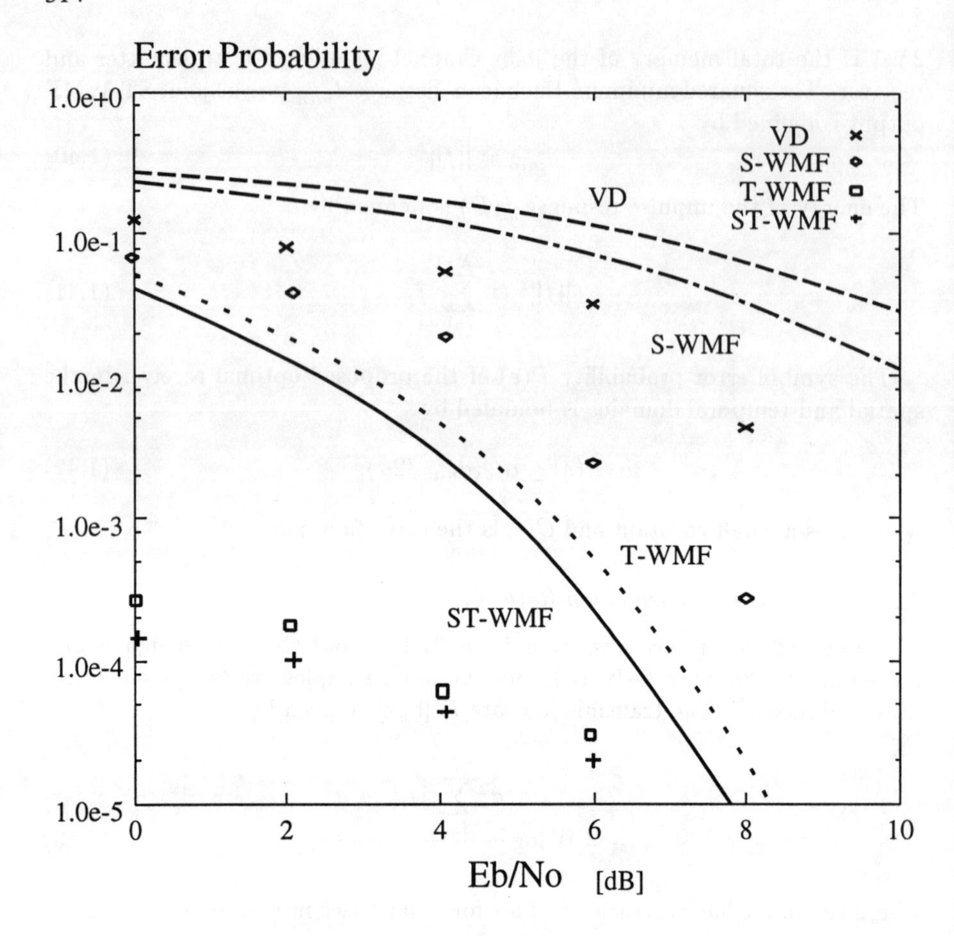


**Figure 1.12** BER Performance

$\mu_n^j(I_n^j)$ at time = n is calculated by using other users' CCI components like this:

$$\mu_n^j(I_n^j) = \mu_{n-1}^j(I_n^j) + \left( z_k^j - \sum_{i=1}^{J} R_{i,j} \hat{z}_k^i \right)^2, \tag{1.44}$$

where $z_k^j$ and $\hat{z}_k^j$ represent the output of the spatial filter bank for the $j^{\text{th}}$ user and the estimated output for the $j^{\text{th}}$ user, respectively. $R_{i,j}$ shows the correlation between user $i$ and user $j$.

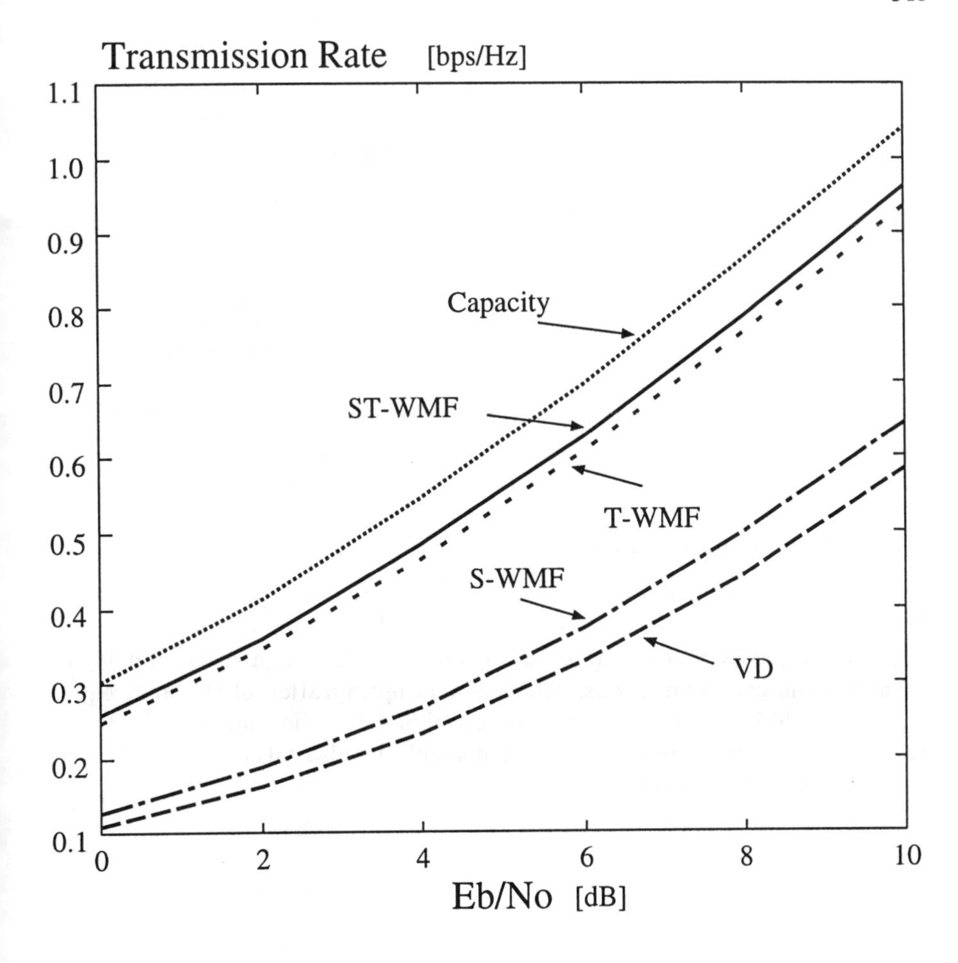


**Figure 1.13** Acheivable Transmission Rate

Fig. 1.15 illustrates that the bit error probability depends on the number of users. In a digital cellular system, an increase in channel capacity is needed and by using the proposed system a high capacity is expected. Moreover, Fig. 1.16 illustrates that the achievable transmission rate of the proposed system can be close to the channel capacity when the number of users is small.

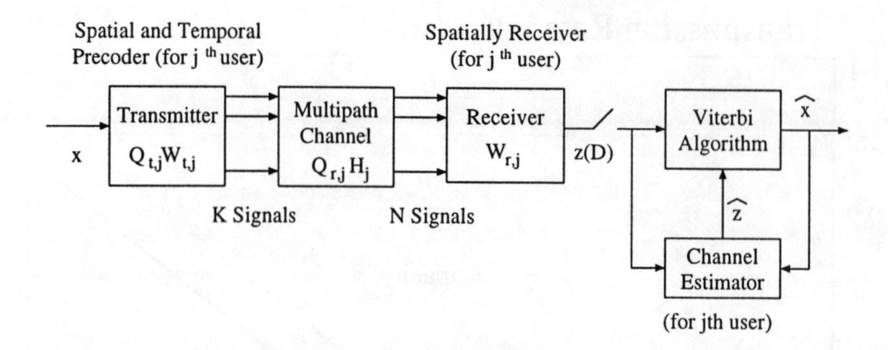


**Figure 1.14** S & T Joint Multiuser Transmitter-Receiver System for CDMA

## 1.7 CONCLUDING REMARKS

Feasibility of implementing an adaptive array antenna is increasing in higher frequency bands such as millimeter wave band. The spatial and temporal communication theory will be more important to achieve high-speed and highly reliable radio communications. Analysis and optimization of the joint equalization should be acheived for more practical mobile radio channels. For future research, modulation and demodulation should be also taken into account for software radio architecture.

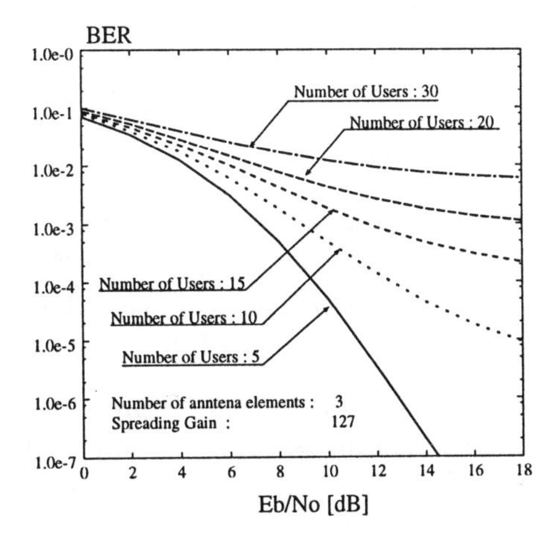


**Figure 1.15** BER as a Function of the Number of Users in S & T Joint Transmitter-Receiver System for CDMA

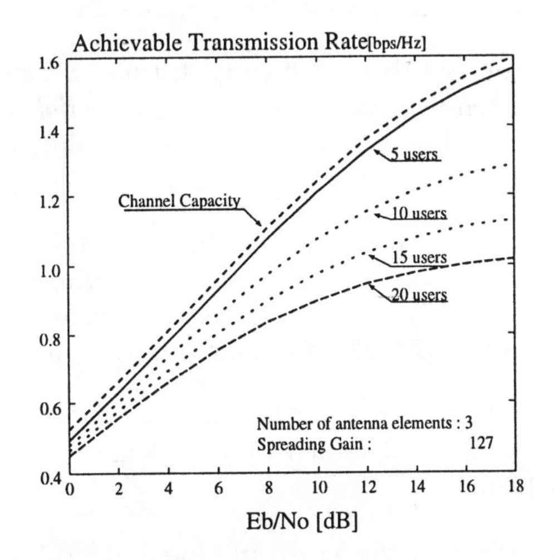


**Figure 1.16** Achievable Transmission Rate as a function of the Number of Users in S & T Joint Transmitter-Receiver System for CDMA

## References


[1] D. L. Schilling, "*Wireless Communications Going Into the 21st Century,*" IEEE Trans. Vehic. Tech., vol. 43, no. 3, pp. 645 – 652, Aug. 1994.

[2] D. D. Falconer, F. Adachi, and B. Gudmundson, "*Time Division Multiple Access Methods for Wireless Personal Communications,*" IEEE Commun. Mag., vol. 33, no. 1, pp. 50 – 57, Jan. 1995.

[3] R. Kohno, R. Meidan, and L. B. Milstein, "*Spread Spectrum Access Methods for Wireless Communications,*" IEEE Commun. Mag., vol. 33, no. 1, pp. 57 – 67, Jan. 1995.

[4] J. S. Thompson, P. M. Grant, and B. Mulgrew, "*Smart Antenna Arrays for CDMA Systems,*" IEEE Pres. Commun., vol. 3, no. 5, pp. 16 –25, Oct. 1996.

[5] B. D. V. Veen and K. M. Buckley, "*Beamforming: A Versatile Approach to Spatial Filtering,*" IEEE ASSP Mag., pp. 4 – 24, Apr. 1988.

[6] R. Kohno, C. Yim and H. Imai, "*Array Antenna Beamforming Based on Estimation on Arrival Angles Using DFT on Spatial Domain,*" Proc. of the 2nd International Symposium on Personal, Indoor and Mobile Radio Communications (PIMRC'91), London, UK, pp. 38 – 43, Sept. 1991.

[7] M. Nagatsuka, N. Ishii, R. Kohno and H.Imai, "*Adaptive Array Antenna Based on Spatial Estimation Using Maximum Entropy Method* ", Proc. of the 4th International Symposium on Personal, Indoor and Mobile Radio Communications (PIMRC'93), Yokohama, Japan, pp.567-671, Sept. 1993.

[8] R. O. Schmidt, "*Multiple Emitter Location and Signal Parameter Estimation* ",IEEE Trans. Antennas and Propagation, vol. AP-34, No.3, pp.276 – 280, March 1986.

[9] J. Mitola, "*The Software Radio Architecture,*" IEEE Commun. Mag., vol. 33, no. 5, pp. 26 – 38, May 1995.

[10] J. Kennedy and M. C. Sulivan, "*Direction Finding and "Smart Antennas" Using Software Radio Architectures,*" IEEE Commun. Mag., vol. 33, no. 5, pp. 62 – 68, May 1995.

[11] Y. Karasawa and H. Inomata, "*Research on Digital and Optical Beamforming Antennas in Japan*," Proc. JINA'96, pp. 159 – 168, Nov. 1996.

[12] P.E. Mogensen, et al., "*A Hardware Testbed for Evaluation of Adaptive Antennas in GSM/UMTS*," Proc. IEEE PIMRC'96, pp. 540 – 544, Oct. 1996.

[13] R.Kohno, "*Spatial and Temporal Filtering for Co-Channel Interferece in CDMA*," in Chapter 3 of *Code Division Multile Access Communications*, edited by S.G.Glisic and P.A.Leppanen, Kluwer Academic Publishers, pp.117–146, 1995.

[14] N. Ishii, "*Signal Design and Detection Theory Based on an Adaptive Array Antenna in Spatial and Temporal Domains*," a doctoral thesis in Yokohama National University, Dec., 1996.

[15] J.J. Blanz, P. Jung and P.W. Baier, "*A Flexibly Configurable Statistical Channel Model for Mobile Radio Systems with Directional Diversity*," AGARD SPP Symposium, Athenes, Greece, pp. 38-1 – 38-11, 1995.

[16] Shahid U.H.Qureshi : "*Adaptive Equalization*," Proceedings of the IEEE, Vol. 73, No. 9 Sep. 1985

[17] R. Kohno, H. Imai, S. Pasupathy : "*Adaptive Array Antenna Based on Combination of Spatial and Temporal Filtering for Channels with Multipath Distortion*," EURASIP Fifth European Signal Conference, Proceedings, Barcelona Sep. 1990

[18] N.Kuroiwa, R.Kohno, H.Imai : "*Design of a Directional Diversity Receiver Using an Adaptive Array Antenna*," Trans. IEICE Japan, J73-B-II, No. 11, pp. 755–763 Nov. 1990

[19] R.Kohno, H.Wang, H.Imai : "*Adaptive Array Antenna Combined with Tapped Delay Line Using Processing Gain for Spread-Spectrum CDMA Systems*," PIMRC'92, Boston, Massachusetts 1992

[20] M.Nagatsuka, N.Ishii, R.Kohno and H.Imai, "*Array Antenna Based on Spatial Spectrum Estimation Using Maximum Entropy Method*," IEICE Trans. Commun., vol.E77-B, No.5, pp.624 – 633, May 1994

[21] N.Ishii and R.Kohno, "*Spatial and Temporal Equalization Based on an Adaptive Tapped Delay Line Array Antenna*," IEICE Trans. Commun., vol.E78-B, No.8, pp.1162 – 1169, August 1995

[22] R.Kohno, "*Spatial and Temporal Filtering for Co-Channel Interferece in CDMA*," IEEE International Symposium on Spread Spectrum Tech. & Appli., pp.51 - 60, July 1994

[23] R. T. Compton. Jr., "*Adaptive Antennas: Concepts and Performance*," Prentice Hall, Englewood Cliffs, NJ 1988

[24] G. D. Forney, Jr., "*Maximum-likelihood Sequence Estimation of Digital Sequences in The Presence of Intersymbol Interference*," IEEE Trans. Inform. Theory, vol.IT-18, pp.363–378, May 1972

[25] F. R. Magee, jr., J. G. Proakis, "*Adaptive Maximum-Likelihood Sequence Estimation for Digital Signaling in the Presence of Intersymbol Interference*," IEEE Trans. Inform. Theory, vol.IT-19, pp. 120 - 124, Jan. 1973

[26] R. Kohno , H. Imai and M. Hatori , "*Cancellation techniques of co-channel intererence in asynchronous spread spectrum multiple access systems*," Trans. IEICE, vol. J66-A, no. 5, pp. 416 - 423, May 1983

[27] S. Verdu, "*Minimum probability of error for asynchronous Gaussian multiple access channels*," IEEE Trans. Inform. Theory, vol. IT-32, pp. 85 - 96, Jan. 1986

[28] R. Kohno, H. Imai, M. Hatori and S. Pasupathy, "*Combination of an Adaptive Array Antenna and a Canceller of Interference for Direct-Sequence Spread-Spectrum Mutliple-Acess System*," IEEE J. Select. Areas Commun., vol. 8, p. 675 - 682, May 1990

[29] R. Kohno, H. Imai, M. Hatori and S. Pasupathy, "*An adaptive canceller of co-channel interference for spread spectrum multiple access communication networks in a power line*," IEEE J. Select. Areas Commun., vol. 8, p. 691 - 699, May 1990

[30] M. Nagatsuka, R. Kohno and H. Imai, "*Optimal Receiver in Spatial and Temporal Domains Using Array Antenna*," International Symposium on Information Theory and Its Applications 1994, pp. 893 - 898, Nov. 1994

[31] Shmuel Y. Miller and Stuart C. Schwartz, "*Integrated Spatial-Temporal Detectors for Asynchronous Gaussian Multiple-Access Channels*," IEEE Trans. on Commun., vol. 43, pp. 396 - 411, Feb./March/April 1995

[32] Rafal Krenz and Krzysztof Wesolowski, "*Comparison of Several Space Diversity Techniques for MLSE Receivers in Mobile Communications*," IEEE International Symposium on Personal Indoor and Mobile Radio Communications (PIMRC'94), vol.II, pp.740–744, Sept. 1994

[33] A. Saifuddin, R. Kohno and H. Imai, "*Integrated Co-Channel Interference Cancellation and Decoding Scheme over Fading Multipath Channel for CDMA*," Trans. IEICE, Vol. J77-B-II, No. 11, pp. 608 – 617, Nov. 1994

[34] A. Saifuddin, R. Kohno and H. Imai, "*Cascaded Combination of Cancelling Co-Channel Interference and Decoding of Error-Correcting Codes for CDMA*," Proc. of IEEE ISSSTA'94, pp. 373 – 377, Oulu, Finland

[35] A. Saifuddin and R. Kohno, "*Performance Evaluation of Near-Far Resistant Receiver for DS/CDMA Cellular System over Fading Multipath Channel*," Trans. IEICE, Vol. E78-B, No. 8, pp. 1136 – 1144, Aug. 1995

[36] A. Saifuddin, R. Kohno and H. Imai, "*Integrated Receiver Structure of Staged Decoder and CCI Canceller for CDMA with Multilevel Coded Modulation*," Europ. Trans. on Telecomm. and Related Technol., Vol. 6, No. 1, pp. 9 – 19, Jan.-Feb., 1995

# 2 REVIEW OF RAY MODELING TECHNIQUES FOR SITE SPECIFIC PROPAGATION PREDICTION

George Liang
Henry L. Bertoni

**Abstract:** A review of modeling techniques that employ ray optics and the Uniform Theory of Diffraction (UTD) for site specific propagation prediction in the UHF band will be given. A number of different methods will be described which include: a two dimensional ray trace in the horizontal plane, a method employing two 2D ray traces in orthogonal planes, ray tracing in full three dimensional space and the Vertical Plane Launch (VPL) technique, which approximates a full 3D ray trace. When employed for microcells with ranges that are less than one kilometer, these techniques can provide accurate predictions of the propagation characteristics in a cluttered urban environment. The strengths and weaknesses of each type of ray tracing implementations will be addressed along with direct comparisons between three of the methods and with measurements in typical urban environments.

## 2.1 HISTORIC PERSPECTIVE ON PROPAGATION MODELING

Propagation modeling within urban environments increased in importance with the advent of commercial mobile communication systems and continues to be an area of research interest. The initial deployment of cellular mobile radio (CMR) systems used cells whose radius exceeded tens of kilometers. In order to support the installation of this system, the goal of the early modeling was to quantify the path loss with a few parameters. Attention was given to the development of empirical models based on a limited set of experimental measurements. Currently, there are major efforts underway to provide ubiquitous mobile communications in the form of Personal Communication Services (PCS) to an ever growing number of users. This has lead to the use of microcells to reduce coverage area, thereby allowing greater frequency reuse in order to accommodate a higher density of users. This in turn has led to a need to understand the radio wave propagation characteristic in a more detailed manner. Over the small ranges used for microcells, statistical prediction models based on measurements can show considerable error, especially in areas having mixed building sizes. In contrast, ray tracing techniques that are able to find the dominant propagation paths can be expected to exhibit accuracy over these small ranges that is superior to the statistical models.

One of the underlying difficulties with developing a propagation model for any environment is that no two areas are identical in the composition of the buildings and terrain. Therefore it is often difficult to apply a model developed for one area to another area with a slightly different environment. Empirical models usually begin with a general qualitative classification into urban, suburban and rural areas with further quantitative characteristics, such as building density, area covered and locations, as well as the average building height, vegetation density and terrain variation being specified within each of these areas [1]. **Okumura** et al [2] developed a series of curves use to predict the field strength based on a set of measurements in and around Tokyo for a very high base station antennas at VHF and UHF bands. In this approach the **path loss** between the base station and mobile in an urban area is found by the adding the free space path loss to a median attenuation factor $A_{mu}$, which is determined from empirical graphs at various frequencies from 100-3000 MHz and distance ranging from 1-100 km [2]. From Okumura's results, Hata developed a series of computationally efficient formulas that calculated propagation loss with small errors for frequencies between 100-1500 MHz, distances in the range of 1-20 km, base station antenna heights between 30-200 m, and mobile antenna height of 1-10 m [3]. This formula for predicting path loss in urban areas is the standard from which correction factors are added to account for large or medium-small cities and for suburban and rural areas.

Although the models developed by Okumura and Hata are widely quoted for propagation prediction in cities, they are useful only for large coverage areas in excess of 1 km, and high base stations. In the case of microcells, where the maximum range of coverage may not exceed 1 km and the base station antenna is near to or below the rooftops of the surrounding buildings, a different model must be used to determine the path loss. In order to support the implementation of microcells there have been a large number of microcellular measurements performed in urban environments. For example, Whitteker conduct a series of measurements for street-lamp height base station antennas and a 3.65 m mobile antenna height at 910 MHz along LOS and non-LOS streets in the urban core of Ottawa, Canada [4]. Although he did not develop a specific microcellular model he did concluded that the path loss was approximately free space along the LOS streets while it was 20 dB or more below free space for non-LOS streets. Harley formulated a model based on measurement conducted in Melbourne, Australia for base station antenna heights between 5-20 m [5] along LOS streets. He found that the measured path loss was considerably lower than the path loss found by extrapolating Okumura's or Hata's results and concluded that it would be unreasonable to extend their models for microcells.

Empirical microcellular models were also formulated for non-LOS regions based on measurements made in Manhattan at 894 MHz [6]. These measurements were the basis for a non-LOS coverage model [7] that was developed for cities with a rectilinear street grid. Although the model agrees well with the measurements for Manhattan, it is uncertain whether the model can be applied to any other environment. Measurements were also conducted for urban and suburban areas around San Francisco Bay by Xia et al [8] at 900 and 1900 MHz and with various base station antenna heights. Regression fits to the measurements were made and compared for LOS and non-LOS streets. It was found that for non-LOS paths, the path loss is significantly effected by the intervening buildings. It was also shown that a propagation model can be developed by incorporating building location parameters and street orientation.

## 2.2 RECENT TRENDS USING THEORETICAL MODELING METHODS

While measurement based models have been the predominate method of characterizing the propagation environment, for microcells it has become apparent that theoretical models can be used to predict the path loss in an urban area. The simplest of these is the multiple ray model [9, 6] for LOS street along urban canyons. This model accounts for interference between the direct ray, the two rays that reflected once from each wall of the buildings lining the street, and the corresponding ground reflected rays, to predict the variation of the signal down a street. As is the case for empirical models, it is desirable to obtain a model that can be applied to both LOS and non-LOS regions. From the measurements conducted by Ikegami et al [10, 11] it was postulated that the field strength could be predicted in non-LOS regions based on a few dominate ray paths that propagate over uniform height buildings down to the mobile. The European research committee, COST 231, has developed a model based on the results of Ikegami [11] and Walfisch [12] for estimating the path loss over a series of buildings of uniform height and separation. The model is expected to provide an average error of around 4 dB and a standard deviation of 6 dB. The COST 231 **Walfisch-Ikegami** formulation has been compared with measurements for a number of European cities including Mannheim and Darmstadt, Germany [13] and Lisbon, Portugal [14] with slightly better than expected prediction results.

Although COST 231-WI provides a good estimate of the path loss, it is valid only for areas with buildings of fairly uniform heights located on a regular street grid, and when the base station antenna is near or above the rooftops. For the areas where the buildings are not uniform, or when the base station is immerse within the surrounding buildings, it is necessary to developed a site specific propagation model based on ray optics and the **Geometrical Theory of Diffraction** [15]. These **ray tracing** models utilize the **shoot and bounce ray (SBR)** method or the method of images and treat each building as a individual scatterer in order to predict the power at any given location. Reflections from the building walls are usually approximated by the Fresnel reflection coefficient [16], while diffraction from the building edges are approximated using the diffraction coefficient for a perfectly absorbing [17], perfectly conducting [18] or a dielectric [19] wedge. The use of ray tracing techniques for propagation prediction stems from indoor propagation modeling [20, 21], where the propagation environment is in a confined area and for the most part requires that only reflection and transmission at walls be accounted for.

### *2.2.1 Two Dimensional Methods*

Outdoor ray tracing models generally need to account for diffraction at the building edges in order to accurately predict the path loss, especially in non-LOS regions. For a low base station antenna in a high rise urban environment, almost all the models assume that the buildings are infinite in height and therefore the ray trace is confined to the horizontal plane. The simplest of these assume a very simple and gross building shapes and then identifies the few dominate rays that reflect (LOS) and/or diffract (non-LOS) to the mobile [22, 23]. A level of sophistication beyond this is to use one of the ray tracing methods to find all ray paths having a number of building interactions (i.e. reflections and edge diffractions) up to some predetermined maximum. These techniques include ray launching at equally spaced interval along the full 360° arc around the transmitter in the horizontal plane [24] or image methods [25, 26, 27], and apply UTD to find the path loss at the mobile.

For these 2D models, the ray trace is confined to the horizontal plane by neglecting propagation over the buildings. The buildings are assumed to be infinitely high and the walls of the buildings are approximated as being vertical. This assumption allows the model to represent the buildings as polygons with an arbitrary number of vertices. Figure 2.1 shows the top view of a simple

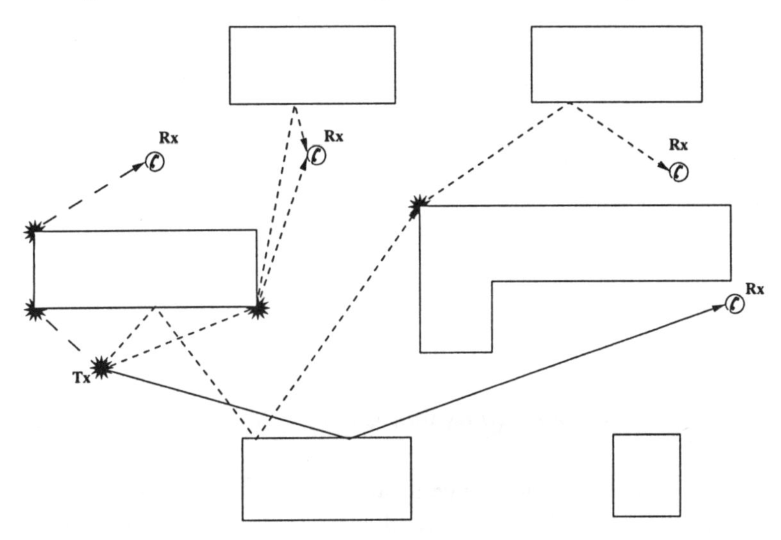


**Figure 2.1** 2D ray tracing in the horizontal plane.

layout of six buildings. A reflected ray is generated by the intersection of a ray launched from the source with a wall. Multiply reflected rays are generated in the same manner by rays reflecting from a wall at the specular direction. Diffractions at the vertical edges (corners) of a building are modeled in a two step process, with the edges being treated as a receiver initially to determine all the rays that illuminate it. After all the rays have been traced from the actual transmitter, each illuminated corner is treated as a secondary source for the subsequent ray tracing. By employing the mechanisms of reflections from the walls and diffraction from the vertical edges of the buildings, the propagation around a dense high rise urban environment can be accurately predicted.

On the other hand when the transmitter is located on the rooftop of a buildings or is near to rooftops of the surrounding low raise buildings, it is necessary to consider the propagation over the buildings. As mentioned previously, the COST 231-WI model provide good predictions for the case when the transmitter is above the rooftop and the building heights and location are nearly uniform, and demonstrates the importance of this mechanism. For irregular building environments, ray tracing techniques can be employed to predict the signal due to propagation over the buildings. The ray trace is normally performed in the vertical plane that contains the transmitter and the receiver and cuts across the intervening buildings to produce a profile of the building edges. Figure 2.2 shows an example of a ray trace in the vertical plane for a transmit-

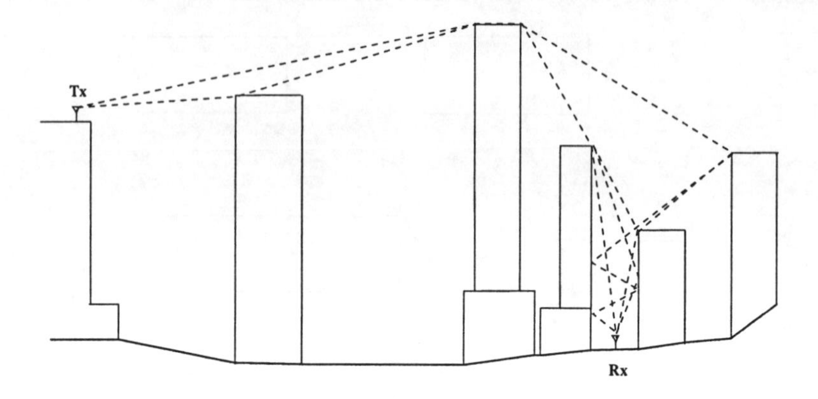


**Figure 2.2** 2D ray tracing in the vertical plane.

ter located on a roof and the mobile located at street level. For this case it is necessary to determine all the horizontal edges that act as a secondary source illuminating the mobile. Although a ray trace in the vertical plane provide reasonable predictions for propagation over buildings, it fails to account for rays that go over low buildings but around tall buildings.

### *2.2.2 Vertical Plane/Transverse Plane Method*

To account for propagation over buildings, a model was developed by a group from the University of Karlsruhe that employs two 2D ray traces, one in the vertical plane and the other in the transverse plane between the base station and mobile antenna. This model also has a third component call the **multipath scattering** mode which accounts for the diffuse reflection from buildings when the surface is in the LOS of both the transmitter and the receiver [28]. Figure 2.3 illustrates the components of this model, with the 2D-VPM accounting for the propagation over the buildings, the 2D-TPM used to determine the ray paths that propagation around the buildings and the 3D-MSM for determining the diffuse scattering paths. Versions of this model has also been implemented by other groups [30] including the Propagation Prediction Group at Polytechnic [31].

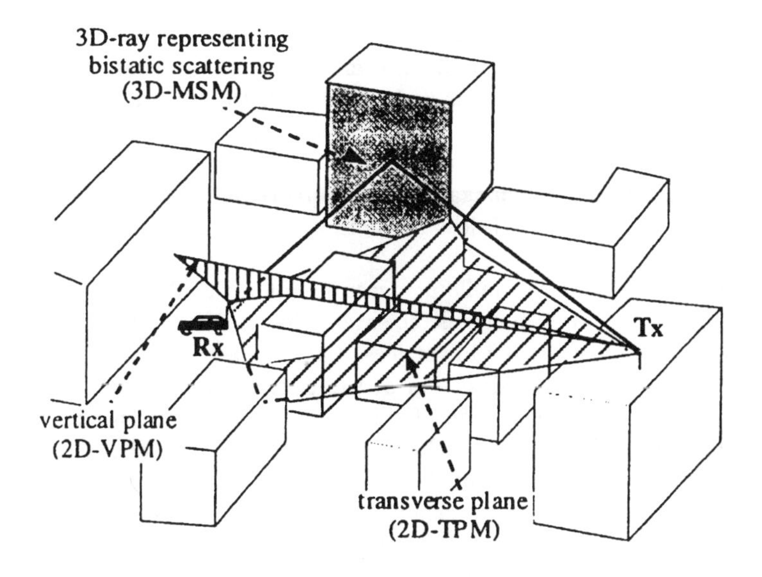


**Figure 2.3** Approximate 3D ray tracing method using vertical planes[29].

The Karlsruhe approach provides good prediction for low base station antennas but was found to be deficient in accounting for the some ray paths. For instance, in the case when a mobile is close to a high base station the transverse plane becomes too steep to account for the proper paths. This error is due to the result of two limitation of a steep transverse plane, the first being that a steep plane will fail to interact with the buildings further away from the source than the mobile and the second reason being that ray paths that undergo multiple reflections do not stay confined to the transverse plane. Finally, this **split plane method** is not able to consider propagation paths that goes over as well as around the buildings which exists in mixed height building environments.

### *2.2.3 Three Dimensional Methods*

Although 2D ray tracing propagation models provide accurate predictions for low base station antennas in a high rise environment, its validity begins to weaken when propagation over the buildings must be accounted for. Therefore it becomes necessary to develop a more robust model that can predict the path loss in this type of environment. The logical step is to implement a full 3D ray trace model that can generate rays for the entire $4\pi$ steradian around the transmitter. This approach has been implemented by many research groups [32]–[33], each of which use some type of simplification to limit the number of rays that are generated. The most severe limitation of a full **3D ray tracing** propagation model is associated with the huge numbers of rays generated by diffractions which usually limits the number of diffractions to 2 at the most. Therefore in a environment where the buildings are of relatively uniform height, so that the propagation path must travel over many consecutive rows of buildings [34, 35], path loss prediction with a full 3D model becomes unmanageable.

The representation of the building features for a full 3D method usually considers each surface to have a specific extent that is defined an edge polygon composed of a arbitrary number of vertices. Each surface also has a orientation in 3D space as given by the direction of the outward normal of the surface. Given the location and orientation of the walls and edges, a **ray launching** (SBR) technique [31, 32] or the **method of images** [36, 37, 38] can be used to determine the 3D physical paths of all reflected and edge diffracted rays. As for the case with the 2D ray trace in the horizontal plane the largest computational effort is associated with the edge diffractions. However, for the 3D case edge diffraction is even more severe due to the facts that: 1) there are more edges in 3D for the same number of buildings, and 2) even for an edge that is finite in length, multiple illuminating rays must be considered at different segments along the edge. Therefore, unlike the 2D case, where only a single secondary ray trace is needed to find the signal due to diffraction at each corner, in the 3D case an edge will contain multiple secondary sources requiring an equal number of ray traces for each segment of the edge. This task becomes completely unmanageable for cascading multiple diffractions.

### *2.2.4 Vertical Plane Launch Method*

As a way to bridge the gap between the computational efficiencies of a full 3D ray trace and the inherent inaccuracy and lack of robustness of the simpler 2D models, a **vertical plane launch technique** (VPL) was devised. The concept of the VPL method for a rooftop antenna is illustrated in Figure 2.4, which shows half planes originating from a vertical line through the transmitter and extending outward in one direction. As an example, the plane between the transmitter and receiver 1 in Figure 2.4 contains a ray that must propagate over the intervening rows of building. The ray reaching receiver 2 is contained in the plane shown, and is reflected from a tall building and travels over a series of lower buildings before arriving at the receiver. For receiver 3, the illuminating ray in the vertical plane undergoes diffraction at the vertical edges of two buildings before traveling over the rooftops of lower buildings to the receiver. Although Figure 2.4 illustrates a case for a rooftop antenna, the VPL

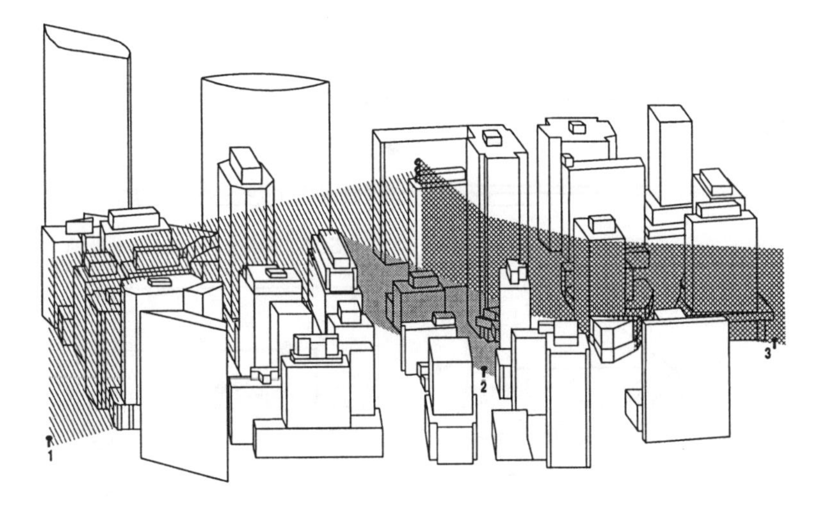


**Figure 2.4** Approximate 3D ray tracing method using vertical planes.

is entirely general and can be applied to base stations that are located well below the rooftops of the surrounding buildings.

Unlike a full 3D SBR method, where rays are launched in a 3D space and all directions are treated in a unified manner, the VPL method takes into account the nearly universal use of vertical walls in building construction and differentiates the horizontal and vertical directions. In the horizontal directions, 2D rays representing the vertical planes are launched from the source in a manner similar to that of the 2D pin cushion method previously used for indoor propagation [20]. This method generates a binary tree at the point where the vertical plane intersects an exterior face of a building wall, with one plane continuing along the incident direction, and a second plane going off in the direction of specular reflection, as shown in Figure 2.5. The plane that continues in the incident direction contains rays that propagate directly over the building and is assumed to contain the rays that are diffracted over the buildings at its horizontal edges. The plane that is spawned in the reflected direction contains rays that are specularly reflected from the building face, and is assumed to contain the rays diffracted at the top horizontal edge of the wall. The path that a ray travels in the vertical direction is found by examining the profile of all the buildings in the unfolded set of vertical plane segments between the source and receiver, and uses deterministic equations to calculate the vertical displacement and received signal strength. Restricting the rays diffracted at horizontal edges to lie in the vertical planes, rather than a cone, is the primary ray tracing approximation made in the VPL method.

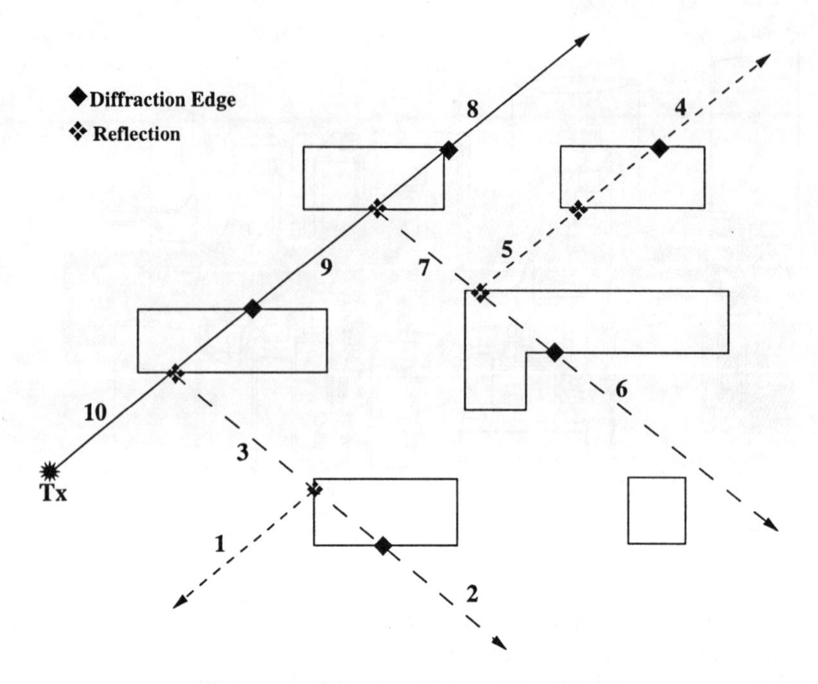


Figure 2.5 Binary ray tree generated from the VPL method

A vertical plane segment is considered to have illuminated the receiver if two conditions are met. The ray must intersect the capture circle of a receiver, as described in [20], and it must lie in the vertical wedge of the illumination. The capture circle is a concept used in conjunction with rays traced at a discrete angular separation to ensure that the signal associated with a continuous family of rays undergoing the same set of reflections and diffractions before illuminating the receiver is counted once and only once. In the vertical plane, the illumination wedge is used to represent the extent in the vertical direction of the region illuminated by a continuous family of rays that have undergone a particular set of reflections and diffractions.

Once the receiver is considered to be illuminated, then all the interactions along the unfolded path can be use to determine the total path loss. In addition to the path length effects, the total path loss depends on the product of the reflection and diffraction coefficients. For a vertically or horizontally polarized source, the reflection coefficient at walls is approximated by that of a TE or TM polarized plane wave, respectively, at a dielectric half space [16] with a dielectric constant of $\epsilon_r = 6$. The use of the reflection coefficient for a dielectric half space with $\epsilon_r \approx 5 - 7$ is suggested by direct measurements [39], and this range of $\epsilon_r$ is found in our predictions to give the least error with the measurements. Diffraction at edges can be approximated as the product of the UTD diffraction coefficients from a dielectric wedge [19], although this is limited to a few edges in cascade. Alternatively, numerical techniques such as discussed in [40, 12, 41] can be employed when more than a few horizontal edges must be accounted for and a more rigorous evaluation of the diffraction loss is necessary.

The VPL technique is able to account for rays that undergo multiple interactions as a result of a combination of forward diffractions at horizontal edges with reflections from building walls and single or double diffraction at a vertical edge of a building. The VPL method neglects rays that are transmitted through the building, diffuse scattering from the walls, and rays that travel under a structure. It also neglects reflections from the rooftop, which travel upward and hence away from the buildings and receivers. These simplifications are made because it is believed that the rays do not significantly contribute to the total received power in a microcellular environment, or that they occur very infrequently, and their inclusion would substantially increase the model's complexity and the computation time.

## 2.3 COMPARISONS BETWEEN METHODS

Each of the four methods, the 2D ray trace in the horizontal plane, the split plane, the full 3D and the VPL methods all have certain advantages and limitations due to the approximations that are made as discussed above. All the different methods, except the full 3D method, have been implemented at Polytechnic University, and enables us to compare the results of each method for a single building environment. Comparison are made between measurements and the predictions from the 2D method for street level base stations since it is well known that propagation around the buildings is important in this case. The predictions made with Vertical/Transverse plane method are compared with measurements for rooftop mounted base stations since propagation over the buildings must be accounted for in this case. For both cases, the predictions from the VPL method will be compared with the different methods as well as with the measurements to demonstrate the robustness and versatility of the technique.

### *2.3.1 Description of the Experimental Area*

Measurements were conducted in a predominately commercial area of Rosslyn Virginia, as shown in Figure 2.6, that is mainly composed of high rise office towers with some buildings of only a few stories intersperse. Since the area is limited, the terrain variation is minimal and can be considered flat throughout the entire area. A set of 400 receivers were located consecutively along each

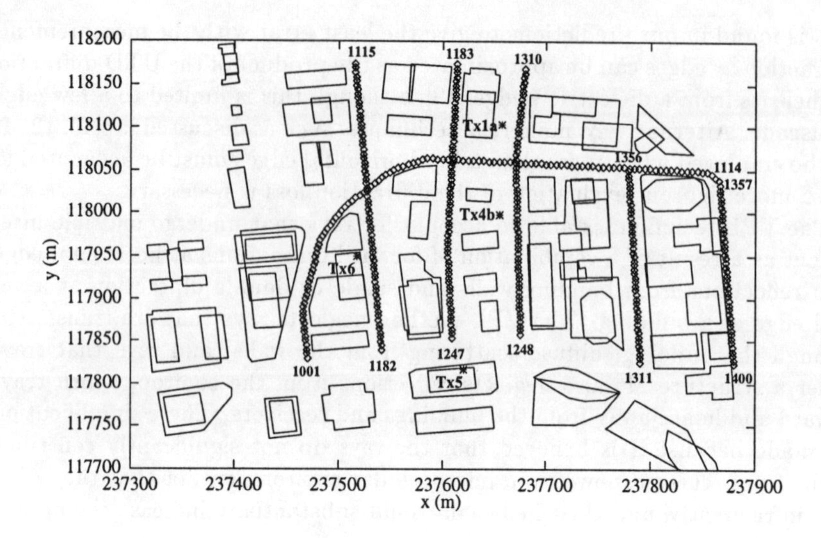


**Figure 2.6** Transmitter and receiver locations for Rosslyn, VA.

street in the experimental area as shown in the Figure 2.6. There were two street level base stations approximately 10 meters above the ground located at Tx 1a and 4b in Figure 2.6. In addition, there were two rooftop mounted antennas, labeled Tx 5 and 6 in Figure 2.6, that were located approximately 42 and 46 meters above street level, respectively. Both rooftop antenna locations employed directional antennas with a half power beamwidth of about 30° in both the E and H planes. The boresight of the antenna was pointing due north or parallel to the y-axis for Tx 5 while the boresight was pointing northeast or 45° counterclockwise from the positive x-axis for Tx 6.

### *2.3.2 Comparisons for Street Level Base Stations*

For street level transmitters that are 10m above the ground, the predictions for the VPL and 2D ray trace in the horizontal plane are expected to be similar when the surrounding buildings are much higher. In a high rise urban environment, propagation over the buildings does not significantly contribute to the total power received. Instead, the dominate propagation paths are expected to lie in a nearly horizontal plane that contains the transmitter and all the receivers. Therefore the propagation mechanisms are associated with specular reflections from the walls, in combination with single or double diffraction at the vertical corners of the buildings. These ray paths can be accurately and readily found by performing a 2D ray trace in the horizontal plane.

For an environment containing buildings of mixed height, where not all the buildings are significantly higher than the source or the receiver, propagation paths that go over as well as around the buildings are needed for some receiver locations. In these cases the 2D ray trace provides pessimistic predictions. On the other hand the VPL method retains the rays propagating over the buildings by accounting for the vertical dimension within the ray trace.

Figure 2.7 shows a comparison of the predictions obtained from a 2D ray

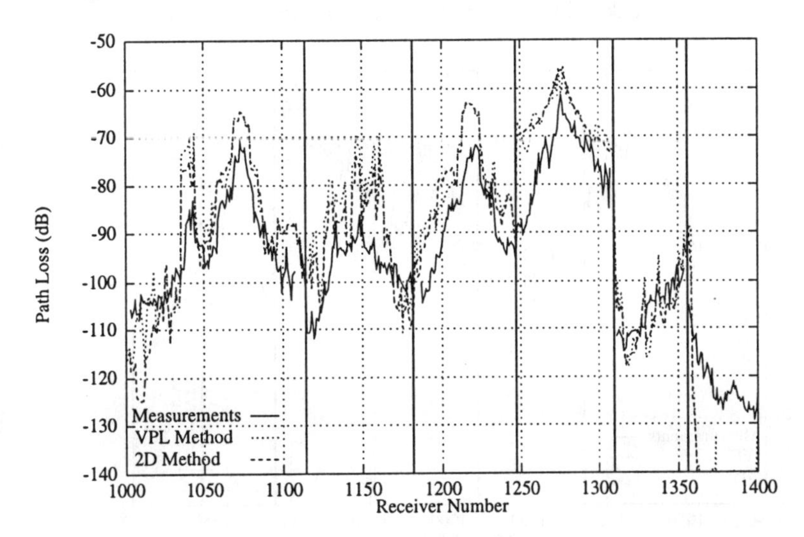


**Figure 2.7** Comparison between 2D and VPL method with measurements for Tx4b at 900MHz.

trace (dashed curve), as well as those obtained from the VPL method (dotted curve), and the measurements (solid curve) for Tx4b, which is located near the center of the area shown in Figure 2.6. The predictions for both methods were made by accounting for one corner diffraction and up to 6 reflections before and after the corner diffraction. In combination with this, the VPL also includes a maximum of up to four diffraction at the horizontal building edges. In general the predictions for both methods are more optimistic than the measurements especially along the line of sight streets. On the other hand, along street that are heavily shadow from the transmitter, the 2D method is very pessimistic while the VPL method provides closer agreement with the measurements. This is particularly evident for the receivers located on streets that are on the extreme left and right margins of the measurements areas (Rx 1001–1021 and 1359–1400). For receiver locations 1001–1021, the 2D method provides slightly pessimistic predictions while the predictions from the VPL method matches more closely with the measurements. On the other side of the prediction area, at locations 1359–1400, heavy shadowing results in 2D predictions that are so weak that they fall below the bottom limit of the graph. On the other hand, the VPL predictions, while still pessimistic, are now within 20dB of the measurements.

The same result can be seem in the comparisons between the two method for Tx1a at 1900MHz, as shown in Figure 2.8. For this case, double diffraction at the vertical edges are taken into account for the predictions with both meth-

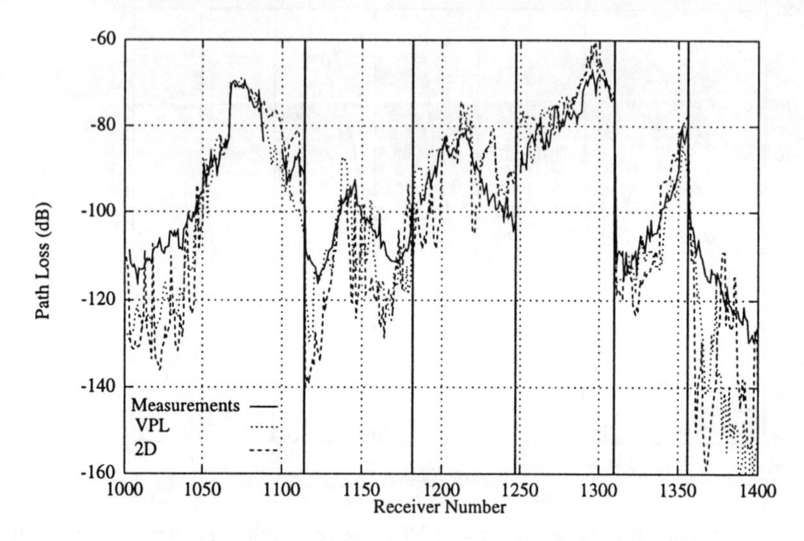


**Figure 2.8** Comparison between 2D and VPL method with measurements for Tx1a at 1900MHz.

ods. In general the predictions from the 2D (dashed curve) and VPL methods (dotted curve) correspond with each other in the LOS and lightly shadowed regions. However, the VPL provides better results in the heavily shadow regions where the predictions from the VPL tend to agree more closely with the measurements than the simple 2D ray trace.

In general, the VPL method is able to provide better results than the 2D method for a mixed height building environment such as exists in Rosslyn. Table 2.1 list the prediction statistics for the VPL and 2D methods for both Tx1a and 4b at 900 and 1900MHz but neglecting the results for the receiver along the extreme right hand side (Rx 1357–1400). In both cases the VPL method gives lower values of $\sigma$, while the value of $\eta$ is generally slightly larger. The cost of improved accuracy of the VPL method comes at a increased computational time due to the additional complexity of considering rooftop diffraction.

| | Tx1a | | | | Tx4b | | | |
|---|---|---|---|---|---|---|---|---|
| | 900MHz | | 1900MHz | | 900MHz | | 1900MHz | |
| Tx | $\eta$ (dB) | $\sigma$ (dB) | $\eta$ (dB) | $\sigma$ (dB) | $\eta$ (dB) | $\sigma$ (dB) | $\eta$ (dB) | $\sigma$ (dB) |
| VPL | 1.04 | 7.75 | -2.12 | 7.66 | 6.05 | 7.58 | 6.05 | 7.65 |
| 2D | 0.51 | 8.77 | 3.95 | 10.15 | 4.72 | 7.77 | 5.52 | 8.30 |

**Table 2.1** Statistical results for the comparison between the 2D and VPL methods.

### *2.3.3 Comparisons for Rooftop Base Stations*

For a rooftop transmitter, the dominated propagation mechanisms are expected to include paths that go over as well as around the buildings. This includes paths that travel over a series of buildings before diffracting at a horizontal edge down to the receiver. There can also be rays that reflect from a building wall in the vicinity of the source before propagating over the buildings, and rays that reflect from a building wall near the receiver after the ray has gone over other buildings. Finally, paths that undergo a single diffraction at a vertical corner of a building in combination with propagation over lower buildings or reflections, will also contribute to the total power received at the receiver.

Predictions for a rooftop antenna made with the SP/VP method (dashed line) and the VPL method (dotted line) are compared in Figure 2.9, along

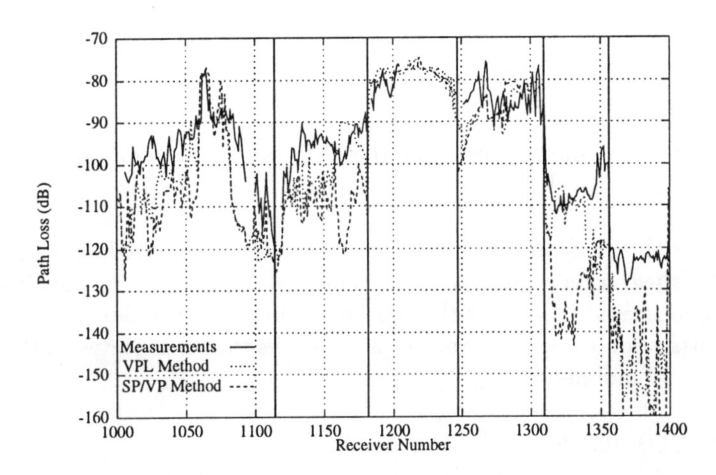


**Figure 2.9** Comparison between Vertical/Transverse Plane and VPL method with measurements for Tx5 at 900MHz.

with the measurements (solid line), for Tx5 at 900MHz. The VPL method is seen to provide better agreement with the measurements than the SP/VP method, especially for receivers close to the transmitter (Rx 1155–1182 and 1315–1325). For these locations the SP/VP method misses some important rays reflected from the buildings because those rays are not confined to either the slant or vertical planes. The VPL method however can account for rays that propagate over buildings in addition to reflecting off a nearby building close to the transmitter or the receiver. The VPL method reduces $|\eta|$ by over 4 dB, from 9.87 dB for the SP/VP method to 5.40 dB, with a corresponding reduction of $\sigma$ from 9.83 to 7.66 dB.

A similar result can be seen in Figure 2.10 for Tx6 at 1900MHz. In this

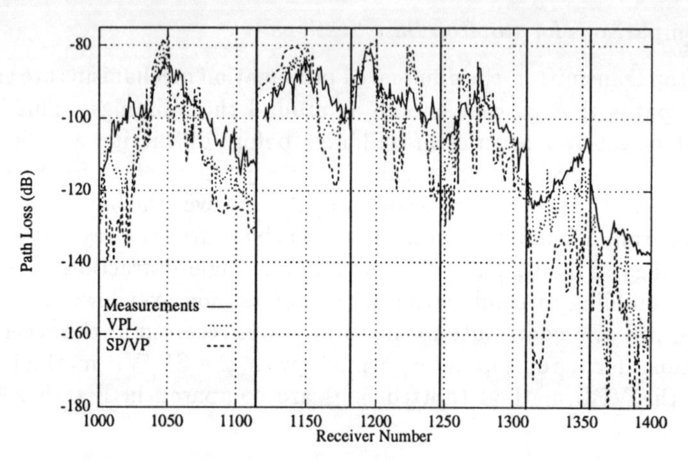


**Figure 2.10** Comparison between Vertical/Transverse Plane and VPL method with measurements for Tx6 at 1900MHz.

case the dielectric constant used for the walls was $\epsilon_r = 3$ for the predictions made with both the SP/VP and VPL method. As shown in Figure 2.10 the predictions made with the VPL (dotted curve) method tend to provide better agreement with the measurements for receivers where the predictions from the SP/VP (dashed curve) tend to be extremely pessimistic. It is evident that for rooftop base stations where the slant plane becomes very steep, the SP/VP method misses some important propagation paths that exist between the transmitter and the mobile

## 2.4 CONCLUSIONS

The deployment of PCS in microcells with a small cell radius has fueled a growing interest in theoretical propagation predictions utilizing ray tracing techniques and UTD. This type of propagation modeling has certain advantages over traditional measurement based models, including the ability to use the same model in different building environments, at different transmitter locations and heights and at different frequencies. When the base station antenna is low in comparison to all of the surrounding buildings, a 2D ray tracing is sufficient to accurately characterizes the propagation environment. On the other hand in a heterogeneous building environment where the building heights are not uniformly high, it becomes necessary to account for the energy that goes over as well as around the buildings.

A split plane approach which considers 2 separate ray traces, one in the vertical and another in transverse plane containing the transmitter and receiver has been proposed to consider propagation paths that go over the buildings. Unfortunately, this method is limited in approximating full 3D ray paths. Alternatively, a full 3D ray tracing can be employed to consider all possible propagation paths, but is limited by its computational inefficiencies to the inclusion of two diffractions at the most. Finally, the new VPL technique overcomes the deficiencies of the other methods. It accomplishes this by including rays paths that are not accounted for in the other methods either because of inherent approximations or because of computational time constraints.

## References

[1] J. D. Parsons, *The Mobile Radio Propagation Channel*, John Wiley & Sons, Inc., New York, NY, 1st edition, 1992.

[2] Y. Okumura, E. Ohmori, T. Kawano, and K. Fukuda, "Field Strength and Its Variability in VHF and UHF Land-Mobile Radio Service", *Review of the Electrical Communication Laboratories*, pp. 826–873, 1968.

[3] M. Hata, "Empirical Formula for Propagation Loss in Land Mobile Radio Services", *IEEE Transactions on Vehicular Technology*, vol. VT-29, no. 3, pp. 317–325, August 1980.

[4] J.H. Whitteker, "Measurements of Path Loss at 910 MHZ for Proposed Microcell Urban Mobile Systems", *IEEE Transactions on Vehicular Technology*, vol. 37, no. 3, pp. 125–129, August 1988.

[5] P. Harley, "Short Distance Attenuation Measurements at 900 MHz and 1.8 GHz Using Low Antenna Heights for Microcells", *IEEE Journal on Selected Areas in Communications*, vol. 7, no. 1, pp. 5–11, January 1989.

[6] A.J. Rustako Jr., N. Amitay, G.J. Owens, and R.S. Roman, "Radio Propagation at Microwave Frequencies for Line-of-Sight Microcellular Mobile and Personal Communications", *IEEE Transactions on Vehicular Technology*, vol. 40, no. 1, pp. 203–210, February 1991.

[7] A.J. Goldsmith and L.J. Greenstein, "A Measurement-Based Model for Predicting Coverage Areas of Urban Microcells", *IEEE Journal on Selected Areas in Communications*, vol. 11, no. 7, pp. 1013–1023, September 1993.

[8] H.H. Xia, H.L. Bertoni, L.R. Maciel, A. Lindsay-Stewart, and R. Rowe, "Microcellular Propagation Characteristics for Personal Communications in Urban and Suburban Environments", *IEEE Transactions on Vehicular Technology*, vol. 43, no. 3, pp. 743–752, August 1994.

[9] A.J. Rustako Jr., N. Amitay, G.J. Owens, and R.S. Roman, "Radio Propagation Measurements At Microwave Frequencies for Microcellular Mobile and Personal Communications", in *Conference Record ICC'89*, 1989, pp. 482–486.

[10] F. Ikegami and S. Yoshida, "Analysis of Multipath Propagation Structure in Urban Mobile Radio Environments", *IEEE Transactions on Antennas and Propagation*, vol. AP-28, no. 4, pp. 531–537, July 1980.

[11] F. Ikegami, S. Yoshida, T. Takeuchi, and M. Umehira, "Propagation Factors Controlling Mean Field Stength on Urban Streets", *IEEE Transactions on Antennas and Propagation*, vol. AP-32, no. 8, pp. 822–829, August 1984.

[12] J. Walfisch and H.L. Bertoni, "A Theoretical Model of UHF Propagation in Urban Environments", *IEEE Transactions on Antennas and Propagation*, vol. 36, no. 12, pp. 1788–1796, December 1988.

[13] K. Löw, "Comparison of Urban Propagation Models With CW Measurements", in *Proceedings of 42nd VTS Conference.* 1992, vol. 2, pp. 936–942, IEEE.

[14] A. Domingues, D. Caiado, N. Goncalves, and L.M. Correia, "Testing the COST231-WI Propagation Model in the City of Lisbon", in *COST 231 TD(96).* 1996, European Cooperation in Field of Scientific and Technical Research.

[15] J.B. Keller, "Geometrical Theory of Diffraction", *Journal of the Optical Society of America*, vol. 52, no. 2, pp. 116–130, February 1962.

[16] C.A. Balanis, *Advanced Engineering Electromagnetics*, John Wiley & Sons, New York, NY, 1989.

[17] L.B. Felsen and N. Marcuvitz, *Radiation and Scattering of Waves*, Prentice-Hall, Inc., Englewood Cliffs, NJ, 1973.

[18] R.F. Kouyoumjian and P.H. Pathak, "A Uniform Geometrical Theory of Diffraction for an Edge in a Perfectly Conducting Surface", *Proceedings of the IEEE*, vol. 62, pp. 1448–1461, November 1974.

[19] R.J. Luebbers, "Finite Conductivity Uniform GTD Versus Knife Edge Diffraction in Prediction of Propagation Path Loss", *IEEE Transactions on Antennas and Propagation*, vol. AP-32, no. 1, pp. 70–76, January 1984.

[20] W. Honcharenko, H.L. Bertoni, J.L. Dailing, et al., "Mechanisms Governing UHF Propagation on Single Floors in Modern Office Buildings", *IEEE Transactions on Vehicular Technology*, vol. 41, no. 4, pp. 496–504, November 1992.

[21] S.J. Fortune, D.M. Gay, B.W. Kernighan, et al., "WISE Design of Indoor Wireless Systems: Practical Computation and Optimization", *IEEE Computational Science and Engineering*, pp. 58–68, 1995.

[22] F. Niu and H.L. Bertoni, "Path Loss and Cell Coverage of Urban Microcells in High-Rise Building Environments", in *Proc. IEEE GLOBECOM'93*, Houston, TX, 1993, pp. 266–270.

[23] C. Bergljung and L.G. Olsson, "Rigorous Diffraction Theory Applied To Street Microcell Propagation", in *Proc. IEEE GLOBECOM '91*, Phoenix, AZ, 1991, pp. 1292–1296.

[24] G. Liang, "Ray Trace Simulation of UHF Propagation in a High Rise Urban Environment", Master's thesis, Polytechnic University, Brooklyn,NY, January 1995.

[25] M.C. Lawton and J.P. McGeehan, "The Application of a Deterministic Ray Launching Algorithm for the Prediction of Radio Channel Characteristics in Small-Cell Environments", *IEEE Transactions on Vehicular Technology*, vol. 43, no. 4, pp. 955–969, November 1994.

[26] P. Daniele, V. Degli-Exposti, G. Falciasecca, and G. Riva, "Field Prediction Tools For Wireless Communications in Outdoor and Indoor Environments", in *Proc. IEEE MTT Symposium*, Turin, Italy, November 1994, IEEE.

[27] K. Rizk, J.-F. Wagen, and F. Gardiol, "Two-Dimensional Ray-Tracing Modeling for Propagatin Prediction in Microcellular Environments", *IEEE Transactions on Vehicular Technology*, vol. 46, no. 2, pp. 508–518, May 1997.

[28] T. Kürner, D.J. Cichon, and W. Wiesbeck, "Concepts and Results for 3D Digital Terrain-Based Wave Propagation Models: An Overview", *IEEE Journal on Selected Areas in Communications*, vol. 11, no. 7, pp. 1002–1012, September 1993.

[29] D.J. Cichon, T. Kürner, and W. Wiesbeck, "Multipath Propagation Modelling in Urban Environments", in *IEEE International Workshop on Personal, Indoor and Mobile Radio Communications*, May 1992.

[30] M. Feistel and A. Baier, "Performance of a Three-Dimensional Propagation Model in Urban Environments", in *Proceedings of 6th International Symposium on Personal, Indoor and Mobile Radio Communications*, Toronto, Canada, 1995, vol. 2, pp. 402–407, IEEE.

[31] T. M. Willis, B.J. Guarino, J.D. Moore, , et al., "UHF and Microwave Propagation Prediction in an Urban Environment", in *AGARD Conference Proceedings 574*, April 1996, pp. 8.1–8.10.

[32] K.R. Schaubach and N.J. Davis IV, "Microcellular Radio-Channel Propagation Prediction", *IEEE Antennas and Propagation Magazine*, vol. 36, no. 4, pp. 25–34, August 1994.

[33] V. Erceg, S.J. Fortune, J. Ling, A.J. Rustako, and R.A. Valenzuela, "Comparisons of a Computer-Based Propagation Prediction Tool with Experimental Data Collected in Urban Microcellular Environments", *IEEE Journal on Selected Areas in Communications*, vol. 15, no. 4, pp. 677–684, May 1997.

[34] L.R. Maciel, H.L. Bertoni, and H.H. Xia, "Unified Approach to Prediction of Propagation Over Buildings for All Ranges of Base Station Antenna Height", *IEEE Transactions on Vehicular Technology*, vol. 42, no. 1, pp. 41–45, February 1993.

[35] H.L. Bertoni, W. Honcharenko, L.R. Maciel, and H.H. Xia, "UHF Propagation Prediction for Wireless Personal Communications", *Proceedings of the IEEE*, vol. 82, no. 9, pp. 1333–1359, September 1994.

[36] A.G. Kanatas, I.D. Kountouris, G.B. Kostaras, and P. Constantinou, "A UTD Propagation Model in Urban Microcellular Environments", *IEEE*

*Transactions on Vehicular Technology*, vol. 46, no. 1, pp. 185–193, February 1997.

[37] S.Y. Tan and H.S. Tan, "UTD Propagation Model in an Urban Street Scene for Microcellular Communications", *IEEE Transactions on Electromagnetic Compatibility*, vol. 35, no. 4, pp. 423–428, November 1993.

[38] S.Y. Tan and H.S. Tan, "Propagation Model Microcellular Communications Applied to Path Loss Measurements in Ottawa City Streets", *IEEE Transactions on Vehicular Technology*, vol. 44, no. 2, pp. 313–317, May 1995.

[39] O. Landron, M.J. Feuerstein, and T.S. Rappaport, "A Comparison of Theoretical and Empirical Reflection Coefficients for Typical Exterior Wall Surfaces in Mobile Radio Environment", *IEEE Transactions on Antennas and Propagation*, vol. 44, no. 3, pp. 341–351, March 1996.

[40] L.E. Vogler, "An Attenuation Function for Multiple Knife-Edge Diffraction", *Radio Science*, vol. 17, no. 6, pp. 1541–1546, Nov-Dec 1982.

[41] H.H. Xia and H.L. Bertoni, "Diffraction of Cylindrical and Plane Waves by an Array of Absorbing Half-Screens", *IEEE Transactions on Antennas and Propagation*, vol. 40, no. 2, pp. 170–177, February 1992.

# 3 FUNDAMENTALS OF SMART ANTENNAS FOR MOBILE RADIO APPLICATIONS

P.W. Baier
J.J. Blanz
R. Schmalenberger

**Abstract:** When developing standards for UMTS and FPLMTS, important issues include capacity and spectrum efficiency, which can be enhanced by antenna diversity techniques. The basic principles of these techniques are explained and, as an example, are applied to a specific UMTS air interface concept. This concept evolves from GSM by the introduction of a CDMA component in addition to FDMA and TDMA.

## 3.1 INTRODUCTION

The age of second generation terrestrial mobile radio systems began in 1992. In that year, after a preparation period of almost one decade, the first digital cellular mobile radio networks entered operation. These networks follow the European GSM standard [1]. In the short time since 1992, GSM has been adopted by ca. 170 operators in ca. 100 countries around the world, and more than 25 million subscribers presently use GSM [2]. This great success of GSM and the interest in some other second generation standards such as IS-95 [3] illustrates the economic importance of cellular mobile radio communications. Market analyses show not only that the demand for second generation mobile radio networks will increase further but that there is a need for a new, third generation of digital wireless networks [4]. Presently, the definition of concepts for such third generation networks is being intensively studied by standardization bodies, manufacturers, operators, university laboratories, etc. In Europe these new concepts are subsumed under the acronym UMTS (Universal Mobile Telecommunications Systems), whereas the designations used on a worldwide scale are FPLMTS (Future Public Land Mobile Telecommunications System) or IMT 2000 (International Mobile Telecommunication 2000).

It is necessary that third generation networks lead to a convergence of fixed and mobile services. These networks will also include a satellite component, even though the terrestrial part will play the predominant role, and they will offer access in the domains of both cellular, fixed wireless and cordless. Besides voice communication, these future networks should provide video and multimedia services as well as bearer services enabling a variety of applications. To fulfil these requirements, it is considered necessary that data rates up to 144 kbit/s for universal wide area coverage and up to 2 Mbit/s for local coverage be supported [5].

For the implementation of third generation mobile radio systems new frequency bands are needed, and it can be expected that the regulators will make adequate spectrum allocations [6]. Nevertheless, spectrum remains a scarce resource. Therefore, when designing third generation standards, a major issue concerning the air interface consists in achieving a capacity as large as possible at an expense as low as possible. The capacity is measured in Erlang per unit area and is closely related to the spectrum efficiency, which relates the total transmissible information to the allocated bandwidth.

In this paper, terrestrial cellular mobile systems are considered. There exist several ways to enhance the capacity of such systems for a given spectrum efficiency, the more conventional ones being the reduction of the cell size - this unfortunately increases cost because more cell sites are required - and the application of efficient source encoding schemes, which reduce the data rates necessary for voice, fax or image transmission. In addition, measures in the areas of signalling and protocols have to be considered because, depending on the adopted signalling procedures, a more or less significant portion of the transmissible data rate has to be set aside for signalling and is therefore lost for traffic data transmission. Other possibilities include increasing spectrum efficiency by the utilization of optimized multiple access schemes or refining modulation and FEC techniques. Last but not least, the employment of special antenna designs in combination with advanced signal processing to exploit specific properties of the terrestrial mobile radio channel and/or the spatial distribution of the base stations (BS) and mobile stations (MS) by antenna diversity techniques should be considered, when striving for high capacity at reasonable cost. The resulting antenna concepts are termed smart or intelligent. A smart antenna system usually disposes of a number of single port antennas or antenna elements for transmitting and/or receiving, with these components cooperating in some constructive way. Of course, smart antennas are not only attractive for third generation mobile radio systems but also for enhancing the capacity of today's second generation mobile radio networks, as long as such antenna concepts can be adopted within the valid standards.

A presupposition of the application of smart antennas is the directional and polarization dispersions of the mobile radio environment as well as the multi-location distribution of the BSs and MSs. These issues are discussed in Section 3.2. In Section 3.3, the principle of antenna diversity techniques is reviewed. Section 3.4 presents a case study. In this section, various intelligent antenna concepts to be applied in combination with the UMTS air interface Joint Detection CDMA (JD-CDMA) [7] are considered. Based on computer simulations it illustrates how the favorable spectrum efficiency of JD-CDMA can be further increased by smart antennas. The conclusions are summarized in Section 3.5.

## 3.2 CONDITIONS RELEVANT FOR THE APPLICATION OF SMART ANTENNAS

### *3.2.1 Directional dispersion and polarization dispersion of the mobile radio channel*

In contrast to free space propagation, the line of sight (LOS) path between transmitter and receiver is generally obstructed in terrestrial mobile radio communications. Even if the BS antenna is at an elevated location, it usually cannot be seen from the MS, because the MS is surrounded by buildings, trees, hills, etc. Therefore, the radio connection of a MS to a BS has to rely on physical effects like reflection, refraction, scattering and diffraction by the various objects in the propagation environment [8]. This means that the electromagnetic waves in a point-to-point connection between a BS and a MS generally propagate along a multitude, or even a continuum, of different paths with each path consisting of several hops. The signals transmitted over these paths are superposed at the receiver. Because of the movement of the MS and possibly other objects in the propagation environment, these paths are time variant and exist for only a limited life time. In addition, the paths differ more or less from each other in the following respects:

- delay time,
- attenuation and phase shift,
- launching direction at the transmitter and direction of incidence at the receiver,
- change of the polarization along the path and dependence of the attenuation and phase shift on the starting polarization chosen at the transmitter,
- time dependence of the path characteristics,
- path life time.

A non-smart antenna system consists of a single port antenna at the transmitter and a single port antenna at the receiver. In a mobile radio propagation environment, the channel between the input of the transmitter antenna and the output of the receiver antenna of such a system, due to these multipath effects, can be described by an impulse response, which originates from a weighted superposition of the impulse responses of a selection of paths, the weighting and selection being determined by the employed antennas. This impulse response is more or less spread along the delay axis - i.e. delay dispersion occurs - and is time variant [8]. It can be described by its lowpass equivalent $\underline{h}(\tau, t)$, where $\tau$ and $t$ denote the delay and the time, respectively. $\underline{h}(\tau, t)$ is related via Fourier transformation to a time variant transfer function

$$\underline{H}(f, t) = \int_{-\infty}^{\infty} \underline{h}(\tau, t) e^{-j2\pi f \tau} \, d\tau, \tag{3.1}$$

which is frequency selective on account of the delay dispersion of $\underline{h}(\tau, t)$. $\underline{h}(\tau, t)$ and $\underline{H}(f, t)$ are lump manifestations of the active paths of the propagation environment as seen through the employed antennas.

If one is interested in a fundamental understanding of the potentials of smart antennas, the description of the mobile radio channel by the above mentioned functions $\underline{h}(\tau, t)$ and $\underline{H}(f, t)$ is not sufficient. Rather, a more basic and differentiated description is required, which is not a priori colored by the chosen antennas, but which describes the electromagnetic properties of the propagation environment between the locations of a transmitter and a receiver in such a way that the channel impulse response can be determined depending on the properties of any given pair of antennas. Such a description should, in addition to taking into account delay dispersion and time variance, also tackle the effects of directional dispersal and polarization dispersal. Directional dispersion means that a unidirectionally radiated wave may impinge at a receiver location from a more or less extended angular range and polarization dispersion means that a wave radiated with a certain polarization may arrive at a receiver location with a set of different polarizations. To obtain such a description, spherical coordinate systems are introduced at the transmitter and receiver locations with the co-elevation angles $\vartheta$ and $\Theta$, respectively, and the azimuthal angles $\varphi$ and $\Phi$, respectively. In order to obtain an unambiguous relation between the angles $\vartheta, \varphi, \Theta, \Phi$ and the four cardinal points at each of the two locations, the following definitions can be adopted: $\vartheta$ equal to zero and $\Theta$ equal to zero indicate the vertical directions at the transmitter and receiver locations, respectively, and $\varphi$ equal to zero and $\Phi$ equal to zero indicate the north directions at the transmitter and receiver locations, respectively. Let us assume that fictitious omnidirectional antennas are used at the transmitter and receiver, with the transmitter antenna generating a wave which has an electrical field component only in $\vartheta$-direction, and with the receiver antenna being sensitive only for electrical field components in $\Phi$-direction. The partial impulse response between the ports of the two antennas resulting from illuminating the scenario in the solid angle $\mathrm{d}\,\vartheta \cdot \mathrm{d}\,\varphi \cdot \sin\vartheta$ centered around $\vartheta$, $\varphi$ and from probing the scenario in the solid angle $\mathrm{d}\,\Theta \cdot \mathrm{d}\,\Phi \cdot \sin\Theta$ centered around $\Theta$, $\Phi$ can be represented in the form $\underline{g}^{(\Phi,\vartheta)}(\vartheta,\varphi,\Theta,\Phi,\tau,t) \cdot \mathrm{d}\,\vartheta \cdot \mathrm{d}\,\varphi \cdot \mathrm{d}\,\Theta \cdot \mathrm{d}\,\Phi \cdot \sin\vartheta \cdot \sin\Theta$, where the delay dispersive and time variant function $\underline{g}^{(\Phi,\vartheta)}(\vartheta,\varphi,\Theta,\Phi,\tau,t)$ is termed differential directional channel impulse response. In order to obtain a comprehensive description of the propagation environment between the transmitter and receiver locations which is independent of the antenna characteristics, four differential directional channel impulse responses $\underline{g}^{(\mu,\nu)}(\vartheta,\varphi,\Theta,\Phi,\tau,t)$, $\mu \in \{\Theta,\Phi\}$, $\nu \in \{\vartheta,\varphi\}$ are needed. Each of these functions describes the transfer characteristics of the scenario for a certain pair of transmitter and receiver polarizations, and these functions cover the dispersions of delay, direction and polarization. Because each of the four functions is constituted by a different set of propagation paths between the transmitter and receiver locations, these functions are more or less independent from each other.

Once the four differential directional channel impulse responses are known for a pair of locations, the channel impulse response between the ports of any transmitter antenna and any receiver antenna being situated at these locations can be determined depending on the antenna characteristics. To accomplish this, for each antenna two characteristics are required, one for the $\vartheta$- and $\Theta$-components, respectively, and the other for the $\varphi$- and $\Phi$-components, respectively, of the electrical fields. These two characteristics are termed $\underline{f}_{\mathrm{t}}^{(\vartheta)}(\vartheta,\varphi)$ and $\underline{f}_{\mathrm{t}}^{(\varphi)}(\vartheta,\varphi)$ for the transmitter antenna and $\underline{f}_{\mathrm{r}}^{(\Theta)}(\Theta,\Phi)$ and $\underline{f}_{\mathrm{r}}^{(\Phi)}(\Theta,\Phi)$ for the receiver antenna. With these four antenna characteristics and the above introduced four differential directional channel impulse responses, the time variant impulse response $\underline{h}(\tau,t)$ between the input of the transmitter antenna and the output of the receiver antenna can be determined. To this purpose, the vectors

$$\underline{f}_{\mathrm{t}}(\vartheta,\varphi) = \left[\underline{f}_{\mathrm{t}}^{(\vartheta)}(\vartheta,\varphi), \underline{f}_{\mathrm{t}}^{(\varphi)}(\vartheta,\varphi)\right]^{T}, \tag{3.2}$$

$$\underline{f}_{\mathrm{r}}(\Theta,\Phi) = \left[\underline{f}_{\mathrm{r}}^{(\Theta)}(\Theta,\Phi), \underline{f}_{\mathrm{r}}^{(\Phi)}(\Theta,\Phi)\right]^{T} \tag{3.3}$$

and the matrix

$$\underline{G}(\vartheta,\varphi,\Theta,\Phi,\tau,t) = \begin{bmatrix} \underline{g}^{(\Theta,\vartheta)}(\Theta,\vartheta,\Theta,\Phi,\tau,t) & \underline{g}^{(\Theta,\varphi)}(\Theta,\varphi,\Theta,\Phi,\tau,t) \\ \underline{g}^{(\Phi,\vartheta)}(\Phi,\vartheta,\Theta,\Phi,\tau,t) & \underline{g}^{(\Phi,\varphi)}(\Phi,\vartheta,\Theta,\Phi,\tau,t) \end{bmatrix} \tag{3.4}$$

are introduced. Then,

$$\begin{aligned} \underline{h}(\tau,t) &= \iint\iint_{\substack{\vartheta,\Theta=0\ldots\pi \\ \varphi,\Phi=0\ldots 2\pi}} \underline{f}_{\mathrm{r}}^{\mathrm{T}}(\Theta,\Phi)\cdot\underline{G}(\vartheta,\varphi,\Theta,\Phi,\tau,t)\underline{f}_{\mathrm{t}}(\vartheta,\varphi) \\ &\times \sin\vartheta\cdot\sin\Theta\cdot d\vartheta\cdot d\varphi\cdot d\Theta\cdot d\Phi \end{aligned} \tag{3.5}$$

holds.

As a presupposition of applying the above introduced differential directional channel impulse responses to determine the channel impulse response $\underline{h}(\tau,t)$, see (5), the transmitter and receiver antennas should not be too closely surrounded by obstacles. Rather, the objects closest to the antennas should lie in the farfield regions. Admittedly, this presupposition restricts the applicability of the above theory because, especially the nearfield regions of the MS, antennas are usually not free from obstacles. Nevertheless, this theory may be helpful in developing a basic understanding of the physical effects being relevant in smart antenna design and application.

Knowledge of the properties of the mobile radio channels, i.e. ideally about the above introduced differential channel impulse responses, is important in two different respects. On the one hand, such knowledge is required for designing mobile radio systems in an optimum way for the environments in which they have to work and for comparing different system concepts on a realistic basis; the information required for these purposes being usually obtained by channel sounders (CS) [9] in expensive measurement campaigns [10] and by evaluation of the measurement results, but also by implementing models which imitate the geometrical and electrical conditions of the considered scenario on the computer [11]. On the other hand, when operating a mobile radio network, the valid channel properties have to be estimated continuously in order to be able to adapt the transmitter and/or receiver including their antennas to the channel and thus optimize transmission with the ultimate aim to achieve a large capacity. To solve this task, channel estimation devices are required which work in real time [12]. Usually, not the total knowledge on the channel, i.e. not the above introduced differential channel impulse responses, but rather a partial knowledge becomes available by channel measurements as well as by channel estimation devices. Channel measurements by channel sounders and channel modelling will now be considered. In the case study presented in Section 3.4 information on the possibilities of real time channel estimation during mobile radio system operation will be given.

A channel sounder is comparable to a bi-static radar system and consists of a mobile station (CS-MS), which is usually the transmitter, and a fixed station (CS-BS), which is usually the receiver. The transmitter emits test signals, which are received and evaluated by the receiver on-line or off-line with the goal of obtaining information on the radio channel between the two stations. In order to obtain information concerning path directions and polarizations, which are needed for the investigation of smart antenna concepts, one or several of the following measures have to be taken:

- Movements of the CS-MS to study the time variance introduced by the motions of the mobile and to determine the directions of relevant wave propagation paths at the location of the CS-MS by Doppler analysis.

- Application of a rotating directional antenna at the CS-BS in order to gain informations on the directions of relevant wave propagation paths at the location of the CS-BS.

- Periodic alternation of the polarization of the waves radiated by the CS-MS antenna to study the dependence of the transmission on the polarization chosen at the CS-MS.

- Probing simultaneously different polarizations of the waves impinging at the CS-BS to study the dependence of the transmission on the polarization chosen at the CS-BS.

A useful extension of the above mentioned principles consists in using one CS-MS and simultaneously two CS-BSs at different locations.

Carrying out channel measuring campaigns is rather expensive. Therefore it is sometimes attractive to establish channel models on the computer to represent the behavior of real world mobile radio channels. Two types of such channel models can be distinguished: geometry-based channel models and algorithmic channel models. In geometry-based channel models, the model of the wave propagation in the real mobile radio channel is based on the geographical, topographical and morphological data of the considered scenario [11]. In algorithmic channel models, the mobile radio channels are represented by tapped

a)

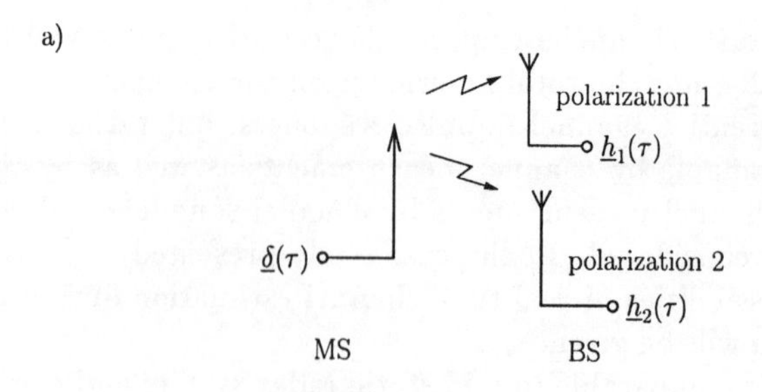


b)

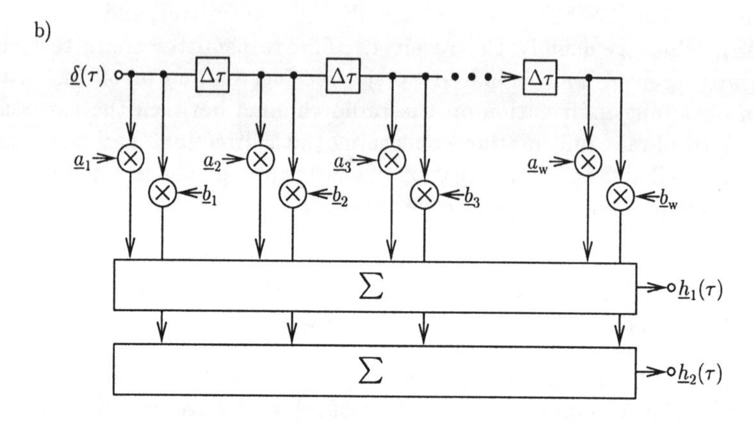


**Figure 3.1** Algorithmic channel modell a) antenna configuration b) tapped delay line circuit

delay line arrangements. An example for an algorithmic channel model is given in Fig.3.1a,b. As indicated in Fig.3.1a, at the MS a transmitter antenna with certain directional and polarization characteristics is used, and at the BS we have two antennas for two different polarizations 1 and 2, with each antenna having a certain directional characteristic. The mobile radio channel between the port of the MS antenna and the ports of the two BS antennas can be mathematically modelled by an algorithm, which includes the effects of polarization and which can be visualized by a tapped delay line circuit as shown in Fig.3.1b. When quantitatively describing mobile radio channels based on channel sounder measurements or on channel models, the channel can be characterized either by a sequence of individual channel impulse responses or by statistical measures, which characterize the channel in the mean. Quantities of the latter kind include the delay spread, the Doppler spread, the coherence time, the coherence bandwidth and the angular or directional spread [13]. The directional spread is particulary interesting with respect to smart antennas.

### *3.2.2 Multi-location distribution of base stations and mobile stations*

In a mobile radio scenario, a multitude of radio links is simultaneously active in the same service area, and a priori no natural isolation between the different links exists. As a consequence of this lack of isolation, the radiated signals arrive not only at those receivers for which they are intended, which means that part of the radiated power is wasted, but also at the other receivers where the signal is undesired. The sum of undesired signals impinging at a receiver location from within and from outside their own cell constitutes the intracell and intercell multiple access interference (MAI), respectively. In most cases cellular mobile radio systems are interference limited, which means that the factor which limits capacity is not thermal receiver noise but MAI. If the multiple access schemes FDMA and/or TDMA are applied, intracell MAI is a priori avoided in conventional system architectures, where each time and frequency slot is allocated only to a single user in each cell. This is not true in the case of CDMA, where intracell MAI is unavoidable but may be afterwards eliminated by joint detection or interference cancellation. A vital task of the receivers consists in detecting the desired signals despite of the presence of MAI with sufficient quality. As an essential feature of mobile radio scenarios, the transmitters and receivers are spatially distributed over the service area. For any pair of transmitter and receiver locations, a matrix $\underline{\boldsymbol{G}}(\vartheta, \varphi, \Theta, \Phi, \tau, t)$ of differential directional channel impulse responses exists, see (4). It is important with respect to the application of smart antennas that, in general, the differential directional channel impulse responses of two pairs of locations differ especially with respect to their dependence on their angles $\vartheta, \varphi, \Theta, \Phi$, if at least one location is not the same for the two pairs, i.e. if at least three different locations are involved. Even in the case of LOS propagation such differences would arise. In addition, in the case of real world mobile radio scenarios effects of the following kinds may be observed:

- Signals emitted by two transmitters $T_1$ and $T_2$, which appear under the same LOS directions at a certain receiver location R, may impinge with different bearings at this receiver location, see Fig.3.2a.

- Signals emitted by two transmitters $T_1$ and $T_2$, which appear under different LOS directions at a certain receiver location R, may impinge with the same bearing at this receiver location, see Fig.3.2b.

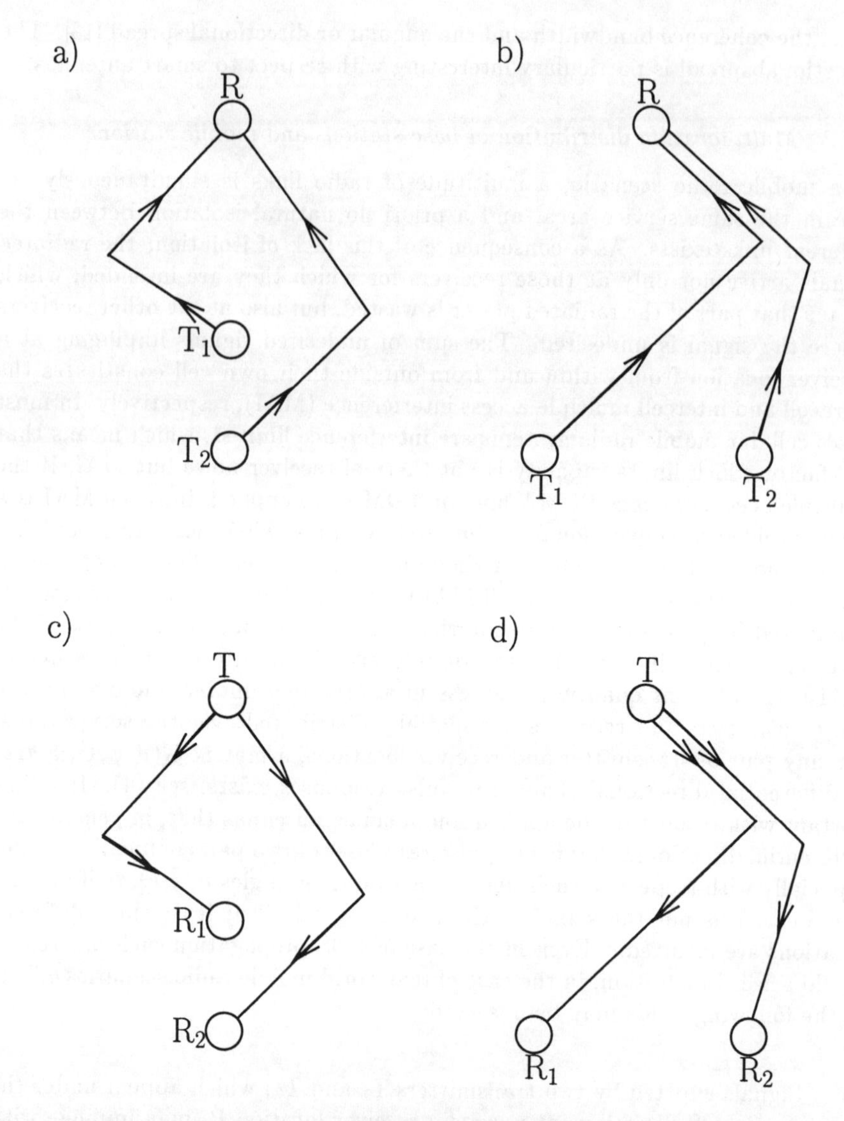


**Figure 3.2** Directional effects in mobile radio channels a) same LOS direction of $T_1$ and $T_2$ b) different LOS direction of $T_1$ and $T_2$ c) same LOS direction of $R_1$ and $R_2$ d) different LOS direction of $R_1$ and $R_2$

- Signals emitted by a transmitter T in two different directions may impinge at two receiver locations $R_1$ and $R_2$ appearing under the same LOS direction from this transmitter location, see Fig.3.2c.
- Signals emitted by a transmitter T in one direction may impinge at two receiver locations $R_1$ and $R_2$ appearing under different LOS directions from this transmitter location, see Fig.3.2d.

Such effects, which result from the spatial distribution of transmitters and receivers, offer the potential to increase the isolation between different radio links in the same service area by smart antennas.

## 3.3 FUNDAMENTALS OF ANTENNA DIVERSITY TECHNIQUES

### *3.3.1 Basic structures and effects*

Both time variance and frequency selectivity mentioned in Section 3.2.1 are undesired because they increase the sensitivity of the receiver to MAI [14]. In the case of time variance, this increase results from the nonlinear dependence of the error probability on the carrier-to-interference ratio $C/I$. Frequency selectivity increases the sensitivity to MAI because equalization is required at the receivers, which always comes along with an SNR degradation. If a certain quality of service has to be guaranteed, the increased sensitivity to MAI has to be compensated by reducing MAI. Such a reduction can be achieved by decreasing the number of users per cell and/or the cluster size, both of which directly lower capacity. It is well known that the detrimental effects of time variance and frequency selectivity can be combatted by diversity, i.e. by transmitting the desired signal not only over one, but over several channels, and by properly combining the channel output signals at the receiver [15]. The more uncorrelated these channels are, the more pronounced the benefits of diversity become. As another positive aspect of diversity, the differences of MAI powers and time variance on the different channels lead to interferer diversity, which can be exploited to improve the resulting MAI statistics. In the case of single port antennas at the transmitter and receiver, the only possible ways to translate the diversity principles into action are time diversity realized e.g. by interleaving in combination with FEC coding, and frequency diversity realized e.g. by frequency hopping or CDMA. In both cases, in fact different channels are not actually used, but one and the same channel is accessed at different instants of time or at different carrier frequencies.

A basically different diversity approach is antenna diversity, which is characterized by employing for each BS and/or MS instead of only one single port antenna a number of such antennas which are separately accessible. If these antennas are closely spaced, an array antenna is produced. According to (3.5), the propagation environment offers the potential to obtain, depending on the chosen antennas, a variety of more or less independent channel impulse responses between each pair of transmitter and receiver locations. Antenna diversity can be applied as the only diversity approach or in addition to time diversity and frequency diversity. However, it should be mentioned that there is a limit to the total diversity benefit achievable. With the antenna numbers $K_{\mathrm{a}}^{(\mathrm{BS})}$ and $K_{\mathrm{a}}^{(\mathrm{MS})}$ at the BS and the MS respectively, the antenna system can be considered as a multiport network. Fig.3.3 shows such a network for the case $K_{\mathrm{a}}^{(\mathrm{BS})}$ equal to three and $K_{\mathrm{a}}^{(\mathrm{MS})}$ equal to two. The multiport network is characterized by $K_{\mathrm{a}}^{(\mathrm{BS})} \cdot K_{\mathrm{a}}^{(\mathrm{MS})}$ impulse responses $\underline{h}^{(\mu,\nu)}(\tau,t)$, $\mu \in \{1 \ldots K_{\mathrm{a}}^{(\mathrm{BS})}\}$, $\nu \in \{1 \ldots K_{\mathrm{a}}^{(\mathrm{MS})}\}$, see Fig.3.3. Each of these impulse responses originates by

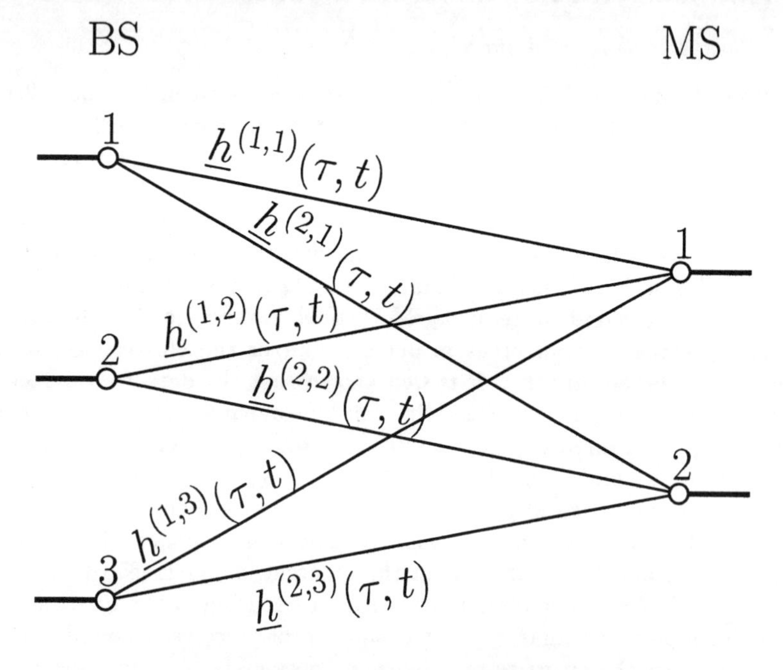


**Figure 3.3** Multiport network

a weighted superposition of the impulse responses of a selection of paths, the weighting and selection being determined by the characteristics of the employed antennas. In contrast to time diversity or frequency diversity with single port antennas at both stations, novel channels are generated by antenna diversity, and a new dimension of diversity is opened by the following basic effects:

- Path separation: Portions of the same signal transmitted over different sets of paths can be separately received and processed.
- Introduction of additional paths: Sets of paths not used for the transmission of the signal when employing single port antennas can be made usable.
- Reuse of paths: Portions of a signal having traveled over a set of paths can be used several times with different complex weightings.

On account of space limitations at the MSs, multiple antennas can be much more easily implemented at the BSs. Therefore only multiple antennas at the BSs are considered in this paper, with the number of BS antennas termed $K_{\mathrm{a}}$. However, it is known that antenna diversity techniques can be also utilized at the MSs [16].

Multiple antenna arrangements can be classified into macro structures and micro structures, the difference between which will be explained by considering the receiver side of multiple antenna transmission system. In the case of macro structures, the $K_{\mathrm{a}}$ antennas are so far apart that at each of the $K_{\mathrm{a}}$ antenna locations different wave fronts impinge, see Fig.3.4. Macro structures enable

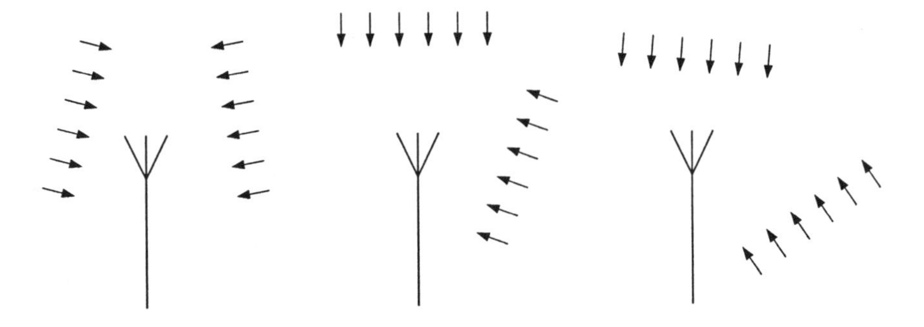

**Figure 3.4** Macro structure: At each of the antenna locations other wave fronts impinge

spatial macro diversity and can be implemented in the form of BS diversity and of remote antennas [17]. By spatial macro diversity additional paths are introduced. In the case of micro structures, the $K_{\mathrm{a}}$ antennas are so closely spaced that each of the antennas is hit by the same impinging wave fronts, although with different delays, see Fig.3.5. Practical concepts which apply micro structures are the following:

- Directional diversity realized by a set of $K_{\mathrm{a}}$ co-located directional antennas covering different solid angles [18], which leads to path separation.

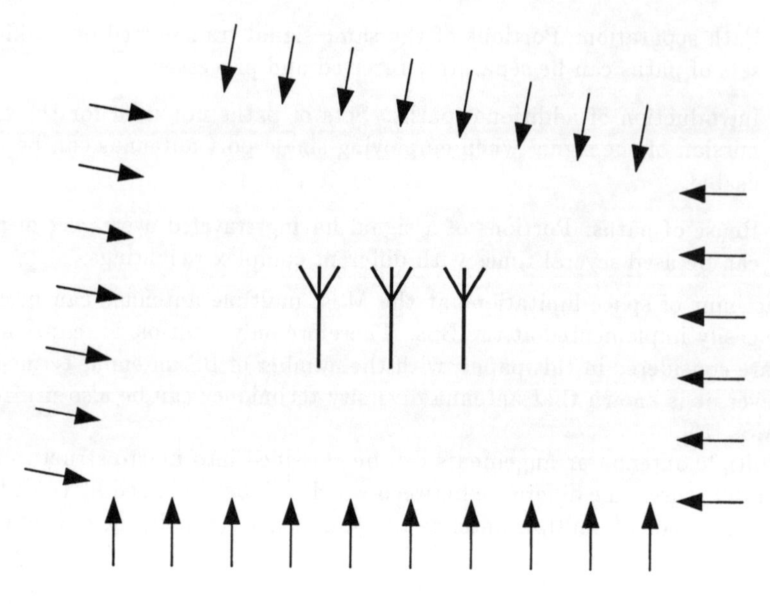

**Figure 3.5** Micro structure: Each of the antennas is hit by the same impinging wave fronts

- Polarization diversity realized by a set of $K_a$ co-located antennas with different orientations of polarization [19], which results in path separation and in the introduction of additional paths.

- Spatial micro diversity realized by arrays of $K_a$ generally equal and equally oriented antennas at closely spaced locations [20], which has the effect of path reuse and path separation.

- Field component diversity realized by a set of $K_a$ antennas, with different antennas having different characteristics for the electric and magnetic field components [21]. This leads to path separation and to the introduction of additional paths.

Also hybrid structures are conceivable, e.g. remote antennas with an array antenna at each of the remote locations connected to a certain BS, or BS diversity in combination with directional antennas at each of the involved BSs. [h] The antenna diversity arrangements have a single port at the MS side and $K_a$ ports at the BS side, see Fig.3.6. From the port of the MS antenna to the ports of the $K_a$ BS antennas, $K_a$ channels with the channel impulse responses $\underline{h}^{(k_a)}(\tau, t)$, $k_a = 1 \ldots K_a$, exist. In what follows, expressions will be presented

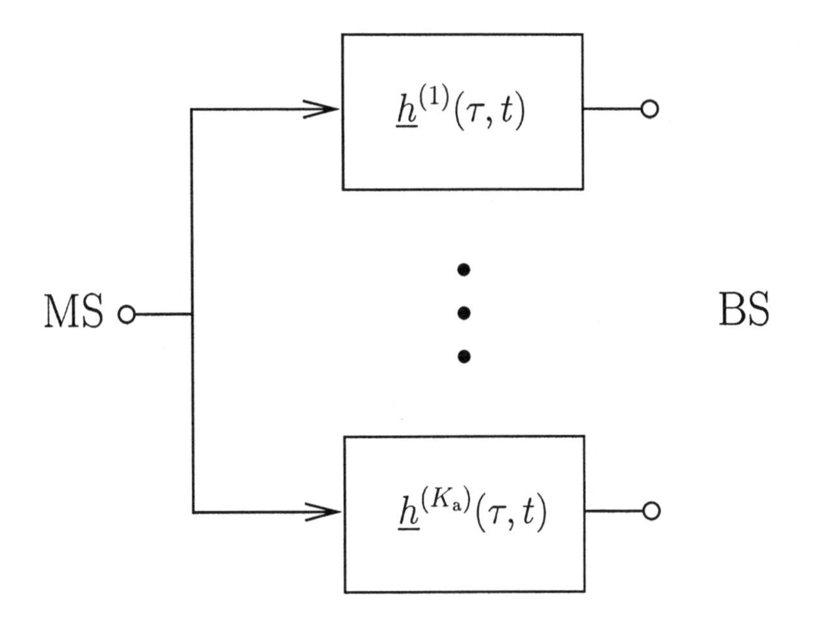


**Figure 3.6** Single port/multiple port channel

which show how the channel impulse responses $\underline{h}^{(k_a)}(\tau, t)$, $k_a = 1 \dots K_a$, result from the differential directional channel impulse responses introduced in Section 3.2.1 and from the specific arrangement under consideration. If directional diversity, polarization diversity or field component diversity are considered, $K_a$ receiver antennas are at virtually the same location, each with a different characteristic, $\underline{f}_r^{(k_a)}(\Theta, \Phi)$. Consequently, the $K_a$ channel impulse responses take the form

$$\begin{aligned} \underline{h}^{(k_a)}(\tau, t) &= \iiiint\limits_{\substack{\vartheta, \Theta = 0 \dots \pi \\ \varphi, \Phi = 0 \dots 2\pi}} \underline{f}_r^{(k_a)T}(\Theta, \Phi) \cdot \underline{G}(\vartheta, \varphi, \Theta, \Phi, \tau, t) \underline{f}_t(\vartheta, \varphi) \\ &\times \sin\vartheta \cdot \sin\Theta \cdot d\vartheta \cdot d\varphi \cdot d\Theta \cdot d\Phi, \qquad (3.6) \\ & k_a = 1 \dots K_a. \end{aligned}$$

In the case of spatial micro diversity, $K_a$, equal and equally oriented antennas are used, which are so closely spaced that a micro structure is obtained, see Fig.3.5. A wave impinging from direction $\Theta$, $\Phi$ arrives at each of the $K_a$ antennas with a different phase shift $\Psi^{(k_a)}(\Theta, \Phi)$. Therefore, the $K_a$ channel impulse responses take the form

$$\underline{h}^{(k_a)}(\tau,t) = \iiiint\limits_{\substack{\vartheta,\Theta=0\ldots\pi\\ \varphi,\Phi=0\ldots2\pi}} \exp\left[j\Psi^{(k_a)}(\Theta,\Phi)\right]\underline{f}_{\mathrm{r}}^{T}(\Theta,\Phi)\cdot\underline{G}(\vartheta,\varphi,\Theta,\Phi,\tau,t) \tag{3.7}$$

$$\cdot\underline{f}_{\mathrm{t}}(\vartheta,\varphi)\cdot\sin\vartheta\cdot\sin\Theta\cdot d\vartheta\cdot d\varphi\cdot d\Theta\cdot d\Phi,$$

$$k_{\mathrm{a}} = 1\ldots K_{\mathrm{a}}.$$

Finally, spatial macro diversity will be considered. In this case, for each of the $K_{\mathrm{a}}$ antenna locations a different set $\underline{G}^{(k_a)}(\vartheta,\varphi,\Theta,\Phi,\tau,t)$ of differential directional channel impulse responses is valid, and the $K_{\mathrm{a}}$ channel impulse responses are

$$\underline{h}^{(k_a)}(\tau,t) = \iiiint\limits_{\substack{\vartheta,\Theta=0\ldots\pi\\ \varphi,\Phi=0\ldots2\pi}} \underline{f}_{\mathrm{r}}^{T}(\Theta,\Phi)\cdot\underline{G}^{(k_a)}(\vartheta,\varphi,\Theta,\Phi,\tau,t)\underline{f}_{\mathrm{t}}(\vartheta,\varphi)$$

$$\times\sin\vartheta\cdot\sin\Theta\cdot d\vartheta\cdot d\varphi\cdot d\Theta\cdot d\Phi, \tag{3.8}$$

$$k_{\mathrm{a}} = 1\ldots K_{\mathrm{a}}.$$

In (3.8) it is assumed that equal and equally oriented antennas are used at each of the $K_{\mathrm{a}}$ locations.

### *3.3.2 Uplink*

In order to determine, which benefits can be achieved by antenna diversity in the uplink, the structure shown in Fig.3.6 is supplemented in such a way that the single input/multiple output channel is transformed into a single input/single output channel. When doing so, the resulting structure must comply with the procedures usually applied in receiver design and operation. This transformation can be performed as indicated in Fig.3.7. Each branch of the structure shown in Fig.3.6 is supplemented by an adder to introduce MAI and a filter matched to the corresponding channel impulse response. The design of the matched filters requires that the channel impulse responses $\underline{h}^{(k_a)}(\tau,t)$, $k_{\mathrm{a}} = 1\ldots K_{\mathrm{a}}$, can be determined at the BS receiver, e.g. based on transmitted training signals. The filter outputs are fed into an adder and then further processed. The depicted combiner unit in Fig.3.7 performs maximal ratio combining (MRC) for the case of uncorrelated MAI with equal power in the two diversity branches. The spectral efficiency and correlation properties of the MAI can also be considered when combining the received signals according to MRC [23]. Other combining techniques are selection combining and equal gain combining [22]. When adjusting the combiner, various sources of information can be utilized. For instance, the responses of the $K_{\mathrm{a}}$ effective channels

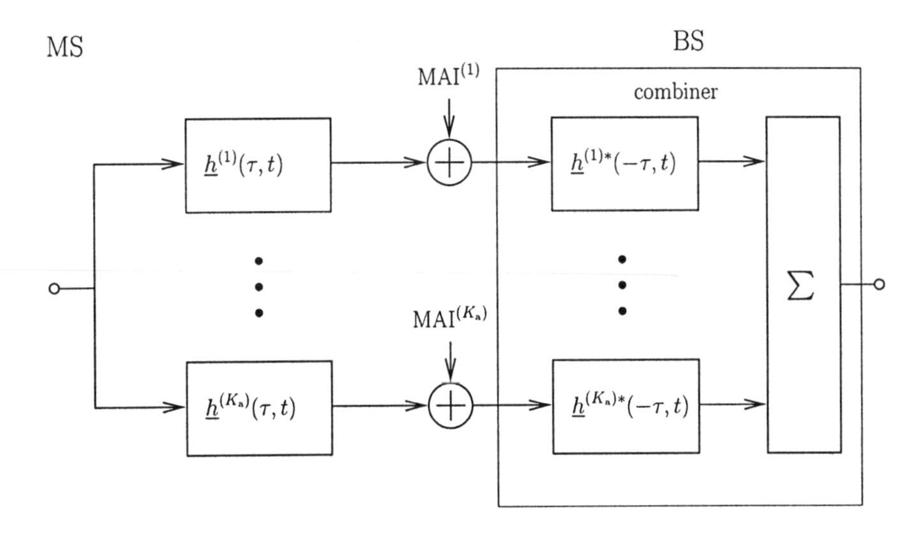


**Figure 3.7** Transformation of the single input/multiple output channel into a single input/single output channel

to transmitted training signals could be used [24] or estimated directions of incidence, which can be obtained by algorithms like MUSIC and ESPRIT [25].

Between the input of the transmitter antenna and the output of each of the $K_{\mathrm{a}}$ matched filters, an effective channel with the channel impulse response

$$\underline{h}_{\mathrm{eff}}^{(k_{\mathrm{a}})}(\tau,t) = \underline{h}^{(k_{\mathrm{a}})}(\tau,t) * \underline{h}^{(k_{\mathrm{a}})*}(-\tau,t), \quad k_{\mathrm{a}} = 1 \ldots K_{\mathrm{a}} \tag{3.9}$$

results. Each of these $K_{\mathrm{a}}$ effective channel impulse responses has a real valued mainlobe and otherwise assumes complex values, which decrease with increasing distance from the mainlobe. Each of the effective channels is corrupted by MAI. The MRC unit composes the $K_{\mathrm{a}}$ effective channel impulse responses $\underline{h}_{\mathrm{eff}}^{(k_{\mathrm{a}})}(\tau,t)$ to obtain the resultant effective channel impulse response $\underline{h}_{\mathrm{eff}}(\tau,t)$, which is valid for the effective channel between the input of the transmitter antenna and the output of the combiner, and which is also corrupted by MAI. The crux of combining consists in enhancing the desired mainlobe signals and in simultaneously hampering the undesired MAI in order to push up the carrier-to-interference ratio $C/I$. In terms of antenna characteristics, proper combining can be interpreted as generating antenna diagrams with relatively high gains into the directions, from which mainly desired signals come, and with relatively low gains into directions, from which mainly MAI impinges. This combining has two other positive effects. First, fading is reduced because, when constructively combining the mainlobes of the effective channel impulse responses $\underline{h}_{\text{eff}}^{(k_a)}(\tau,t)$, the fluctuations of the individual mainlobes are equalized. Secondly, the delay dispersion is reduced because the mainlobes of the effective channel impulse responses $\underline{h}_{\text{eff}}^{(k_a)}(\tau,t)$ are constructively combined whereas, apart from the mainlobes, the combination is performed in a more or less randomly way. The beneficial effects of antenna diversity in the uplink can be exploited to increase the capacity by accommodating more users per cell [26] and/or decreasing the cluster size. It may be even possible to separate the users of a cell merely based on the different directions of incidence of their signals at the BS. Such concepts are termed SDMA (Space Division Multiple Access) and would allow to reuse the same time and frequency slots of an FDMA/TDMA system within a cell [27, 28].

Concerning the variance of the MAI power, which also should be small, this quantity decreases in relation to its mean value, if the number of affecting MAI sources is increased by the antenna concept, i.e. if the interferer diversity is increased. Consequently, the variance of the MAI power in relation to its mean value increases in the case of directional diversity because the total number of relevant sources of MAI decreases. In the case of polarization diversity and macro diversity, the number of relevant MAI sources increases and the variance of the MAI power in relation to its mean value decreases. In the case of spatial micro diversity it can be expected that interferer diversity remains approximately the same as in the case of a single output receiver antenna. In Table 3.1, the various types of uplink antenna diversity, their mechanisms and their pros and cons are summarized.

### *3.3.3 Downlink*

In the downlink, an antenna diversity arrangement appears as a multiple input/single output channel, see Fig.3.6. As a prerequisite to enjoying the benefits of antenna diversity in the uplink, see Section 3.3.2 , the channel impulse responses $\underline{h}^{(k_a)}(\tau,t)$, $k_a = 1 \ldots K_a$, have to be known in the receiver. The knowledge of these uplink channel impulse responses does not imply the knowledge of the corresponding downlink channel impulses. In the case of time division multiplexing (TDD), the time variance causes the channel impulse responses valid in the downlink to deviate from the corresponding channel impulse responses of the uplink; in the case of frequency division multiplexing (FDD), the channel impulse responses in the uplink differ from those in the downlink on account of the different transmission frequencies. For these reasons, it is not practical to use the channel impulse responses known from the uplink to design the downlink signals in such a way that transmission is optimized. Nevertheless, it

| diversity type | mechanism | delay dispersion | fading fast | fading slow | interferer diversity | $C/I$ |
|---|---|---|---|---|---|---|
| directional | path separation | + | + | | − | + |
| polarization | path separation, additional paths | + | + | | + | + |
| spatial micro | path reuse | + | + | | | + |
| field component | path separation, additional paths | + | + | | | + |
| spatial macro | additional paths | + | + | + | + | + |

**Table 3.1** Pros and cons of different types of antenna diversity in the uplink

can be presumed at the BS that uplink channels with a low attenuation lead to downlink channels which have a low attenuation as well. This knowledge can be exploited to concentrate the transmitter power of the downlink on those channels which show low attenuation. By doing so, the power is concentrated towards the MSs for which it is meant [29]. This simultaneously reduces the MAI power impinging at non-addressed MSs. Fig.3.8 shows a corresponding arrangement. The feeding signals of the $K_{\mathrm{a}}$ antennas at the BS are weighted by complex factors $\underline{c}_{k_{\mathrm{a}}}$, $k_{\mathrm{a}} = 1 \ldots K_{\mathrm{a}}$ in such a way that radiation is concentrated in those directions, which are made up of low attenuation paths towards the addressed MS, and in which low MAI is caused for non-addressed MSs.

## 3.4 CASE STUDY: JD-CDMA WITH SMART ANTENNAS

### *3.4.1 System description*

In the early 1990s the development of a novel air interface began, in which a combination of the multiple access methods FDMA and TDMA is supplemented by a CDMA component and in which the increased interference typical of CDMA is combatted by joint detection in the receivers [7, 30]. This air interface concept is termed JD (Joint Detection)–CDMA and has been verified by extensive computer simulations [31]. Presently, a JD-CDMA hardware demonstrator is being implemented in a joint project of a European manufacturer and the authors. Elements of JD–CDMA are also considered within ACTS.

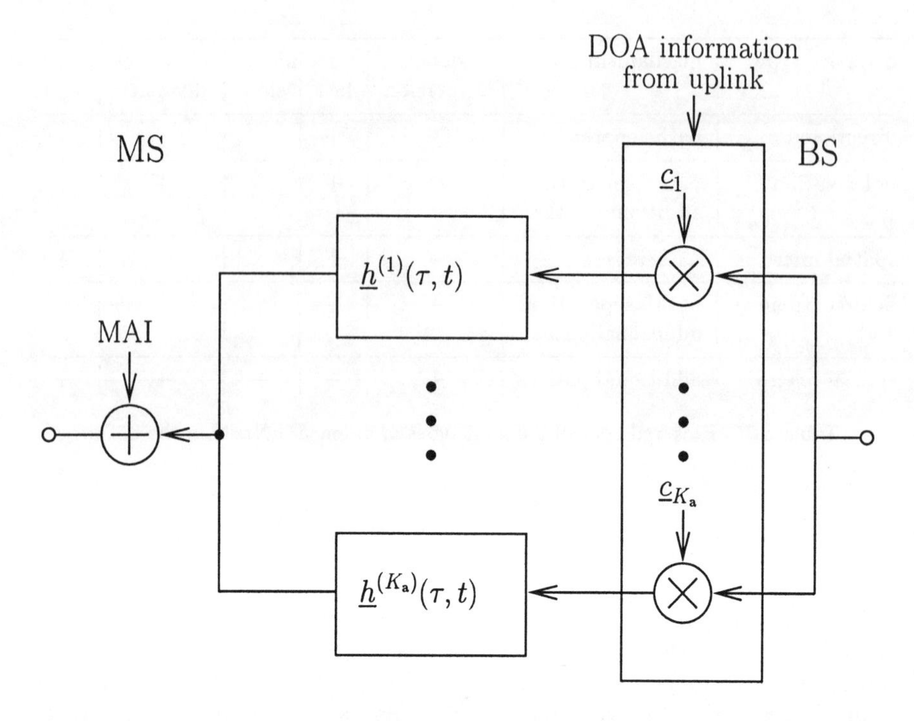


**Figure 3.8** Downlink beamforming based on uplink DOA information

In the JD-CDMA air interface concept $K$ mobile users are simultaneously active in the same frequency and time slot, each using a user specific spreading code to allow signal separation in the receiver. The frame structure of this time-slotted CDMA concept is illustrated in Fig.3.9, where $B, T_{\text{fr}}, N_{\text{fr}}, T_{\text{bu}}$, and $K$ denote the bandwidth of a frequency slot, the duration of a TDMA frame, the number of bursts per TDMA frame, the burst duration, and the number of users per frequency and time slot, respectively. A burst consists of two data blocks separated by a user specific midamble, which is used for channel estimation, and a guard interval, see Fig.3.9. The assumed frame and burst structures are similar to those used in GSM and facilitate backward compatibility.

In what follows, the beneficial effects of receiver antenna diversity techniques in the uplink of a JD–CDMA air interface will be demonstrated. Three different antenna configurations are investigated: a spatial micro diversity concept using omnidirectional antennas with rather large distances between each other; a spatial micro diversity concept with an array of closely spaced omnidirec-

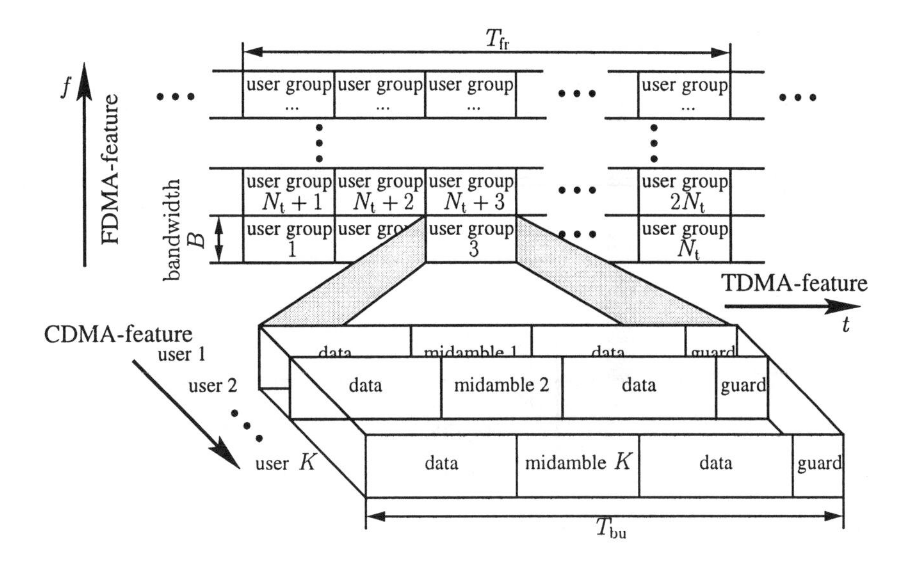


**Figure 3.9** JD-CDMA frame structure

tional antennas; and a directional diversity concept using a set of co-located sector antennas. The performance enhancements achievable by these antenna configurations are determined by link level and system level simulations. As a measure of performance, the spectrum efficiency mentioned in Section 3.1 is chosen. An intelligent antenna concept for the JD-CDMA downlink will also be considered. In this concept, the transmitter powers are directionally transmitted in such a way that the carrier-to-interference ratio $C/I$ is maximized for all users. The benefits achievable by this technique will be demonstrated by means of the cumulative distributions of the $C/I$.

### *3.4.2 Antenna diversity concepts for the JD-CDMA Uplink*

In Fig.3.10 the structure of the JD-CDMA uplink receiver is depicted for the case of spatial micro diversity with two omnidirectional receiver antennas. For each antenna output, a channel estimator (CE #1, CE #2) estimates the channel impulse responses for the $K$ connections between the corresponding receiver antenna and the $K$ mobile users according to the maximum likelihood channel estimation algorithm presented in [12]. Based on the estimated impulse responses and the knowledge of the user specific spreading codes, joint detection is performed using the JD algorithms presented in [30].

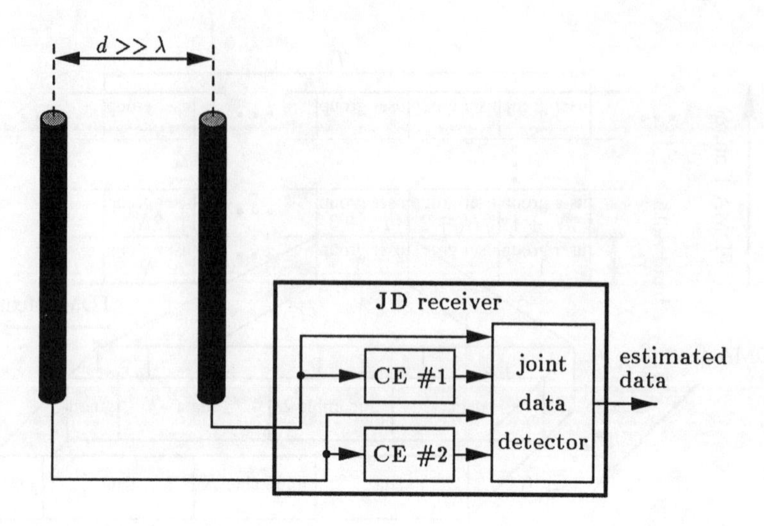


**Figure 3.10** JD-CDMA uplink receiver: Spatial micro diversity with omnidirectional antennas

The second antenna diversity concept for the JD-CDMA uplink considered is a spatial micro diversity concept based on an array of omnidirectional antennas [32]. As depicted in Fig.3.11, the directions of arrival (DOAs) of waves carrying desired as well as undesired signals are estimated in the receiver by a DOA estimator, e.g. by using ESPRIT-like algorithms [25]. According to the maximum likelihood channel estimation algorithm presented in [12], which is extended in order to take the estimated DOAs into account, see [25], estimates of the channel impulse responses valid for different DOAs are determined. Finally, the transmitted data are estimated using a JD technique which is an extension of the JD technique described in [30], and which explicitly takes into account the DOAs of waves carrying both desired and undesired signals. As shown in [32], this JD technique can be split up into a beam forming part and a signal separating part. As a first step, beam formers are implemented for the DOAs associated with each desired signal. The beam formers maximise the power ratio of desired and undesired signal components. Therefore, the resulting antenna pattern for such a beam former effects low gains for DOAs associated with strong undesired signals and high gains for the DOAs of the desired signals. As a second step, the outputs of the beam formers pertaining to a certain user are combined by MRC. The outputs of the MRC units are fed into a JD unit, which removes intersymbol interference (ISI) as well as MAI

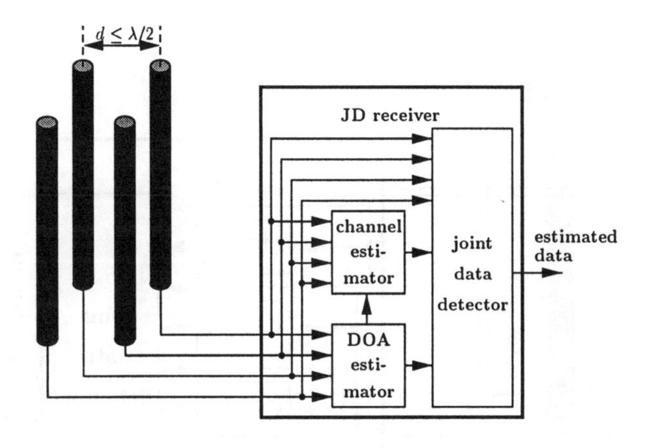


**Figure 3.11** Spatial micro diversity with an array of omnidirectional antennas

originating from users assigned to the considered BS, [30]. The ISI and MAI canceller exploits information about the user specific CDMA codes, the DOAs of both desired and undesired signals, and the channel impulse responses associated with the DOAs of desired signals in order to determine estimates of the transmitted data.

The last antenna diversity concept for the JD-CDMA uplink applies directional micro diversity [32]. A number of $K_a$ sector antennas is placed at approximately the same location and covering different sectors of the total space. Such an antenna configuration is depicted in Fig.3.12 for the case of four identical sector antennas. As in the case of spatial micro diversity, a channel estimator as described in [12] is implemented for each antenna output, see CE #1 ... CE #4 in Fig.3.12. These channel estimators determine the channel impulse responses for the connections of the corresponding sector antenna to all $K$ users, which are assigned to the considered BS. Based on the estimated channel impulse responses and on the user specific CDMA codes, the transmitted data are estimated by JD according to [30]. Due to the fixed diagrams of the $K_a$ sector antennas, no information about DOAs is necessary for receiver operation.

With the above introduced quantities $B$, $T_{fr}$, $T_{bu}$ and $K$, with the information rate $R$ per user and with the cluster size $r$, the spectrum efficiency is given by [33]

$$\eta = \frac{K \cdot T_{fr} \cdot R}{r \cdot T_{bu} \cdot B}. \tag{3.10}$$

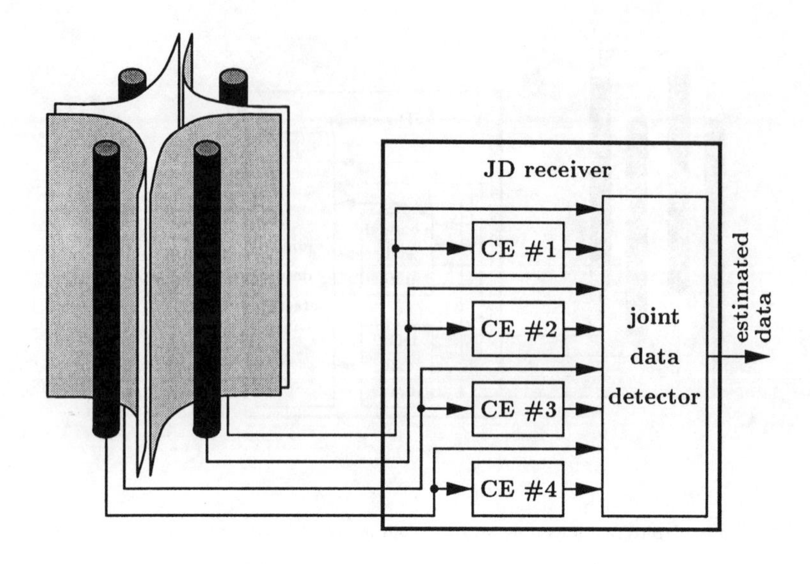


**Figure 3.12** Third antenna concept: Directional diversity with sector antennas.

This expression shows that, for given values of $T_{\text{fr}}$, $R$, $T_{\text{bu}}$ and $B$, $\eta$ is determined by $r$ and $K$. Concerning $r$, the minimum possible value which is one, should be chosen to make $\eta$ large. Concerning $K$, at first glance, $\eta$ could be made arbitrarily large by increasing $K$ more and more. However, the choice of the values for $r$ and $K$ has an impact on the performance of the system: the smaller $r$, the more severe the effect of intercell MAI, and the larger $K$, the larger the SNR degradation experienced when separating the user signals by JD at the receiver [30]. This means that, when choosing $r$ and $K$ such that $\eta$ is maximized, the side condition of keeping the system performance sufficiently high has to be observed.

To quantify the system performance, a quality of service (QoS) criterion is chosen which is defined as follows: the QoS criterion is met, if the bit error probability $P_{\text{b}}$ exceeds a given upper bound $P_{\text{b}}^{\text{M}}$ with a probability not greater than a given value $P_{\text{out}}$. $P_{\text{out}}$ is termed outage probability. The values $P_{\text{b}}^{\text{M}}$ and $P_{\text{out}}$ determine the required QoS. For each pair $(r, K)$, a cumulative distribution function (cdf) $\text{Prob}\{P_{\text{b}} \leq \Gamma\}$ holds. In order to test, whether the QoS criterion specified by $P_{\text{b}}^{\text{M}}$ and $P_{\text{out}}$ is met, the value $\text{Prob}\{P_{\text{b}} > P_{\text{b}}^{\text{M}}\} = 1 - \text{Prob}\{P_{\text{b}} \leq P_{\text{b}}^{\text{M}}\}$ has to be determined. Then the following decision rule holds:

$$\begin{aligned} &\text{Prob}\{P_{\text{b}} > P_{\text{b}}^{\text{M}}\} > P_{\text{out}} : \text{ QoS criterion not fulfilled,} \\ &\text{Prob}\{P_{\text{b}} > P_{\text{b}}^{\text{M}}\} \leq P_{\text{out}} : \text{ QoS criterion fulfilled.} \end{aligned} \tag{3.11}$$

If the QoS criterion is not fulfilled, $r$ has to be increased and/or $K$ has to be decreased until the QoS criterion is met. On the other hand, if the QoS criterion is met with a certain pair $(r, K)$, one should try to increase $K$ and/or decrease $r$ with the goal to arrive at another pair $(r, K)$, for which the QoS criterion is still fulfilled, however, with a larger $\eta$.

The above described procedure to determine $\eta$ cannot be performed in a closed analytical way. Rather, extensive computer simulations are required to reach this goal. To determine $\eta$, the authors have developed the simulation tool ADAMO (Antenna Diversity And MOre). ADAMO is based on Monte Carlo simulations of the data transmission in the JD-CDMA uplinks of a reference cell, which is embedded into a cellular environment. Based on a Hata-like model for slow fading [33], an inhomogeneous directional distribution of the intercell MAI is generated for the reference cell. In this, both the average powers and the DOAs of the interferers are taken into consideration. Fast fading models, which also include the distribution of DOAs, are used to generate time varying and frequency selective channel impulse responses for the connections between the MSs and the BS of the reference cell. A detailed description of these channel models can be found in [34].

In what follows, values of $\eta$ determined by simulations with ADAMO will be presented. The simulations performed with the system parameters are listed in Table 3.2. Further details of the simulation conditions are:

- Slow frequency hopping from burst to burst,
- MS speed 90 km/h,
- GMSK-like spreading modulation,
- rural propagation area [34],
- $P_{\mathrm{b}}^{\mathrm{M}} = 10^{-3}$, $P_{\mathrm{out}} = 5 \cdot 10^{-2}$,
- $K_{\mathrm{a}} = 8$ antennas.

The obtained values of $\eta$ are shown in Table 3.3. The results presented in Table 3.3 show that all considered antenna diversity concepts offer an increase of $\eta$ as compared to single omnidirectional antennas. The largest $\eta$ is obtained for the uniform linear array. The antenna diversity concepts with omnidirectional antennas and with a uniform linear array have the further advantage that a cluster size $r$ equal to one can be chosen, which facilitates frequency planning.

### *3.4.3 Antenna diversity concept for the JD-CDMA downlink*

The spatial micro diversity concept for the JD-CDMA uplink, see Fig.3.11, allows determination of the directions of incidence of the desired signals at the

| | | |
|---|---|---|
| user bandwidth | $B$ | 1.6 MHz |
| burst structure | | |
| burst duration | $T_{\mathrm{bu}}$ | 0.5 ms |
| TDMA frame duration | $T_{\mathrm{fr}}$ | 6 ms |
| midamble chips | $L_{\mathrm{mid}}$ | 268 |
| guard interval | $T_{\mathrm{g}}$ | 30 $\mu$s |
| data symbols per data block | $N$ | 24 |
| symbol duration | $T_{\mathrm{S}}$ | 7 $\mu$s |
| chips per symbol | $Q$ | 14 |
| chip duration | $T_{\mathrm{c}}$ | 0.5 $\mu$s |
| size of data symbol alphabet | $M$ | 4 (4PSK) |
| convolutional encoder | | |
| constraint length | $K_{\mathrm{c}}$ | 5 |
| rate | $R_{\mathrm{c}}$ | 1/2 |
| general | | |
| interleaving depth | $I_{\mathrm{d}}$ | 4 bursts |
| resulting data rate per user | $R$ | 8 kbit/s |

**Table 3.2** Parameters of the JD-CDMA uplink considered in the simulations

| | diversity type | $(r, K)$ | $\eta / \left(\frac{\text{bit/s}}{\text{Hz}} \text{ per cell}\right)$ |
|---|---|---|---|
| $K_{\text{a}} = 8$ | spatial macro | (1,3) | 0.18 |
| | spatial micro | (1,4) | 0.24 |
| | directional micro | (3,8) | 0.16 |
| $K_{\text{a}} = 1$ | single omnidirectional antenna | (4,3) | 0.045 |

**Table 3.3** Spectrum efficiencies $\eta$ of the JD-CDMA uplink for different BS antenna concepts

uplink receivers. As mentioned in Section 3.4.3, such directional information can be exploited in the downlink to increase the $C/I$ by directionally radiate the transmitter powers. It is assumed that the downlink waves mainly propagate along the LOS bearing under which the MSs appear from the BSs, which is a reasonable assumption in rural and urban macro cells. It is further assumed that at each BS of the scenario the bearings of all MSs of the scenario are known, an assumption which can be only partially fulfilled in reality. A spatial micro diversity structure with $K_{\text{a}}$ antennas arranged in a linear array with the distance $\lambda/2$ between adjacent antennas is used for downlink transmission. For each connection, BS/MS, the complex steering weights of the $K_{\text{a}}$ antennas are chosen in such a way that the ratio of the desired power at the addressed MS and the sum of the MAI powers at all other MSs becomes maximum. Further, the transmitter powers of all BSs are adjusted such that the $C/I$ becomes equal at all MSs. Following this strategy, which is explained in more detail in [35], cumulative distribution functions of the $C/I$ were determined for the JD-CDMA downlink by extensive computer simulations by the authors. In these simulations, intracell interference can be omitted because it is eliminated by joint detection. The simulation results are shown in Fig.3.13a–d. The parameter of the curves is the number $K$ of users per time slot. As is expected, the curves for a certain $K$ are shifted to the right with increasing $K_{\text{a}}$. In order to illustrate the benefit of directional transmission, let us assume that a $C/I$ of 0 dB is required to obtain a sufficient transmission quality, and that this quality is maintained or exceeded for 95% of the time. Under these presuppositions depending on $K_{\text{a}}$ the user numbers, $K$, shown in Table 3.4, can be accommodated. These results show that the spectrum efficiency, $\eta$, can be increased by a factor of 6 if the number $K_{\text{a}}$ of antennas is increased from two to eight.

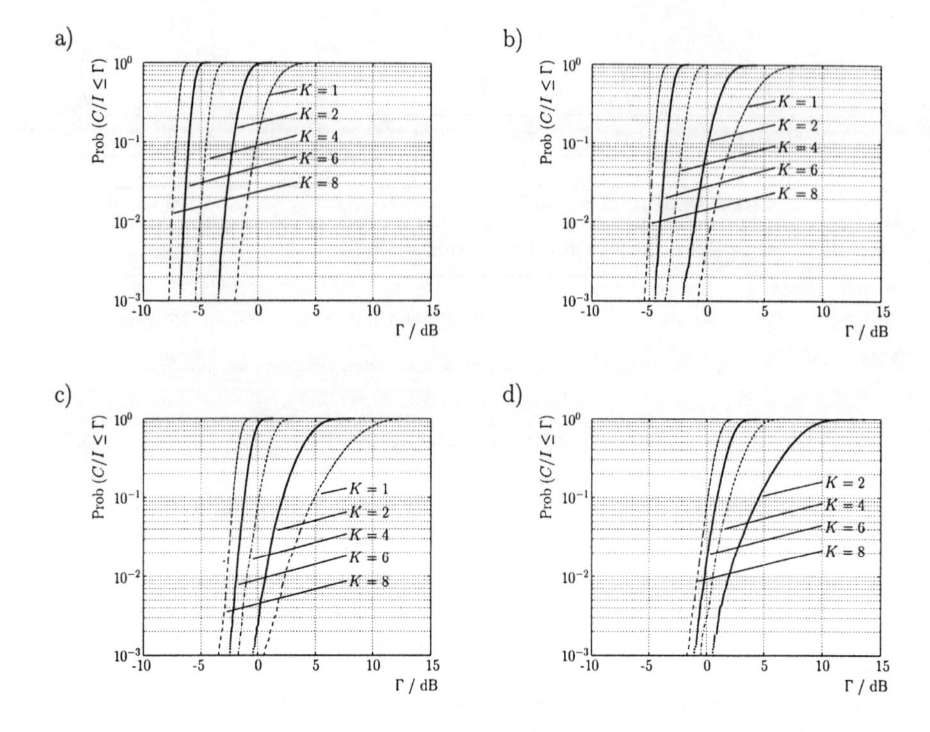


**Figure 3.13** JD-CDMA downlink with directional transmission at the BSs: Cumulative distribution functions of the $C/I$ a) $K_\mathrm{a} = 1$ b) $K_\mathrm{a} = 2$ c) $K_\mathrm{a} = 4$ d) $K_\mathrm{a} = 8$

| $K_\mathrm{a}$ | 1 | 2 | 4 | 8 |
|---|---|---|---|---|
| max. allowable $K$ | 0 | 1 | 2 | 6 |

**Table 3.4** Maximum allowable number $K$ of users per time slot depending on the number $K_\mathrm{a}$ of antennas at each BS

## 3.5 CONCLUSIONS

It has been shown that the exploitation of the properties of the mobile radio channels and the spatial distribution of the BSs and MSs offer a new dimension of diversity. The impacts of MAI, delay dispersion and frequency selectivity can be reduced by smart antennas, which offer the potential to increase the capacity. By means of the UMTS air interface JD-CDMA the beneficial effects of smart antennas can be demonstrated.

## ACKNOWLEDGEMENT

The sponsoring of part of this work by Deutsche Forschungsgemeinschaft as well as the invaluable support of the supercomputer staff at the Regionales Hochschulrechenzentrum Kaiserslautern (RHRK) are gratefully acknowledged.

## References

[1] M. Mouly, M.B. Pautet: The GSM system for mobile communications. Published by the authors, 1992. ISBN 2-9507190-0-78.

[2] F. Hillebrand: Defining the UMTS vision. Conference Proceedings. Next generation mobile, evolution or revolution? IBC UK Conferences, London (1996).

[3] B. Frison, M. Woinsky, A. Kripalani: PCS standards. IEEE Personal Communications, vol.1 (1994), 10–11.

[4] Acts 96 Programme Guide. Brussels (1996).

[5] J. Rapeli: UMTS-targets, system concept and standardization in a global framework. IEEE Personal Communications, vol. 2 (1995), 20–28.

[6] E. E. Reinhart: Mobile communications. WARC-92 Special Report, IEEE Spectrum, vol. 29 (1992), 27-29.

[7] P. Jung: Joint detection CDMA. Invited Paper. In Ramjee Prasad: "CDMA for Wireless Personal Communications,"Artech House, Boston (1996), Section 11.5, 348–353.

[8] D. Parsons: The mobile radio propagation channel. Pentech Press. London, 1992.

[9] T. Fehlhauer, P.W. Baier, W. König, W. Mohr: Optimized wideband system for unbiased mobile radio channel sounding with periodic spread spectrum signals. IEICE Transactions on Communications, vol. E76-B (1993), 1016-1029.

[10] T. Fehlhauer, P.W. Baier, W. König, W. Mohr: Wideband characterization of fading outdoor radio channels at 1800 MHz to support mobile radio system design. Wireless Personal Communications, vol. 1 (1995), 137-149.

[11] T. Kuerner, D.C. Cichon, W. Wiesbeck: Concepts and results for 3D digital terrain-based wave propagation models: an overview. IEEE Journal on Selected Areas in Communications, vol. 11 (1993), 1002–1012.

[12] B. Steiner, P. Jung: Optimum and suboptimum channel estimation for the uplink of CDMA mobile radio systems with joint detection. European Transactions on Telecommunications and Related Technologies, vol. 5 (1994), 39–50.

[13] P.C.F. Eggers: Angular Dispersive Mobile Radio Environments sensed by highly Directive Base Station Antennas. Proc. IEEE Sixth International symposium on Personal Indoor and Mobile Radio Communications (PIMRC'95), Toronto (1995), 522–526.

[14] G. C. Hess: Land-mobile radio system engineering. Artech House, Boston, 1993.

[15] A. Klein, B. Steiner, A. Steil: Known and novel diversity approaches in a JD-CDMA system concept developed within COST 231. Proc. International Symposium on Personal, Indoor and Mobile Radio Communications (PIMRC'95), Toronto (1995), 512-516.

[16] J. Fuhl, E. Bonek, P. Balducci, P. Nowak, H. Garn: Internal antenna arrangements for personal communication systems. Proc. First European Personal and Mobile Communications Conference (EPMCC'95), Bologna (1995), 62-67.

[17] P. Jung, B. Steiner, B. Stilling: Exploitation of intracell macrodiversity in mobile radio systems by development of remote antennas. Proc. IEEE Fourth International Symposium on Spread Spectrum and Applications (ISSSTA'96), Mainz (1996), 302-307.

[18] J.J. Blanz, P. Jung, A.J. Steil, P.W. Baier: Coherent receiver antenna diversity with directional antennas. IEEE Third International Conference on Telecommunications (ICT'96), Istanbul (1996), 410–417.

[19] P.C.F. Eggers, J. Toftgård, A.M. Opera: Antenna systems for Base Station Diversity in Urban Small and Micro Cells. IEEE Journal on Selected Areas in Communications, vol. 11 (1993), 1046–1056.

[20] C. Farsakh, M. Haardt, J.A. Nossek, K. Pensel: Adaptive antenna arrays in mobile radio systems. Proc. ITG Conference on Mobile Communications, Neu-Ulm (1995), 101–108.

[21] W.C.Y. Lee: Mobile Communications Engineering. Mc Graw–Hill Book Company.

[22] D.G. Brennan: Linear diversity combining techniques. IRE, vol.47 (1959), 1075–1102.

[23] J.J. Blanz, M. Haardt, A. Papathanassiou, I. Furió and P. Jung: "Combined Direction Of Arrival and Channel Estimation for a Time-Slotted CDMA. "In Proceedings of the Sixth International Conference on Telecommunications (ICT'97), Melbourne, Australia, April 1997.

[24] J. Fuhl, E. Bonek: Space–Time Decomposition: Exploiting the Full Information of a Training Sequence for an Adaptive Array. Electronics Letters, vol. 32 (1996), 1938–1940.

[25] M. Haardt: Efficient One-, Two- and Multidimensional High resolution Array Signal Processing. Shaker Verlag, Aachen (1996), ISBN 3-8265.2220-6.

[26] J. S. Thompson, P.M. Grant and B. Mulgrew: Smart Antenna Arrays for CDMA Systems. IEEE Personal Communications Magazine, vol. 3 (1996), 16–25.

[27] J.Fuhl, A.F. Molisch: Space Domain Equalization for Second and Third Generation Mobile Radio Systems. Proc. ITG Conference on Mobile Radio, Neu-Ulm (1995), 85–92.

[28] M. Tangemann, C. Hoek, R. Rheinschmitt: Introducing Adaptive Array Antenna Concepts in Mobile Communications Systems. RACE Mobile Telecommunications Workshop, Amsterdam (1994), 714–727.

[29] C. Farsakh, J.A. Nossek: Channel allocation and downlink beamforming in an SDMA mobile radio system. Proc. IEEE Sixth International symposium on Personal Indoor and Mobile Radio Communications (PIMRC'95), Toronto (1995), 687–691.

[30] J.J. Blanz, A. Klein, M.M. Naßhan, A. Steil: Performance of a cellular hybrid C/TDMA mobile radio system applying joint detection and coherent receiver antenna diversity. IEEE Journal on Selected Areas in Communications, "Special Issue on CDMA networks", vol. 12 (1994), 568–579.

[31] P. Jung, J.J. Blanz, M.M. Naßhan, P.W. Baier: Simulation of the uplink of JD–CDMA mobile radio systems with coherent receiver antenna diversity. Wireless Personal Communications — An International Journal (Kluwer), vol. 1 (1994), 61–89.

[32] J.J. Blanz, R. Schmalenberger, P. Jung: Smart antenna concepts for time–slotted CDMA. Proc. IEEE 47th Vehicular Technology Conference (VTC'97), Phoenix (1997).

[33] A.J. Steil, J.J. Blanz: Spectral efficiency of JD-CDMA mobile radio systems applying coherent receiver antenna diversity with directional antennas. Proc. IEREE Fourth International Symposium on Spread Spectrum Techniques & Applications (ISSSTA'96), Mainz (1996), 313–319.

[34] J.J. Blanz, P.W. Baier, P. Jung: A flexibly configurable statistical channel model for mobile radio systems with directional diversity. Proc. ITG Conference on Mobile Radio, Neu-Ulm (1995), 93–100.

[35] R. Schmalenberger, J.J. Blanz: Multi antenna C/I balancing in the downlink of digital cellular mobile radio systems. Proc. IEEE 47th Vehicular Technology Conference (VTC'97), Phoenix (1997).

# Part 4

# Advanced Systems & Technology

# 1 MOBILE AND PERSONAL COMMUNICATIONS: ACTS AND BEYOND[1]

J. Schwarz da Silva,
B. Arroyo-Fernández,
B.Barani,
J. Pereira
D. Ikonomou

**Abstract**

In Europe, Mobile and Personal Communications have always been considered a key driver for growth and innovation, as well as being a necessary building block of the Wireless Information Society. Since 1988, European Union (EU)-funded R&D projects have been working towards the development of the next generations of mobile communication concepts, systems and networks.

The ACTS (Advanced Communications Technologies and Services) Programme, launched in 1995 and extending until 1998, provides a first opportunity to master and trial mobile and personal communications services and technologies, involving service providers, communications operators and equipment manufacturers.

From the user's perspective, the ACTS programme strives to ensure that current mobile services are extended to include multimedia and broadband services, that access to services is possible without regard to the underlying networks, and that convenient, light weight, compact, and power efficient terminals adapt automatically to whatever air-interface parameters are appropriate to the user's location, mobility, and desired services.

Well into the last half of the ACTS programme, it is time to make a first assessment of the progress relating to three aspects of R&D on Third Generation Mobile and Personal Communication Systems, namely future mobile/wireless services, mobile/wireless platforms, and enabling technologies.

[1] These views expressed in this article are those of the authors and do not engage the European Commission

At the same time, it becomes necessary to start looking ahead towards the next R&D programme, specially in the area of Mobile and Personal Communications. In this regard, the European Commission has recently put forward its position regarding the next Framework Programme, highlighting the need to continue if not intensify its support to Mobile and Personal Communications and the associated advanced technologies, services and applications.

## 1.1 INTRODUCTION

Mobile and Personal Communications are recognised as a major driving force of socio-economic progress and are crucial for fostering the European industrial competitiveness and for its sustained economic growth, and its balanced social and cultural development.

The emergence of GSM (Global System for Mobile Communications) as a world standard shows the potential of a concerted action at an European level, covering all aspects from R&D to political and regulatory to spectrum availability. Resulting notably from extensive research in the scope of the RACE (Research in Advanced Communications in Europe) programme, GSM is a demonstration of the European know-how and the basis for the European competitiveness in the world telecommunications scene.

The unprecedented growth of world-wide mobile/wireless markets, coupled with advances in communications technology and the accelerated development of services taking place in fixed networks, points now to the urgent introduction of a flexible and cost effective Third Generation Mobile Communication System. In this context, UMTS (Universal Mobile Telecommunications System), as such system is commonly referred to in Europe, has been the subject of extensive research carried out primarily in the context of the European Community R&D programmes RACE and ACTS (Advanced Communications Technologies and Services).

The next phase, to be pursued in the scope of the upcoming Information Society (IS) programme, will add a different dimension. Concurrently with the much needed and anticipated integration of fixed and mobile, terrestrial and satellite systems, and the continued development of Mobile/Wireless Broadband Communication systems, a dramatic paradigm shift is envisioned with the adoption of Software Communication concepts. At the same time, research in basic technology will insure that the necessary developments are in place (from smart batteries to high temperature superconductivity to flat panel displays) to allow for a true mass market of telecommunication products. With those, a myriad of personalised telecommunication services will emerge, requiring the development of new service engineering tools and the extension to the telecommunications arena of Just-in-Time service definition and provisioning. To stress the point, a user-centric perspective will refocus the work towards the seamless offer of personalised, Just-What-the-Customer-Wants services, and provide the user with simple, user-friendly interfaces to the service providers, from profile definition to service selection to billing.

The success of mobile communication systems, as well as its consequences in terms of need of better, more efficient systems, and of additional spectrum, is analysed in Section 1.2. In Section 1.2.1 the success stories in Europe, US and Asia, each reflecting different approaches and the weight of legacy systems, are reviewed.

Due to the recent explosive growth of mobile communications, in both cellular and cordless-based systems, after the first steps in deregulation, Japan (and namely NTT) is targeting 2000 for the deployment of Third Generation systems, with an eye already on a second phase to be deployed by 2005 with much higher data rates (10 Mbps versus 2 Mbps). Europe, also bracing for spectrum shortage due to the accelerated uptake of its digital systems, is targeting 2002 for basic UMTS service, and 2005 for full UMTS capabilities (2 Mbps); a second phase is envisioned, conditioned upon market needs. In the US, however, the huge First Generation legacy, and the very recent, heavy investments in Second Generation systems, has led to a somewhat relaxed approach, targeting 2005 or even later. Enhancements to Second Generation systems are envisioned, to meet the market's needs.

Adding to the varying degrees of urgency to deploy Third Generation systems, different expectations, as well as different degrees of market liberalisation, lead to substantially distinct approaches to standardisation in Europe, Japan, and the US. In Section 1.2.2 the work ongoing at ETSI (European Telecommunications Standards Institute) and ITU (International Telecommunications Union) is discussed.

Critical for the development of the new Third Generation systems, was the WARC '92 and WRC '95 recommendations of spectrum allocation for FPLMTS/IMT-2000 systems. Spectrum for UMTS has been identified in Europe, coinciding with W(A)RC's recommendations. Section 1.2.3 discusses the present spectrum recommendations for Third Generation systems.

Section 1.2.4 discusses the role of he UMTS Forum in setting the industry strategy and identifying the regulatory requirements for the success of UMTS.

In the context of the ACTS programme, R&D in advanced mobile and personal communications services and networks, is called upon to play an essential role. The specific objectives that are addressed by the current ACTS projects include the development of Third Generation platforms for the cost effective transport of broadband services and applications, aiming at responding to the needs of seamless services provision across various radio environments and under different operational conditions. Since the scope of future mobile communications encompasses multimedia, far beyond the capabilities of current mobile/wireless communication systems, the objective is to progressively extend mobile communications to include multimedia and high performance services, and enable their integration and inter-working with future wired networks. The achievements thus far of the many projects in what is called the Mobile domain are analysed in detail in Section 1.3. Projects in the Mobile domain address three aspects of R&D on Third Generation Mobile and Personal Communication Systems: future mobile/wireless services, mobile/wireless platforms, and enabling technologies.

Section 1.4 discusses the 5FWP, its priorities and underlying criteria. It also provides a discussion of the major challenges and choices that are facing Europe if its leadership is to be maintained in the emerging 21st Century Information Society.

Section 1.5 presents the conclusions, emphasising the importance of R&D relating to the converging information and telecommunication technologies given the pressure to expand the services and capabilities of mobile/wireless systems, and the consequent need for enhanced, more efficient systems, and mainly the need for a complete overhaul of the sector towards a mass market perspective. From an R&D point of view, the challenges ahead are significant. However, the area of Mobile and

Personal Communications will no doubt be given the necessary priority to insure continued success.

## 1.2 AN EXPLODING GLOBAL MARKET

The tremendous success of mobile/wireless communication systems in Europe, US and Asia, as well as its consequences in terms of need of better, more efficient systems, and of additional spectrum, is analysed in this section. A perspective is also provided relating to the substantially distinct approaches to standardisation in Europe, Japan, and the US. Spectrum recommendations for Third Generation systems are analised. Finally, the role of the UMTS Forum in setting the industry strategy and identifying the regulatory requirements for UMTS' successful uptake in Europe is discussed.

### *1.2.1 Three Showcases: Europe, Japan and US*

This section analyses the mobile and personal communication success stories in Europe, US and Japan, each reflecting different approaches to and schedules for deregulation, as well as the weight of legacy systems.

#### *1.2.1.1 The European Success Story*

Europe has witnessed in recent years a massive growth of mobile communications, ranging from the more traditional analogue based systems to the current generation of digital systems such as GSM (Global System for Mobile Communications), DCS-1800 (Digital Communication System at 1800 MHz), ERMES (European Radio Messaging System), and to a lesser extent DECT (Digital European Cordless Telephone), and TETRA (Trans European Truncked Radio). The GSM family of products (GSM + DCS-1800), which represents the first large scale deployment of commercial digital cellular system ever, enjoys world wide success, having already been adopted by over 230 operators in more than 110 countries. As it can be seen in Figure 1, in a very short period of time, the percentage of European cellular subscribers using GSM or DCS-1800 has already exceeded 50%. In addition, the figure portrays the penetration rates of the combined analogue and digital cellular systems for the same time-frame. A snapshot of the situation for each of the EU countries, as well as the rest of Europe, in terms of number of subscribers and penetration of analogue/digital cellular for November 1996 is depicted in Figure 2. It is worth noticing that the biggest markets of Europe in terms of subscribers (i.e., UK, Italy and Germany) are not the markets with the largest penetration rates. In this respect, the largest penetration rates are found in the Nordic countries, already exceeding 25% of the population.

At current growth rates, it is envisaged that the total number of subscribers will reach some 200 million, in Europe alone, by the turn of the century. By then it is expected that some 320 GSM networks will be on the air in some 130 countries. It should be noted that while the telephone density for fixed telephones is not expected to exceed the 50% mark (i.e., at most one phone for every two persons, or approximately two per family), personal mobile communications, in all forms, promises to reach nearly 80% of Europe's population. Other systems, catering for more specialised applications or markets, such as wireless local loop, private mobile radio for police and safety systems and paging, are also called upon to contribute very strongly to the development of the market and the economic growth of Europe.

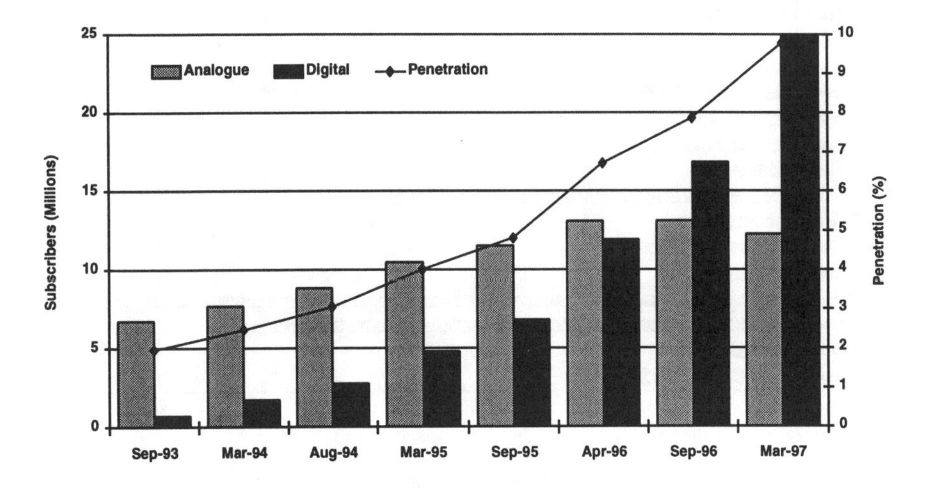


**Figure 1** Mobile Cellular subscribers/penetration in Europe in March 1997

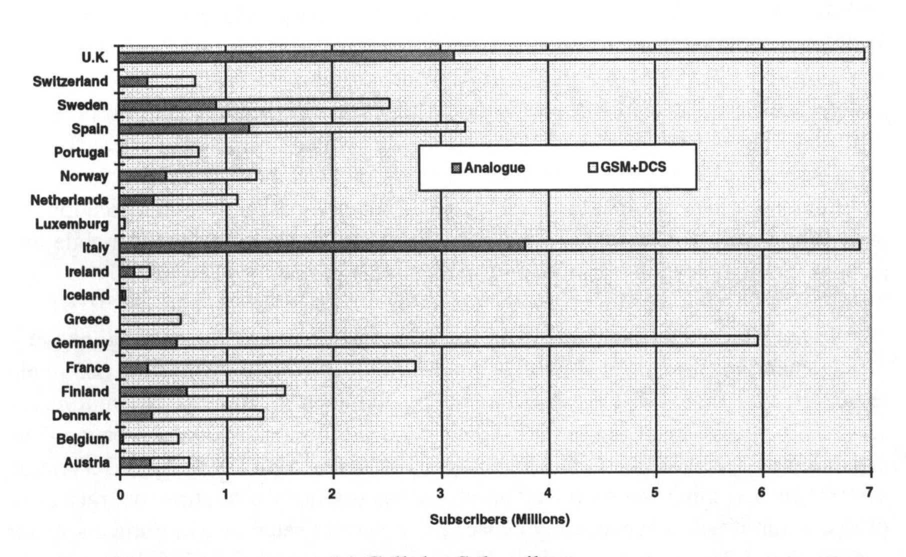


(a) Cellular Subscribers

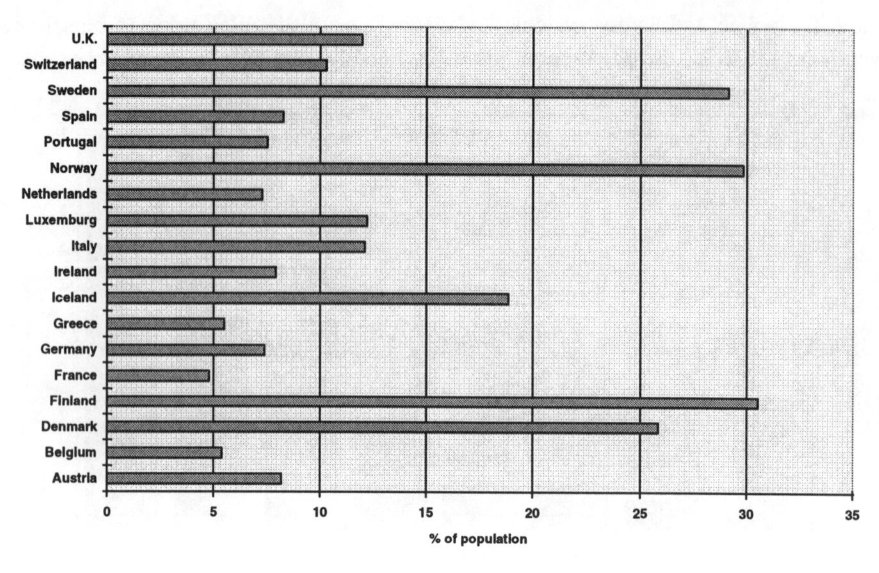


(b) Cellular Penetration

**Figure 2** Cellular Subscribers and Penetration in Europea in March 1997

Third Generation systems and technologies are being actively researched world wide. In Europe, such systems are commonly referred under the name UMTS (Universal Mobile Telecommunications Systems) while internationally, and particularly in the ITU context, they are referred to as FPLMTS (Future Public Land Mobile Telecommunications Systems) or more recently IMT-2000 (International Mobile Telecommunications 2000).

In this context, but also in a world-wide perspective, with many competing mobile and personal communication technologies and standards being proposed to fulfil the users' needs, the essential questions, to which no immediate, conclusive, firm answers can be given, are: "To what extent, and how fast, will the users' requirements evolve beyond the need for voice and low data rate communications?", and "Which will be the technologies that will meet the requirements for mobile and personal communications services and applications beyond the year 2000?".

The rapid advance of component technology; the pressure to integrate fixed and mobile networks; the developments in the domains of service engineering, network management and intelligent networks; the desire to have multi-application hand-held terminals; and above all the increasing scope and sophistication of the multimedia services expected by the customer; all demand performance advances beyond the capability of Second Generation technology. The very success of Second Generation systems in becoming more cost-effective and increasingly cost-attractive raises the prospect that it will reach an early capacity and service saturation in Europe's major conurbations. These pressures will lead to the emergence of Third Generation systems representing a major opportunity for expansion of the global mobile marketplace rather than a threat to current systems and products.

The ground work for UMTS [1] started in 1990, and some early answers can already be provided regarding its requirements, characteristics and capabilities, with the initial standards development process already under way at ETSI (European

Telecommunications Standards Institute). The basic premise upon which work is being carried out, is that by the turn of the century, the requirements of the mobile users will have evolved and be commensurate with those services and applications that will be available over conventional fixed or wireline networks. The citizen in the third millennium will wish to avail himself of the full range of broadband multimedia services provided by the global information infrastructure, whether wired or wireless connected.

### *1.2.1.2 The Mobile Explosion in Japan*

While in Europe the need for a full fledged Third Generation system is anticipated for 2005, the situation in Japan is such that the Japanese Ministry of Posts and Telecommunications (MPT) felt compelled to accelerate the standardisation process to have a new, high capacity system deployed by the year 2000.

Users of mobile communication systems, including cellular (analogue, TACS, and digital, PDC) and cordless (PHS, Personal Handyphone System) have increased dramatically in Japan (Figures 3 and 4), coinciding with the first, tentative steps at deregulation, at the rate of more than one million new users per month. By the end of March 1997 there were already almost 27 million users, out of which around 6 million were PHS users. (By the end of April 1997 they were, respectively, more than 28 million and 6.4 million.)

Digital overtook analogue at the beginning of 1996, coinciding with the peak of analogue, which since then has started loosing subscribers (the same kind of substitution phenomenon that is observed in most European countries).

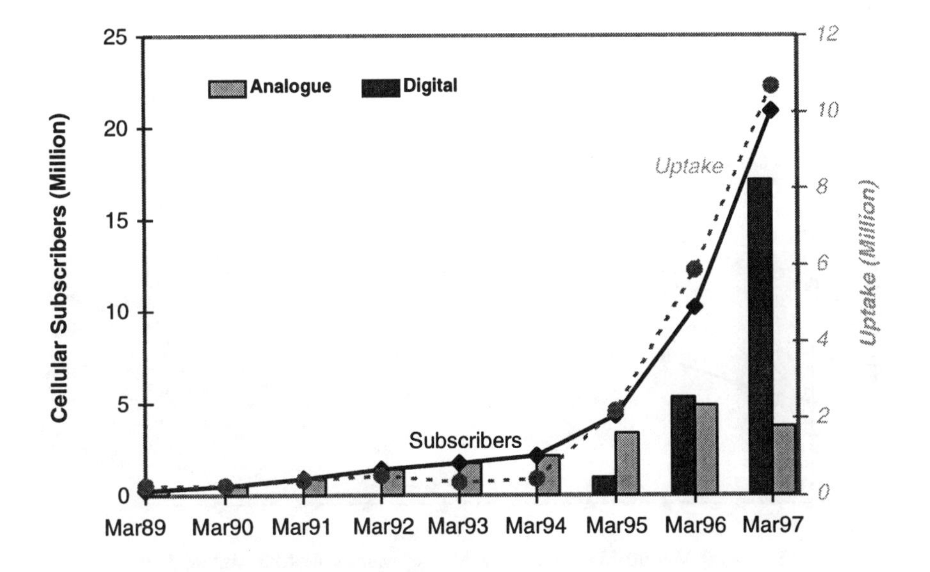


**Figure 3** Growth of (Digital) Cellular Subscribers

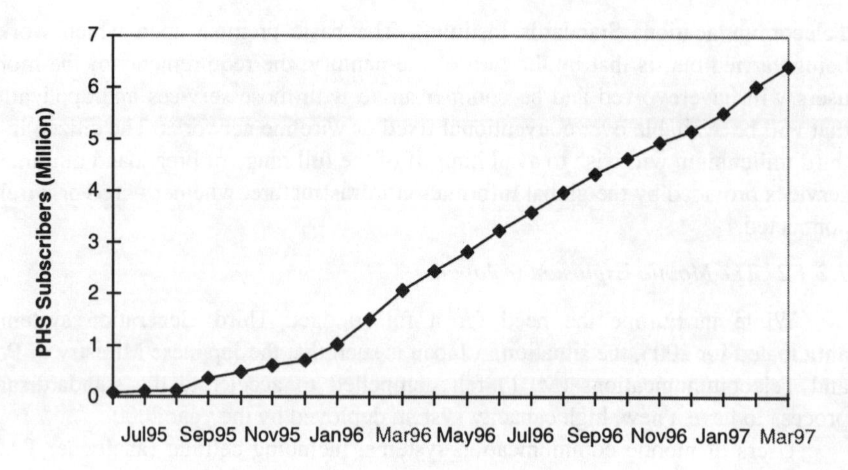


**Figure 4** Growth of PHS Subscribers

A very interesting trend, however, is that of the paging business (Figure 5). The growth of cellular did not seem to affect the growth of paging, much like it is the case in the US and like it is shaping up in Europe. However, as soon as a cheap and versatile cordless solution became established, the number of subscribers started to diminish, at a rate of more than half a million per year. This bodes nothing good for single product service providers in a time were integration is the rule. In any case, it must be said that even the paging industry is evolving, namely towards two-way messaging, or at least providing message delivery acknowledgement, and becoming definitely alphanumeric.

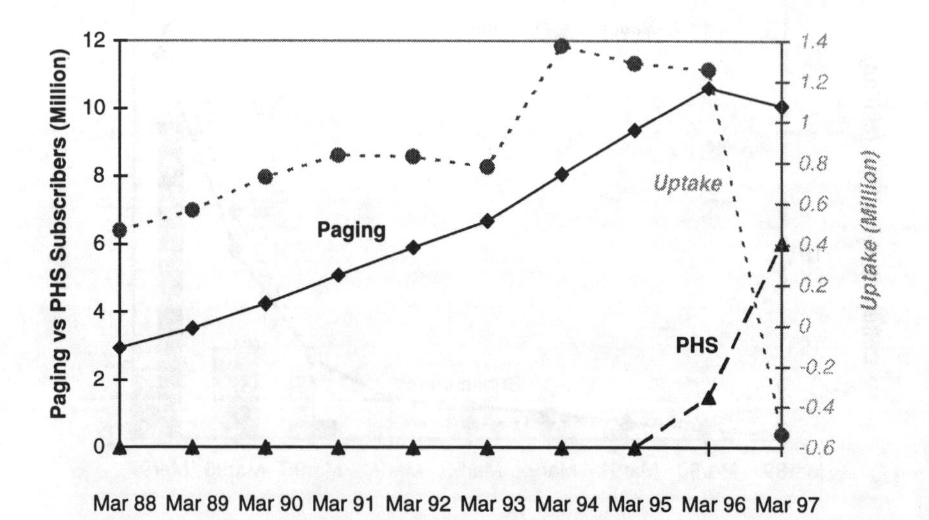


**Figure 5** Paging Subscribers in Japan

In an effort to meet the increasing demand for mobile communication systems, measures for promoting more efficient utilisation of frequencies are being studied and implemented, and a significant impetus has been given to the early introduction of Third Generation systems.

The MPT created in 1996 the Study Group on the Next-Generation Mobile Communication Systems, and tasked it to prepare by April 1997 a concrete vision of next generation systems, and examine related technical issues including present status and trends, identification of the technological requirements, and measures to be taken towards implementation. The report has been recently published: it calls for a concerted effort towards Wideband-CDMA, although it stops short of a clear endorsement of such a solution. Interest groups will keep studying other solutions like Wideband-TDMA and OFDMA.

In the meantime, NTT came out openly for its own Wideband-CDMA system, which they wish to deploy by the year 2000. For that purpose they have called upon a world-wide partnership with European and US companies.

*1.2.1.3 Developments in Korea*

The development of FPLMTS in Korea was launched by the Ministry of Information and Communications, and put under the supervision of ETRI (Electronics & Telecommunications Research Institute), with the aim of being specified by the year 2000 and deployed no later than 2001.

As for the air interface technology, it has already been decided it will be wideband-CDMA, a technology already extensively researched in Korea, starting as an adaptation of IS-95, and ending up as a full fledged wideband-CDMA system. It must be noted here that the number of CDMA users in Korea has already passed (in April 1997) the one million mark.

In terms of the viability of a Korean-only system, talks are apparently ongoing with Japan to arrive at a common approach, or at least to maximise the points in common.

*1.2.1.4 The US Situation*

The most important factor to take into consideration in the US case is the significant AMPS (analogue, First Generation system) cellular base. On top of that, the recently deployed Second Generation systems represent huge investments that need to bring significant returns before replacement plans can be considered. That leaves out new entrants and some muscled operators that would anyway wish for a modular deployment of Third Generation systems, building upon their existing investments.

Even if the full set of capabilities expected from FPLMTS/IMT-2000 may not be needed in the US until well after 2005, some subset may be needed earlier, and are already being actively pursued (see Section 1.2.2.3). The exact deployment time-frame will be dictated by the demands of the market.

It is furthermore considered essential not to over-prescribe the Third Generation solution in terms of "required" data rates. The approach is one of gradual availability of increasingly higher speed service offerings, as the market demands them, instead of leapfrogging into the final, high performance solution (market push, not technology pull).

In any case, US companies are already positioning themselves to provide an evolutionary path from IS-95 to a fully FPLMTS/IMT-2000 compliant IS-95-based

Wideband CDMA system (Wideband-95). A few major companies have jointly announced that they intend to work together with the CDMA Development Group (CDG) to develop specifications for next generation wireless communications standards, while stating that they would also support developments of other Third Generation standards around the globe.

They propose an evolutionary path building upon and being compatible with today's IS-95 systems and IS-41 based networks. The new specification would enable the full re-use of the IS-41 network elements and existing PSTN interfaces and administrative elements, including billing and network management systems, thus enabling integration with existing systems, and the maximization of the return on investment.

### *1.2.2 Standardisation Work*

Adding to the varying degrees of urgency to deploy Third Generation systems, different expectations, as well as different degrees of market liberalisation, lead to substantially distinct approaches to standardisation in Europe, Japan, and the US. They are briefly analysed in this section.

#### *1.2.2.1 The Role of ETSI*

SMG (Special Mobile Group) within ETSI is a business and service oriented organisation responsible for two ETSI Projects:

- GSM (including GSM and DCS-1800);
- GSM-UMTS (including UMTS services concept and the UMTS evolution of the core network), a name still under consideration;

and one horizontal task

- UMTS GRAN (Generic Radio Access Network).

The responsibilities of the SMG include, among others: definition of the services and features offered; selection and specification of the most efficient radio techniques, including the co-ordination of validation tasks; elaboration of specifications for network architecture, signalling protocols and conditions of inter-working with other networks; elaboration of roaming aspects with GSM and selected other systems, namely satellite systems; selection of the appropriate security procedures; elaboration of highly efficient, timely available tools for management of the network and of network evolution; development of test specifications for the type approval of mobile terminals and other network elements; and maintenance of the specifications and controlled release of updated standards.

The specific objective of the GSM-UMTS project is "to develop UMTS as a world-wide competitive and versatile mobile multimedia communication standard with capabilities extending beyond a fully evolved GSM" by developing and maintaining the specifications of UMTS in the following areas: UMTS objectives and requirements; specification of UMTS service aspects; and specification of the evolution of the GSM core transport network to enable a GSM network operator to support UMTS GRAN.

Such an extended scope of activities led to the formation of subgroups which function as working groups of the ETSI GSM-UMTS project, namely: SMG2 Radio aspects; SMG1 Services and Facilities; SMG3 Network aspects; SMG4 Data services; and SMG6 Operation and Maintenance.

*1.2.2.2 Standards Work in Japan*

As early as August 94, MPT convened the Multimedia Mobile Communications Study Committee with the purpose of defining the concept, the market, the applications, and the milestones for the introduction of Mobile Multimedia Communication systems. The committee identified the following subjects as essential for implementing Mobile Multimedia Communications: 1) technological development, including radio technologies, hardware and software; 2) spectrum for Mobile Communications; 3) developments in the communications infrastructure; and 4) market adjustment towards Mobile Multimedia Communications, including user-friendly interfaces, and quality content.

The milestones for Mobile Multimedia Communication systems were split in two phases: one around the year 2000, involving the implementation of FPLMTS as defined usually (data rates up to 2 Mbps), and a second phase, around the year 2010, for FPLMTS Phase 2, with data rates up to 10 Mbps. Standardisation work began in September 1994, with the establishment of ARIB (Association of Radio Industries and Businesses). The action plan involved the comparative study of CDMA, TDMA and other radio technologies, in two rounds, with down selection.

*1.2.2.3 Expediting Standards Work — an US experiment*

The US wireless marketplace is an interesting topic for study in terms of competition and free market economics, with the government and regulatory authorities committed to allowing industry to compete in an open wireless market, from the licensing of spectrum through the choice of technologies deployed. The FCC (Federal Communications Commission), in particular, wants to ensure that only a minimum set of technical requirements be mandated such that new technologies and innovations are allowed to flourish based upon the demands of the market. As was seen in the recent PCS (Personal Communication Systems) auctions, the FCC tends to favour allowing market competition to determine what technologies will prevail, in contrast to mandated standards as is the case in other parts of the world. Therefore, from the US point of view, the challenge for global Third Generation standardisation is to be as open and flexible as possible in order to gain acceptance in the market.

Given the increased complexity, as well as the multiplicity of the standards for mobile communication systems, the telecommunications industry felt that the TIA (Telecommunications Industry Association) and the EIA (Electronic Industry Association) could no longer keep up with demand. As a result, many industry organisations have flourished with the single purpose of coming up with a standard proposal, as complete as considered appropriate, and to achieve this in a timely manner. At that point, when there is enough industry momentum, EIA/TIA comes in to articulate the proposal with existing standards (still more than just rubber-stamping the proposal).

As an illustration, mention can be made of the CDPD Forum, and the CDMA Development Group (CDG). The former managed to arrive at the specification for CDPD, a digital packet data overlay on AMPS, and get to commercial operation in less than two years - worring about making CDPD a standard came much later.

The CDG has actively worked to define, among other things, High Speed Data Services (HSDS), both circuit- and packet-switched, over IS-95, capable of providing, in phase 1, up to 64 kbps symmetrical data capabilities. The result of that work is now being incorporated into the TIA standard IS-95 Revision B (High Speed IS-95), the next revision of the CDMA Standard. In the scope of its Advanced Systems initiative, the CDG expects to continue evolving the system towards full FPLMTS/IMT-2000 capabilities.

#### *1.2.2.4 UMTS and FPLMTS/IMT-2000 Standardisation Plans*

At the global level, the work carried out within ITU, and particularly within TG 8/1, has been instrumental in the definition of FPLMTS/IMT-2000.

Due to the already existing differences in the use of FPLMTS bands by PCS in the US, PHS in Japan, and DECT in Europe, there will most likely be different air interfaces as well as different channel assignments in different regions of the world at least for the terrestrial component of FPLMTS. In any case, the goal in ITU-R TG 8/1 is still "to enable world-wide compatibility of operation and equipment, including international and intercontinental roaming."

The minimum performance capabilities for circuit and packet switched data that have been agreed upon by TG 8/1 are:

- Vehicular Test Scenario: 144 kbps
- Pedestrian, and outdoor to indoor test scenario: 384 kbps
- Indoor office test scenario: 2048 kbps

The agreed upon time-plans of the different standardisation bodies in the different regions is put together below to show the difference in pace resulting from the anticipated need for an advanced Third generation system in each region, as well as the need to respond to the challenge raised by Japan. Even if there is already a very active group working towards Third Generation (CDG), the US preparation is left far more open than the others, given the prevailing philosophy of letting the market decide when and how to act: other proposals are indeed expected, although their schedule is still unknown. (The year 2000 target for Wideband 95 was dictated by NTT's deadline, certainly not by the US market.)

### *1.2.3 Spectrum Issues*

Critical for the development of the new Third Generation systems are the recommendations of spectrum allocation for FPLMTS/IMT-2000 systems. The present spectrum recommendations for Third Generation systems are briefly analised below.

**Table 1** UMTS and FPLMTS/IMT-2000 Standardisation Schedules

| | ITU-R TG 8/1 | US | | Europe | Japan |
|---|---|---|---|---|---|
| Basic Technology selection | | IS-95 | ? | end 1997 | 1995 |
| Candidate proposals | end 1997 | | | | 1995 |
| Evaluation of candidates | March 1998 | | | | 1996 |
| Key choices for the radio interface | March 1999 | | | | end 1996 |
| Final Recommendation | end 1999 | | | end 1999 | 3Q 1997 |
| Deployment | | ~2000 | ? | Basic 2002<br>Full 2005 | ~2000 |

### *1.2.3.1 WARC '92 Spectrum Assignments for FPLMTS*

The 1992 World Administration Radio Conference (WARC) of the International Telecommunications Union (ITU) targeted 230 MHz, in the 2 GHz band (1885-2025 MHz and 2110-2200 MHz), on a world-wide basis, for FPLMTS/IMT-2000, including both terrestrial and satellite components. The objective was identified as that of establishing, through the appropriate Global Standards and the co-ordinated assignment of spectrum by the various National and Regional Authorities, the future, truly ubiquitous personal mobile communications system, creating a seamless radio infrastructure capable of offering a wide range of services, in all radio environments, with the quality we have come to expect from the fixed networks.

The availability of the spectrum assigned to FPLMTS/IMT-2000 varies from region to region, the same happening with the strategies followed to insure its availability when time comes. Another aspect that is region specific, is how that spectrum can, or cannot, be shared with other systems. The whole 230 MHz of spectrum identified by WARC-92 was reserved in Europe to Third Generation, UMTS, technology. This even if part of the band allocated for DECT falls into the FPLMTS band.

The issue of technology migration from Second to Third Generation via the use of spectrum in the FPLMTS/UMTS bands has been repeatedly raise in many fora. This may result in the spectrum being allocated, in some parts of the world, in an inefficient piecemeal fashion to evolved Second Generation technologies and potentially many new narrow-application systems, thereby impeding the development of broadband mobile multimedia services.

Terminal, system and network technology as researched within the EU-funded ACTS projects, may alleviate to a large extent the complexity of the sharing of the spectrum between the Second and Third Generation systems.

### *1.2.3.2 Spectrum for UMTS in Europe*

The European Radiocommunications Office (ERO), has recently prepared a draft decision which advocates a two-phase approach, with phase 1 aiming to safeguard the spectrum already allocated to Third Generation mobile communications systems, and phase 2 being an effort for allocation of additional spectrum taking into account the ever increasing market demand for multimedia mobile services being catered by mobile communications networks. Some preliminary indications of a possible way forward are:

- The total amount of spectrum needed by Third Generation mobile communications systems is yet unclear. It is estimated that the requirements will be between 300 MHz and 500 MHz. These figures assume a multimedia service provision environment with competing operators. It is further noted that the minimum bandwidth required for one public operator is in the order of 2x20 MHz, and that each country will have to accommodate at least three public networks.
- Further R&D is needed in technologies that will enable a more efficient use of spectrum (e.g., continuous dynamic channel selection, smart antennas, adaptive radio access and coding)
- The possibility of using the Mobile Satellite Systems (MSS) frequency bands for terrestrial indoor use under a frequency sharing scenario needs to be further studied.
- There is no immediate need (before year 2000) for additional spectrum, but additional spectrum will become a necessity for the second phase of the UMTS introduction.

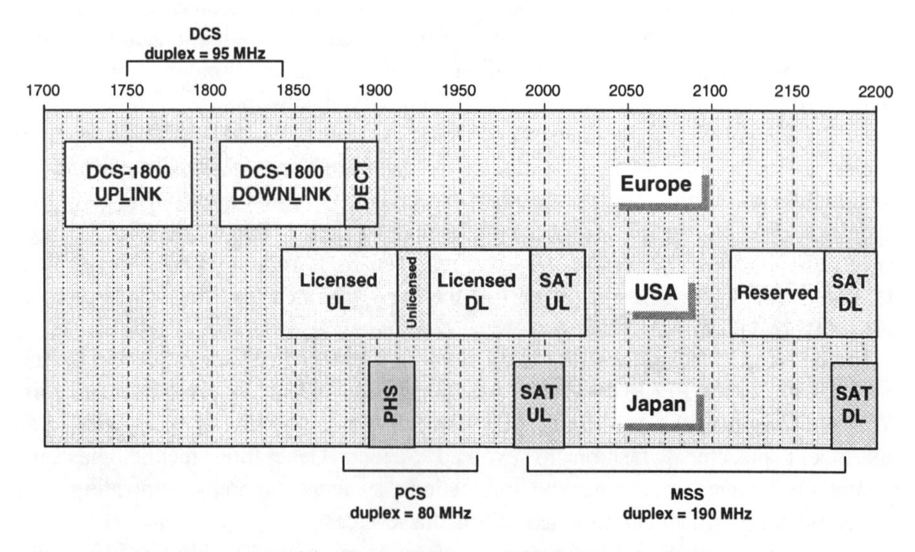


**(a)** Current Usage of Spectrum

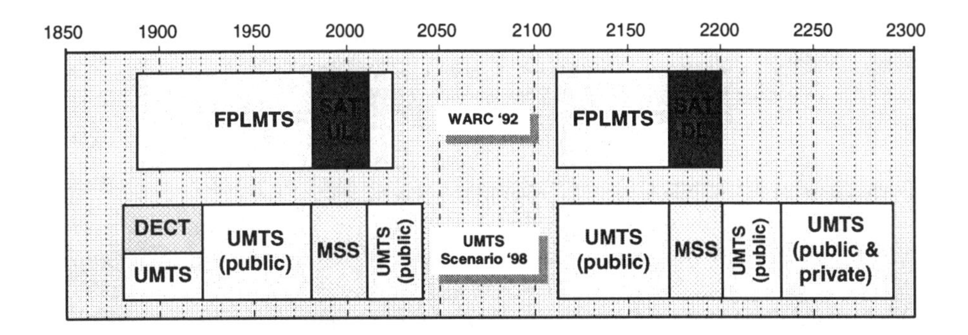


(**b**) Planned UMTS Bands in Europe

**Figure 6** FPLMTS/IMT-2000 and UMTS Spectrum versus Current Use

*1.2.4 The UMTS Forum*

In response to the imperatives of the internal European market, specific measures were taken, as early as 1987, to promote the Union-wide introduction of GSM, DECT, and ERMES. European Council Directives were adopted to set out common frequency bands to be allocated in each Member State to ensure pan-European operation, together with European Council Recommendations promoting the co-ordinated introduction of services based on these systems.

In 1994, the European Commission adopted a Green Paper on Mobile and Personal Communications [2] with the aim of establishing the framework of the future policy in the field of mobile and personal communications. The Green Paper proposed to adapt, where necessary, the telecommunications policy of the European Union to foster a European-wide framework for the provision of mobile infrastructure, and to facilitate the emergence of trans-European mobile networks, services, and markets for mobile terminals and equipment.

Based on the Green Paper, the European Commission set out general positions on the future development of the mobile and personal sector, and defined an action plan which included actions to pursue the full application of competition rules; the development of a Code of Conduct for service providers; and the agreement on procedures for licensing of satellite-based personal communications. The action plan also advocated the possibility of allowing service offerings as a combination of fixed and mobile networks in order to facilitate the full-scale development of personal communications; the lifting of constraints on alternative telecommunications infrastructures and constraints on direct interconnection with other operators; the adoption and implementation of Decisions of the ERC (European Radiocommunications Committee) on frequency bands supporting DCS-1800 and TETRA; the opening up of an Europe-wide Numbering Space for pan-European services including personal communications services; and continuing support of work towards UMTS.

The combination of these regulatory changes will contribute to a substantial acceleration of the EU's mobile communications market and speed the progress towards Third Generation mobile/personal communications. It will however be necessary to encourage potential operators and manufacturers to invest in the required technology, by setting out a clear calendar for the adoption of the required new standards and the re-farming of the necessary spectrum. The applicable licensing regimes and rules for flexible sharing of the available spectrum need also to be adopted at an early stage so as to permit the identification of novel market opportunities commensurate with the broadband multimedia requirements of the Third Generation mobile telecommunications systems.

The continued evolution of Second Generation systems has been recognised as an issue of great societal and economic importance for Europe and the European industry. To facilitate and crystallise such an ambition, and in accordance with the political mandate given by the European Parliament and the European Council, an ad-hoc group called the UMTS Task Force was convened by the European Commission and was charged with the task of identifying Europe's mobile communications strategy towards UMTS. The report of the UMTS Task Force and its recommendations [3] have been largely endorsed by the European mobile industry, and as a result the UMTS Forum has now been created with the mandate to provide an on-going high level strategic steer to the further development of European mobile and personal communications technologies. High on the priorities of the UMTS Forum are the issues of technology, spectrum, marketing and regulatory regimes. Drawing participation beyond the European industry, the UMTS Forum is expected to play an important role in bringing into commercial reality the UMTS vision.

As part of its mandate and in response to the growing need for a coherent approach towards the development of UMTS, the UMTS Forum has recently issued a report [4] which highlights the enabling factors -- political leadership, spectrum availability, fair licensing arrangements and availability of standards -- which are considered crucial for the success of UMTS.

Finding the solution to the problem of evolution and migration path from Second (GSM, DCS-1800, DECT) to Third Generation systems (FPLMTS/UMTS), particularly from a service provision point of view (see Figure 7), is also the subject of intense research carried out in the context of ACTS projects. Some of the key questions that are addressed include a detail consideration of the feasibility, as well as the cost effectiveness and attractiveness of the candidate enhancements. In this context, the ACTS projects will develop a set of guidelines aiming at reducing the uncertainties and associated investment risks regarding the new wireless technologies, by providing the sector actors and the investment community with clear perspectives on the technological evolution and on the path to the timely availability to the user of advanced services and applications. In particular, the Mobile ACTS projects interact directly with the Technology Aspects Group (WG4) of the UMTS Forum, many times being able to provide timely inputs to the strategic discussions ongoing there.

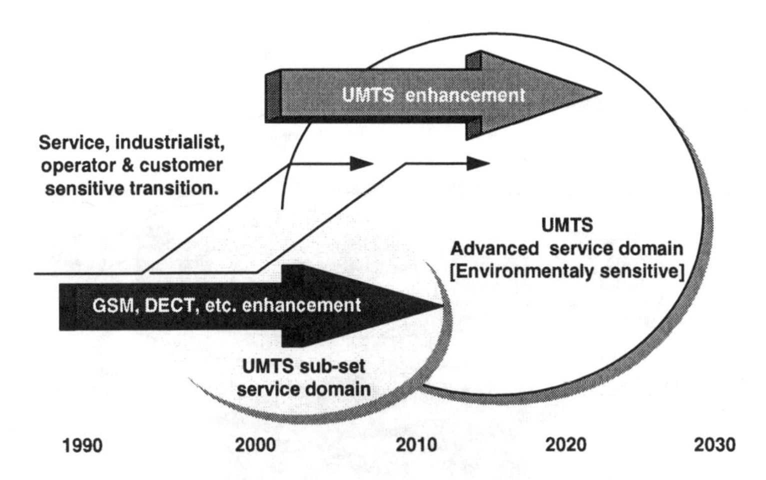


**Figure 7** Evolution from Second to Third Generation

## 1.3 ASSESSING ACTS

R&D in advanced mobile and personal communications services and networks, in the context of the ACTS programme, is called upon to play an essential role. The specific objectives that are addressed by the current ACTS projects include the development of Third Generation platforms for the cost effective transport of broadband services and applications, aiming at responding to the needs of seamless services provision across various radio environments and under different operational conditions. Since the scope of future mobile communications encompasses multimedia, far beyond the capabilities of current mobile/wireless communication systems, the objective is to progressively extend mobile communications to include multimedia and high performance services, and enable their integration and inter-working with future wired networks.

Third Generation mobile communication systems aiming also at integrating all the different services of Second Generation systems, provide a unique opportunity for competitive service provision to over 50% of the population, and cover a much wider range of broadband services (voice, data, video, multimedia) consistent and compatible with the technology developments taking place within the fixed telecommunications networks. The progressive migration from Second to Third Generation systems, expected to start at the turn of the century, will therefore encourage new customers while ensuring that existing users will perceive a service evolution that is relatively seamless, beneficial, attractive and natural.

Figure 8 illustrates the range of service environments, from in-building to global, in which the Third Generation of personal mobile communication systems will be deployed. Appropriately positioned are all projects in the Mobile domain, and some relevant "invited" ACTS projects from other domains.

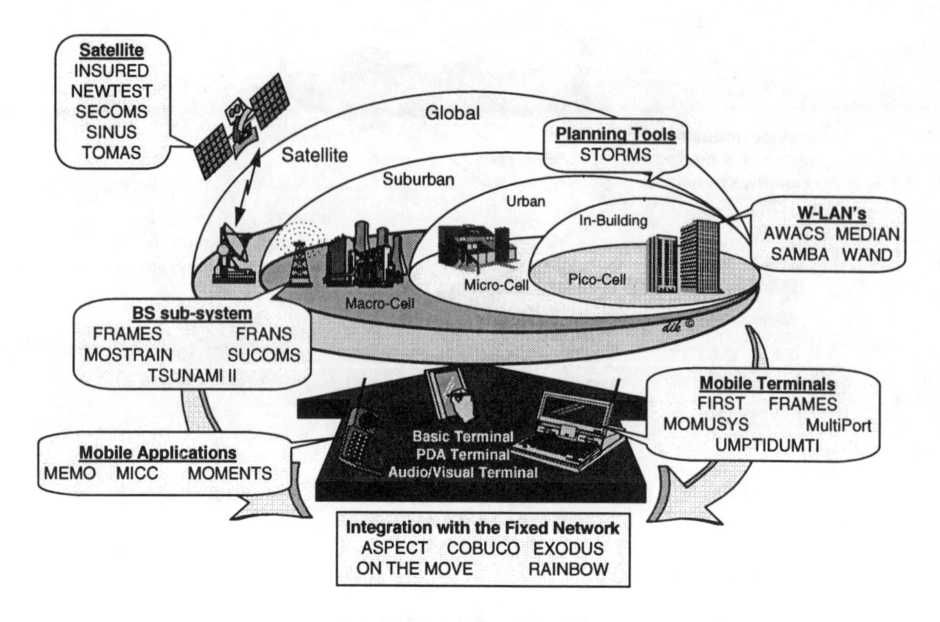


**Figure 8** System environments

Figure 9 portrays the technological capabilities of Third Generation systems, measured in terms of terminal mobility and required bit rates as compared to those of Second Generation platforms such as GSM.

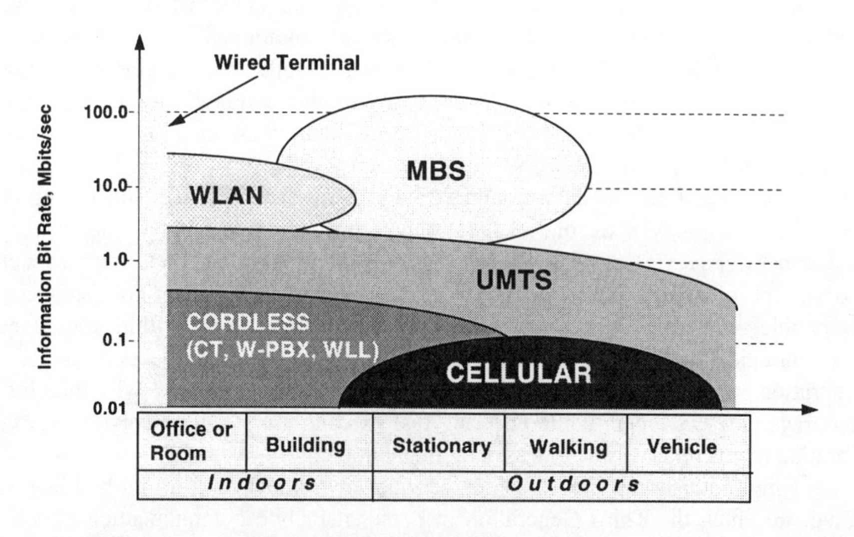


**Figure 9** Mobility versus bit rates

In Europe, R&D on Third Generation technology, commonly referred to as UMTS, falls under the European Community ACTS Programme. UMTS is conceived as a multi-function, multi-service, multi-application digital mobile system that will provide personal communications at rates ranging from 144 kbps up to 2 Mbps according to the specific environment, will support universal roaming, and will provide for broadband multimedia services. UMTS is designed to have a terrestrial and a satellite component with a suitable degree of commonality between them, including the radio interfaces. The R&D effort concentrates on the development of technical guidelines regarding, in particular, the level of UMTS support of ATM transmission technology of IBC networks, the compatibility of UMTS and fixed-network architecture, the allocation of intelligent functionality (UPT — Universal Personal Telecommunications, and IN — Intelligent Network), the level of integration of the satellite component of UMTS, and the multi-service convergence philosophy of the UMTS radio interface.

MBS (Mobile Broadband Systems), including their W-LAN (Wireless Local Area Network) dimension, are an extension to the wired B-ISDN, with the ability to provide radio coverage restricted to a small area (e.g., sports arenas, factories, television studios, etc.) allowing communication between MBS mobile terminals and terminals directly connected to the B-ISDN at rates up to 155 Mbps. Mobility and the proliferation of portable and laptop computers together with potential cost savings in the wiring and re-wiring of buildings are also driving forces in the introduction of broadband wireless customer premises networks in a picocell environment supporting the requirements for high-speed local data communication up to and exceeding 155 Mbps.

It is estimated that some 24.000 person-months of effort will be devoted, until the end of 1998, to R&D in mobile and personal communication in the context of the ACTS programme. The activities of the ACTS projects relate to services, network platforms, terminals, and technologies. The demonstration and assessment of novel services and applications take into account the full implications of user environment, system characteristics, and service provision and control. Projects aim at proving the validity of novel components or sub-system technologies, including multi-mode transceivers as well as tools for network planning. The common factor to all projects is that they contain a strong and clearly addressed thread of user involvement and innovative technology, and that the R&D effort is undertaken in the context of technology, service and application trials. For more information on the ACTS mobile projects, the interested reader should refer to the WWW pages of the ACTS projects that are linked to the ACTS Mobility and Personal Communications Domain home page [5]. Regarding the now terminated RACE Programme, and in particular its mobile dimension, the reader is directed to the proceedings of the annual RACE conferences [5,6,7].

Given the global nature of the mobile communications markets, the EU-sponsored ACTS R&D programme has been open to international co-operation and to the participation of non-European companies in ACTS projects on a mutual benefit basis. Indeed different world regions need to co-operate, particularly in the area of standards development and frequency allocations to ensure not only the widespread availability of advanced and affordable wireless services and applications, but also global roaming and inter-operability. National and regional borders must be transcended, inter-connection of networks and inter-operability of services and applications must be encouraged, all with the objective of ensuring, through the convergence of telecommunications, computer and multimedia, the fulfilment of the potential of the upcoming wireless information age.

*1.3.1 Scope and Objectives of the Mobile Domain*

Projects in the Mobile domain address three aspects of R&D on Third Generation Mobile Systems: future mobile/wireless services, mobile/wireless platforms, and enabling technologies. This is portrayed in Figure 10, including the interrelation with other ACTS domains.

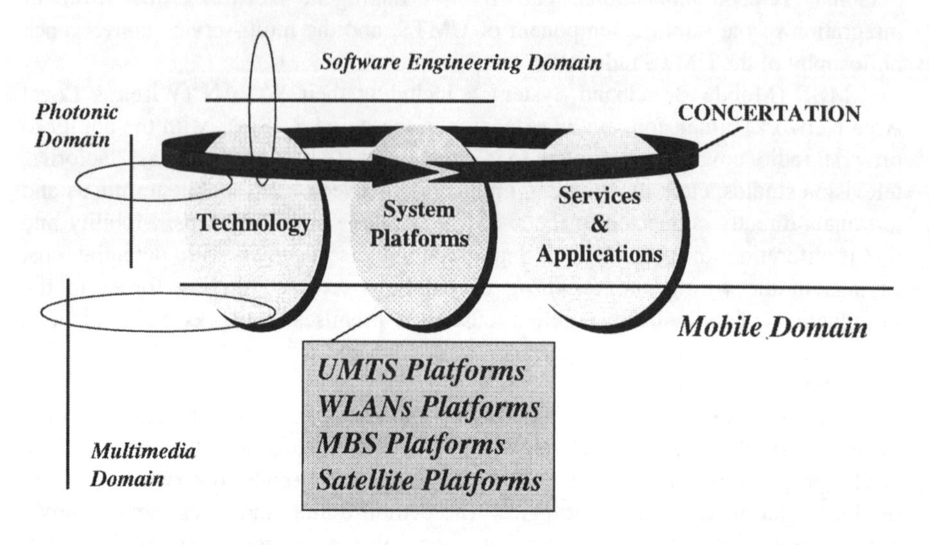


**Figure 10** Scope of the Mobile Domain

Figure 11 below illustrates the communication environments addressed by the various projects of the Mobile domain as well as the main areas covered; the expected interaction with the other ACTS domains and related activities is also depicted.

*1.3.2 Enabling Technologies*

At the core of the future mobile communication networks are a series of enabling technologies whose role is essential in permiting such networks to meet the capacity and quality of service requirements at a cost/performance level attractive to both operators, service providers and users. These enabling technologies range from those usually related to hardware issues (antennas, adaptive wideband radio front-ends, batteries, etc.) to those related to software issues (service creation, authentication, network planning and control, etc.).

Software radio concepts for adaptive and flexible transmission schemes is the main subject of *FIRST*. Its aim is to demonstrate that it is feasible and cost effective to develop and deploy Intelligent Multi-mode Terminals capable of operation with UMTS as well as with multiple standards and with the ability to deliver multi-media services to mobile users.

Project *TSUNAMI* seeks to further develop the technology for smart antennas and aims to demonstrate that it is feasible and cost effective to deploy adaptive antennas within the infrastructure of third generation mobile systems such as UMTS.

The introduction of High Temperature Superconductivy (HTS) components in the Base Stations of mobile communications systems has become lately a hot topic for research world wide. HTS applied to the sub-systems of mobile communications networks will offer not only enhanced performance as a result of the low microwave losses in filters, delay lines and resonators, but also the possibility of novel architectures and functions that support adaptation and reconfiguration. Project *SUCOMS* is involved in the design, fabrication an assessment of communications transceiver components and sub-system items built to a pre-defined specification.

Finally, the *NEWTEST* project aims to study and trial advanced algorithms based on Neural Network techniques to combat the signal impairments traditionally encountered on satellite non linear channels.

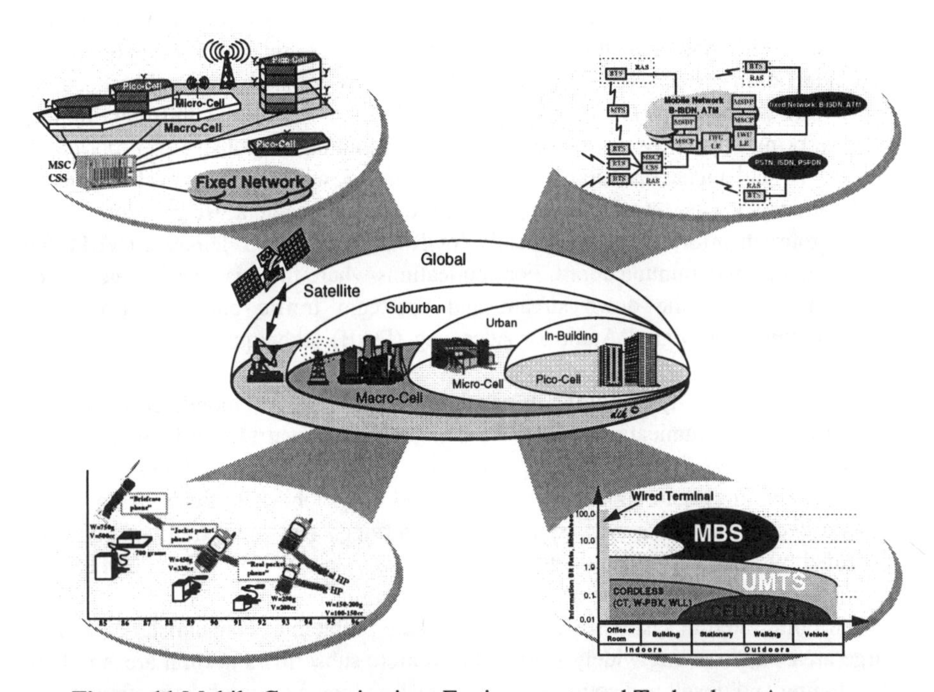


**Figure 11** Mobile Communications Environments and Technology Aspects

### *1.3.3 Mobile/Wireless Platforms*

The essential platforms of relevance to mobile communications are those meeting the requirements in a variety of environments (public, private, rural, office) and at different levels of service and availability. These can be broadly classified as cellular public networks (e.g., UMTS), private and public local area networks (W-LANs and MBS) and global low mobility networks (satellites)

#### *1.3.3.1 Universal Mobile Telecommunications Systems (UMTS)*

The vision of UMTS, as it has emerged from work undertaken within RACE [8], calls for UMTS to support all those services, facilities and applications which customers presently enjoy, and to have the potential to accommodate, yet undefined, broadband multimedia services and applications with quality levels commensurate to those of the fixed IBC networks. In this context, projects will identify the cardinal services that UMTS must support, the "future-proofing" UMTS bearer requirements in macro-, micro- and picocell environments, and the applications likely to be supported by UMTS. A considerable effort will be devoted to determining how best to ensure that UMTS will be designed so as to be perceived by the customers as a broadband service evolution of Second Generation technologies (evolution, not revolution), while ensuring a competitive service provision in a multi-operator environment.

UMTS represents a new generation of mobile communications systems in a world where personal services will be based on a combination of fixed and wireless/mobile services providing a seamless end-to-end service to the user. Bringing this about will require unified offering of services to the user in wireless and wired environments, mobile technology that supports a very broad mix of communication services and applications, flexible, on-demand, bandwidth allocation for a wide variety of applications, and standardisation that allows full roaming and inter-working capability, where needed, while remaining responsive to proprietary, innovative and niche markets. In particular, UMTS will support novel telematics applications such as dynamic route guidance, fleet management, freight control and travel/ tourism information, specifically for Road Transport Telematics (RTT) and high speed train communications. For applications where there is a very large degree of asymmetry in the down stream and upstream traffic channels, UMTS in combination with Digital Audio Broadcasting (DAB) techniques, can provide cost effective solutions.

With UMTS fully exploiting its capabilities as the integral mobile-access part of B-ISDN, telecommunications will make a major leap forward towards the provision of a technically integrated, comprehensive, consistent, and seamless personal communication system supported by both fixed and mobile terminals. As a result, mobile access networks will begin to offer services that have traditionally been provided by fixed networks, including wideband services up to 2 Mbps. UMTS will also function as a stand-alone network implementation.

Mobile satellite systems are a solution to the problem of economically covering large areas, and serving widely scattered or remote subscribers in rural areas in both developing and developed countries, as well as for ship and aircraft communications. It is intended that the satellite component of UMTS be integrated with its terrestrial component. To keep the inter-working between the two transparent to the end user, three levels of integration are identified as follows: network integration where the space and terrestrial networks can operate as separate entities; equipment integration that requires common service standards, consistent transmission parameters and common radio interface between satellite and terrestrial implementations; and system integration in which the satellite is an integral part of the terrestrial network and able to support handover between terrestrial and satellite megacells. Clearly the satellite component of UMTS raises technical and economic issues that could impinge on the performance of the UMTS terrestrial service and, consequently, convergence must be carefully weighed.

At the time UMTS reaches service, ATM will be an established transmission technique; hence the UMTS environment must support ATM-cell transmission through to the user's terminal. This compatibility will enable service providers to offer a homogeneous network, where users can receive variable bit-rate services regardless of their access media (mobile or fixed, including wireless local loop). The flexibility of service provision across Europe, in multi-operator environment, also demands close attention to spectrum-sharing techniques, charging, billing and accounting, numbering, network security, privacy, etc., all of which may have regulatory implications. To these issues, one must add the degree of compatibility between UMTS and fixed network functionality, and the form of the multi-service capability at the radio interface.

UMTS will require a revolution in terms of radio air-interface design, and continued evolution of intelligent network (IN) principles. The arrival of a fully capable UMTS does not preclude the extension of such developments into those bands currently open to Second Generation technology. The resulting parallel process of UMTS design and Second Generation enhancement will call for careful market management and co-existence between UMTS and Second Generation service standards to ensure a smooth, customer sensitive transition. Indeed, multi-mode/multi-band transceiver technology may be used to provide multi-standard terminal equipment, particularly between UMTS, GSM/DCS-1800, and DECT.

*UMTS Platform Projects*

The main distinction of UMTS relative to second generation systems is the hierarchical cell structure designed for gradated support of a wide range of multimedia broadband services within the various cell layers by use of advanced transmission and protocol technologies. The three dimensional hierarchical cell structure in UMTS aims to overcome second generation problems by overlaying, discontinuously, pico and micro cells over the macro cell structure with wide area coverage. Global/Satellite cells can be seen in the same sense to support area coverage where macro cell constellations are not economic to build up or/and to support wide distance traffic. In the centre of Figure 11 such a hierarchical cell structure configuration is illustrated.

The choice of an air interface parameter set corresponding to a multiple access scheme is a critical issue as far as spectral efficiency is concerned. Within the framework of the ACTS Programme, project *FRAMES* is in charge of defining a hybrid multiple access scheme based on the adaptive air interface concepts that were earlier developed in RACE. After a comparative assessment of a dozen of candidate techniques, the FMA (FRAMES Multiple Access) System has been proposed to ETSI for the process of definition of the UMTS air interface. The FMA concept lends itself to the introduction of novel technologies, such as joint detection, multi user detection for interference rejection, advanced coding schemes as turbo codes and adaptive antennas which contribute to more efficient utilisation of the UMTS frequency bands.

The introduction of third generation mobile systems require an effective software tool able to assist and support Operators to design and plan the UMTS network. The tool being designed by *STORMS* is conceived as an open software platform where external or newly developed modules can be easily plugged. A Topographical and Geographical Information System (TGIS) has been developed in the *radio coverage engineering* module. Following the evaluation of the coverage pattern associated to each single cell, a radio *coverage optimisation* module has also been developed. It allows to optimise the number of radio base stations necessary to ensure a given coverage percentage with a certain confidence degree.

The fixed infrastructure (links between the UMTS network nodes) *resources optimisation* module considers the wide variety of UMTS network architectures tailored to the environment constraints (Public, Business, Domestic). A possible three dimensional cell planning layout is demonstrated in the top left part of Figure 11.

Design and development of sustainable transmission systems, with very high capacity is a major challenge for the realisation of Universal Personal Telecommunication (UPT). Adaptive antennas such as the ones developed by *TSUNAMI-2* are a new technology that have the potential to provide large increases in capacity. The top level design of the field trial equipment has been completed, and the project is in the process of constructing a single fully adaptive BTS that will be deployed and used in a DCS 1800 network. Performance studies suggest that the gain in the uplink direction will be greater than downlink and the project is now studying ways of correcting this imbalance.

The study a "generic" UMTS access infrastructure, able to cope with different innovative radio access techniques (long term view) and, at the same time, to guarantee a soft migration from second to third generation systems (short term view), is addressed by *RAINBOW*. The main achievement on the issue of network integration is the proposal by *RAINBOW* of a reference configuration of the UMTS system introducing a *single generic functional interface* between the UMTS access part and the core network and a separation of the *radio technology dependent* parts from the *radio technology independent* parts. These two new concepts have been presented and promoted in ETSI SMG and ITU-R TG8/1 with the purpose of obtaining a unified view of the evolution of the standards towards the UMTS.

A fully operational example of system migration from 2nd generation to 3rd generation, as proposed by *COBUCO,* is in progress. The phased approach adopted by *COBUCO* provides a realistically wide scale of application, experiments, migration studies and research, and usage. It means a practical trial of UMTS features, protocols and data base models defined in RACE II (e.g. MONET). The protocols are already specified and their implementation will follow subsequently.

To facilitate and promote the development of a wide spectrum of mobile multimedia application a standardised mobile Application Programme Interface (API) is being developed by *ON THE MOVE*. The project has succeeded in defining the MASE middleware architecture and its services offered to "mobile-aware" applications. These include among others location awareness, messaging and multiparty services, cost control, security, disconnected operation, terminal adaptation, profile management and support for distributed environments. The application designed and implemented for this experiment included stock portfolio information, news telegrams, travel information and access to news magazines as well as a dashboard providing feedback to the users about their location and available bandwidth and the results will allow the project to focus on the essential requirements both with respect to the MASE and the applications to be designed for the upcoming field trials.

*1.3.3.2 Wireless-LAN (W-LAN) and Mobile Broadband Systems (MBS)*

The strategic importance of mobile broadband communications systems catering for different mobility requirements ranging from stationary (for wireless local loops), through quasi-stationary (office and industrial environments), to fully mobile was recognised at an early stage in the context of RACE. The objectives of the work were namely to develop a quasi-mobile wireless system for bit-rates of up to 155 Mbps throughput (in the 40 or 60 GHz bands), and to create the industrial capacity to produce the necessary system components (RF, IF and baseband systems, antennas, terminals). In the context of MBS applications, the investigation and definition of system aspects, radio access schemes, network management issues, integration with IBC, etc., is critical. MBS systems will cater for novel multimedia and video mobile telecommunication applications, including those appropriate to W-LAN and WLL (wireless local loop) broadband systems.

As wireless terminal extensions to the B-ISDN, MBS systems concepts are being actively researched in the context of ACTS with emphasis on specific objectives including the demonstration of mobile broadband applications, video distribution, interactive video, audio and data communication service at bit-rates up to 155 Mbps on a mobile terminal connected to an IBC network; the demonstration of ATM compatibility between mobile and fixed terminals with implementation of the necessary mobility management-functions (especially for handover) and the required signalling, control and service-provision protocols; and finally the validation of quality-of-service parameters corresponding to the evaluated applications. As with UMTS, a satellite component of MBS will be investigated to determine the optimum frequency band (Ka: 20-30 GHz, or 60 GHz bands).

The widespread availability of portable and laptop computers dictates the need for third-generation mobile systems to incorporate an integrated W-LAN capability to maintain its "universality". Application areas include mobile systems for offices, industrial automation, financial services, emergency and medical systems, education and training, with network connection for portable computers and personal digital assistants as well as ad hoc networking. The specific nature of each of the above related environments do, however, influence security, range, defined working area, transmission rate, re-using of frequencies, cost, maintenance, penetration potential, etc.

In creating a high-speed (up to 155 Mbps) local data communication link, significant research is required to identify a suitably reliable system and associated air interface. Important issues include frequency allocation and selection, choice of bandwidth, efficient coding schemes, specification of medium access procedures, definition of link control protocols, as well as connectivity aspects related to connection to other wired or wireless communications networks. In brief, ongoing W-LAN R&D activities seek solutions that recognise application, environment, cost, performance, networking and system architecture requirements.

ERO is already looking into spectrum allocation for MBS [9]. The main tasks identified are:

- to review the currently designated frequency bands for MBS (terrestrial and satellite) and the applications which they may be able to support in terms of capacity, user bit rates, etc.;

*1.3.3.2 Wireless-LAN (W-LAN) and Mobile Broadband Systems (MBS)*

The strategic importance of mobile broadband communications systems catering for different mobility requirements ranging from stationary (for wireless local loops), through quasi-stationary (office and industrial environments), to fully mobile was recognised at an early stage in the context of RACE. The objectives of the work were namely to develop a quasi-mobile wireless system for bit-rates of up to 155 Mbps throughput (in the 40 or 60 GHz bands), and to create the industrial capacity to produce the necessary system components (RF, IF and baseband systems, antennas, terminals). In the context of MBS applications, the investigation and definition of system aspects, radio access schemes, network management issues, integration with IBC, etc., is critical. MBS systems will cater for novel multimedia and video mobile telecommunication applications, including those appropriate to W-LAN and WLL (wireless local loop) broadband systems.

As wireless terminal extensions to the B-ISDN, MBS systems concepts are being actively researched in the context of ACTS with emphasis on specific objectives including the demonstration of mobile broadband applications, video distribution, interactive video, audio and data communication service at bit-rates up to 155 Mbps on a mobile terminal connected to an IBC network; the demonstration of ATM compatibility between mobile and fixed terminals with implementation of the necessary mobility management-functions (especially for handover) and the required signalling, control and service-provision protocols; and finally the validation of quality-of-service parameters corresponding to the evaluated applications. As with UMTS, a satellite component of MBS will be investigated to determine the optimum frequency band (Ka: 20-30 GHz, or 60 GHz bands).

The widespread availability of portable and laptop computers dictates the need for third-generation mobile systems to incorporate an integrated W-LAN capability to maintain its "universality". Application areas include mobile systems for offices, industrial automation, financial services, emergency and medical systems, education and training, with network connection for portable computers and personal digital assistants as well as ad hoc networking. The specific nature of each of the above related environments do, however, influence security, range, defined working area, transmission rate, re-using of frequencies, cost, maintenance, penetration potential, etc.

In creating a high-speed (up to 155 Mbps) local data communication link, significant research is required to identify a suitably reliable system and associated air interface. Important issues include frequency allocation and selection, choice of bandwidth, efficient coding schemes, specification of medium access procedures, definition of link control protocols, as well as connectivity aspects related to connection to other wired or wireless communications networks. In brief, ongoing W-LAN R&D activities seek solutions that recognise application, environment, cost, performance, networking and system architecture requirements.

ERO is already looking into spectrum allocation for MBS [9]. The main tasks identified are:

- to review the currently designated frequency bands for MBS (terrestrial and satellite) and the applications which they may be able to support in terms of capacity, user bit rates, etc.;

- to examine the trends for further applications in different domains of public and private interest and their expected growth in terms of capacity, user bit rate, and bandwidth requirements, in order to provide advice on the appropriateness of the existing frequency allocations;
- to examine and identify the application sectors in different domains of public and private interest which are more likely to the demands for spectrum, taking into account the increasing use of multimedia services in a multi-operator environment.

*W-LAN / MBS Platforms Projects*

Proliferation of portable and laptop computers, wide acceptance of mobility, and the potential cost savings in avoiding the wiring or re-wiring of buildings arc driving forces for broadband wireless-access in an in-building environment. Consequently third-generation mobile systems must incorporate an integrated WLAN capability to maintain "universality". Wireless ATM will allow users to transmit and receive data at data rates (>20 Mbps) and controlled service levels that match those of the wired ATM world. In creating a high-rate (up to 155Mb/s) local data communication link significant research is required to identify a suitably reliable system and associated air interface. Important issues include frequency allocation and selection, choice of bandwidth efficient coding schemes, specification of medium access procedures, definition of link control protocols as well as connectivity aspects related to connection to other wired or wireless communications networks.

Figure 12 positions the various ACTS /MBS projects with respect to frequency band under consideration and user data rate capabilities over the air interface.

The current achievements of *WAND* include the complete functional system specification on the demonstrator which has been specified with the SDL (Specification and Description Language) and verified with the simulation model, the demonstrator implementation work has been started. Project *WAND* aims to develop and evaluate in realistic user environments a wireless ATM transmission facility (at 5.2 GHz) that expands the reach of ATM technology to the ambulant user of premises communications networks.

*WAND* has been active in its liaison and standardisation activities by contributing and harmonising the work between ATM forum and ETSI RES10. The stochastic radio channel model for channel simulations was developed and verified by measurements on 5 and 17 GHz frequency bands.

The user requirement study has been concluded and the main user and application characteristics of potential wireless applications have been identified.

The main objective of the *MEDIAN* project is to evaluate and implement a high speed Wireless Local Area Network pilot system for multimedia applications and demonstrate it in real-user trials. The pilot system relies on a multi-carrier modulation scheme which is adaptive to the transmitted data rates and channel characteristics, and on wireless ATM network extension.

The system design of the *MEDIAN* Demonstrator is finished. The hardware design and implementation phase started. Some hardware components are already available.

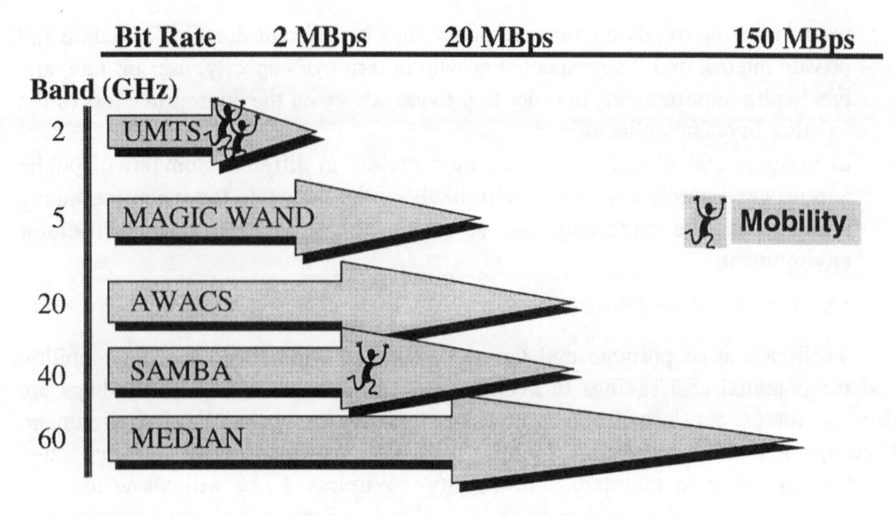


**Figure 12** ACTS Broadband Access Projects in the Mobile Domain

The *SAMBA* project is developing an Mobile Broadband Systems (MBS) trial platform comprising of two Base Stations and two Mobile Stations operating in the 40 GHz frequency band each offering transparent bearer services of up to 34 Mbps, that aims to demonstrate the feasibility of the MBS concept by offering all the essential functionalities expected from MBS. The trials performed on the *SAMBA* trial system will demonstrate the provision of multimedia services (such as high resolution video), the integration of MBS to a fixed broadband network via standard ATM interfaces and prove the enabling technologies developed for operation in the 40 GHz frequency band.

The main result of the *SAMBA* project will be the proof of the cellular mobile broadband concept. For that, an air interface providing reliable transmission of ATM cells and capable of handover, location management, medium access control and radio resource management will be validated by the Trial Platform. In addition the work will include the development of a portable mobile station. So far the user requirements are defined and the target quality of service parameters set.

The Trial Platform is specified. This includes the air interface, the cell shapes and the antennas, the millimeterwave transceiver, baseband processing unit, control unit as well as the ATM switch and ATM mobility server. The development of ASICs, MMICs and modules as well as protocol software has started.

The *AWACS* project aims at the development of a low mobility system operating in the 19 GHz frequency band offering user bit rates of up to 34 Mbps and with a radio transmission range of up to 100 m.

During the first phase of the project, the equipment and channel sounder have been configured for 19 GHz propagation characteristics and propagation simulation techniques in order to measure the propagation data efficiently. The project has defined wireless ATM applications to be used for trials towards the final stage of the project and the demonstration system has been configured for this purpose in order to achieve real time experience for further *AWACS* development.

In system studies, initial link level simulations (BER performance of radio transmission chain) and system level simulations (users, capacity and coverage) have been conducted. Some preliminary statistical propagation models from ETSI-RES10 have been integrated for the simulations. Spatial propagation model proposed for wireless ATM. Further, reference models and performance parameter sets from ETSI-RES10 and ATM Forum have been analysed and first step towards a definition of MAC/LLC procedures for the project have been taken.

The real time trials are expected to indicate the capacity of the available system in a real user environment. The trials results will contribute to the development of common specifications and standards such as ETSI-RES10, HIPERLAN specifications, ITU and ARIB in Japan.

*1.3.3.3 Satellite Communications*

A considerable technological and market debate is ongoing regarding the role that satellite communications may be called to play in the overall mobile communications scene. Broadly speaking, the applications that may be foreseen for future satellite mobile communications are those relating to providing the satellite component of an integrated global system of which UMTS is the terrestrial component, as well as those looking ahead to multimegabit broadband services mainly for low mobilify applications.

The satellite component of UMTS will most probably be positioned as the successor of the many global S-PCS systems currently proposed to serve narrow band services such as voice, fax and low rate data (4.8-9.6 Kbps). Although some of these systems are partly developed with European industry, none of them has got a European leadership. Considering the expected growing importance of satellite global communication systems in the overall satellite communication market as well as the strong US push in this sector, it will be important for European industry to be well positioned on this market segment through relevant R&D preparatory work. Figure 13 shows some of the service configurations offered by satellite personal communication systems. Through ACTS, a number of projects are also dealing with S-UMTS development.

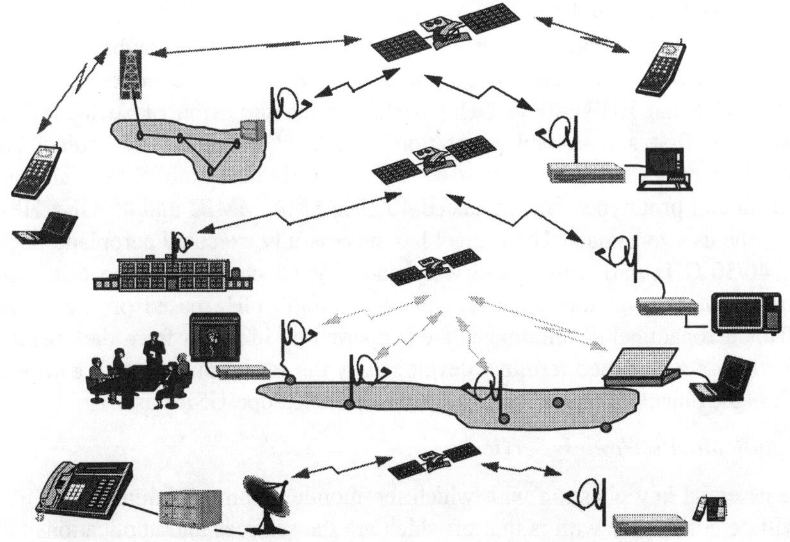

**Figure 13** Satellite Personal Communications

*Satellite Platform Projects*

Clearly the satellite component of UMTS raises technical and economic issues that could impinge on the performance of the UMTS terrestrial service and consequently convergence must be carefully weighed. Project *SINUS* aims at the development of a software simulator allowing to validate complex specifications such as the satellite air interface with the aim of identifying maximum commonalties with *FRAMES*, and the handover procedure using channel performance results derived from *RAINBOW*. The software demonstrator will also allow to simulate different satellite constellations and to associate to each one of those a performance result. The project has defined the services and applications for UMTS space segment and in particular the services to be implemented in the testbed for trials. The standard interfaces and specification for terminal application is now completed as well as the definition of generic upper layer protocols able to manage different radio interfaces.

In parallel, project *INSURED* has completed the first release of the S-UMTS Reference Model, to be updated at the end of the project in line with the result of the trials and simulations;, especially for integration with the terrestrial network.. Insofar as possible, this model will be validated through field trials using the first of its kind IRIDIUM satellite together with a terrestrial network infrastructure based on an IRIDIUM gateway and a modified GSM network. The architecture for the INSURED Demonstrator has been defined.

The specification and validation work of the above two satellite projects will be complemented by project *TOMAS*, which by using previously developed equipment (terminal developed in RACE project MOEBIUS) together with specific satellite facilities primarily procured through the Inmarsat operator to propose an open platform allowing host projects to test equipment developed within the Programme using real satellite channels (GEO). Example of such equipment could be an advanced modem or an advanced video codec. First integration trials with *MOMUSYS* have been completed successfully

Higher frequency bands are also investigated to provide broadband mobile satellite services with low mobility. The *SECOMS/ABATE* project investigates the Ka (20/30 GHz) and EHF (40/50 GHz) bands for the provision of such services respectively for first and second generation broadband mobiles. The project has developed the target satellite system concept, the 20 GHz land mobile array antenna and RF front end prototypes, the advanced 40 GHz LNA MMIC and 45 GHz HPA MMIC for the user terminals. The project has successfully executed aeroplane flight trials for 40/50 GHz narrowband and wideband channel characterisation tests, and the campaign of propagation measurements at 20/40/50 GHz based on the Italsat beacon This aeronautical developments are supported by Inmarsat from the operator side and will use a modified terminal developed by the Jet Propulsion Laboratory in the US. It is the objective to use trial results on a joint Europe-USA basis.

### *1.3.4 Future Mobile/Wireless Services*

The essential key question with which the mobile communications sector as a whole will be confronted with is that of which are the services and applications that will be the key drivers of the new technological generation. While it is widely accepted that "voice communications" will certainly continue to be one of the key requirements to be met by a mobile/wireless system, the debate is still wide open regarding the nature, type and main characteristics of future advanced multimedia mobile services, as well as its likely date of "take-off".

In conjunction to the above system platform concepts and enabling technologies projects, there are a number of application development projects, expected to provide the 'proof of concept' as well as an insight on what the users should expect from future generations of mobile communications networks.

The trials dimension for this projects is oriented to user trials, and when feasible, trials are based on platforms developed by the system projects of the Domain.

The distribution of advanced multimedia products by means of a wireless media highway is the main subject of *MOMENTS*, which also covers the related billing and charging issues. So far the configuration for the Service Platform has been defined, and the HW and the SW for the Basic Service Platform and terminals for 100 end-users have been delivered and installed to each trial network. Contributions to standardisation bodies (ITU-T and MPEG4) have been submitted regarding the video coding schemes

The wireless communications requirements of the health care sector is addressed by *MULTIPORT*, with the aim to demonstrate PDA access to databases as well as remote consultation in a UMTS environment with intermediate stages that incorporate wireless LAN/GSM trials, and UMTS simulations. The first prototypes of the different components have been integrated into the first prototype of the MultiPort system which is being trialed in a Hospital.

The same sector is also trialed for wireless services in a hospital environment by *WAND*, the underlying platform is a WCPN operating in the 5 GHz frequency band.

The construction sector will benefit from the work of *MICC* which will bring to construction site workers completely new communications capabilities through mobile communications. *MICC* will publish a set of tested European wide recommendations. It is seen by the consortium as a major step to organise industry-wide solutions for on-site telecommunications for Construction.

Every user of telecommunications services has individual special needs, by taking account of the needs of users with disabilities the usability requirements of the population as a whole can be accommodated. This is the basic theme of *UMPTIDUMPTI* whose primary objective is to verify that emerging broadband and mobile services and equipment are usable by all. Within ETSI TC Human Factors the project has initiated two new work items concerned with standardisation of text-telephony (a key service for the deaf) and the further analysis of requirements for mobile video telephony.

### *1.3.5 Concertation Activities*

Specific issues of common interest to Mobile projects are tackled by SIGs (Special Interest Groups) with the aim of reaching results applicable to the widest possible set of projects while seeking specific contributions to the work of standards bodies. Table 2 below illustrate the basic objectives, and expected contributions of each of the ongoing Special Interest Groups.

**Table 2** Objectives and Expected Outputs from the SIGs of the Mobile Domain

| | Objectives | Achievements |
|---|---|---|
| **Air Interface (SIG 1)** | To investigate the aspects affecting the choice and design of air interfaces for 3rd generation of mobile communications systems (UMTS/FPLMTS, MBS, WLANs), including satellite aspects. | Common deliverable on Air-interface options. |
| **Joint Trials Cluster**<br><br>**(SIG 2)** | To provide a forum for the exchange of views with the aim of using the maximum possible of H/W system resources provided by the projects developing H/W platforms in order to support the largest possible number of application demonstrators. | Joint trials. |
| **Mobile Terminals (SIG 3)** | To determine the implication of the user requirements and technologies implications on the specification and design of terminals for the 3rd generation of mobile communication systems. | Common deliverables in preparation, will be submitted to ETSI. |
| **Network Issues**<br><br>**(SIG 4)** | To perform work on all aspects affecting the network architecture of future mobile systems such as UMTS, MBS, WLANs and their satellite components. This work will primarily cover network architecture and planning, service definition and provisioning, impact of radio interface features on network aspects, protocol architecture and interworking aspects. | -Analysis of the UMTS transport chain extended to satellite-based services, wireless ATM systems and transport chain for signalling information.<br><br>-New ATM Adaptation Layer type 2, proposed by ITU-T for the transport of low bit rate services considered. A comparison with the use of AAL5 has been performed.<br><br>-First version of the deliverable "Preliminary Framework for the UMTS Network, "released |

In the context of the Domain's concertation activities, a number of workshops have been organised on the following subjects: Mobile User Requirements, Mobility aspects, Radio Access Techniques, Mobile Systems Evolution and Migration, Mobile Terminals, and Trials and Linkages to Guidelines.

A related activity, but intended for a wider audience, is the organisation of a series of workshops addressing key issues for the evolution of mobile and wireless communications and that will play an important role in future collaborative R&D actions in the framework of EU-funded programmes.

## 1.4 BEYOND ACTS

After the previous assessment of the ACTS programme, it is just natural to start looking ahead towards the Fifth Framework Programme (5FWP), specially in the area of Mobile and Personal Communications.

The section starts with a brief review of the process that has led to the present Commission proposal for the 5FWP, highlighting the most distinguishing features of the new approach to R&D. Then the Information Society (IS) programme will be discussed in certain detail, specially in what concerns Mobile and Personal Communications.

### *1.4.1 A New Integrated and Co-ordinated Approach to R&D*

Taking into account the final conclusions and recommendations of the general 5-year assessment report on the research activities of the Union, and through consultation with European *R&D organisations* and member states, the Commission has concluded for the need to focus efforts to a greater extent, improve the consistency of the overall approach, and pay greater attention to the exploitation of results.

The 5FWP will be marked by a particular effort of selectivity and concentration on a limited number of areas and objectives. Its content will be determined based upon three categories of criteria: economic development and scientific and technological potential, furtherance of social objectives, and European value added according to the subsidiarity principle.

In terms of economic development and scientific and technological potential, the selected areas must correspond to areas in which European firms can and must become more competitive, areas which are expanding and have good growth potential, and areas in which significant technological progress is anticipated. Employment, quality of life, health and a clean environment are obvious priorities and a serious concern of the EU citizens. As such, they must be factors in deciding the areas of concentration. Finally, subsidiarity leads to the selection of areas where a "critical mass" in human and financial terms needs to be established and a mixture of complementary expertise found in the various countries is needed, and those that concern European problems and the development of the European area or involve aspects of standardisation.

Based upon all the above, the Commission has proposed organising the 5FWP on the basis of six programmes, three "thematic" and three "horizontal". "Thematic" and "horizontal" programmes are necessarily complementary and interrelated.

The "horizontal" programmes cover 1) co-operation in the field of R&D with third countries and international organisations with the objective of *confirming the international role of European research*, 2) the dissemination and optimisation of results through *innovation and participation of SMEs*, and 3) *improving human potential* through stimulation of training and mobility of researchers.

The "thematic" programmes are R&D programmes focusing on the following priorities: 1) unlocking the resources of the living world and the ecosystem, 2) creating a user-friendly information society, and 3) promoting competitive and sustainable growth. The work carried out in ACTS will be continued mainly in the scope of the Information Society programme.

### *1.4.2 The Information Society (IS) Programme*

The most important aspect of the Information Society is its potential for renewed growth and the creation of new forms of work: several million jobs will result in Europe alone from the development of the Information Society, provided the challenges of technical excelence and preservation of cultural diversity can be met.

The full potential of the Information Society can not be achieved on the basis of today's systems and technologies. Continued R&D, side by side with a technology adoption effort, is therefore necessary. The aim of the Information Society (IS) programme should be to have the advanced goods, services and methodologies made possible by the Information Society contribute to creating new jobs and strengthening the competitiveness of European companies, stimulating the development of new markets and services and strengthening the role of the public in society, improving the position of Europe as a place for investment, reserach and innovation, and strengthening the scientific and technological base of the EU with the goal of reinforcing global competitiveness. The need for accessibility and interoperability at all levels, from technologies and tools to systems and applications, is paramount.

Among the Key Actions under consideration are 1) Systems and Services for the Citizen, 2) Electronic trade and new methods of work, 3) Multimedia content, and 4) Essential technologies and infrastructures. Mobile and Personal Communications, including in particular satellite-based services, are specifically being considered as part of key action 4.

### *1.4.3 Mobile and Personal Communications in the Future*

The impact of the technologies associated with the Information Society far exceeds that of the industries directly involved: they have implications in terms of fundamental industrial (mass market perspective) and social (new activities, new relations) change. It is therefore essential that carefull consideration is given to R&D in Mobile and Personal Communications, as an essential enabler, in the scope of an integrated R&D programme.

The unprecedented growth of world-wide mobile/wireless markets analysed in Section 1.2, coupled with advances in communications technology and the accelerated development of services taking place in fixed networks, has made the introduction of a flexible and cost effective Third Generation Mobile Communication System a priority.

The next phase, to be pursued in the scope of the IS programme, will add a different dimension. Concurrently with the much needed and anticipated integration of fixed and mobile, terrestrial and satellite systems, the continued development of Mobile/Wireless Broadband Communication systems, the systematic search for performance improvements through the (optimum) combination of technologies (from adaptive antennas to modulation/coding schemes to per-survivor processing), a dramatic paradigm shift is envisioned with the adoption of Software Communication concepts.

At the same time, research in basic technology will insure that the necessary developments are in place (from smart batteries to high temperature superconductivity to flat panel displays) to allow for a true mass market of telecommunication products.

With those, a myriad of personalised telecommunication services will emerge, requiring the development of new service engineering tools and the extension to the telecommunications arena of Just-in-Time service definition and provisioning. To stress the point, a user-centric perspective will refocus the work towards the seamless offer of personalised, Just-What-the-Customer-Wants services, and provide the user with simple, user-friendly interfaces to the service providers, from profile definition to service selection to billing.

A particular aspect of the development of the Mobile and Personal Communications sector is that relating to its satellite communications dimension. It has indeed been recognized that Europe must take a more proactive and consistent approach given that the developments in the satellite communications sector are of a global strategic importance, and that new requirements are emerging which demand high performance access and space-based backbone infrastructures.

As a result of a number of initiatives of an industrial and political nature, a *Satellite Action Plan* has been adopted by the European Commission [10], specifically calling for the elaboration of a strategic vision relating to the strengthening of the role of the European services and manufacturing industry in global, advanced broadband multimedia satellite systems, services and applications, and to advancing of the European position in global satellite personal communication systems, services and applications particularly relating to UMTS.

On issues relating to spectrum availability, licensing, standardization and international co-operation, the European Commission has recently adopted a Communication [11]. In so doing it hopes to stimulate a debate and raise the awareness of the sector actors to the challenges and choices which Europe will face in the near future.

## 1.5 CONCLUSIONS

This paper, together with the current perspective on the European mobile and wireless telecommunication sector, presents a first assessment of the achievements of the ACTS projects in the Mobile Domain.

Supported by telecommunications operators, equipment manufacturers, service providers, research institutes, universities and leading-edge users, a number of EU-funded R&D projects have proved to be instrumental in the development of Third Generation systems. The participation of representative users reflects the policy that

R&D must be demand-driven to ensure that the developed technologies, services and applications are responsive to market requirements.

The case is also made, in the scope of the discussion of the 5 Framework Programme, for the urgent need for enhanced, more efficient mobile communication systems, as well as for the spectrum required by the anticipated multimedia services that will be offered. R&D in Mobile and Personal Communications is believed to be essential, due to its multiplicative, enabling effect in all areas of the Information Society.

## References

[1] *The European Path towards UMTS*, Special Issue of the IEEE Personal Communications Magazine, February 1995

[2] Green Paper on A Common Approach in the Field of Mobile and Personal Communications in the EU, European Commission, April 1994

[3] UMTS Task Force Final Report

[4] UMTS Forum Report

[5] http://www.infowin.org/ACTS/IENM/CONCERTATION/MOBILITY

[6] Proceedings of the RACE Mobile Telecommunications Summit, European Commission, Cascais, Portugal, 22-24 November 1995

[7] http://www.infowin.org

[8] RACE Vision of UMTS, Workshop on Third Generation Mobile Systems, DG XIII-B, European Commission, Brussels, January 1995

[9] Mobile Broadband Systems (MBS), Draft Final Report, ERO, November 1996

[10] Communication on "EU Action Plan: Satellite Communications in the Information Society", COM(97) 91, European Commission, March 1997

[11] Communication on "The further development of mobile and wireless communications - challenges for the European Union", COM(97) ??, European Commission, June 1997

# 2 OVERVIEW OF RESEARCH ACTIVITIES FOR THIRD GENERATION MOBILE COMMUNICATION

Tero Ojanperä

**Abstract**

This paper gives an overview of worldwide air interface research activities towards the third generation mobile communications (UMTS and FPLMTS/IMT-2000). The status of standardization for each main region (Europe, Asia, USA) is discussed. Third generation research activities are described focusing into the air interface concept developments. Details of different CDMA, TDMA, OFDM and hybrid air interface designs are given. Furthermore, the implications of Time Division Duplex operation are covered.

## 2.1 INTRODUCTION

Since the work started in the standardization bodies ITU TG8/1 (International Telecommunication Union Task Group) committee for FPLMTS/IMT-2000 (Future Public Land Mobile Telephone System) 1986 and SMG5 (Special Group Mobile) subtechnical committee for UMTS (Universal Mobile Telecommunication System) 1991, the third generation activities have formed an umbrella for advanced radio system developments. During the recent years, standardization activities towards UMTS and FPLMTS/IMT-2000 have accelerated towards concrete specifications. According to present plans ETSI will select UMTS Radio Access Concept by the end of 1997, and the main system parameters will be frozen by the end of 1998. Moreover, the FPLMTS/IMT-2000 radio transmission techniques evaluation process has started in ITU. Submissions of candidate technologies are expected latest by October 1998 and the completion of ITU selection process by early 1999.

To support the standardization activities several research programs throughout the world have developed third generation air interface concepts. In addition, laboratory and field trials have been performed. In this paper we focus into those air interface research activities that currently are considered as candidates for UMTS and FPLMTS/IMT-2000 air interface standards.

The paper is organized as follows. Section 2 introduces the UMTS system concept and in section 3 the radio access system research activities throughout the world are described. Section 4 introduces the FRAMES multiple access evaluation campaign. Wideband CDMA based radio interfaces are dealt in section 5. Sections 6 and 7 assess TDMA and hybrid schemes, respectively and section 8 presents OFDM based air interface schemes. Furthermore, section 9 covers Time Division Duplex (TDD) concepts and related problems. The conclusions are presented in the last section.

## 2.2 RADIO ACCESS SYSTEM CONCEPT FOR UMTS

UMTS is the third generation mobile radio system offering to the mass multimedia services on the move, world-wide roaming in a flexible and efficient way. UMTS will be based on the concept of Generic Radio Access Network (GRAN), capable of connecting into several core networks. The GRAN concept, depicted in Figure 1, can be connected into GSM/UMTS, N-ISDN, B-ISDN and packet data network. GSM BSS can also be connected into the GSM/UMTS core network. This connection between GSM and UMTS is important, especially in the beginning of the UMTS life time, in order to facilitate a spectrum efficient utilization of GSM900, GSM1800 and UMTS frequency bands and seamless coverage using multimode GSM/UMTS terminals. The evolved GSM core network facilitates also service portability and full utilization of the existing investments and customer base.

In addition to the UMTS activities, GSM continues evolution towards third generation. Enhanced GSM carrier, currently under feasibility study in ETSI, will offer wide area coverage up to 144 kbit/s and even higher bit rates for local coverage using the 200 kHz carrier and higher order modulation. This will facilitate introduction of a subset of UMTS services already into second generation systems. Of course, a new revolutionary wideband access provides then the full set of UMTS services with increased spectrum efficiency and flexibility.

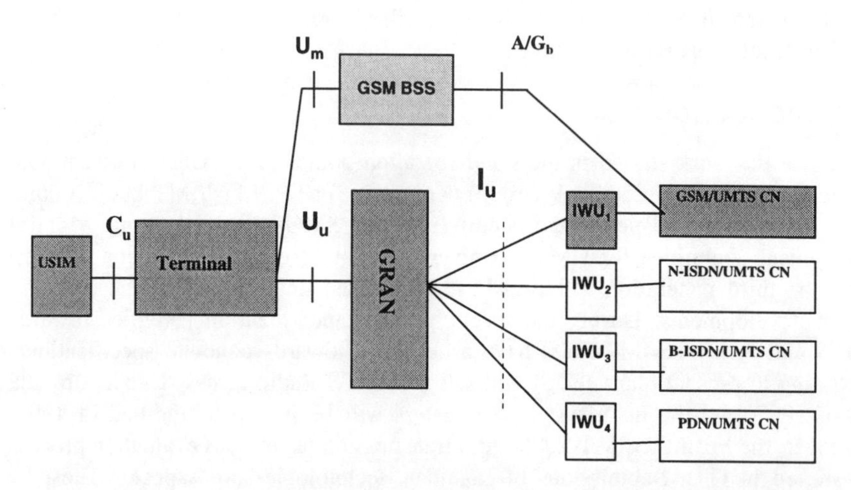


**Figure 1** GRAN concept.

As an example of UMTS services, we can imagine a UMTS user in outdoors scenario. He would like to go into movies and therefore he needs to find out what movies are being shown in the city. A voice recognition software translates his query into a format the computer can understand. An intelligent search agent initiates a search for appropriate information about the movies. After the information has been found it will be downloaded into user's terminal and shown in the terminal screen. After looking at movie titles, he might want to see a preview of a film which can then be downloaded at 100 kbit/s bit rate. Next he wants to know how to go into the movie theater and downloads a digital map and picture of the movie theater. It is clear that the user could get this kind of idea everywhere and does not understand coverage limitations imposed by the system.

In the above example several features of UMTS were used. To set up the connection, the application negotiated the bearer service attributes such as bearer type, bit rate, delay, BER, up/down link symmetry and protection including none or unequal protection. During the session, parallel bearer services (service mix) were active, i.e., UMTS is able to associate several bearers with a call and bearers can be added and released during a call. Furthermore, both real-time and non-real-time communication modes were used. Browsing WWW is a typical asymmetric, non-real time application and the video component that was added during the session is an example of a real-time application. UMTS is also able to adapt the bearer service bit rate, quality and radio link parameters depending on the link quality, traffic, network load, and radio conditions within the negotiated service limits.

Mobile office is a UMTS application already enabled by the GSM data service. The applications include email, fax, file transfer etc. A typical engineering document might exceed 1 Mbytes (10 Mbits) if it contains a large number of pictures. It would thus take approximately five seconds to transfer it with UMTS but 100 seconds with 100 kbit/s GPRS connection. The higher user bit rate of UMTS will increase the total traffic because users will now use applications that they would not if only lower bit rates were available. Higher carrier bit rate will enable more efficient sharing of resources between the users and higher multiplexing gains.

Since the UMTS applications are not yet know, we cannot specify the supported services only based on today's needs. We need a flexible and broad approach in order to have a future proof radio interface. Therefore, based on analysis of large number of existing services and future needs a more generic radio requirements have been derived (data rate, BER, data rate variation, max delay). These application requirements can be mapped into generic radio bearers as shown in Table 1. The exact values are currently under discussion in standardization bodies.

## 2.3 WORLDWIDE RESEARCH ACTIVITIES FOR UMTS AND FPLMTS/IMT-2000 RADIO

### *2.3.1 European activities*

In Europe, RACE I program, launched 1988 and lasted until June 1992, started the third generation research activities. Main studies of RACE concentrated into individual technologies such as cellular propagation studies, handover, dynamic resource allocation; modulation, equalization, coding, channel management and fixed network mobile functions. Many of these individual technologies were later used as basis for the system concept development in RACE II.

During 1992 and 1995, in the RACE II program, CODIT (Code Division Multiple Testbed) and ATDMA (advanced TDMA) projects developed air interface specifications and testbeds for UMTS radio access [36]. The ATDMA test bed provided up to 64 kbit/s user data rate with rate 1/4 coding and the CODIT test bed provided user data rate up to 128 kbit/s. Laboratory tests were performed for both systems. Later, the field tests for 128 kbit/s transmission were concluded for the CODIT test bed [37]. SIG5 (Special Interest Group) compared the CODIT and ATDMA air interfaces using both quantitative and qualitative criteria [1]. Depending on the selected radio environment and service scenario one of them showed advantage but final conclusion about superiority of one scheme over the other, however, could not be reached.

European Research Program ACTS (Advanced Communication Technologies and Services) was launched at the end of 1995 to support collaborative mobile research and development. Within ACTS the project FRAMES (Future Radio Wideband Multiple Access System) has been set up with an objective to define a

**Table 1** UMTS Bearer Service Requirements

| | **Real Time/Constant Delay** | | **Non Real Time/Variable Delay** | |
|---|---|---|---|---|
| **Operating environment** | **Peak Bit Rate** | **BER/Max Transfer Delay** | **Peak Bit Rate (note 3)** | **BER /Max Transfer Delay** |
| **Rural outdoor (terminal speed up to 250 km/h)** | at least 144 kbit/s (preferably 384 kbit/s) granularity appr. 16 kbit/s | delay 20 - 300 ms BER $10^{-3}$ - $10^{-7}$ | at least 144 kbit/s (preferably 384 kbit/s) | BER = $10^{-5}$ to $10^{-8}$ Max Transfer Delay 150 ms or more |
| **Urban/ Suburban outdoor (Terminal speed up to 150 km/h)** | at least 384 kbit/s (preferably 512 kbit/s) granularity appr. 40 kbit/s | delay 20 - 300 ms BER $10^{-3}$ - $10^{-7}$ | at least 384 kbit/s (preferably 512 kbit/s) | BER = $10^{-5}$ to $10^{-8}$ Max Transfer Delay 150 ms or more |
| **Indoor/ Low range outdoor (Terminal speed up to 10 km/h)** | 2 Mbit/s granularity appr. 200 kbit/s | delay 20 - 300 ms BER $10^{-3}$ - $10^{-7}$ | 2 Mbit/s | BER = $10^{-5}$ to $10^{-8}$ Max Transfer Delay 150 ms or more |

proposal for UMTS radio access system. The first target of FRAMES was to investigate hybrid multiple access technologies and based on thorough evaluation of several candidate schemes select the best combination as a basis for further detailed development of UMTS radio access system. The comparison results have been presented in [18][19][20]. The FRAMES multiple access (FMA) proposal consisting of harmonized concept of wideband TDMA with and without spreading (FMA1) and wideband CDMA (FMA2) has been submitted to ETSI for UMTS radio access evaluation process [21]. The FRAMES project is currently developing detailed concepts for the physical layer, medium access and radio link control layer (MAC/RLC) and Radio Network Layer.

In addition to the RACE and ACTS programs, several industrial projects have developed technologies for UMTS and FPLMTS/IMT-2000. During 1992 - 1995, Nokia developed a wideband CDMA concept reported in [5][8][43][44] Furthermore, Nokia wideband CDMA testbed was developed during 1992 -1995 with transmission capabilities up to 128 kbit/s for video application [6][7]. Field tests for the system were carried out during 1996. The test system bandwidth is configurable up to 30 MHz. The Nokia wideband CDMA (MUD-CDMA) [5] was submitted to FRAMES, where it has been used as a basis for FRAMES FMA2 development. Furthermore, the FRAMES FMA1 without spreading is partly based on the wideband TDMA scheme presented in [24] and on the experience from ATDMA project. The FMA1 with spreading is based on JD-CDMA concept developed in University of Kaiserslautern[62]. In addition, a testbed has been developed for JD-CDMA. In UK, the LINK program developed also a wideband CDMA concept and testbed [55]. Telia has developed an OFDM (Orthogonal Frequency Division Multiple Access) concept and testbed [56].

ETSI SMG2 started the definition process of UMTS Terrestrial Radio Access (UTRA) in UMTS workshop held in Sophia Antipolis, December 1996. Main milestones of the process are

- M1 6/97: Definition of a limited number of UMTS Terrestrial Radio Access concepts based on the proposed access technologies or combinations of them.
- M2 12/97: Selection of one UMTS Terrestrial Radio Access concept
- M3 6/98: Definition of key technical aspects of the UMTS terrestrial radio access (including carrier bandwidth, modulation, channel coding, channel types, frame structure, access protocols, channel allocation and handover mechanisms, cell selection mechanisms etc.)

So far, the submitted proposals into ETSI can be grouped under the following main categories

- wideband CDMA,
- TDMA,
- hybrid CDMA/TDMA,
- OFDM and
- ODMA (Opportunity Driven Multiple Access).

In addition to the UMTS activities, GSM continues evolution towards third generation. Enhanced GSM carrier, currently under feasibility study in ETSI, will offer wide area coverage up to 144 kbit/s and even higher bit rates for local coverage using the 200 kHz carrier and higher level modulation [2][42]. This will facilitate the introduction of a subset of UMTS services already into second generation systems.

Of course, a new revolutionary wideband access provides then the full set of UMTS services with increased spectrum efficiency and flexibility.

*2.3.2 Japanese activities*

FPLMTS Study Committee in ARIB (Association of Radio Industries and Business) was established in April 1993 to coordinate the Japanese research and development activities for FPLMTS system. In October 1994, it established Radio Transmission Special Group for radio transmission technical studies and production of draft specification for FPLMTS. The Special Group consists of two Ad Hoc groups: CDMA and TDMA [54].

Several companies have developed wideband CDMA air interface proposals. Originally, 13 different wideband CDMA/FDD (Frequency Division Duplex) radio interfaces were presented to FPLMTS Study Committee. In the beginning of 1995, they were merged into three CDMA/FDD proposals, Core A, B and C, and into one TDD proposal. The Core B system is based on the wideband CDMA standardized for US PCS, known as IS-665. In the end of 1996, the four schemes were further combined into a one single proposal where the main parameters are from the Core-A. In the following we focus into the Core-A proposal which has been presented more detail in [27][28][29][30][31]. Three of the original wideband CDMA proposals from Japan, 2 FDD and 1 TDD concept, have been contributed also to ETSI by respective companies.

Both laboratory tests and field trials have been conducted for the Japanese wideband CDMA proposals during 1995 and 1996. E.g. the Core A has tested speech and 384 kbit/s video transmission in field trials [29]. 2 Mbit/s transmission has also been performed in laboratory conditions using 20 MHz bandwidth. In the second phase of Core-A trials, currently on-going, interference canceller is also tested. The Core-B proposal has only been validated for speech service in the frame work of PCS standardization in USA [25]. Core C has been tested for data rates up to 32 kbit/s. Furthermore, the CDMA/TDD proposal was measured in laboratory conditions [25].

For TDMA, there were seven proposals for pedestrian environment, six for office environment and another six for vehicular environment. From these proposals the group compiled a single carrier TDMA system, MTDMA - Multimode and Multimedia TDMA. A prototype equalizer for MTDMA system with carrier bit rate of 1.536 Mbit/s and user bit rate of 512 kbit/s has been build and tested [25]. Since one of the TDMA proposals was based on multicarrier OFDM technology (BDMA - Band Division Multiple Access proposal), it was decided to continue studies of this technology. BDMA has also been proposed into ETSI.

Main conclusion of the FPLMTS Radio Transmission Technology Special Group is that detailed studies on CDMA shall be started, i.e., wideband CDMA is the main technology choice of Japan for FPLMTS/IMT-2000. Further studies will be made for MTDMA high speed transmission in office and pedestrian environments [54].

*2.3.3 US activities*

The focus of US research activities has been in PCS. Recently, also third generation has gained more attention. Since there exists three main second generation technologies, namely GSM, IS-95 and IS-136, most likely we will see several paths towards FPLMTS/IMT-2000. The FPLMTS frequency band is already partially used for PCS systems and thus evolution of the PCS systems within that frequency band is important.

GSM1900 (also known as PCS1900) standard can already provide bit rates up to 115.2 kbit/ using GPRS and HSCSD. Furthermore, the enhanced GSM carrier can provide bit rates up to 256 kbit/s within the 200 kHz carrier. For IS95 the higher bit rates beyond 9.6 kbit/s are currently standardized. First step is Phase 1 providing bit rates up to 64 kbit/s. Phase 2 will go up to 144 kbit/s. Phase 3, will be wideband CDMA with carrier spacing of 5 MHz. Several companies have made wideband CDMA proposals including Nokia, Lucent, Motorola, Nortel and Qualcomm. IS-136 is currently supporting data rates up to 14.4 kbit/s. Further evolution is expected up to 57.8 kbit/s. However, the evolution path towards FPLMTS/IMT-2000 is unclear. The US PCS standard IS-656, wideband CDMA with bandwidth of 5 MHz, has been used as a basis for the Core-B proposal in Japan [25].

*2.3.4 Korean activities*

Electronics and Telecommunications Research Institute (ETRI) has established an R&D consortium to define and refine the Korean proposal for FPLMTS/IMT-2000 during 1997 and 1999. A wideband CDMA proposal has been developed within ETRI[52].

KMT (Korea Mobile Telecom) decided 1994 to develop wideband CDMA air interface for FPLMTS/IMT-2000. Focus was low-tier system for pedestrian users [51]. Common Air Interface (CAI) was written and a test system build. The test system provides data rate of 36 kbit/s using 5 MHz bandwidth. Final CAI was completed at end of 1996. Currently, pre-commercial system development is under way [51].

## 2.4 FRAMES MULTIPLE ACCESS EVALUATION

Several studies have compared the second generation air interfaces. For the third generation, SIG5 performed preliminary evaluation of ATDMA and CODIT [1] and 2 Mbit/s transmission with TDMA and CDMA air interfaces has been evaluated in [43]. However, the most extensive air interface comparison carried out is FRAMES multiple access evaluation [18][19]. FRAMES evaluation campaign consisted of two stages. At the first stage, several candidate schemes were compared and schemes with similar characteristics were combined. The results of this stage were 2 two multiple access schemes with three options for both [19][20]:

- multicarrier TDMA (multiples of 200 kHz), single wideband TDMA (WB-TDMA, bandwidth 1-2 MHz) and hybrid CDMA/TDMA (bandwidth 1.6 MHz)
- asynchronous CDMA (WB-CDMA, bandwidth 6 MHz), OFDM/CDMA and synchronous CDMA

At the second stage these schemes were extensively evaluated with respect to the following criteria :

- Radio related properties which covered aspects such as provision of various data rates in different environments and bearer service flexibility, spectrum efficiency (capacity), coverage, support for adaptive antennas, hierarchical cell structures

(HCS), duplex method, support for public and private environments and handover.

- Terminal impacts (power consumption and complexity, GSM/UMTS dual mode terminals)
- BSS impacts (evolution from existing systems, BSS complexity and cost)

The main results of the FRAMES multiple access analysis have been summarized in [20]. WB-TDMA combined with TDD provides flexibility for asymmetric and packet oriented services and it has been optimized for high bit rates. Main concern is the complexity of the equalizer in large cells. Hybrid CDMA/TDMA provides good performance both for voice and data. The spreading eases somewhat the delay spread handling in tough radio propagation conditions. However, CDMA/TDMA has high complexity due to joint detection for low bit rate services. WB-CDMA is flexible for circuit switched variable rate services due to its good multirate capabilities. The unpredictable nature of packet switched services will create some concerns especially for power control. WB-CDMA requires large parts of continuous spectrum even for low bit rate services. WB-CDMA seems less suitable for asymmetric services in TDD mode.

The above considerations led the FRAMES project to combine WB-TDMA and CDMA/TDMA into Mode 1 (FMA1, WB-TDMA with and without spreading) and to harmonize WB-CDMA parameters with that to form the Mode 2 of the FRAMES Multiple Access (FMA2). In addition, backwards compatibility with GSM/DCS was kept as a key cornerstone of the FMA concept facilitating easy implementation of GSM/UMTS dualmode terminals [20].

Three multiple access options were dropped from further consideration due to complexity and performance reasons. Multicarrier TDMA has too complex RF. OFDM/CDMA considered only for downlink had low performance with reuse one [11] but could be considered as a modulation method for FMA1. Synchronous CDMA had low performance due to the lack of fast power control [20].

Even though multicarrier TDMA was considered too complex, a single carrier enhanced GSM, i.e. a 200 kHz carrier with linear modulation instead of GMSK, was considered as a viable option [20]. An evolved 200 kHz mode can offer higher data rates 150-200 kbit/s with binary modulation and 300-400 kbit/s with quaternary modulation. Advantages are:

- Can co-exist on GSM carriers - Easy evolution
- Easy dual mode terminals implementation
- Well harmonized with GSM and FMA
- Low spectrum granularity - Efficient with hierarchical cell structures (HCS)

## 2.5 CDMA BASED PROPOSALS

In this section we analyze the similarities and differences of wideband CDMA schemes proposed for third generation systems. We cover five different wideband CDMA designs: FRAMES FMA2 which has been presented in more detail in [20][21][22][23], CODIT air interface proposal [34][35], Core-A [27][28][29][30][31] and two Korean air interfaces (ETRI [52] and KMT [51]). Table 4 gives the main parameters of the schemes.

Most of the proposals utilize the following new capabilities that characterize the third generation wideband CDMA:

- fast power control in downlink
- coherent uplink
- seamless interfrequency handover
- optional multiuser detection
- user dedicated pilot symbols both in up and downlink

All the other systems except the ETRI system are asynchronous and thus do not need external synchronization, such as GPS. This is especially important for low cost base station.

*2.5.1 Carrier spacing and chip rate*

For FMA2 a single RF bandwidth for all services was a target. In [43] it has been shown by simulation that 2 Mbit/s transmission rate is possible to achieve even with 6 MHz carrier spacing. However, better performance is achieved if a larger carrier spacing e.g 10 MHz is used. Therefore, FMA2 employs two carrier spacings 5 MHz and 10 MHz [22].

In contrast to the FMA2, CODIT assumed that with DS-CDMA handling wide selection of services with information bit rates ranging from few kbit/s up to 2 Mbit/s can hardly be achieved with a single RF bandwidth. Therefore, three different RF bandwidths were selected: 1 MHz, 5 MHz and 20 MHz. Different RF bandwidths could be even overlaid as long as the interference can be controlled in proper way but no such control scheme is presented in [34][35]. Later, impact of such overlay scheme has been analyzed for DS-CDMA scheme proposed by ETRI for carrier spacings of 1.25 MHz and 5 MHz .

The Core-A proposes also several RF bandwidths: 1.25 MHz, 5 MHz, 10 MHz and 20 MHz with chip rates of 1.024, 4.096, 8.192, and 16.384 Mchips/s, respectively [25]. However, earlier also chip rates of 0.96 Mchip/s, 3.84 Mchip/s, 7.68 Mchip/s, 15.36 Mchip/s have been considered based on the test bed implementation [29] Bandwidth of 1.25 MHz is employed up to 32 kbit/s, 5 MHz up to 384 kbit/s and higher bandwidths up to 2 Mbit/s. Both Korean CDMA proposals have 5 MHz bandwidth. The chip rates of KMT and ETRI system are 4.608 Mcps and 3.6864 Mchip/s, respectively [52][49].

*2.5.2 Modulation and detection*

The data modulator maps the m incoming coded data bits into one of M ($=2^m$) possible real or complex valued transmitted data symbols, i.e., data modulated. The M-ary data symbol is fed to the spreading circuit where after the resulting signal is filtered and mixed with the carrier signal (i.e., spreading modulation). Typical data modulation schemes are BPSK and QPSK. Spreading circuit can be either binary, balanced quaternary, dual channel quaternary or complex DS-spreading. Typical spreading modulations are BPSK used with binary spreading circuit, QPSK and O-QPSK used with quaternary spreading circuit. Roll-off factor depends on several factors such as power amplifier linearity and adjacent channel attenuation requirements.

FMA2 applies dual channel QPSK spreading circuit [61], i.e., symbol streams on I and Q channels are independent of each other as depicted in Figure 2. The I channel is used for data and the Q channel for control. Core-A, Codit, KMT and ETRI apply balanced quaternary spreading [61].

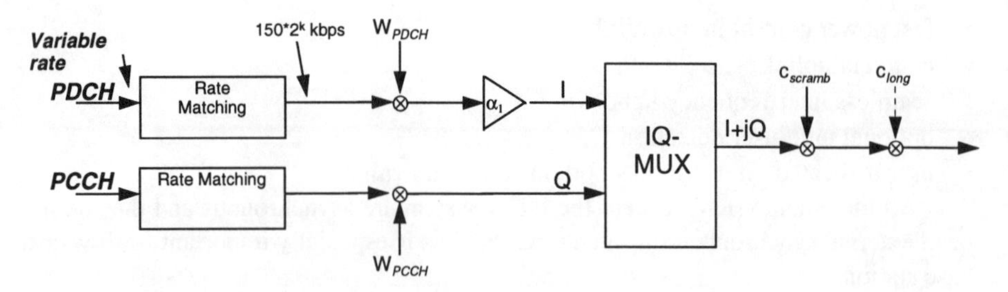


**Figure 2** FMA spreading scheme (PDCH=physical data channel, PCCH=physical control channel, W=Walsh code).

FMA2, CODIT, KMT and ETRI schemes apply QPSK data modulation in both up and downlink. Spreading modulation for them is QPSK in the downlink and O-QPSK in the uplink. The data modulation of the Core-A is QPSK both in up and downlink and the spreading modulation is QPSK in the uplink and BPSK in the downlink [31]. The BPSK spreading in this case means that the same code is used in I and Q channel to save the number of orthogonal codes in the downlink. However, this leads to performance degradation due to crosstalk as a consequence of the phase error in the receiver. The roll-off factors of FMA2 and Core-A are 0.2 and 0.22, respectively.

All schemes apply coherent detection both in up and downlink to enhance the performance. Furthermore, FMA2 and Core-A employ user dedicated pilot symbols in both up and downlinks to obtain the channel estimate for coherent detection. The user dedicated downlink pilot symbol facilitate channel estimation with adaptive antennas. Channel estimate cannot be obtained from a common pilot proposed in [5],[18] and [35] since it experiences a different radio channel as the downlink beam for a specific user. FMA2 pilot symbols can be transmitted at variable rate from 0.5 to 2 kHz depending on the radio channel conditions. Core-A transmits four consecutive symbols every 0.625 ms, i.e., with 1.6 kHz [25]. CODIT performs coherent detection using the demodulated control channel as a pilot [35]. The control channel has differentially coherent detection. Since the CODIT data and control channels are transmitted in parallel, the samples of the 10 ms data frame has to be buffered before despreading and demodulation. In both KMT and ETRI schemes continuous code multiplexed pilot facilitates coherent detection. In the downlink the pilot signal is common to all users and in the uplink it is user dedicated [50][53].

All schemes employ antenna diversity in the uplink. In addition, Core-A employs antenna diversity in mobile station to improve the inferior downlink performance [29]. Other potential methods to improve the downlink performance are investigated in [17].

**Table 2** Features of wideband CDMA proposals (UL=uplink, DL=downlink)

| | FRAMES FMA2 | Core-A | CODIT | ETRI System | KMT System |
|---|---|---|---|---|---|
| **Duplexing method** | FDD | FDD | FDD | FDD | FDD |
| **Channel spacing** | 5 MHz / 10 MHz[1] | 5 MHz/1.25 MHz/20 MHz | 1 MHz/5 MHz / 20 MHz | 5 MHz | 5 MHz |
| **Chip rate** | 3.84 Mcps / 7.68 Mcps[1] | 16.384 Mcps / 4.096 Mcps | | 3.6864 Mchip/s, | 4.608 Mcps |
| **Frame length** | 10 ms | 10 ms | 10 ms | 20 ms | 10 ms |
| **Multirate concept** | UL: Variable spreading<br>DL: Multicode | UL & DL: Variable spreading and/or multicode | UL & DL: Variable spreading | UL & DL: multicode | UL & DL: Variable spreading multicode (over 32 kbit/s) |
| **Spreading** | Spreading factor from 2 up to 256, short, optional long code | Short orthogonal tree structured codes + long code | Long codes | Walsh + Long codes | Walsh + Long codes |
| **Interleaving** | Intra-frame/inter-frame interleaving | Intra-frame interleaving | Intra-frame interleaving | Intra-frame interleaving | Intra-frame interleaving |
| **Data modulation** | QPSK | QPSK | QPSK | QPSK | QPSK |
| **Spreading circuit** | Dual Channel QPSK[2] | Balanced QPSK | Balanced QPSK | Balanced QPSK | Balanced QPSK |
| **Spreading Modulation** | UL: O-QPSK<br>DL: QPSK | UL: O-QPSK<br>DL: QPSK | UL: O-QPSK<br>DL: QPSK | UL: O-QPSK<br>DL: QPSK | UL: O-QPSK<br>DL: QPSK |
| **Pulse shaping** | Root Raised Cosine, roll-off = 0.2 | Root Raised Cosine, roll-off =0.22 | Root Raised Cosine | Root Raised Cosine | Not given |
| **Detection** | UL&DL: Coherent detection (Pilot symbols based) | UL&DL: Coherent detection<br>(Pilot symbols based) | UL: Coherent detection (pilot channel based)<br>DL: Coherent detection (pilot code based) | UL: Coherent detection (pilot channel based)<br>DL: Coherent detection (pilot code based) | UL: Coherent detection (pilot channel based)<br>DL: Coherent detection (pilot code based) |
| **Power control** | UL: Open loop and fast closed loop<br>DL: Fast closed loop | UL: Open loop and fast closed loop<br>DL: Fast closed loop | UL: Open loop and fast closed loop<br>DL: Slow quality based | UL: Open loop and fast closed loop<br>DL: Slow quality based | UL: Open loop and fast closed loop<br>DL: Slow quality based |
| **Diversity** | RAKE, UL antenna diversity | UL & DL: RAKE and antenna diversity | RAKE, UL antenna diversity | RAKE, UL antenna diversity | RAKE, UL antenna diversity |
| **Handover** | Mobile controlled soft handover | Soft handover supported | Soft handover supported | Soft handover supported | Soft handover supported |
| **IF handover** | Supported | Supported | Supported | Not mentioned | Not mentioned |
| **Base station synchronization** | Asynchronous operation | Asynchronous operation | Asynchronous operation | Synchronous operation | Asynchronous operation |
| **Interference reduction** | Short codes supports multi-user detection | Interference canceller (option) | - | - | Synchronized uplink |

[1] Harmonization with FMA Mode 1 and GSM has been taken into account in the choice of these parameters

[2] Studies are ongoing of modifying FMA2 to complex spreading

*2.5.3 Spreading and scrambling codes*

In the downlink FMA2 employs short orthogonal codes to separate physical channels. Base station specific scrambling code is an extended Gold code of length 256. Also in the uplink data and control channel are separated by Walsh codes and the scrambling code is an extended VL-Kasami sequence of length 256. The spreading scheme of FMA2 is illustrated in Figure 3.

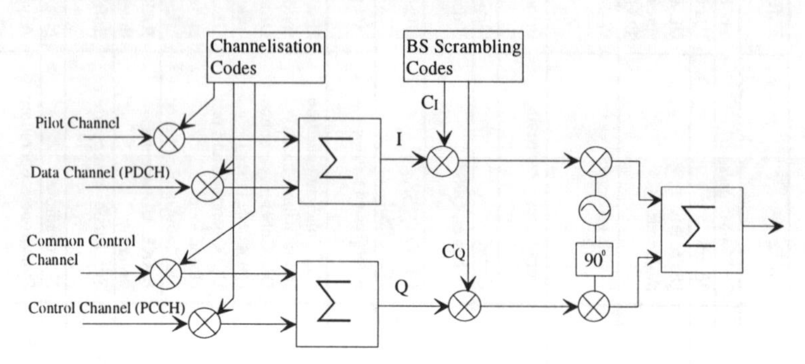


**Figure 3** FMA2 Spreading scheme.

Short codes ease the multi-user detection and allow use of orthogonal signal sets if the data symbol length is multiple of power of two chips. The DS-CDMA air interface designs presented in [5] (MUD-CDMA) and [18] (FMA2) shows that the limitations of short codes pointed out by [35] can be circumvented by a smart system design. Flexibility is not sacrificed because unequal puncturing or repetition coding can be used to obtain any user data rate regardless of the spreading factor. Inter-cell synchronisation is avoided by using a special synchronisation scheme during the soft handover [5]. Code planning is negligible because the number of VL-Kasami sequences, used in the uplink, is more than one million and in the downlink same orthogonal code set can be reused in each cell and sector since it is overlaid by a cell specific scrambling sequence. However, in certain cases, the use of short codes may lead to bad correlation properties, especially with very small spreading factors. In the uplink, multi-user detection can be used to overcome this problem. In case the multi-user detection would not be used, e.g. due to complexity reasons, adaptive code allocation could be used to change the spreading code so that sufficiently good correlation properties are restored. As an alternative, an optional long-code scrambling for FMA2 is considered, where the long code is removed in cells that utilise multi-user detection.

Core-A has a two-layered spreading code allocation [29]. Similar to FMA2, short codes are used for channel separation., i.e., all users are allocated the same set of orthogonal short spreading codes. In addition, cells are distinguished by different long spreading codes and hence a special arrangement is required for cell search [32]. In [25] and [30] tree structure Walsh codes are proposed as short codes. The generation of such codes is described in [33]. Earlier, short orthogonal Gold sequences and Gold sequences length of 2^33-1 were proposed [29].

CODIT has long spreading codes of length $2^{41}$-1 for data channels and Gold codes of length 1023 for control channels. Reasons for selecting long codes were: large number of codes available and flexibility to multiple bit rates, variable spreading factors and requirements for inter-channel and inter-cell synchronization are relaxed [35]. According to [35], the problems of short spreading codes are code planning, restrictions in applicable spreading factors and timing control.

KMT system uses short orthogonal codes of length 128 to separate the physical channels both in the up and downlink. Cells and users in the uplink distinguished by long spreading codes of length 16384 ($2^{14}$). Note: if the uplink is not operated in synchronous mode the long code of length $2^{32}$ is used to separate the users. Two pilots, cluster and cell pilot, are needed to avoid long synchronization time due to long codes. A cluster consists of several cells and under each cluster the long spreading sequence pilots are reused. Each cluster has a cluster pilot which is also a long spreading sequence. There are sixteen cluster pilots and each cluster can have 32 cell sequences. Thus maximum 48 sequence need to be searched. A cluster pilot can be transmitted by the center cell of a cluster only or by each cell. The former technique is suited for hierarchical cell system [51].

Like all the other systems, the ETRI system uses also short orthogonal Walsh codes to separate the physical channels. The length of the code is 256. In the downlink, all other physical channels except the signalling channel for each user are distinguished by Walsh codes and scrambled by a cell/sector specific 20 ms long PN code. Signalling channel for each user uses the same Walsh code as the data channel and it is scrambled by a different PN-sequence[52][53]. In the uplink, data channel, pilot channel and signalling channel are distinguished by three different Walsh codes and scrambled by a common PN-code [53]. Since the system employs the same PN codes with different phase offset, similar to IS-95 standard, careful PN offset planning analogue to frequency planning is needed [65]

### *2.5.4 Multirate*

Multirate design means that how different services with different quality of service requirements are multiplexed together in a flexible and spectrum efficient way. The provision of flexible data rates with different quality of service requirements can be divided into three subproblems: how to map different bit rates into the allocated bandwidth, how to provide desired quality of service, and how to inform the receiver about characteristics of the received signal. The first problem concerns issues like multicode transmission and variable spreading. The second problem concerns coding schemes and the third problem control channel multiplexing and coding.

FMA2 applies multicode transmission in the downlink and variable spreading gain in the uplink [20]. Variable spreading scheme eases the linearity requirements of power amplifier. There is also option to apply multicode transmission in the uplink for highest bit rates. Multicode transmission also facilitates control of quality of service by power allocation instead of coding only. An arbitrary small granularity is achieved by a unequal repetition coding/puncturing scheme mapping the symbol stream after channel coding into the final, predefined symbol rate [11].

Originally, the Core-A multirate scheme was based on multicode transmission for bit rates higher than 128 kbit/s and discontinuous transmission (on-off gating) for bit rates of 128 kbit/s or less [29]. However, this scheme would lead into audible interference due to pulsed transmission (e.g. 8 kbit/s has to be gated on off using 1.6 kHz frequency). Due to these problems and desire to relax power amplifier linearity requirements, variable spreading scheme was introduced as a primary transmission scheme and multicode is employed only for the most highest data rates [31].

CODIT uses variable spreading both in the up and downlink. Long codes facilitate arbitrary spreading ratios. For the KMT scheme multicode technique is used for bit rates higher than 32 kbit/s. Bit rates less than 32 kbit/s are mapped into the symbol rate of 72 ksps by repetition coding. The ETRI scheme employs variable spreading up to 128 kbit/s and multicode for higher bit rates [52].

In all schemes the different services of a same user are time multiplexed into the frame. Furthermore, Core-A and KMT time multiplex also the control information. FMA2, CODIT and ETRI transmit the control information on a parallel code channel. In addition, FMA2 and CODIT have possibility to control the data rate on frame-by-frame basis. The FMA2 scheme is the most flexible and, as already said, facilitates arbitrary small granularity. The other scheme adjust the bit rate in upper layers. The FMA2 data and control channel transmission scheme is described in Figure 4.

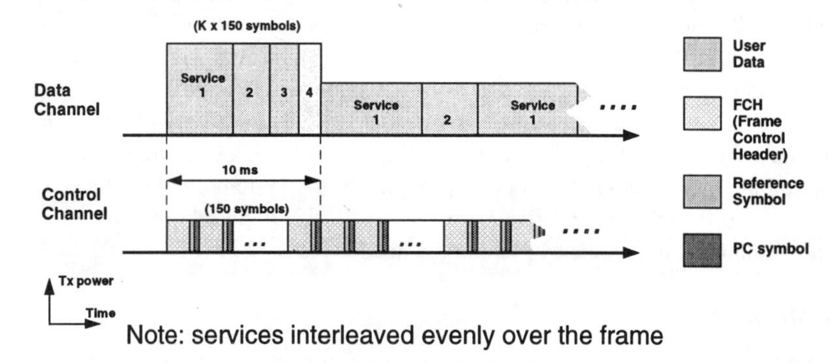


**Figure 4** FMA multirate scheme.

Since FMA2 transmits the FCH header within the same frame as the data, the receiver needs to buffer the traffic channel before the control channel is decoded. Since the spreading ratios are multiples of each other it is possible to despread with lowest spreading ratio and to buffer the subsymbols after despreading instead of the samples before the despreading as in CODIT which has arbitrary spreading ratios.

The frame length of FMA2, CODIT and Core-A is 10 ms, while the frame length of ETRI and KMT systems is 12 ms. The frame length is also the basic interleaving period. Furthermore, FMA2 has also possibility for inter-frame interleaving. All schemes employ convolutional code for services with BER $10^{-3}$ and concatenated convolutional code and RS code for services with BER $10^{-6}$. The detailed parameters of the codes are given in Table 3.

**Table 3** Encoding parameters.

| | Downlink | | Uplink | |
|---|---|---|---|---|
| | CC code rate, constraint length | RS code rate | CC code rate, constraint length | RS code rate |
| **FMA2** | 1/2, K=9, (punctured or repeated) | k/n, varies | 1/3 or 1/2, K=9, (punctured or repeated) | k/n, varies |
| **Core-A** | 1/3, K=7 | k/n, varies | 1/3, K=7 | k/n, varies |
| **CODIT** | 1/3 (punctured) for speech, K=5<br>1/2 for data, K=7 | k/n, varies | k/n, varies | k/n varies |
| **ETRI** | 1/3, K=3 | Not available | 1/3, K=3 | Not available |
| **KMT** | 1/2, K=7 | 3/4 | 1/2,K=7 | 3/4 |

*2.5.5 Power Control*

All schemes use open loop and fast closed loop power control in the uplink. In addition, FMA2 and Core-A apply fast closed loop power control also in the downlink. The power target is decided based on the required quality which is impacted by the channel quality and interference conditions. FMA2 transmits the power control information at a variable rate from 0.5 kbit/s up to 2 kbit/s. The variable rate power control avoids unnecessary waste of capacity for power control signalling, especially for low bit rate services [44]. Variable rate transmission was also tested for the KMT system [48]. Power control parameters are given in Table 4.

**Table 4** Power control parameters

| | Uplink | | |
|---|---|---|---|
| | command rate kHz | dynamic range | step size |
| FMA2 | 0.5-2 | 80 dB | 1.0 dB |
| Core-A | 1.6 kHz | 80 dB | 1.0 dB |
| CODIT | 2 kHz | 80 dB | 1.0 dB |
| ETRI | 1.6 kHz | 80 dB | 0.5 dB |
| KMT | 2 - 6 ms | Not given | 0.5 - 2 dB |
| | **Downlink** | | |
| | command rate | dynamic range | step size |
| FMA2 | 0.5-2 | 20 | 1.0 |
| Core-A | 1.6 kHz | 30 | 1.0 |
| CODIT | slow quality based | | |
| ETRI | No details given, open loop slow PC | | |
| KMT | No details given | | |

Typically, The SIR target for the power control is based on the frame error rate calculated from the Viterbi decoder output. The measured frame error rate is compared to the target FER (frame error ratio) and a SIR target is determined based on this. The target SIR is compared to the measured SIR, which e.g. Core-A measures from the time multiplexed pilot.

The impact of the fast power control in the downlink is twofold. On the other hand, it improves the performance against fading multipath channel. On the other hand, it increases the multiuser interference variance within the cell since orthogonality between users is not perfect due to multipath channel [15]. The net effect, however, is increased performance.

### *2.5.6 Handover*

The all proposals employ soft handover. They all except ETRI also assume unsynchronized network which has to be taken into account in the handover synchronization scheme. FMA2 employs short codes impacting the soft handover planning. However, using synchronization channel and adjustment of transmission time from a new base station the seamless soft handover can be carried out [5]. CODIT employs long codes and thus the mobile station can measure the timing uncertainty between the base stations involved in the soft handover. Core-A uses also long codes.

### *2.5.7 Interfrequency handover*

For interfrequency handover there are two implementation alternatives: compressed mode/slotted mode and dual receiver. The dual receiver has expensive receiver but simple system operation. However, if there is already a diversity receiver (two receiver chains not selection diversity) then it is possible to use the other receiver for measurements on other carriers [21]. Of course, diversity gain is lost during the measurement period. Compressed mode, on the other hand, has simple receiver but more complex system operation and the receiver algorithms has to operate on burst nature during slotted mode, i.e., poorer performance and more control.

For the compressed mode there exists number of alternatives how to implement it. In [45] four different methods are presented: variable spreading factor, code rate increase, multicode and higher order modulation. Variable rate spreading and coding rate increment (puncturing) cause 1.5 - 2.5 dB loss in Eb/No performance and higher order modulation even higher loss(5 dB). This is due to breaks in power control and less coding. A further drawback of variable spreading ratio is that simple terminals have to be able to operate with different spreading ratios (e.g. speech terminal normally operates with spreading ratio of e.g. 256 now it should operate also with spreading ratio of 128).

For FMA2 dual receiver was proposed in [20]. However, currently also compressed mode is under study [21]. Both Core-A and CODIT use compressed mode. Core-A employs 16-QAM during the compressed mode operation [30] and CODIT variable spreading ratio. The KMT and ETRI proposals do not consider interfrequency handover.

### *2.5.8 Multiuser detection and other interference reduction methods*

Even though multiuser detection is only a receiver technique, it impacts the overall system design. E.g. short codes ease the implementation of multiuser detection. Furthermore, multiuser detection will change the relation between system load and coverage. Since multiuser detection reduces the intra cell interference, cell size does not shrink as fast as without multiuser detection when the system load increases [16]. Due to complexity reasons, MUD cannot be used in similar way in the downlink than in the uplink. In addition, terminal must only receive its' own signal in contrast to the base station. Therefore, a simpler interference suppression scheme could be applied in terminal.

FMA2 employs a optional interference cancellation scheme in the uplink. In simulations multistage scheme has been used. Performance results for this scheme have been presented in [9][10][12][14]. For Core-A a pilot symbol assisted coherent multistage interference cancellation has been proposed [30]. CODIT does not employ multiuser detection. Downlink interference cancellation has been studied for MUD-CDMA, scheme which was the basis for FMA2 [15]. The downlink multiuser detection improves performance and allows more accurate power control since it reduces the intra cell interference [15]. Even though orthogonal codes are used in the downlink, there is gain from the multiuser detection since the multipath channel partially destroys the orthogonality [15].

To reduce the intra cell interference, the KMT scheme time synchronizes all users in the uplink with accuracy of 1/8 chip [48]. This is done by measuring the timing in base station and signaling timing adjustment commands with a rate of 2 kbit/s to the mobile station. However, multipath still results to intra cell interference and gain from the orthogonal uplink depends on the channel profile. In addition, the signalling traffic reduces the downlink capacity for each user by 2 kbit/s.

### *2.5.9Packet data*

Packet access procedure in CDMA should minimize the interference to other users. If no connection is established the initial access is not power controlled thus we need to minimize time and information transmitted during this period. FMA2 packet access scheme is similar to Core-A except that it does not allow information transmission in the initial access burst. When there is nothing to send, FMA2 either cuts off the transmission or keeps the physical connection transmitting power control and reference symbols only. In the former case a virtual connection (authentication and other information) is retained in order to establish the link fast again in case of new transmission. Selection between these two alternatives is a tradeoff between the resource spending for synchronization and power control information and resources spend for random access.

Core-A employs adaptive mode selective packet data transmission system where the initial access burst contains information on the data length and data volume [31]. Data can either be transmitted on a common channel or dedicated channel. Closed loop power control is not possible for the common channel so only small volumes of information should be transmitted on that channel On the other hand, dedicated channels are inefficient since resources are tied up even when there is no data to be transmitted. Most appropriate mode is selected according the traffic type. In case there is nothing to send on a dedicated channel, closed-loop power control is still applied [31].

## 2.6 TDMA BASED SCHEMES

Table 2 presents the main characteristics of three TDMA based air interfaces proposed for UMTS/FPLMTS: FMA1 without spreading (wideband TDMA)[22],

ATDMA [39] and MTDMA [25]. The schemes have been organized into columns according to radio environments since typically the burst structure of TDMA schemes depends on the radio environment. Furthermore, Table 3 presents a modified ATDMA scheme, where the carrier bandwidth and frame structure have been aligned with GSM [40]. In the following, we focus into comparison of the schemes with parameters presented in Table 5.

### *2.6.1 Modulation*

Both ATDMA and FMA1 use the same linear modulation methods: B-O-QAM and Q-O-QAM. The modulation method of the MTDMA is QPSK and 16-QAM in pico cell environments. For FMA1 coded modulation methods are under consideration for quaternary modulation. Linear modulation was selected since it is found to be bandwidth efficient, i.e. it provides a high bit rate per bandwidth unit. The Q-O-QAM modulation is a flexible way of extending the chosen B-O-QAM modulation to higher bit rates without increasing the bandwidth requirement. Offset modulation is selected, since it reduces amplitude variation, which reduces the back-off requirement at the transmitter amplifier, giving more efficient transmission. In ATDMA the GSM modulation scheme GMSK is used in long macro cells to extend the maximum range of macro cells. It uses a lower carrier bit rate (360 kbit/s) to maintain the same channel spacing as in the short macro cells [38].

The bandwidth efficiency for FMA1 is with B-O-QAM modulation 1.625 bits/s/Hz and for ATDMA almost the same with few percentages variation. Q-O-QAM has double the bandwidth efficiency i.e. 3.25 bits/s/Hz. The bandwidth efficiency of MTDMA is 1.92 bits/s/Hz which is considerably higher than for the two other schemes. The receiver filter for FMA1 and MTDMA is square root raised cosine filter with roll-off factor of 0.35.

### *2.6.2 Carrier spacing and symbol rate*

In ATDMA the carrier spacing and symbol rate are adapted according to different cell types. The largest channel spacing is approximately 1107 MHz. This results into maximum user bit rate of 1.16 Mbit/s using same coding rate as in FMA1. Thus a user bit rate of 2 Mbit/s would need multicarrier transmission with 2 carriers. The carrier bit rate of 1800 kbit/s was motivated by the requirement of 100 m range for pico cell environment. With the used propagation model it was considered not to be possible to achieve a higher carrier bit rate within the given range requirements. However, in the modification of ATDMA for better compatibility with GSM, presented in [40], the carrier spacing was defined to be 1.6 MHz in pico cells providing 2 Mbit/s user bit rate. In micro cells the carrier spacing and user bit rate of ATDMA were halved into 800 kHz and 1 Mbit/s, respectively.

FMA1 without spreading and MTDMA have both carrier spacing of 1.6 MHz and both can provide 2 Mbit/s user data peak rate. For them the driving force has been the provision of 2 Mbit/s transmission rate in all environments using a single RF bandwidth. The reasoning for the FMA1 carrier spacing is as follows. For circuit switched users we dedicate the entire frame to one user to achieve at minimum a raw bit rate of 3.38 Mbit/s. To provide 2 Mbit/s user data rate, it is possible to use e.g. a punctured rate 1/2 convolutional coding. For packet switched users 10 % of the raw data bits have to be reserved for the control information of the packet protocol.

**Table 5** Comparison of TDMA based schemes

| | FMA1 - wideband TDMA | | ATDMA | | | MTDMA | | |
|---|---|---|---|---|---|---|---|---|
| | FMA1 data burst/regular burst | FMA1 micro cell subburst | ATDMA macro | ATDMA micro | ATDMA pico | MTDMA macro | MTDMA Office | MTDMA Pedestrian |
| Carrier Spacing | 1.6 MHz | 1.6 MHz | 270.92307 kHz | 1107.692308 kHz | 1107.692308 kHz | 400 kHz | 1.6 MHz | 1.6 MHz |
| Carrier Bit Rate | 2.6 Mbit/s | 2.6 Mbit/s | 450 kbit/s | 1.8 Mbit/s | 1.8 Mbit/s | 384/768 kbit/s | 3.072 Mbit/s | 3.072 Mbit/s |
| Modulation | B-O-QAM Q-O-QAM | B-OQAM Q-O-QAM | B-OQAM Q-O-QAM | B-OQAM Q-O-QAM | B-OQAM Q-O-QAM | QPSK 16-QAM | QPSK 16-QAM | QPSK 16-QAM |
| Bandwidth efficiency | 1.625 bit/s/Hz | 1.625 bit/s/Hz | 1.66 bit/s/Hz | 1.62 bit/s/Hz | 1.62 bit/s/Hz | 0.96 bit/s/Hz | 1.92 bit/s/Hz | 1.92 bit/s/Hz |
| Maximum Gross Bit Rate | 2.0 Mbit/s/ 1.69 Mbit/s | 2.37 Mbit/s | 273 kbit/s | 1.1 Mbit/s | 1.382 Mbit/s | 256/512 kbit/s | 2.48 Mbit/s | 2.48 Mbit/s |
| Frame Length | 4.165 ms | 4.165 ms | 5 ms | 5 ms | 5 ms | 10 ms | 10 ms | 10 ms |
| Number of Slots/Frame | 16/64 | 64 | 18 | 72 | 72 | 8 | 64 | 64 |
| Payload (bits) | 684/122 | 144 | 66 (+10 or 6 for in-band signalling) | 66 (+10 or 6 for in-band signalling) | 86 (+10 for in-band signalling) | 320 +40 (control) | 320 +40(control) | 320 +40 (control) |
| Training Sequence (symbols) | 49 (18.8 µs) | 27 (10.3µs) | 29 (64.4µs) / 33(73.3 µs) | 29 (16µs)/ 33 (18µs) | 15 (8.3µs) | 8+48 | 8+48 | 8+48 |
| Tail Bits | 3 | 3 | 8 | 8 | 6 | 8 | 8 | 8 |
| Guard Time (symbols/us) | 11 (4.2 us)/ 10.5 (4us) | 10.5 (4µs) | 12 (26.7 µs) | 12 (6.7 µs) | 9 (5 µs) | 16 (83.3 µs) | 16(10.42 µs) | 16 (10.42 us) |
| Max. Excess Delay | 7 µs | 2.7 µs | 20 µs | 5 µs | 1 µs | 10,4µs | 1.3 µs | 1.3 µs |
| Overhead | 9% / 35 % | 23 % | 39 % | 39 % | 23%/12% (double burst) | 25 % | 25 % | 25 % |

This reduces the available raw data rate. In packet mode, some raw bits have to be reserved also for the error detection (e.g. CRC). Then, for the error correction, e.g. rate 2/3 punctured convolutional code can be used.

Relation of channel spacing and carrier bit rate depends also on adjacent channel suppression requirements. GSM requires adjacent channel suppression of 18 dB, which means that commonly adjacent carriers cannot be used within the same cell and normally also not in the neighboring cells. For UMTS a higher value is desired. The power amplifier linearity aspects also influence the adjacent channel interference since power spillover can occur when power amplifiers are not sufficiently linear.

**Table 6** ATDMA scheme compatible with GSM

| | Long Macro | Short Macro | Micro | Pico |
|---|---|---|---|---|
| Carrier Spacing | 200 kHz | 200 kHz | 800 kHz | 1600 kHz |
| Carrier Bit Rate | 271 kbit/s | 325 kbit/s | 1.3 Mbit/s | 2.6 Mbit/s |
| Frame Length | 4.165 ms | 4.165 ms | 4.165 ms | 4.165 ms |
| Number of Slots/Frame | 8 | 10 | 40 | 80 |
| Payload | 114+2 | 114+2 | 114+2 | 114+2 |
| Training Sequence | 26 | 19 | 19 | 19 |
| Tail Bits | 6 | 6 | 6 | 6 |
| Guard Time (symbols/us) | 8.25(4us) | 9(26.7 μs) | 9 (6.7 μs) | 9 (5 μs) |
| Max. Excess Delay | 2.7 μs | 20 μs | 5 μs | 1 μs |

### *2.6.3 Burst and frame structures*

Burst and frame structure are designed based on

- range and characteristics of services to be supported
- constraints based on multipath propagation and propagation environment
- constraints based on terminal hardware complexity

The frame and slot structure of FMA1 without spreading was optimized for micro and pico cell coverage for high bit rates up to 2 Mbit/s. As shown in [24], range of approximately one kilometer was achieved for WB-TDMA even though the carrier bit rate was 6.5 Mbit/s (carrier spacing of 2 MHz) compared to 5.2 Mbit/s (1.6 MHz) in FMA1. FMA1 without spreading has two different slot sizes 1/16 slot and 1/64 slot. The larger slot reduces overhead for medium and high bit rate services. Frame length was selected to be same as GSM since this supports interleaving depth at least of four for services with a tight delay requirement. The frame and slot structure of FMA1 and ATDMA facilitate both frame-by-frame and slot-by-slot hopping. The latter is used for high bit rate services occupying several slots within a frame. Both FMA1 and ATDMA use rate compatible punctured codes to achieve different coding rates in a flexible way. For FMA1 also Turbo codes are under study [66][67]

In ATDMA, the frame length and the slot length design was based on transmission of speech service with gross bit rate of 13 kbit/s in macro cell environment. A minimum interleaving depth of four was expected to be required to achieve good performance. This led to specification of payload for burst of 66 bits with 10 bits for inburst signalling and frame length of 5 ms. Furthermore, when identifying the macro cell carrier bandwidth, requirement of maximum user bit rate of 64 kbit/s was used in definition of number of slots required within frame. The quality of service requirement was thought to require rate 1/4 coding and thus this led into 450 kbit/s and 18 slots/frame, taking into account one extra slot for measurement of other carriers. Then micro cell carrier bitrate was taken to be fourfold compared to macro cell since delay spread is four times less. And this resulted into 72 slots/frame. Pico cell carrier bit rate was taken to be the same as in micro cell due to expected peak power limitations to achieve the specified range of 100 m. With the carrier bit rate of 1800 kbit/s, maximum user bit rate is 1.16 Mbit/s with Q-O-QAM taking into account the 40 % overhead.

MTDMA is clearly targeted for data transmission. The frame length and interleaving depth is 10 ms which is too long for a for low rate, low delay speech transmission.

### *2.6.4 Training sequence design*

As depicted in Figure 5, a typical training sequence is composed of L precursor symbols followed by P reference symbol. The training sequence is placed into the middle of the burst to get better estimate of the channel over the whole burst. Length of the training sequence depends on the carrier symbol rate and the desired length of delay spread that needs to be estimated. The number of precursor symbols has to be at least as long as the length of the channel impulse response in symbols. After channel delay spread exceeds the specified limit performance starts to degrade gradually and hence the length depends also about the design paradigm: average or worst case conditions. The number of required reference symbols in the training sequence depends on length of the channel's impulse response, the required signal to noise ratio, the expected maximum Doppler frequency shift and the number of modulation levels. The number of reference symbols should not be too large so that the channel characteristics remain practically stable within the correlation window[38]. The training sequence has to exhibit good correlation properties. Typically, training sequences have been optimized for good autocorrelation properties. However, FMA1 utilizes also interference cancellation and thus good cross correlation properties are desirable for good performance [4]. For the FMA1 nonspread data bursts the longer sequence can handle about 7 μs of time dispersion and the shorter one 2.7 μs. It should be noted that if the time dispersion is larger, the drop in performance is slow and depends on the power delay profile.

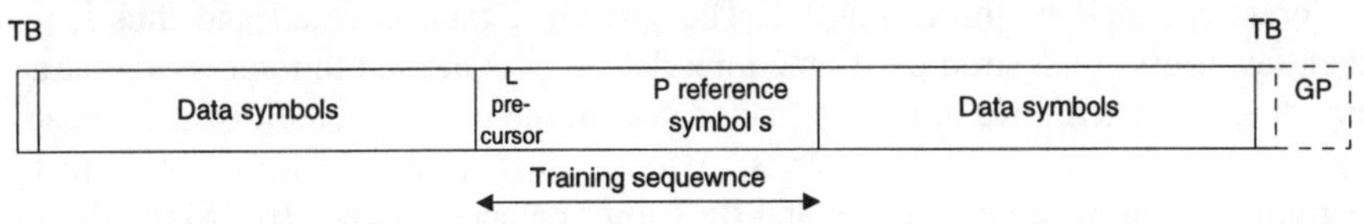


**Figure 5** Training sequence (TB=tail bits, GP=guard period)

MTDMA training sequence is in the beginning of the burst. This facilitates direct processing of data without need to buffer it as in the case of midample. On the other hand, performance might degrade for the last bits in the burst. In MTDMA there is 8 precursor symbols which facilitate delay spread estimation up to 2.6 μs. However, the guard time limits the performance to 1.3 μs.

The guard bits provide a protection interval between the bursts for time alignment uncertainty, time dispersion and power ramping and the tail bits can be used to assist the detection process. Faster ramping leaks more power into the adjacent channels and hence the power ramp-up at the beginning of the bursts and power ramp down at the end of the burst must be smooth enough to avoid spreading of the output spectrum of the transmitter. According calculations for GSM in [14] about three bits ramp time would be sufficient. Furthermore, one symbol is needed for the timing alignment uncertainty. Thus for FMA1 micro cell subburst we have 5.5 symbols left for covering the length of the impulse response which means 2.1 μs excess delay before the soft degradation in performance starts. For data burst/regular burst this is 2.3 μs. For ATDMA micro cell we have 6.7 μs guard time and 3 bits have been reserved for power ramping and 1 bit for timing alignment uncertainty this leaves 4.5 μs for excess delay spread. MTDMA ramp time 4 bits which leaves 4 bits for channel impulse response i.e. 1.3 μs. ATDMA can combine two burst into double bursts to reduce the overhead for higher bit rates. However, a slightly longer training sequence needed to guarantee receiver performance at the burst ends. Contrary to ATDMA, FMA1 employs the guard times between bursts for user data to avoid use of two training sequence types.

*2.6.5 Radio resource management*

ATDMA transport chain has been designed around so called transport modes, which is a characterized by a certain configuration of a set consisting of modulation, error control code and amount of radio resources to guarantee a given performance for a given level of signal-to-noise + interference. Link adaptation controls this configuration [38].

An improved packet access, PRMA++, have been developed for ATDMA [39]. For FMA1, even more advanced scheme is under development which can multiplex both real time and non-real time services in a flexible way.

Both ATDMA and FMA1 use dynamic channel allocation for channel assignments.

## 2.7 HYBRID CDMA/TDMA

FRAMES FMA1 with spreading is based on hybrid CDMA/TDMA concept, also called joint detection [62][63][64]. The role of CDMA there is to multiplex the different channels within a timeslot. The spreading ratio is small and thus if more than few users are desired per frame joint detection is needed to remove the intracell interference. Therefore, in a fully loaded system there is no benefit of the spreading against interference from other cells. Also the slow power control results to high variation in the received signals and thus joint detection is needed. Since the joint detection is a mandatory feature, it is more critical compared to wideband CDMA. However, since the number of users is small the complexity of joint detection might not be not excessive.

In the following we explain the main differences between the FMA1 without spreading, covered in section 6, and FMA1 with spreading. Original parameters of joint detection CDMA, frame length of 6 ms and 12 slots per frame[62], have been aligned with FMA1 without spreading to be backward compatible to GSM, i.e.,

4.615 ms and 8 slots per frame, respectively. The main difference to FMA1 without spreading is, of course, spreading. Short orthogonal spreading codes of length 16 are used. Spreading modulation method is GMSK and data modulation QPSK. The training sequence length of Mode 1 is adapted to the expected channel impulse response length. There are bursts with two different training sequence lengths. The longer training sequence is suited for estimating the 8 different uplink channel impulse responses of 8 users within the same time slot with a time dispersion of up to about 15 μs. If the number of users is reduced, the tolerable time dispersion is increased about proportionally. The shorter training sequence is suited for estimating the 8 uplink channel impulse responses with a time dispersion of up to about 5.5 μs. Furthermore, it is suited for estimating the downlink channel impulse response with a time dispersion of up to about 25 μs, independent of the number of active users and the uplink channel impulse response with this same time dispersion in case all bursts within a slot are allocated to one and the same user.

## 2.8 OFDM BASED SCHEMES

Introduction of OFDM into cellular world has been driven by two main benefits:

- flexibility: each transceiver has access to all subcarriers within a cell layer
- easy equalization: OFDM symbols are longer than the maximum delay spread resulting into flat fading channel which can be easily equalized.

Also the introduction of Digital Audio Broadcasting (DAB) based on OFDM and research of OFDM for HIPERLAN type II and Wireless ATM have increased the interested towards OFDM[47].

Main drawback of OFDM is the high peak to average power. This is especially severe for the mobile station and for long range applications. Different encoding techniques have been investigated to overcome this problem. Furthermore, the possibility to access to all resources within the system bandwidth results into an equally complex receiver for all services, regardless of the bit rate. Of course, a partial FFT for only one OFDM block is possible for low bit rate services, but this would require a RF synthesizer for frequency hopping.

For UMTS and FPLMTS/IMT-2000 two OFDM air interface concepts have been presented: BDMA and OFDM by Telia (Table 7). BDMA concept has actually been proposed both in ETSI and in ARIB FPLMTS Study Committee. More information from Telia's OFDM concept can be found from [56][58][59][57] Main difference between the Telia concept and BDMA is the detection method which impacts the overall design of the systems. Telia OFDM uses coherent detection and BDMA differentially coherent detection. Telia OFDM employs coherent detection due to two reasons: performance gain and opportunity to use arbitrary signal constellations such as 16-QAM [60].

Table 7 **Main features of OFDM proposals.**

| | Telia OFDM | BDMA |
|---|---|---|
| Bandwidth | 5 MHz | 5 MHz |
| number of subcarriers | 1024 | 800 |
| OFDM subcarrier bandwidth | 5 kHz | 6.25 kHz |
| OFDM symbol length | 200 μs | 200 μs |
| OFDM block size | 25 carriers × 3 symbols | 24 carriers × 1 symbol |
| Frame length | 15 ms | 5 ms, divided into 4 subframes |
| Detection | coherent | differentially coherent |

*2.8.1 Bandwidth*

Both Telia OFDM and BDMA have 5 MHz bandwidth. The reasoning for the bandwidth choice is that totally two times 60 MHz FDD (Frequency Division Duplex) spectra has been allocated for FPLMTS/IMT-2000. So, assuming four operators and three cell layers results into 5 MHz system bandwidth [60]. Each transceiver will have an access to the whole 5 MHz bandwidth which is divided into subcarriers. This facilitates very high bit rates and flexible allocation of resources. Telia OFDM has within the 5 MHz bandwidth 1024 subcarrier each 5 kHz. BDMA has 800 subcarriers with bandwidth of 6.25 kHz [25].

*2.8.2 Frame design*

The Telia OFDM frame length is 15 ms and the BDMA has frame length of 5 ms. However, the total interleaving period for BDMA is 20 ms. BDMA frame consists of 4 subframes and each of them is further divided into six timeslots. This frame design facilitates the hopping scheme for BDMA described below.

The OFDM block design principles are different for the schemes since Telia OFDM applies coherent detection and BDMA differential detection. The block size for Telia OFDM is 25 OFDMA carrier and 3 OFDM symbols. The time and frequency correlation function impacts the block shape. The shape should be selected to reflect the peak of the time-frequency correlation function. However, since there is large variation in the radio channels, the shape should be designed for worst case channels. For BDMA the slot size is 24 OFDMA carriers and 1 OFDM symbol slot.. In Telia OFDM pilot symbols occupy 18.7 % of a block. The downlink of Telia OFDM has smaller overhead than the uplink since common pilot can be utilized. However, this is not true for cheap speech terminals if only partial FFT is carried out to demodulate only the own block of a user. In BDMA, one subcarrier is allocated to be used as reference for differential detection. The Telia scheme results into finer granularity, 3.9 kbit/s against 11.2 kbit/s, but seems to offer less diversity for low bit rates compared to BDMA.

For the purpose of coherent detection the OFDM block has to be large enough to allow efficient use of channel correlation in estimation but the number of pilots should be minimal. In the uplink, blocks belonging to the same user should be kept together to utilize the same timing and to better maintain orthogonality. But, on the other hand, information should be transmitted on different places in the time-frequency space to utilize diversity. The placement of the blocks is a trade-off between these two contradicting requirements.

*2.8.3 Coding , interleaving and frequency hopping*

Frequency hopping together with coding, interleaving improves the performance through increased diversity. Telia OFDM hops on frame-by-frame basis, i.e., every 15 ms (67 hops/s) being effective against long fading dips. BDMA hops on burst-by-basis (800 hops/s) and thus better diversity is achieved.

Telia OFDM interleaves the symbols over all transmitted block within one frame. BDMA interleaves over 20 ms i.e. 4 TDMA frames and thus also in this respect better diversity is achieved. Of course, the delay is also longer. The 20 ms interleaving delay might be too long with respect to the 30 ms minimum delay requirements of UMTS bearer services.

In case BDMA receiver is based on partial FFT and hopping synthesizers, the synthesizer settling times are very tight and are even tighter for multislot services. Therefore, full utilization of all six slots for one user might be difficult.

*2.8.4 Handover*

Handover principles are the same for both schemes. In case the base stations are synchronized handover within the same cell layer does not require any extra measures but measurements are performed during idle slots. On the other hand, if the basis stations are not synchronized the mobile station needs to release the existing link and switch to a new frequency band and to perform measurements. Another possibility is to have a dual receiver [60].

*2.8.5 Uplink synchronization*

The transmitted signals from all mobiles must reach the base station with a certain synchrony in order to maintain their orthogonality. Both time and frequency offsets need to be estimated. Number of studies have been carried out to solve this critical aspect for OFDM [58].

*2.8.6 Power control*

For Telia OFDM the power control scheme is still open. In BDMA a closed loop power control based on quality measurements is applied [25]. The power control period is 1.2 ms and step size 1 dB. Also open loop power control is used. In case the frequency hops are correlated, which is the case for slow moving mobiles, the power control can compensate fading and thus improves performance. For higher mobile speeds it will minimize the transmission power but cannot follow the fast fading. However, the frequency hopping mitigates the impact of fast fading.

*2.8.7 Dynamic channel allocation (DCA)*

The Telia OFDM employs a distributed CIR measurement based DCA scheme [57]. Channel rearrangement per frame basis is used to improve the performance.

The relation between frequency hoping (random vs. cyclic hopping) and DCA needs still further investigations.

## 2.9 TIME DIVISION DUPLEX (TDD)

Main discussion about UMTS air interface has been around technologies for frequency division duplex (FDD). However, there are several reasons why also time division duplex (TDD) should be used. First of all, there will be most likely a dedicated frequency bands for TDD within the identified UMTS frequency bands. Furthermore, FDD requires exclusive paired bands and spectrum for such systems is therefore hard to find. On the other hand, TDD can make use of individual bands which do not need to be mirrored for the return path, and hence spectrum is more easily identified. With a proper design including powerful FEC TDD can be used even in outdoor cells. It has been argued that the TDD guard interval would result to excessive overhead in large cells. However, in a cell with a range of three kilometers we would need a 20 us guard time to prevent Tx and Rx time slots to overlap, i.e., an overhead of ca. 4 % assuming frame length of 4.615 ms. When the propagation delay exceeds the guard period, soft degradation of performance occurs. Thus, UMTS TDD mode need not to be restricted into unlicensed indoor solutions but perhaps even most short-range UMTS should be TDD, even that used by the traditional cellular operators, e. g. for high-capacity microcells. Second reason for using TDD is the flexibility in radio resource allocation, i.e., bandwidth can be allocated by changing the number of time slots for up and downlink.

The asymmetric allocation of radio resources leads into two interference scenario that will impact the overall spectrum efficiency of TDD scheme:

- asymmetric usage of TDD slots will impact the radio resource in neighboring cells
- asymmetric usage of TDD slots will lead into blocking of slots in adjacent carriers within own cell

Figure 6 depicts the first scenario. MS2 is transmitting at full power at the cell border. Since the MS1 has different asymmetric slot allocation than MS2 its' downlink slots received at the sensitivity limit are interfered by MS1 causing blocking. On the other hand, since the BS1 can have much higher EIRP (effective isotropically radiated power) power than MS2, it will interfere BS2 receiving MS2. Hence the radio resource algorithm needs to avoid this kind of situation.

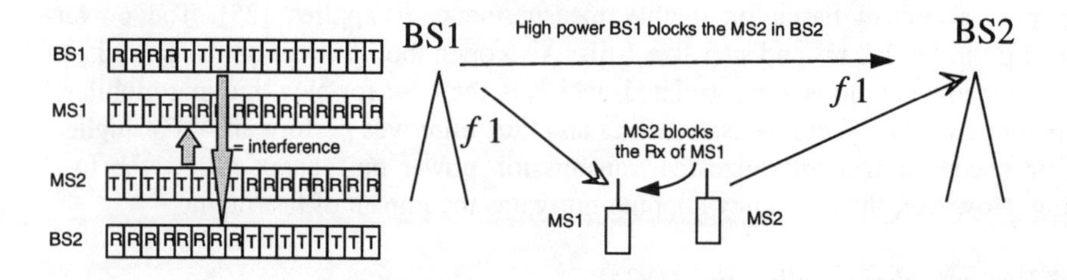


**Figure 6** TDD interference scenario.

In the second scenario, two mobiles would be connected into same cell but using different frequencies. The base station is receiving MS1 on the frequency f1 using the same time slot it uses on the frequency f2 to transmit into MS2. As shown in the Table 8, the transmission will block the reception due to irreducible noise floor of the transmitter regardless of the frequency separation between f1 and f2.

**Table 8** Adjacent Channel Interference Calculation.

| | |
|---|---|
| BTS transmission power for MS2 in downlink 1W | 30 dBm |
| Received power for MS1 | - 100 dBm |
| Adjacent channel attenuation due to irreducible noise floor | 50 -70 dB |
| Signal to adjacent channel interference ratio | -60 - -80 dB |

Third scenario, where the above described blocking effect exists, is an FDD system where if at any moment traffic in a cell is unbalanced between the up-link and down-link then the spare capacity in the low-traffic direction might momentarily be used for two-way operation, i.e., TDD.

DECT is a second generation TDD system. with carrier spacing of 1728 kHz and carrier bit rate of 1152 kbit/s. DECT frame length is 10 ms and each frame is divided into 24 slots. DECT will provide bit rates up 512 kbit/s half duplex and 256 kbit/s full duplex. The fundamental difference between DECT and UMTS are the bit rate capabilities, operating point and channel coding. Since UMTS TDD mode has powerful channel coding, the required C/I and hence also reuse factor is smaller compared to DECT. Therefore, assumption of similar up and downlink interference situation does not hold anymore and the DECT Dynamic Channel Selection (DCS) is not suitable. Also the performance of DECT in high delay spread environments is not very good which limits the outdoor cell range.

Both TDMA and CDMA based scheme have been proposed for TDD. Most of the TDD aspects are common to TDMA and CDMA based air interfaces. However, in CDMA based TDD systems we need to change symmetry of all codes within one slot in order to prevent interference situation where high power transmitter would block another receiver. Thus, TDMA based solutions have higher flexibility. In FRAMES Multiple Access only FMA1 has a TDD option which can be used both with wideband TDMA without spreading and with spreading [20]. However, with spreading option above mentioned drawback of code allocation exists. CDMA/TDD has been proposed in [25].

## 2.10 CONCLUSIONS

We have reviewed the third generation standardization related radio access research activities. Furthermore, a technical overview of the air interface designs for UMTS and FPLMTS/IMT-2000 covering wideband CDMA. TDMA, OFDM and hybrid schemes was presented.

Asynchronous wideband CDMA has recently received lot of attention in the research community. The most widely considered bandwidth seems to be 5 MHz providing bit rates up to 384 kbit/s. Given that Japanese and European approaches to the wideband CDMA are very similar, it is possible that a common air interface standard can be achieved. The new wideband CDMA air interface can be integrated to the GRAN concept. This provides future proof evolution path from the GSM core network into UMTS.

For bit rates above the 384 kbit/s several alternatives are considered: wideband CDMA, wideband TDMA, hybrid CDMA/TDMA and OFDM. In practice wideband CDMA needs very large bandwidth to support 2 Mbit/s transmission. The critical question for OFDM and hybrid approaches is the maturity of the technology. Therefore, it seems that wideband TDMA would offer well proven, high performance technology for the high bit rate services up to 2 Mbit/s, especially for TDD mode, as a complement to 5 MHz wideband CDMA.

## References


[1] J.Jiménez et, al., "Preliminary Evaluation of ATDMA and CODIT System Concepts", SIG5 deliverable MPLA/TDE/SIG5/DS/P/002/b1, September 1995.

[2] H.Honkasalo, "The technical evolution of GSM", Proc. of Telecom 95, Geneva, October 1995.

[3] P.Ranta, A.Lappeteläinen, Z-C Honkasalo, "Interference cancellation by Joint Detection in Random Frequency Hopping TDMA Networks", Proceedings of ICUPC96 conference, Vol 1 , pp.428-432.

[4] M.Pukkila and P. Ranta, "Simultaneous Channel Estimation for Multiple Co-channel Signals in TDMA Mobile Systems", Proceedings of IEEE Nordic Signal Processing Symposium (NORSIG'96), Helsinki 24-26th September, 1996.

[5] T.Ojanperä et.al., "Design of a 3rd Generation Multirate CDMA System with Multiuser Detection, MUD-CDMA", Proceedings of ISSSTA96 conference, Vol 1 pp.334-338. , Mainz, Germany,1996.

[6] K.Pajukoski and J.Savusalo, "Wideband CDMA Test System", Proceedings of PIMRC97, Helsinki, September 1997.

[7] I.Helakorpi, K.Pajukoski, "Performance Measurements of a Wireless Wideband Digital Communications Link", XIV IMEKO World Congress, Tampere, June, 1997.

[8] Westman Tapani and Holma Harri, "CDMA System for UMTS High Bit Rate Services", Proceedings of VTC97, Vol., pp. 824-829, Phoenix, U.S.A, May 1997.

[9] A.Hottinen, H.Holma, A.Toskala, "Performance of Multistage Multiuser Detection in a Fading Multipath Channel", Proceedings of PIMRC'95, pp. 960-964, Toronto, Canada, September 27-29, 1995.

[10] H. Holma, A. Toskala, A.Hottinen "Performance of CDMA Multiuser Detection with Antenna Diversity and Closed Loop Power Control", Proceedings of VTC'96, Atlanta, Georgia, USA, April 1996.

[11] A.Hottinen and K.Pehkonen, "A Flexible Multirate CDMA Concept with Multiuser Detection", Proceedings of ISSSTA96 conference, Vol 1 pp.556-560, Mainz, Germany,1996.

[12] S.Hämäläinen, H.Holma, A.Toskala "Capacity Evaluation of A Cellular CDMA Uplink with Multiuser Detection", Proceedings of ISSSTA96 conference, Vol 1, pp.556-560, Mainz, Germany,1996.

[13] H.Holma, A.Toskala, A.Hottinen "Performance of CDMA Multiuser Detection with Antenna Diversity and Closed Loop Power Control", Proceedings of VTC'96, Atlanta, USA, pp. 362-366, April/May 1996.

[14] A. Hottinen, H.Holma and A.Toskala, "Multiuser Detection for Multirate CDMA Communications", Proc. of ICC'96, Dallas, USA, June 1996.

[15] R.Wichman and A.Hottinen, "Multiuser Detection for Downlink CDMA Communications in Multipath Fading Channels", Proceedings of VTC97, Vol.2, pp.572-576, Phoenix, USA, May 1997.

[16] H.Holma, A.Toskala, T.Ojanperä, "Cellular Coverage Analysis of Wideband MUD-CDMA System", Proceedings of PIMRC97, Helsinki, September 1997.

[17] S.Hämäläinen, H.Holma, A.Toskala, M.Laukkanen, "Analysis of CDMA Downlink Capacity Enhancements", Proceedings of PIMRC97, Helsinki, September 1997.

[18] T.Ojanperä Tero et al., "FRAMES - Hybrid Multiple Access Technology ", Proceedings of ISSSTA96 conference, Vol 1, pp. 320 - 324, Mainz, Germany, 1996.

[19] T.Ojanperä et.al., "A Comparative Study of Hybrid Multiple Access Schemes for UMTS", Proceedings of ACTS Mobile Summit Conference, Vol 1., pp. 124-130, Granada, Spain, 1996.

[20] T.Ojanperä et.al, "Comparison of Multiple Access Schemes for UMTS", Proceedings of VTC97, Vol.2, pp. 490-494, Phoenix, U.S.A, May 1997.

[21] T. Ojanperä, A.Klein and P.O.Andersson, "FRAMES Multiple Access for UMTS", IEE Colloquium on CDMA Techniques and Applications for Third Generation Mobile Systems, London, May 1997.

[22] F.Ovesjö, E.Dahlman, T.Ojanperä, A.Toskala and A.Klein, "FRAMES Multiple Access Mode 2 - Wideband CDMA", Proceedings of PIMRC97, Helsinki, September 1997.

[23] A.Klein, R.Pirhonen, J.Sköld and R.Suoranta, "FRAMES Multiple Access Mode 1 - Wideband TDMA with and without Spreading" Proceedings of PIMRC97, Helsinki, September 1997.

[24] E.Nikula and E.Malkamäki, "High Bit Rate Services for UMTS using wideband TDMA carriers", Proceedings of ICUPC'96, Vol.2, pp. 562 - 566, Cambridge, Massachusetts, September/October 1996.

[25] ARIB FPLMTS Study Committee, "Report on FPLMTS Radio Transmission Technology SPECIAL GROUP, (Round 2 Activity Report)", Draft v.E1.1, January 1997.

[26] F.Muratore and V.Palestini, "Burst transients and channelization of a narrowband TDMA mobile radio system", Proceedings of the 38th IEEE

Vehicular Technology Conference, 15-17 June, 1988, Philadelphia, Pennsylvania.

[27] F.Adachi et.al., "Multimedia mobile radio access based on coherent DS-CDMA", Proc. 2nd International workshop on Mobile Multimedia Commun., A2.3, Bristol University, UK Apr. 1995.

[28] K.Ohno et.al., "Wideband coherent DS-CDMA", Proc. IEEE VTC'95, pp.779 - 783, Chicago, U.S.A, July 1995.

[29] T.Dohi et.al., "Experiments on Coherent Multicode DS-CDMA:", Proc. IEEE VTC'96, pp.889 - 893, Atlanta GA, USA.

[30] F.Adachi et.al., "Coherent DS-CDMA: Promising Multiple Access for Wireless Multimedia Mobile Communications", Proc. IEEE ISSSTA'96, pp.351 - 358, Mainz, Germany, September 1996.

[31] S.Onoe et.al., "Wideband-CDMA Radio Control Techniques for Third Generation Mobile Communication Systems", Proceedings of VTC97, Vol.2, pp.835-839, Phoenix, USA, May 1997.

[32] K.Higuchi et.al., "Fast Cell Search Algorithm in DS-CDMA Mobile Radio Using Long Spreading Codes", Proceedings of VTC97, Vol.3, pp.1430-1434, Phoenix, USA, May 1997.

[33] F.Adachi, M.Sawahashi and K.Okawa, "Tree-Structured generation of orthogonal spreading codes with different lengths for forward link of DS-CDMA mobile radio" , Electronics Letters, Vol.33, No.1., pp.27-28, January 1997.

[34] Baier et.al., "Design Study for a CDMA-Based Third Generation Mobile Radio System , IEEE JSAC Selected Areas in Communications, Vol.12, No.4, pp. 733 - 743, May 1994.

[35] Andermo (ed.), "UMTS Code Division Testbed (CODIT)", CODIT Final Review Report, September 1995.

[36] P-G.Andermo and L-M.Ewerbring, "A CDMA-Based Radio Access Design for UMTS", IEEE Personal Communications, Vol.2, No.1, pp.48-53, February 1995.

[37] M.Ewerbring et.al., Performance Evaluation of Wideband Testbed based on CDMA", Proceedings of VTC'97, pp. 1009 - 1013, Phoenix, Arizona, USA.

[38] A.Urie et.al., "ATDMA System Definition", ATDMA deliverable R2084/AMCF/PM2/DS/R/044/b1, January 1995.

[39] A.Urie et.al., "An Advanced TDMA Mobile Access System for UMTS", IEEE Personal Communications, Vol.2, No.1, pp. 38-47, February 1995.

[40] A.Urie, "Advanced GSM: A Long Term Future Scenario for GSM", Proceedings of Telecom 95, Vol.2 pp.33-37, Geneva, October 1995.

[41] J.Jiménez et, al., "Preliminary Evaluation of ATDMA and CODIT System Concepts", SIG5 deliverable MPLA/TDE/SIG5/DS/P/002/b1, September 1995.

[42] J.Sköld et.al, "Cellular Evolution into Wideband Services", Proceedings of VTC97, Vol.2, pp. 485 - 489, Phoenix, USA, May 1997.

[43] K.Pehkonen et.al., "A Performance Analysis of TDMA and CDMA Based Air Interface Solutions for UMTS High Bit Rate Services", Proceedings of PIMRC97, Helsinki, September 1997.

[44] T.Ojanperä, P.Ranta, S.Hamalainen and A.Lappetelainen, "Analysis of CDMA and TDMA for 3rd Generation Mobile Radio Systems", Proceedings of VTC97, Vol.2, pp.840-844, Phoenix, USA, May 1997.

[45] M.Gustafsson. et.al, "Different Compressed Mode Techniques for Interfrequency Measurements in a Wide-band DS-CDMA System" submitted to PIMRC´97.

[46] M.Thornberg, "Quality based power control in the CODIT UMTS Concept", Proc. of RACE Mobile Telecommunications Summit Vol.2 pp.308-312, Cascais, Portugal November 1995.

[47] J.Mikkonen and J.Kruys, "The Magic WAND: a wireless ATM access system", Proceedings of ACTS Mobile Summit Conference, Vol 2., pp. 535 - 542, Granada, Spain, 1996.

[48] Y-W.Park et.al.,"Radio Characteristics of PCS using CDMA", Proc. IEEE VTC'96, pp.1661-1664, Atlanta GA, USA.

[49] E-K.Hong et.al., "Radio Interface Design for CDMA-Based PCS", Proceedings of ICUPC´96, pp.365-368, 1996.

[50] J.M.Koo et.al. , "Implementation of prototype wideband CDMA system", Proceedings of ICUPC´96, pp.797-800, 1996.

[51] J.M.Koo, "Wideband CDMA technology for FPLMTS", The 1st CDMA International Conference, Seoul Korea, November 1996.

[52] S.Bang, et al., Performance Analysis of Wideband CDMA System for FPLMTS, Proceedings of VTC'97, pp. 830 - 834, Phoenix, Arizona, USA.

[53] H-R.Park, "A Third Generation CDMA System for FPLMTS Application", The 1st CDMA International Conference, Seoul Korea, November 1996.

[54] A.Sasaki, "A perspective of Third Generation Mobile Systems in Japan", IIR Conference Third Generation Mobile Systems, The Route Towards UMTS, London, February 1997.

[55] S.C.Wales, "The U.K. LINK Personal Communications Programme: A DS-CDMA Air Interface for UMTS", Proceedings of RACE Mobile Telecommunications Summit, Cascais, Portugal, November 1995.

[56] B.Engström and C.Österberg, "A System for Test of Multi access Methods based on OFDM", Proc. IEEE VTC'94, Stockholm, Sweden, 1994.

[57] M.Ericson et.al., "Evaluation of the mixed service ability for competitive third generation multiple access technologies", Proceedings of VTC'97, Phoenix, Arizona, USA.

[58] M.Wahlqvist, R.Larsson and C.Österberg, "Time synchronization in the uplink of an OFDM system", Proc. IEEE VTC'96, pp.1569 - 1573, Atlanta GA, USA.

[59] R.Larsson, C.Österberg and M.Wahlqvist, "Mixed Traffic in a multicarrier System", Proc. IEEE VTC'96, pp.1259-1263, Atlanta GA, USA.

[60] ETSI SMG2, "Description of Telia's OFDM based proposal" TD 180/97 ETSI SMG2, May 1997.

[61] R.L.Peterson, R.E.Ziemer and D.E.Borth, "Introduction to Spread Spectrum Communications", Prentice Hall, 1995.

[62] A. Klein, P.W. Baier: Linear unbiased data estimation in mobile radio systems applying CDMA. IEEE Journal on Selected Areas in Communications, vol. SAC-11 (1993), pp. 1058-1066.)

[63] M.M.Naßhan, P. Jung, A. Steil, P.W. Baier, "On the effects of quantization, nonlinear amplification and band limitation in CDMA mobile radio systems using joint detection" Proceedings of the Fifth Annual International Conference on Wireless Communications WIRELESS'93, Calgary/Canada (1993), pp. 173-186.

[64] P.Jung, J.J.Blanz, M.M.Naßhan, P.W. Baier, "Simulation of the uplink of JD-CDMA mobile radio systems with coherent receiver antenna diversity", Wireless Personal Communications, An International Journal (Kluwer), vol. 1 (1994), pp. 61-89.

[65] J.Yang et.al., "PN Offset Planning in IS-95 based CDMA system" Proceedings of VTC'97, Vol.3, pp.1435-1439, Phoenix, Arizona, USA.

[66] P.Jung, J.Plechinger, M.Doetsch and F.Berens, "Pragmatic approach to rate compatible punctured Turbo-Codes for mobile radio applications", 6th International Conference on Advances in Communications and Control: Telecommunications/Signal Processing, Grecotel Imperial, Corfu, Greece, 23-27 June 1997.

[67] P.Jung, J.Plechinger, M.Doetsch and F.Berens, "Advances on the application of Turbo-Codes to data services in third generation mobile networks" International Symposium on Turbo-Codes, Brest, September 3-5, 1997.

# 3 ADVANCED CDMA FOR WIRELESS COMMUNICATIONS

Markku Juntti

Savo Glisic

**Abstract:** Within this paper some selected topics currently being subject of extensive research for applications in future code-division multiple-access (CDMA) networks are reviewed. The presentation focuses on capacity aspects of advanced CDMA systems utilizing multiuser receivers. A comprehensive review of the multiuser demodulation theory and techiques is presented. Special emphasis is on optimal, suboptimal linear equalizer type and interference cancellation multiuser receivers. Future trends in practical system development are also discussed.

## 3.1 INTRODUCTION

Capacity evaluation of a code-division multiple-access (CDMA) network and comparison of CDMA and time-division multiple-access (TDMA) systems has become an important and controversial issue. One of the reasons for such situation is a lack of a systematic, easy to follow mathematical framework for this evaluation. The situation is complicated by the fact that several parameters are involved and some of the system components are rather complex. Because of that often the problem has been tried to analyze by simulations. The results are controversial. Although there is always a description of the simulation environment it is not easy to comment simulation results due to lack of a complete insight into the simulation process. The conventional CDMA is an interference limited system and each increase in the interference level will reduce the system capacity. The interference coming from other users in the same cell (intracell interference) will be the major limiting factor in achieving high capacity in the system.

Due to multipath propagation each signal will be received as a number of mutually delayed replicas propagated over the paths with different intensity coefficients. This will degrade the code orthogonality. Channel capacity will be increased if a RAKE receiver is used. In addition to this, in advanced CDMA systems, multiple-access interference (MAI) cancellation will be used. The efficiency of the canceler is characterized by so called canceler efficiency which has maximum value one if the MAI is completely eliminated. In the base station any of the known multiuser receivers might be assumed [1, 2, 3] while in the mobile unit some form of simplified adaptive linear minumum mean squared error (LMMSE) detector is expected to be used for these purposes [4, 5, 6, 7]. In any of these cases there will be some residual interference. In general case canceler efficiency will be a function of the interfering signal level and the number of users in the network. A closed form expression for this efficiency can be derived in the case of linear detectors. For most detectors such a derivation is not available at the moment. Even for the linear detector the result is in the for

The interference coming from other cells will depend on interference cancellation factor, cell isolation efficiency and signal isolation efficiency, defined earlier. For land mobile communications the cell isolation is due to large propagation losses and intercell MAI cancellation might be impractical. For multibeam satellite communication this efficiency depends on antenna directivity only and might not be enough to suppress intercell interference to the level needed to eliminate an excessive influence among the cells. For this reason additional action for intercell MAI cancellation might be needed. This could result in unacceptable complexity so that signal isolation might be increased by using different frequencies for each beam (FDM). In this case for a limited frequency band, code chip rate must be reduced leading to reduced capabilities of multipath resolution and less efficient RAKE receiver. Overlay type internetwork interference can also be characterized by cancellation efficiency among the signal More common terminology in this case would be interference suppression gain. A number of papers have been published in the field of narrowband interference suppression for applications in CDMA overlay networks [8, 9, 10, 11, 12, 13, 14, 15, 16, 17, 18, 19, 20, 21, 22, 23, 24]. Capacity gains are discussed in [25, 26]. The major drawback of these schemes is the need for 5-10 times wider CDMA signal bandwidth than the bandwidth of narrowband users.

Further research to find solutions for successful interference suppression in the case when two overlaid networks occupy the same bandwidth should increase feasibility of practical implementations of this technique.

A concept of an advanced CDMA network should in general take all these elements into account and come up with a proposal that is a reasonable trade off between the system performance and its complexity. Within this paper we first discuss relative impact of the some of the above issues on the system capacity and then concentrate more on multiuser receiver theory and techniques. Signal model is presented in Section 2. Short discussion of the relevant system parameters from the capacity point of view is presented in Section 3. A comprehensive review of multiuser demodulation is presented in Section 4. Evolution of CDMA systems towards the advanced concepts including multiuser receivers is discussed in Section 5. A conclusion with a prediction of the future trends in the field is given in Section 6. A comprehensive list of references is intended for the readers more interested in different details.

As it was already pointed out the overall interference level can be reduced by increasing space isolation factor. Spatial filtering and adaptive antennas are considered within separate contributions in this book by R. Kohno and P. W. Baier. In general, the system capacity is increased by using demodulation techniques that offer good performance with lower signal-to-noise ratios. For these purposes sophisticated error correcting should be used. These techniques are reviewed within the papers by B. Vucetic and E. Biglieri. Overlay type CDMA networks are discussed by D. L. Schilling and numerous new evolving applications are discussed by T. Ojanperä.

## 3.2 SYSTEM MODEL

A general multiuser CDMA system is illustrated in Fig. 3.1. So called multiple-access channel model [27, Chap. 14] is used. In this model $K$ users share the same communication media, and the signals transmitted by the users pass through separate and independent channels. The outputs of the channels are added to a common noise process. The transmitted data is demodulated in a centralized multiuser receiver, which makes a joint decision of the data of all users. In a cellular mobile communication system, for example, the setup is valid for the uplink. The mathematical formulation of the transmission system presented in this section has been used in a number of papers, e.g., [28, 29, 30, 31, 32, 33].

A user $k \in \{1, 2, \ldots, K\}$ transmits in the $n$th symbol interval $t \in \Big[(n-1)T, nT\Big)$ a complex signal

$$b_k^{(n)} A_k s_k^{(n)}(t - \tau_k), \tag{3.1}$$

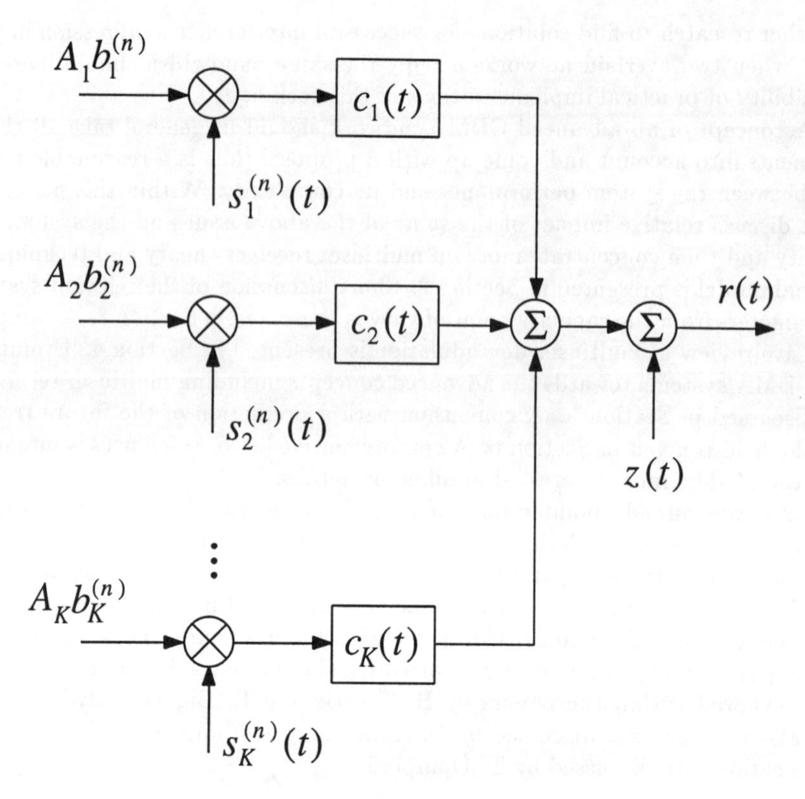


**Figure 3.1** CDMA system.

where $T$ is the length of the symbol period, $b_k^{(n)} \in \Xi$ is the transmitted data symbol, $\Xi$ is the modulation symbol alphabet, $A_k = \sqrt{E_k}$ is the transmitted amplitude of user $k$ (assumed to be constant over the transmission), $E_k$ is the energy per symbol of the corresponding real bandpass signal, $\tau_k \in [0, T)$ is the delay of $k$th user's transmitted signal, and $s_k^{(n)}(t)$ is the signature waveform of user $k$. For convenience $s_k^{(n)}(t)$ is assumed to be real (the analysis can be straightforwardly generalized to complex case) and normalized so that $s_k^{(n)}(t) = 0$, if $t \notin [0, T)$, and $\int_0^T \mid s_k^{(n)}(t) \mid^2 dt = 1$. In a DS-CDMA system the signature waveforms are of the form

$$s_k^{(n)}(t) = \sum_{m=0}^{N_c-1} s_{k,m}^{(n)} \psi(t - mT_c), \tag{3.2}$$

where $s_{k,m}^{(n)}$ is the $m$th chip of user $k$ on symbol interval $n$, $T_c$ is the length of the chip period, $N_c = T/T_c$ is the processing gain, and $\psi(t)$ is the chip waveform.

It is assumed that the channel of user $k$ appears as a linear filter with impulse response $c_k^{(n)}(t)$ (Fig. 3.1). It is assumed that the channel impulse responses consists of discrete multipath components [34, Chap. 14] so that they can be expressed as

$$c_k^{(n)}(t) = \sum_{l=1}^{L} c_{k,l}^{(n)} \delta\left(t - \tau_{k,l}^{(n)}\right), \tag{3.3}$$

where $L$ is the number of multipath components[1] of the channel, $c_{k,l}^{(n)}$ is the complex coeffcient (gain) of $l$th multipath of user $k$ at symbol interval $n$, $\tau_{k,l}^{(n)} \in [0, T_m)$ is the delay of $l$th multipath component of user $k$ at symbol interval $n$, $T_m$ is the delay spread of the channel and $\delta(t)$ is the Dirac's delta function. The delays will be marked by $\tau_{k,l}$ in the forthcoming analysis. Furthermore, it is assumed that the delay spread of the channel is less than the symbol interval, i.e., $T_m < T$.

The received CDMA signal is the convolution of the transmitted signal (3.1) and the channel impulse response (3.3) plus the additive channel noise. Thus, the complex envelope of the received signal can be expressed as

$$\begin{aligned} r(t) &= \sum_{n=0}^{N_b-1} \sum_{k=1}^{K} b_k^{(n)} A_k s_k^{(n)}(t - nT - \tau_k) * c_k^{(n)}(t) + z(t) \\ &= \sum_{n=0}^{N_b-1} \sum_{k=1}^{K} b_k^{(n)} A_k \sum_{l=1}^{L} c_{k,l}^{(n)} s_k^{(n)}(t - nT - \tau_k - \tau_{k,l}) + z(t), \end{aligned} \tag{3.4}$$

where $N_b$ is the number of symbols in the data packet, the asterix $*$ denotes convolution, $z(t)$ is complex zero mean additive white Gaussian noise process with two-sided power spectral density $\sigma^2$.

It is known that the set of matched filter (MF) outputs sampled once in symbol interval forms the sufficient statistics for the detection of the transmitted data[2] [29, 32]. The sampled output of the filter matched to the $k$th users $l$th multipath component is

$$y_{k,l}^{(n)} = \int_{nT+\tau_k+\tau_{k,l}}^{(n+1)T+\tau_k+\tau_{k,l}} r(t) s_k^{(n)}(t - nT - \tau_k + \tau_{k,l}) dt. \tag{3.5}$$

Let the vectors of MF output samples for the $n$th symbol interval be defined as

$$\mathbf{y}_k^{(n)} = (y_{k,1}^{(n)}, y_{k,2}^{(n)}, \ldots, y_{k,L}^{(n)})^\top \in \mathbb{C}^L \tag{3.6}$$

$$\mathbf{y}^{(n)} = (\mathbf{y}_1^{\mathsf{T}(n)}, \mathbf{y}_2^{\mathsf{T}(n)}, \ldots, \mathbf{y}_K^{\mathsf{T}(n)})^{\mathsf{T}} \in \mathbb{C}^{KL} \tag{3.7}$$

and their concatenation over the whole data packet

$$y = \left(\ \mathbf{y}^{\mathsf{T}(1)}\ \ \mathbf{y}^{\mathsf{T}(2)}\ \ \cdots\ \ \mathbf{y}^{\mathsf{T}(N_b)}\ \right)^{\mathsf{T}} \in \mathbb{C}^{N_b KL}. \tag{3.8}$$

Let $\mathbf{R}^{(n)}(i) \in (-1,1]^{KL\times KL}$ be a crosscorrelation matrix[3] with partition

$$\mathbf{R}^{(n)}(i) = \begin{pmatrix} \mathbf{R}_{1,1}^{(n)}(i) & \mathbf{R}_{1,2}^{(n)}(i) & \cdots & \mathbf{R}_{1,K}^{(n)}(i) \\ \mathbf{R}_{2,1}^{(n)}(i) & \mathbf{R}_{2,2}^{(n)}(i) & \cdots & \mathbf{R}_{2,K}^{(n)}(i) \\ \vdots & \vdots & \ddots & \vdots \\ \mathbf{R}_{K,1}^{(n)}(i) & \mathbf{R}_{K,2}^{(n)}(i) & \cdots & \mathbf{R}_{K,K}^{(n)}(i) \end{pmatrix} \in \mathbb{R}^{KL\times KL}, \tag{3.9}$$

where matrices $\mathbf{R}_{k,k'}^{(n)}(i) \in \mathbb{R}^{L\times L}$, $\forall\, k,k' \in \{1,2,\ldots,K\}$ have elements

$$\left(R_{k,k'}^{(n)}(i)\right)_{l,l'} = \int_{-\infty}^{\infty} s_k^{(n)}(t-\tau_k-\tau_{k,l})\, s_{k'}^{(n-i)}(t+iT-\tau_{k'}-\tau_{k',l'})\,dt,$$

$$\forall\, l,l' \in \{1,2,\ldots,L\} \tag{3.10}$$

The vector (3.7) can be expressed as [30]

$$\begin{aligned} \mathbf{y}^{(n)} &= \mathbf{R}^{(n)}(2)\mathbf{C}^{(n-2)}\mathbf{A}\mathbf{b}^{(n-2)} + \mathbf{R}^{(n)}(1)\mathbf{C}^{(n-1)}\mathbf{A}\mathbf{b}^{(n-1)} \\ &\quad + \mathbf{R}^{(n)}(0)\mathbf{C}^{(n)}\mathbf{A}\mathbf{b}^{(n)} + \mathbf{R}^{(n)}(-1)\mathbf{C}^{(n+1)}\mathbf{A}\mathbf{b}^{(n+1)} \\ &\quad + \mathbf{R}^{(n)}(-2)\mathbf{C}^{(n+2)}\mathbf{A}\mathbf{b}^{(n+2)} + \mathbf{w}^{(n)}, \end{aligned} \tag{3.11}$$

where

$$\mathbf{A} = \mathrm{diag}\,(A_1, A_2, \ldots, A_K) \in \mathbb{R}^{K\times K} \tag{3.12}$$

is a diagonal matrix of transmitted amplitudes,

$$\mathbf{C}^{(n)} = \mathrm{diag}\left(\mathbf{c}_1^{(n)}, \mathbf{c}_2^{(n)}, \ldots, \mathbf{c}_K^{(n)}\right) \in \mathbb{C}^{KL\times K}, \tag{3.13}$$

is the matrix of channel coefficient vectors

$$\mathbf{c}_k^{(n)} = \left(c_{k,1}^{(n)}, c_{k,2}^{(n)}, \ldots, c_{k,L}^{(n)}\right)^{\mathsf{T}} \in \mathbb{C}^{L}, \tag{3.14}$$

$$\mathbf{b}^{(n)} = \left(b_1^{(n)}, b_2^{(n)}, \ldots, b_K^{(n)}\right)^{\mathsf{T}} \in \Xi^{K}, \tag{3.15}$$

is the vector of the transmitted data and $\mathbf{w}^{(n)} \in \mathbb{C}^{KL}$ is the output vector due to noise. It is easy to show that $\mathbf{R}^{(n)}(i) = \mathbf{0}_{KL}$, $\forall\, |i| > 2$ and

$\mathbf{R}^{(n)}(-i) = \mathbf{R}^{\top(n+i)}(i)$, where $\mathbf{0}_{KL}$ is a all-zero matrix of size $KL \times KL$. Thus, the concatenation vector of the matched filter outputs (3.8) has the expression

$$\boldsymbol{y} = \mathcal{RCA}\boldsymbol{b} + \boldsymbol{w} = \mathcal{RC}\boldsymbol{h} + \boldsymbol{w}, \tag{3.16}$$

where

$$\mathcal{R} = \begin{pmatrix} \mathbf{R}^{(0)}(0) & \mathbf{R}^{\top(1)}(1) & \mathbf{R}^{\top(2)}(2) & \cdots & \mathbf{0}_{KL} \\ \mathbf{R}^{(1)}(1) & \mathbf{R}^{(1)}(0) & \mathbf{R}^{\top(2)}(1) & \cdots & \mathbf{0}_{KL} \\ \mathbf{R}^{(2)}(2) & \mathbf{R}^{(2)}(1) & \mathbf{R}^{(2)}(0) & \cdots & \mathbf{0}_{KL} \\ \vdots & \vdots & \vdots & \ddots & \vdots \\ \mathbf{0}_{KL} & \mathbf{0}_{KL} & \mathbf{0}_{KL} & \cdots & \mathbf{R}^{(N_b-1))}(0) \end{pmatrix}$$

$$\in \mathbb{R}^{N_bKL \times N_bKL}, \tag{3.17}$$

$$\mathcal{C} = \mathrm{diag}\left(\mathbf{C}^{(0)}, \mathbf{C}^{(1)}, \ldots, \mathbf{C}^{(N_b-1)}\right) \in \mathbb{C}^{N_bKL \times N_bK}, \tag{3.18}$$

$$\mathcal{A} = \mathrm{diag}\,(\mathbf{A}, \mathbf{A}, \ldots, \mathbf{A}) \in \mathbb{R}^{N_bK \times N_bK}, \tag{3.19}$$

$$\boldsymbol{b} = \left(\mathbf{b}^{\top(0)}, \mathbf{b}^{\top(1)} \ldots, \mathbf{b}^{\top(N_b-1)}\right)^{\top} \in \Xi^{N_bK}, \tag{3.20}$$

$\boldsymbol{h} = \mathcal{A}\boldsymbol{b}$ is the data-amplitude product vector, and $\boldsymbol{w}$ is the Gaussian noise output vector with zero mean and covariance matrix $\sigma^2\mathcal{R}$.

### 3.3 CDMA SYSTEM CAPACITY

The starting point in the evaluation of CDMA system capacity [35, 36, 37, 38, 39, 40, 41, 42, 43, 44, 45, 46, 47, 48, 49, 50, 51, 52, 53, 54] is parameter $E_b/N_0$, the received signal energy per bit per overall noise density in a given reference receiver. For the purpose of this paper we can represent this parameter in general case as

$$\frac{E_b}{N_0} = \frac{PT}{I_{oc} + I_{oic} + I_{oin} + \sigma^2} \tag{3.21}$$

where $I_{oc}$, $I_{oic}$, and $I_{oin}$ are power densities of intracell, intercell and overlay type internetwork interference, respectively, and $\sigma^2$ is noise power density of the thermal additive white Gaussian noise (AWGN). Parameter $P$ is the overall received power of the useful signal and $T$ is the information symbol duration. In the case of a RAKE receiver $E_b$ is the overall power of the corresponding signal at the output of the combiner. In general, the parameters $I_{oc}$, $I_{oic}$ and $I_{oin}$, can be represented in the form

$$I_o = \sum_{i=1}^{N_{cells}} \sum_{k=1}^{K} \sum_{l=1}^{L} (1 - \zeta_{ikl})(i - \theta_{ikl})(1 - \vartheta_{ikl}) P_{ikl} \tag{3.22}$$

where $\zeta_{ikl} < 1$ is the space isolation efficiency between the reference receiver and signal coming through path $l$ from user $k$ in cell $i$ having power $P_{ikl}$. Parameter $\vartheta_{ikl}$ is the efficiency of interference canceler for the same signal. After despreading in the reference receiver the residual interfering signal component will be proportional to $(1 - \theta_{ikl})$ where $\theta_{ikl}$ is so called signal isolation factor. For orthogonal codes $\theta_{ikl}$ would be equal to one, only for zero delay between the reference and other codes used in the network (Walsh functions). For m-sequences or Gold codes $1 - \theta_{ikl} = \rho_{ikl}^2$, where $\rho_{ikl}$ is the crosscorrelation between the reference signal and the signal component with index $ikl$. Very often in the analysis this is approximated with $1 - \theta_{ikl} = 1/G$, where $G$ is the system processing gain.

In land mobile communications the space isolation factor for signals coming from different cells, is high due to high propagation losses and additional interference cancellation might not be necessary. For example, the most often used model for macrocell reverse link with power control [53] would suggest

$$1 - \zeta_{ikl} = \frac{10^{\xi_0/10}}{r_0^n} \frac{r_m^n}{10^{\xi_m/10}} = (\frac{r_m}{r_0})^n 10^{(\xi_0 - \xi_m)/10}, \tag{3.23}$$

where $r_0$ is the distance of the interfering mobile set from the reference base station, $r_m$ its distance from its own base station and attenuation factor $n$ has value from 2 to 4 even 5 in rural and urban environments, respectively. The first term in (3.23) takes into account the interfering signal attenuation due to distance $r_0$ and shadowing which is modeled as lognormal function with the Gaussian variable $\xi_0$ of zero mean and variance. The second term models the system power control.

For internetwork interference $P_{ikl}$ might be significantly higher than the power of the reference signal so that additional interference suppression scheme is needed. In this case parameter $1 - \vartheta_{ikl}$ is given as

$$1 - \vartheta_{ikl} = \frac{1}{G_{ikl}}, \tag{3.24}$$

where $G_{ikl}$ is suppression gain of the interference suppression scheme [8].

In multiple beam satellite CDMA network $\zeta_{ikl}$ is controlled by antenna beam width and might not be enough to reduce the interference level down to the desired level so that additional intercell interference cancellation might be needed.

In this case

$$1 - \zeta_{ikl} = g_{ikl}(\phi_{ikl})g_r(\phi_r), \tag{3.25}$$

where $g(\phi)$ is antenna gain in direction $\phi$ and $r$ stands for "reference receiver". Contributions of $I_{oic}$ and $I_{oin}$ to $N_0$ has been discussed in a number of papers [35, 36, 37, 38, 39, 40, 41, 42, 43, 44, 45, 46, 47, 48, 49, 50, 51, 52, 53, 54]. If internetwork interference is not present, relative contribution of $I_{oic}$ to $N_0$ is about 60% of $I_{oc}$ [35]. If $\sigma^2$ is neglected then the maximum improvement in $N_0$ when $I_{oc}$ is completely eliminated would be $(I_{oc} + I_{oic})/I_{oic} \cong 1.6/0.6 \leq 2.8$. Due to presence of AWGN and imperfections of interference cancellation this improvement will be in the range 2-2.5. Experimental and simulation results for IS-665 are 3.15 dB which is slightly more than factor 2. Power control imperfections can also severely degrade the system capacity [36, 54, 48]. As an extreme case in the presence of severe near-far effect CDMA can completely fail. These effects are eliminated in multiuser receiver, that makes these schemes very attractive for next generation of CDMA systems. Heaving this in mind in the next section we review multiuser techniques and discuss feasibility of practical implementations.

## 3.4 MULTIUSER DEMODULATION

The concept of multiuser receivers has been known for about two decades. Schneider [55] studied the zero-forcing decorrelating detector. Later Kashihara [56] and Kohno *et al.* [57] studied multiple-access interference cancellation receivers. Both Schneider and Kohno had also suggested the use of Viterbi algorithm for optimal detection in asynchronous multiuser communications. The real trigger to the increasing interest in multiuser demodulation was the Verdú's work on optimal multiuser detection [29, 58], where the application of Viterbi algorithm for optimal detection was developed, and the performance of the optimal detection was analyzed. Verdú showed that the CDMA systems are inherently neither interference nor near-far limited, but both are actually limitations of the conventional single-user receiver. Since the optimal multiuser detection is prohibitively complex to implement in most practical applications, numerous suboptimal schemes have been investigated. Earlier tutorial reviews of multiuser demodulation techniques can be found in [1, 2, 3] and an overview in [59].

Our purpose here is to present an updated overview of the multiuser demodulation literature relevant from the advanced CDMA system point of view. The main emphasis is on those multiuser demodulation techniques for DS-CDMA systems[4], which are most important either from practical or theoretical point of view. A centralized[5] multiuser receiver can be illustrated as in Figs. 3.2(a) or 3.2(b). The multiuser procssing can be performed either before the multipath combining, by processing the matched filter bank output vector $\boldsymbol{y}$, or after the multipath combining, by processing the maximal ratio combined matched filter bank output vector

$$\boldsymbol{y}_{[MRC]} = \mathcal{C}^{\mathrm{H}} \boldsymbol{y}. \tag{3.26}$$

It should be noted, however, that the block diagrams in Figs. 3.2 are simplified and cannot fit all multiuser receivers into their framework. Most multiuser signal processing can also be implemented before matched filtering, i.e., by processing the received spread-spectrum signal samples. It should also be noted that most multiuser receivers alleviate not only the decrimental effects of multiple-access interference, but the intersymbol interference as well.

Optimal multiuser receivers are considered in Section 3.4.1. Due to prohibitive computational complexity of the optimal multiuser receivers suboptimal solutions, reviewed in Sections 3.4.2–3.4.4, have been studied extensively. They represent an approximation of the optimal maximum likelihood sequence detector (MLSD). Most receivers can process either the matched filter bank output (Fig. 3.2(b)) or its maximal ratio combined version (Fig. 3.2(a)). The latter receivers do not eliminate the effect of MAI on channel estimation. Therefore, the multiuser detectors processing the MF bank output are often more desirable in practice, and our discussion will focus on such receivers. Section 3.4.2 concentrates on linear equalizer type receivers, whereas interference cancellation receivers are considered in Section 3.4.3. Other suboptimal receivers are reviewed in Section 3.4.4.

### *3.4.1 Optimal Multiuser Demodulation*

The centralized multiuser receiver minimizing the bit error probability of one symbol of a particular user has been studied by Verdú [70] for the known channel case. The minimum error probability receiver must find the most probably transmitted data symbol for all users for all symbol intervals. In other words, $N_b K$ separate minimizations should performed. Each minimization computes a metric for all possible $|\Xi|^{N_b(K-1)}$ interfering data symbol combinations, where $|\Xi|$ denotes the cardinality of the symbol alphabet $\Xi$. Although a dynamic programming algorithm can be devised to implement the minimum probability of error detector, the required number of operations is an exponential function of the number of users. Furthermore, the performance of the minimum probability of error detector is difficult to analyze. Thus, similarly to the single-user ISI channels [34], the maximum likelihood sequence detector is often considered to be an optimal multiuser detector.

The MLSD multiuser receiver minimizes the probability of an erroneous decision on the bit vector $\boldsymbol{b}$ including the data symbols of all users on all symbol intervals. If the channel is known, the decision can be expressed in the

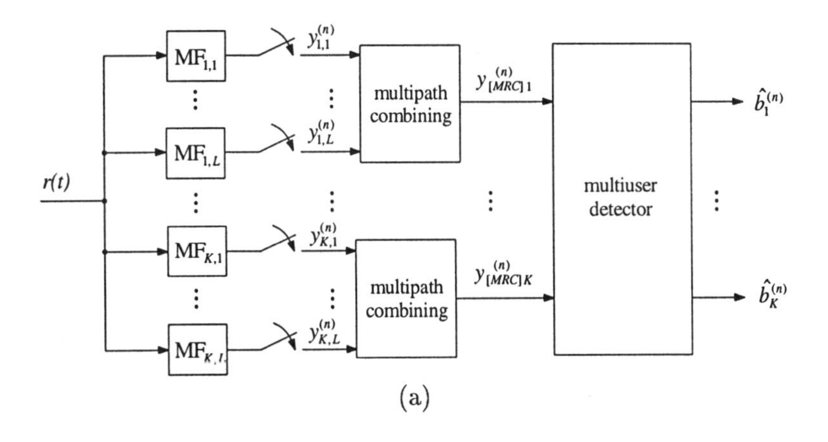


(a)

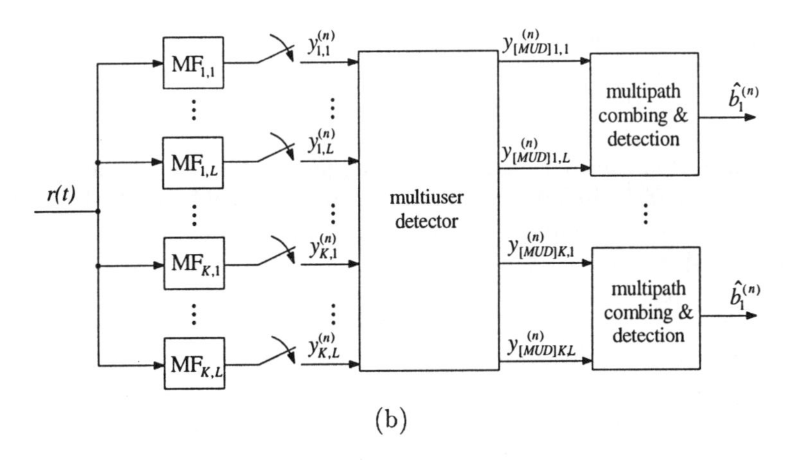


(b)

**Figure 3.2** Multiuser receiver structures.

form [29, 71, 72]

$$\hat{b}_{[MLSD]} = \arg \min_{\mathbf{b} \in \Xi^{N_b K}} \Omega(b), \tag{3.27}$$

where the log-likelihood function $\Omega(b)$ is

$$\Omega(b) = 2\mathrm{Re}(b^{\mathrm{H}} \mathcal{A}^{\mathrm{H}} \mathcal{C}^{\mathrm{H}} y) - b^{\mathrm{H}} \mathcal{A}^{\mathrm{H}} \mathcal{C}^{\mathrm{H}} \mathcal{R} \mathcal{C} \mathcal{A} b. \tag{3.28}$$

The maximum likelihood detector admits the structure of the receiver in Fig. 3.2(a). If the signature waveforms are time-invariant, the minimization can be implemented by a dynamic programming algorithm so that the implementation complexity depends exponentially on the number of users only, not on the data packet length [70, 29]. The implementation complexity makes, however, the MLSD infeasible for many practical applications. The asymptotic multiuser efficiency of the MLSD has been analyzed in [58, 73]. MLSD has been extended for trellis-coded modulated CDMA transmissions in AWGN channels in [74], and for convolutionally encoded transmissions in [75]. The effect of delay estimation errors to MLSD has been considered in [76]. Joint maximum likelihood sequence detection and amplitude estimation in AWGN channels has been analyzed in [77, 78].

The performance of the MLSD is analyzed in [29, 58]. It turned out to be impossible to derive a closed form bit error probability expression for the MLSD. Upper and lower bounds, although complicated to calculate, were found. The simplest lower bound is the *single-user bound* (or matched filter bound), which is the performance of a communication system with one active user ($K = 1$). The performance results on the MLSD demonstrate that significant performance gains can be obtained over the conventional single-user receiver. It has been demonstrated that the CDMA systems are not inherently interference limited, but that is the limitation of the conventional detector.

The maximum likelihood sequence detection for relatively fast fading channels has also been analyzed. Flat Rician fading channels with synchronous CDMA have been considered in [79] and two path Rician fading channels with asynchronous CDMA in [80]. MLSD for synchronous CDMA in Rayleigh fading channels has been presented in [81, 82]. The resulting MLSD receiver consists of channel estimator for all possible data sequences and a correlator, which multiplies the received signal with the estimated channel (estimator-correlator receiver [83]).

Optimal MLSD receiver for channels with unknown delays is significantly more difficult to derive. The reason is the fact that the received signal depends nonlinearly on the delays, and the MLSD receiver does not admit a simple estimator-correlator interpretation. One way to approximate the MLSD for the reception of a signal with unknown delays is to perform joint maximum likelihood estimation on the data, received complex amplitude, and the delays [84]. The joint ML estimation has clearly extremely high computational complexity, which is exponential in the product of the number of users $K$, number of propagation paths $L$, and number of samples per symbol interval $N_s$.

Optimal decentralized multiuser detectors for AWGN channels have been considered in [85], where the multiple-access interference was modeled as non-Gaussian noise. The optimal decentralized multiuser detectors can also admit the utilization of the knowledge of a subset of the $K - 1$ interfering signature waveforms. The optimal decentralized multiuser detector has also computational complexity which depends exponentially on the number of users.

### 3.4.2 Linear Equalizer Type Multiuser Demodulation

Linear equalizer type multiuser receivers process the matched filter output vector $\boldsymbol{y}$ (or the maximal ratio combined vector $\boldsymbol{y}_{[MRC]}$) by a linear operation. In other words, the output $\boldsymbol{y}_{[LIN]}$ of a linear multiuser detector $\mathcal{T} \in \mathbb{C}^{N_b KL \times N_b KL}$ is

$$\boldsymbol{y}_{[LIN]} = \mathcal{T}^{\mathrm{H}} \boldsymbol{y}. \tag{3.29}$$

Different choices of the matrix $\mathcal{T}$ yield different multiuser receivers. The identity matrix $\mathcal{T} = \mathbf{I}_{N_b KL}$ is equivalent to the conventional single-user receiver. The linear equalizer type receivers apply the principles of linear equalization, which has been used in ISI reduction [34].

The decorrelating or zero-forcing linear equalizer type receiver, which completely removes the MAI, corresponds to the choice [30, 86]

$$\mathcal{T} = \mathcal{R}^{-1}. \tag{3.30}$$

Performance of the decorrelating detector in AWGN channels has been analyzed in [87, 30, 88, 89]. It has been shown that the decorrelating detector is optimally near-far resistant in the sense it achives the same NFR as the MLSD. The performance of the decorrelator in known, slowly fading channels has been analyzed in [86, 71, 90, 72, 91]. The corresponding analysis for estimated, relatively fast fading channels has been presented in [79, 80, 81, 92, 93]. The performance of the decorrelator utilizing the matched filter bank output $\boldsymbol{y}$ or the maximal ratio combined MF bank output $\boldsymbol{y}_{[MRC]}$ has been compared in [94, 95]. The principle of decorrelating receiver has been extended to receivers utilizing antenna arrays [96, 97, 98, 99, 100, 101, 95], multiple base stations [102, 103, 104], or multiple data rates [105, 106]. Adaptive implementations of the decorrelating receiver for synchronous CDMA systems have been considered in [107, 108] and for asynchronous CDMA systems in [109]. The decorrelating receiver for convolutionally encoded CDMA transmissions in AWGN channels has been studied in [110]. The same receivers for quasi-synchronous CDMA systems in AWGN channels without precise delay estimation has been proposed in [111, 112, 113, 114], and for code acquisition in quasi-synchronous CDMA in [115]. The effect of delay estimation errors to the decorrelator performance has been analyzed in [116, 117]. The impact of quantization has been considered in [118, 119].

A partial decorrelator, which also makes the additive channel noise component white, so called noise-whitening detector is defined as

$$\mathcal{T} = \mathcal{L}^{-1}, \tag{3.31}$$

where $\mathcal{L}$ is lower triangular Cholesky factor of $\mathcal{R}$ such that $\mathcal{R} = \mathcal{L}^{\mathrm{H}} \mathcal{L}$ [120][6]. The noise-whitening detector forces the MAI due to past symbols to zero. The

MAI due to future symbols may be supressed by interference cancellation utilizing decision-feedback [121, 120] or the MAI may be handled by some suboptimal tree-search algorithm [123, 124, 125].

If the information symbols $b_k^{(n)}$ are independent and uniformly distributed and the channel is known, the linear receiver which minimizes the mean squared errors at the detector outputs (so called LMMSE detector) is [126]

$$\mathcal{T} = \left[\mathcal{R} + \sigma^2 (\mathcal{C}\mathcal{A}\mathcal{A}^{\mathrm{H}}\mathcal{C}^{\mathrm{H}})^{-1}\right]^{-1}. \tag{3.32}$$

The LMMSE receiver is equal to the linear receiver maximizing the signal-to-interference-plus-noise ratio (SINR) [6]. The centralized LMMSE receivers have been proposed for AWGN channels in [127], for fading channels in [91, 128, 129], and for antenna array receivers [98, 99, 130, 131]. Bounds for the NFR and SINR of the LMMSE receiver in AWGN channels have been derived in [132], and the bit error probability has been analyzed in [133].

The LMMSE receivers have attracted most interest due to their applicability to decentralized adaptive implementation, where other users' spreading sequences are not needed. Decentralized LMMSE multiuser receivers for AWGN or slowly fading channels suitable for adaptive implementation based on training have been considered in [4, 5, 6, 7]. The convergence of the adaptive algorithms for the LMMSE multiuser receivers has been considered in [134, 135, 136]. A modified adaptive multiuser receiver applicable to relatively fast fading frequency-selective channels with channel state information has been proposed in [137]. CDMA system capacity with LMMSE receivers has been studied in [138, 139], where the *spreading-coding tradeoff*[7] has been addressed for systems with multiuser receivers. An improved LMMSE receiver, less sensitive to the time delay estimation errors, has been proposed in [140]. Receivers suitable for blind adaptation utilizing the minimum output energy (MOE) criterion have been studied in [141, 142, 143, 144]. It has been shown that the linear filter optimal in MOE sense is equal to the linear filter optimal in the MMSE sense [141]. A blind receiver performing both the MOE filtering and timing estimation has been studied in [145].

Decentralized linear receivers include MAI-whitening filters, which model the multiple-access interference as colored noise. The filters are then designed to whiten the colored MAI-noise plus the AWGN. The MAI-whitening filters have been studied in [146, 147, 148, 149, 150]. The implemention of the MAI-whitening filters is difficult, since it requires information on the MAI covariance. Approximate implementation results in adaptive receivers similar to their LMMSE counterparts [148].

The linear multiuser receivers process ideally the complete received data block, length of which approaches infinity in asynchronous CDMA systems. In other words, the memory-length of the linear equalizer type receivers is infinite. In [30] it was shown that as $N_b \to \infty$ the decorrelating detector approaches a time-invariant, stable digital multichannel infinite impulse response (IIR) filter. Truncation of the IIR filters to obtain finite impulse response (FIR) filters has been suggested in [30] for the decorrelating detector. However, the effect of such a truncation on the detector performance was not analyzed. The FIR detector designs have been considered in [124, 151, 152]. Several other ways to obtain finite memory-length multiuser detectors have also been proposed. The most natural way is to leave symbol intervals regularly without transmission. This will result in finite blocks of transmitted symbols and obviously the detectors would then have finite memory-length [153, 154]. The drawback is that the introduced redundancy will reduce the effciency of the transmission. Other approaches to obtain finite memory-length multiuser detectors include nonlinear subtraction of estimated multiple-access interference ("edge correction") [155], and hard decision approximation of decorrelator [127], which ends up in the decision directed, nonlinear MAI canceler. One-shot detection [1, 156] has also been studied.

In addition to the infinite memory-length, the linear multiuser receivers have relatively high implementation complexity due to the matrix inversion as in (3.30), (3.31), or (3.32). An approximate update algorithm has been proposed in [157, 155]. However, the algorithm is restricted to track only small changes in the correlations caused by minor delay changes. Approximate multistage linear equalizer type detectors have been proposed in [154]. Their computational requirements are still a cubic function of the number of users. Another approach called $\delta$-adjusted multiuser detection has been proposed in [158], but its near-far resistance is an open problem. Application of iterative algorithms has been proposed in [159, 160].

### *3.4.3 Interference cancellation*

The idea of *interference cancellation* (IC) receivers is to estimate the multiple-access and multipath induced interference and then subtract the interference estimate from the MF bank (or MRC) output. The interference cancellation can be derived as an approximation of the MLSD receiver with the assumption that the data, amplitude, and delays of the interfering users (or a subset of them) are known [31]. There are several principles of estimating the interference leading to different IC techniques. The interference can be canceled simultaneously from all users leading to *parallel* interference cancellation (PIC), or on a user by user basis leading to *serial* (successive) interference cancellation (SIC). Also *groupwise* serial (GSIC) or parallel (GPIC) interference cancellation are possible. The interference estimation can utilize tentative data decisions.

The scheme is called hard decision (HD) interference cancellation. If tentative data decisions are not used, the scheme is called soft decision (SD) interference cancellation. The interference cancellation can also iteratively improve the interference estimates. Such a technique is utilized in multistage receivers.

Multistage hard-decision parallel interference cancellation (HD-PIC) output at $m$th stage can be presented as [31]

$$\mathcal{y}_{[HD-PIC]}(m) = \mathcal{y} - (\mathcal{R} - \mathbf{I}_{N_b KL})\hat{\mathcal{C}}(m-1)\mathcal{A}\hat{\boldsymbol{b}}(m-1), \tag{3.33}$$

where $\hat{\mathcal{C}}(m-1)$ and $\hat{\boldsymbol{b}}(m-1)$ denote the tentative channel and data estimates provided by the stage $m-1$ of the multistage HD-PIC receiver. The multistage PIC can be initialized by any linear equalizer type receiver. In the soft-decision parallel interference cancellation (SD-PIC) the amplitude-data product is estimated linearly without making explicit data decision, or a tentative data decision with a soft nonlinearity (such as hyberbolic tangent of linear clipper) is made. In other words, the product $\hat{\mathcal{C}}(m-1)\mathcal{A}\hat{\boldsymbol{b}}(m-1)$ of the estimates $\hat{\mathcal{C}}(m-1)$, $\hat{\boldsymbol{b}}(m-1)$, and $\mathcal{A}$ in (3.33) is replaced by an estimate $(\widehat{\mathcal{CA}\boldsymbol{b}})(m-1)$ of the product $\mathcal{CA}\boldsymbol{b}$. Contrary to the linear equalizer type multiuser receivers, the PIC receivers have inherently finite memory. The output of the HD-PIC receiver for $n$th symbol interval is

$$\mathbf{y}^{(n)}_{[PIC]}(m) = \mathbf{y}^{(n)} - \hat{\boldsymbol{\Psi}}^{(n)}_{[PIC]}(m). \tag{3.34}$$

The multiple-access interference estimate $\hat{\boldsymbol{\Psi}}^{(n)}_{[PIC]}(m)$ has the form

$$\hat{\boldsymbol{\Psi}}^{(n)}_{[PIC]}(m) = \sum_{i=-P_{PIC}}^{P_{PIC}} (\mathbf{R}(-i) - \delta_{i,0}\mathbf{I}_{KL})\, \hat{\mathbf{C}}^{(n+i)}(m-1)\mathbf{A}\hat{\mathbf{b}}^{(n+i)}(m-1), \tag{3.35}$$

where $\delta_{i,j}$ is the Kronecker delta, $P_{PIC} = \lceil \frac{T+T_m}{T} \rceil$, and $\lceil x \rceil$ denotes the smallest integer larger than or equal to $x$. The tentative estimates and decisions may be replaced by final ones at those symbol intervals for which they are available [127]. The result is decision-feedback HD-PIC receiver.

The multistage HD-PIC receiver has been proposed and analyzed for AWGN channels in [161, 31, 162, 163, 164, 165, 166]. The corresponding receivers for slowly fading channels have been studied in [167, 168, 169, 170, 171, 172, 173], and for relatively fast fading channels in [174, 175, 176, 33]. The HD-PIC receivers for transmissions with diversity encoding has been analyzed in [177], and for systems with multiple data rates has been studied in [178]. HD-PIC receivers for trellis-coded modulated CDMA systems in AWGN channels have been studied in [74]. The application of the HD-PIC to multiuser delay estimation in relatively fast fading channels has been considered in [179, 180, 33]. The

SD-PIC receivers with linear data-amplitude product estimation for slowly fading channels have been considered in [181, 182], and for multicellular systems in [183]. The SD-PIC receivers with soft nonlinearity have been considered for AWGN channels in [184, 185, 186]. Modifications of the PIC receiver have also been presented. Replacing the matrix $(\mathcal{R} - \mathbf{I}_{N_bKL})\hat{\mathcal{C}}(m-1)\mathcal{A}$ in (3.33) by an adaptively controlled weighting matrix for AWGN channels has been proposed in [184, 187, 188, 189, 190]. A PIC receiver with a weighting matrix in front of the matrix $(\mathcal{R} - \mathbf{I}_{N_bKL})\hat{\mathcal{C}}(m-1)\mathcal{A}$ in (3.33) has been proposed in [191, 192]. The weights were chosen in an ad hoc manner according to the reliability of the interference estimates. By applying expectation-maximization (EM) or space alternating generalized EM (SAGE) algorithm a class of iterative multistage receivers is obtained [193, 194, 195, 196]. The iterative EM or SAGE based receivers lead to apply different modified interference cancellation principles. The effect of delay estimation errors on the performance of the HD-PIC receiver has been considered in [76], and to SD-PIC receiver in [181].

The serial interference cancellation is performed on user by user basis [197, 198]. In the SIC, the amplitude and data of user 1 are estimated first. Using the obtained estimates the MAI estimate of user 1 is subtracted from the MF outputs of the rest of the users. Then the amplitude and data of user 2 are estimated, and the MAI estimate of user 2 is subtracted from the MF outputs of the users $k = 3, 4, \ldots, K$ etc. The cancellation should start with the user with the largest average power (indexed as user 1), the second powerful user (indexed as user 2) should be canceled next etc. The ordering is a problem in relatively fast fading channels, since it must be updated frequently. SD-SIC has been considered in [198, 199, 182], and HD-SIC in [197, 200, 201]. The SIC for multirate CDMA communications has been studied in [202, 203]. The SIC has the inherent problem that in asynchronous CDMA systems the processing-window of user 1 must be $K$ symbols so that the MAI caused by users $1, 2, \ldots, K-1$ can be canceled from the MF output of user $K$ [3]. Another problem with SIC is that it may not yield good enough performance in heavily loaded CDMA systems, where the performance of the conventional receiver is poor. The reason for that is that the SIC is initialized by a conventional receiver for user 1. If the MAI estimate of the signal of user 1 is poor in the cancellations, the estimation errors propagate to all users. SIC has good performance in systems where the powers of users differ significantly. This cannot be the case in systems with very large number of users. The effect of delay estimation errors to SD-SIC has been considered in [204], and to HD-SIC in [205].

The groupwise interference cancellation receivers detect the symbols of users within some group and form an estimate of the MAI caused by the users within that group based on the symbol decisions. The MAI estimate is then subtracted from the other users' MF outputs. The groupwise interference cancellation can be performed either serially or in parallel. The groupwise serial interference cancellation has been proposed in [206], and the groupwise parallel interference cancellation in [207]. The grouping can also be performed on consecutive symbols of a particular user in time [207]. The groupwise SIC has been proposed also for multiple data rate CDMA systems utiling multiple processing gains [208, 209]. The detector for a group of users can, in principle, apply any known multiuser detector, such as the conventional detector, the decorrelating detector, the PIC detector, or the maximum likelihood sequence detector. The groupwise interference cancellation is a special case of more general groupwise multiuser receivers [210, 211].

The interference cancellation can also be combined to linear equalizer type reception. Most often that is based on decision-feedback of the detected symbols to perform a subtractive cancellation of part of the MAI. The DF-IC in conjunction with the noise-whitening detector has been considered in [121, 120], and in conjunction with adaptive equalizer in [212, 213]. The DF-IC receiver for convolutionally encoded CDMA transmissions in AWGN channels has been studied in [110].

### *3.4.4 Other multiuser receivers*

Most suboptimal multiuser receivers fit into the categories presented in Sections 3.4.2 and 3.4.3. The other most interesting techniques are reviewed in short in this section.

In addition to the linear equalization or interference cancellation, the MLSD can be approximated by partial trellis-search algorithms. The log-likelihood metric (3.28) is computed for a subset of all possible data vectors $\boldsymbol{b}$. Different criteria to choose the subsets result in different partial trellis-search algorithms. The application of sequential decoding has been proposed in [214]. Some of the groupwise multiuser receivers discussed above in Section 3.4.3 can also be interpreted as partial trellis-search algorithms [207]. Other partial tree-search algorithms have been proposed in [215, 216]. A partial trellis-search algorithm for trellis-coded modulated CDMA transmissions in AWGN channels has been studied in [74], and for convolutionally encoded transmissions in [75].

The multiuser parameter estimation, i.e., the complex amplitude and delay estimation has gained increasing interest. Since there is usually no a priori distribution available for the delays, maximum likelihood estimation is usually considered to be the optimal technique for delay acquisition and tracking [217, 218, 84]. This approach has also been considered for amplitude estimation [219]. Suboptimal techniques include subspace estimators [218, 220, 221, 222, 223], a hierarchic ML estimation [224], large sample mean ML estimation [225], an extended Kalman filter [226, 227], recursive least squares algorithm [227], and sequential estimation [228, 229]. The amplitude estimation in AWGN channels with unknown delays has been the topic in [230]. The estimation of number of active users in AWGN channels has been studied in [231, 232, 233, 234, 235].

Multiuser detection based on empirical distribution of the MAI has been proposed in [236, 237]. The distribution of the MAI is estimated by forming a corresponding histogram, and the received symbol is selected so that it matches best into the histogram.

Neural networks have been proposed to approximate the decision regions of the optimal receivers. Multilayer perceptron networks both for centralized and decentralized detection in AWGN channels have been proposed in [238] and single-layer perceptron networks in [239]. Self-organizing maps for centralized detection in AWGN channels have been studied in [240]. Radial basis functions for decentralized detection in AWGN channels have been considered in [241]. Hopfield networks for centralized detection in AWGN channels have been proposed in [242, 243].

## 3.5 EVOLVING CDMA SYSTEMS

The IS-95 cellular system [244, 245] is a second generation cellular wireless communication system applying CDMA technology. Since the bandwith of IS-95 is relatively narrow (1.25 MHz), IS-95 is often called a narrowband CDMA system. There are several third generation systems under development, which are proposed to utilize DS-CDMA technique. One of them is a CDMA system utilizing multiuser detection [246, 247] proposed in a European research project FRAMES [248]. Another developing DS-CDMA system is so called wideband CDMA (W-CDMA) system proposed for Japan [249, 250]. The modified IS-95 standard IS-665 introduces also a wideband CDMA system [49]. All the above mentioned third generation CDMA systems are proposed to utilize multiuser receivers, namely some form of multiple-access interference (MAI) cancellation.

## 3.6 CONCLUSIONS

In this paper we summarize existing work in the field of multiuser detectors and discuss relative contribution of this technique to the overall system capacity. It was pointed out that in a cellular mobile network, with perfect power control, expected improvement factor of the system capacity, when multiuser detectors are used, is 2-3. Power control imperfections can also severely degrade the system capacity As an extreme case in the presence of severe near-far effect CDMA can completely fail. These effects are eliminated in multiuser receiver, that makes these schemes very attractive for next generation of CDMA systems. A comprehensive review of multiuser detectors is presented in separate subsections mainly: optimal multiuser demodulation, linear equalizer type multiuser demodulation and interference cancellation covering parallel, serial and groupwise receivers. What used to be theory few years ago is becoming reality these days. This is discussed within Section 5 – Evolving CDMA systems. In this section a short summary of new standards and projects, proposing employment of multiuser techniques in the third generation of the mobile communications, is presented. Further evolution of this technique is expected in years to come. Especially intensive research is expected in the field of further simplification of these schemes to make practical massive implementation more feasible. It is expected that the third generation of the mobile communications will be using this technique. It was also clearly emphasized that capacity improvement in the system can be effectively achieved by using other steps. Some of these steps like error correcting coding, adaptive antennas are also covered within this book by other authors, as mentioned in the introduction of the paper.

## Notes

1. The number of propagation paths is assumed to be equal for all users for notational simplicity.

2. This is true due to the key assumption that the channel noise $z(t)$ has Gaussian complex amplitude distribution and the delays are known so that the MF outputs can be sampled at correct times.

3. For notational compactness, the discrete-time index $n$ will be left out from $\mathbf{R}^{(n)}(i)$, when possible without confusion.

4. Although the multiuser receivers have gained most interest in conjunction with DS-CDMA systems, they can be applied in any non-orthogonal multiple-access scheme. They have been considered for TDMA [60], hybrid DS-CDMA/TDMA [44, 61], FH-CDMA [62, 63], MC-CDMA [64, 65, 66, 67], wavelet packet CDMA [68], and spread-signature CDMA [69] communications.

5. *Centralized* multiuser detectors make a joint detection of the symbols of different users. *Decentralized* multiuser detectors (called sometimes also single-user detectors) demodulate a signal of one desired user only.

6. The definition of Cholesky factorization used in this paper is an *upper* triangular matrix times a *lower* triangular matrix [121, 120] as opposed to the usual lower triangular times upper triangular matrix [122].

7. The spreading-coding tradeoff deals with the question how much on the bandwith expansion should be invested in forward error-correcting encoding and how much should be invested in spreading.

## References

[1] S. Verdú, "Multiuser detection," in *Advances in Statistical Signal Processing*, vol. 2, pp. 369–409. JAI Press Inc., Greenwich, CT, 1993.

[2] A. Duel-Hallen, J. Holtzman, and Z. Zvonar, "Multiuser detection for CDMA systems," *IEEE/ACM Pers. Commun.*, vol. 2, pp. 46–58, Apr. 1995.

[3] S. Moshavi, "Multi-user detection for DS-CDMA communications," *IEEE Commun. Mag.*, vol. 34, no. 10, pp. 124–137, Oct. 1996.

[4] P. B. Rapajic and B. S. Vucetic, "Adaptive receiver structures for asynchronous CDMA systems," *IEEE J. Select. Areas Commun.*, vol. 12, no. 4, pp. 685–697, May 1994.

[5] P. B. Rapajic and B. S. Vucetic, "Linear adaptive transmitter-receiver structures for asynchronous CDMA systems," *European Trans. Telecommun.*, vol. 6, no. 1, pp. 21–27, January-February 1995.

[6] U. Madhow and M. L. Honig, "MMSE interference suppression for direct-sequence spread-spectrum CDMA," *IEEE Trans. Commun.*, vol. 42, no. 12, pp. 3178–3188, Dec. 1994.

[7] S. L. Miller, "An adaptive direct-sequence code-division multiple-access receiver for multiuser interference rejection," *IEEE Trans. Commun.*, vol. 43, no. 2/3/4, pp. 1746–1755, Feb./Mar./Apr. 1995.

[8] L. B. Milstein, "Interference rejection techniques in spread spectrum communications," *Proc. IEEE*, vol. 76, no. 6, pp. 657–971, June 1988.

[9] J. W. Ketchum and J. G. Proakis, "Adaptive algorithm for estimating and suppressing narrowband interference in PN spread spectrum systems," *IEEE Trans. Commun.*, vol. 30, no. 5, pp. 913–924, May 1982.

[10] L. B. Milstein and R. A. Iltis, "Signal processing for interference rejection in spread spectrum communications," *IEEE Acoust. Speech Sign. Proc. Mag.*, vol. 44, no. 1, pp. 18–31, Apr. 1986.

[11] A. K. Gupta, "On suppression of sinusoidal signal in broad band noise," *IEEE Trans. Acoust. Speech Sign. Proc.*, vol. 33, no. 4, Aug. 1985.

[12] L. Li and L. B. Milstein, "Rejection of narrow band interference in PN spread spectrum signals using transversal filters," *IEEE Trans. Commun.*, vol. 30, no. 5, pp. 925–928, May 1982.

[13] L. Li and L. B. Milstein, "Rejection of pulsed CW interference in PN spread spectrum signals using complex adaptive filters," *IEEE Trans. Commun.*, vol. 31, no. 1, pp. 10–20, Jan. 1983.

[14] E. Masry, "Closed form analytical results for the rejection of narrow band interference in PN spread spectrum systems–part I: Linear prediction filters," *IEEE Trans. Commun.*, vol. 32, no. 8, pp. 888–896, Aug. 1984.

[15] E. Masry, "Closed form analytical results for the rejection of narrow band interference in PN spread spectrum systems–part II: Linear interpolator filters," *IEEE Trans. Commun.*, vol. 33, no. 1, pp. 10–19, Jan. 1985.

[16] R. Vijyan and H. V. Poor, "Nonlinear techniques for interference suppression in spread spectrum systems," *IEEE Trans. Commun.*, vol. 38, pp. 1060–1065, July 1991.

[17] L. Garth and H. V. Poor, "Narrowband interference suppression in impulsive channels," *IEEE Trans. Aeros. Electron. Syst.*, vol. 28, no. 1, pp. 15–34, Jan. 1992.

[18] C. J. Masreliez, "Approximate with non-gaussian filtering with linear state and observation realtions," *IEEE Trans. Autom. Contr.*, pp. 107–110, Feb. 1975.

[19] M. A. Hasan, J. C. Lee, and V. K. Bhargava, "A narrowband interference canceller with adjustable center weight," *IEEE Trans. Commun.*, vol. 42, no. 2/3/4, pp. 877–880, Feb./Mar./Apr. 1994.

[20] J. F. Doherty, "Linearly constrained direct-sequence spread-spectrum interference rejection," *IEEE Trans. Commun.*, vol. 42, no. 2/3/4, pp. 865–871, Feb./Mar./Apr. 1994.

[21] I. Ansari and R. Viswanathan, "Performance study of maximum likelihood receivers and transversal filters for the detection of direct-sequence spread-spectrum signal in narrowband interference," *IEEE Trans. Commun.*, vol. 42, no. 2/3/4, pp. 1939–1946, Feb./Mar./Apr. 1994.

[22] L. A. Rusch and H. V. Poor, "Narrowband interference suppression in CDMA spread spectrum communications," *IEEE Trans. Commun.*, vol. 42, no. 2/3/4, pp. 1769–1779, Feb./Mar./Apr. 1994.

[23] S. G. Glisic, A. Mämmelä, P. Kaasila, and M. D. Pajkovic, "Rejection of a FH signal in a DS spread-spectrum system using complex adaptive filters," *IEEE Trans. Commun.*, vol. 43, no. 1, pp. 136–145, Jan. 1995.

[24] S. G. Glisic and M. D. Pajkovic, "Rejection of a frequency sweeping signal in a DS spread-spectrum system using complex adaptive filters," *IEEE Trans. Commun.*, vol. 43, no. 5, pp. 1982–1991, May 1995.

[25] D. M. Grieco, "The capacity achievable with a broadband CDMA microcellular underlay to an existing cellular macrosystem," *IEEE J. Select. Areas Commun.*, vol. 12, no. 4, pp. 744–750, May 1994.

[26] L. B. Milstein, D. L Schilling, R. L. Pickholtz, V. Erceg, M. Kullback, E. G. Kanterakis, D. S. Fishman, W. H. Biederman, and E. D. Salermo, "On the feasibility of a CDMA overlay for personal communication networks," *IEEE J. Select. Areas Commun.*, pp. 655–667, May 1992.

[27] T. M. Cover and J. A. Thomas, *Elements of Information Theory*, John Wiley, New York, 1991.

[28] M. B. Pursley, "Performance evaluation for phase-coded spread-spectrum multiple-access communication–Part I: System analysis," *IEEE Trans. Commun.*, vol. 25, no. 8, pp. 795–799, August 1977.

[29] S. Verdú, "Minimum probability of error for asynchronous Gaussian multiple-access channels," *IEEE Trans. Inform. Th.*, vol. 32, no. 1, pp. 85–96, Jan. 1986.

[30] R. Lupas and S. Verdú, "Near-far resistance of multiuser detectors in asynchronous channels," *IEEE Trans. Commun.*, vol. 38, no. 4, pp. 496–508, Apr. 1990.

[31] M. K. Varanasi and B. Aazhang, "Multistage detection in asynchronous code-division multiple-access communications," *IEEE Trans. Commun.*, vol. 38, no. 4, Apr. 1990.

[32] Z. Zvonar, *Multiuser Detection for Rayleigh Fading Channels*, Ph.D. thesis, Northeastern University, Boston, MA, Sept. 1993.

[33] M. Latva-aho and J. Lilleberg, "Parallel interference cancellation in multiuser CDMA channel estimation," *Wireless Pers. Commun., Kluwer, to appear*, June 1998.

[34] J. G. Proakis, *Digital Communications*, McGraw-Hill, Inc., New York, 3rd edition, 1995.

[35] K. S. Gilhousen, I. M. Jacobs, R. Padovani, A. J. Viterbi, L. A. Weaver, and C. E. Wheatley III, "On the capacity of a cellular CDMA system," *IEEE Trans. Vehic. Tech.*, vol. 40, no. 2, pp. 303–312, May 1991.

[36] A. M. Viterbi and A. J. Viterbi, "Erlang capacity of a power controlled CDMA system," *IEEE J. Select. Areas Commun.*, vol. 11, no. 6, pp. 892–899, Aug. 1993.

[37] Jr. J. C. Liberti and T. S. Rappaport, "Analytical results for capacity improvements in CDMA," *IEEE Trans. Vehic. Tech.*, vol. 43, no. 3, pp. 680–690, Aug. 1994.

[38] C. Y. Huang and D. G. Daut, "Evaluation of capacity for CDMA systems on frequency-selective fading channels," in *Proc. IEEE Int. Conf. Universal Personal Communications*, Boston, MA, Sept. 29 –Oct. 2 1996, pp. 975–979.

[39] A. F. Naguib, A. Paulraj, and T. Kailath, "Capacity improvement with base-station antenna arrays in cellular CDMA," *IEEE Trans. Vehic. Tech.*, vol. 43, no. 3, pp. 691–698, Aug. 1994.

[40] R. Kohno and L. B. Milstein, "Spread spectrum access methods for wireless communications," *IEEE Commun. Mag.*, pp. 58–67, Jan. 1995.

[41] W.C.Y. Lee, "Overview of cellular CDMA," *IEEE Trans. Vehic. Tech.*, vol. 40, no. 2, pp. 291–302, May 1991.

[42] S. Kondo and L. B. Milstein, "Performance of multicarrier DS-CDMA systems," *IEEE Trans. Commun.*, vol. 44, no. 2, pp. 238–246, Feb. 1996.

[43] P. Jung, P. W. Baier, and A. Steil, "Advantages of CDMA and spread spectrum techniques over FDMA and TDMA in cellular mobile radio applications," *IEEE Trans. Vehic. Tech.*, vol. 42, no. 3, pp. 357–364, Aug. 1993.

[44] J. Blanz, A. Klein, M. Nasshan, and A. Steil, "Performance of a cellular hybrid C/TDMA mobile radio system applying joint detection and coherent receiver antenna diversity," *IEEE J. Select. Areas Commun.*, vol. 12, no. 4, pp. 568–579, May 1994.

[45] P. Newson and M. R. Heath, "The capacity of a spread spectrum CDMA system for cellular mobile radio with consideration of system imperfections," *IEEE J. Select. Areas Commun.*, vol. 12, no. 4, pp. 673–684, May 1994.

[46] K. I. Kim, "CDMA cellular engineering issues," *IEEE Trans. Vehic. Tech.*, vol. 42, no. 3, pp. 345–350, Aug. 1993.

[47] A. J. Viterbi, A. M. Viterbi, K. S. Gilhousen, and E. Zehavi, "Soft handoff extends CDMA cell coverage and increases reverse link capacity," *IEEE J. Select. Areas Commun.*, vol. 12, no. 8, pp. 1281–1288, Oct. 1994.

[48] A. J. Viterbi, A. M. Viterbi, and E. Zehavi, "Performance of power-controlled wideband terrestrial digital communication," *IEEE Trans. Commun.*, vol. 41, no. 4, pp. 559–569, Apr. 1993.

[49] A. Fukasawa, T. Sato, Y. Takizawa, T. Kato, M. Kawabe, and R. E. Fisher, "Wideband CDMA system for personal radio communications," *IEEE Commun. Mag.*, vol. 34, no. 10, pp. 116–123, Oct. 1996.

[50] T. Eng and L. B. Milstein, "Comparison of hybrid FDMA/CDMA systems in frequency selective rayleigh fading," *IEEE J. Select. Areas Commun.*, vol. 12, no. 5, pp. 938–951, June 1994.

[51] M. G. Jansen R. Prasad and A. Kegel, "Capacity analysis of a cellular direct sequence code division multiple access system with imperfect power control," *IEICE Trans. Commun.*, vol. E76-B, no. 8, pp. 894–905, Aug. 1993.

[52] M. G. Jansen and R. Prasad, "Capacity, throughput and delay analysis of a cellular DS CDMA system with imperfect power control and imperfect sectorization," *IEEE Trans. Vehic. Tech.*, vol. 44, no. 1, pp. 67–75, Feb. 1995.

[53] W. Huang and V. K. Bhargava, "Performance evaluation of a DS/CDMA cellular system with voice and data services," in *Proc. IEEE Int. Symp. Personal, Indoor and Mobile Radio Commun.*, Taipei, Taiwan, Oct. 15–18 1996, pp. 588–592.

[54] C.-C. Lee and R. Steele, "Closed-loop power control in CDMA systems," *IEE Proc. — Series F*, vol. 143, no. 4, pp. 231–239, Aug. 1996.

[55] K. S. Schneider, "Optimum detection of code division multiplexed signals," *IEEE Trans. Aeros. Electron. Syst.*, vol. AES-15, no. 1, pp. 181–185, January, 1979.

[56] T. K. Kashihara, "Adaptive cancellation of mutual interference in spread spectrum multiple access," in *Proc. IEEE Int. Conf. Commun.*, 1980, pp. 44.4.1–44.4.5.

[57] R. Kohno, H. Imai, and M. Hatori, "Cancellation technique of co-channel interference in asynchronous spread-spectrum multiple-access systems," *IEICE Trans. Commun.*, vol. 65-A, pp. 416–423, May 1983.

[58] S. Verdú, "Optimum multiuser asymptotic efficiency," *IEEE Trans. Commun.*, vol. 34, no. 9, pp. 890–897, Sept. 1986.

[59] P. Jung and P. D. Alexander, "A unified approach to multiuser detectors for CDMA and their geometrical interpretations," *IEEE J. Select. Areas Commun.*, vol. 14, no. 8, pp. 1595–1601, Oct. 1996.

[60] P. A. Ranta, A. Hottinen, and Z.-C. Honkasalo, "Co-channel interference cancelling receiver for TDMA mobile systems," in *Proc. IEEE Int. Conf. Commun.*, Seattle, WA, June 18-22 1995, vol. 1, pp. 17–21.

[61] G. Kramer, U. Loher, J. Ruprecht, and P. Jung, "A comparison of demodulation techniques for code time division multiple access," in *Proc. IEEE Glob. Telecommun. Conf.*, London, U.K., Nov. 18-22 1996, vol. 1, pp. 525–529.

[62] T. Mabuchi, R. Kohno, and H. Imai, "Multiuser detection scheme based on canceling cochannel interference for MFSK/FH-SSMA system," *IEEE J. Select. Areas Commun.*, vol. 12, no. 4, pp. 593–604, May 1994.

[63] K. W. Halford and M. Brandt-Pearce, "Performance of a multistage multiuser detector for a frequency hopping multiple-access system," in *Proc. Conf. Inform. Sciences Systems*, Princeton University, Princeton, NJ, Mar. 20–22 1996, vol. 1, pp. 605–610.

[64] A. Haimovich and Y. Bar-Ness, "On the performance of a stochastic gradient-based decorrelation algorithm for multiuser multicarrier CDMA," *Wireless Pers. Commun., Kluwer*, vol. 2, no. 4, pp. 357–371, 1996.

[65] L. K. Rasmussen and T. J. Lim, "Detection techniques for direct sequence & multicarrier variable rate broadband CDMA," in *Proc. IEEE Int. Conf. Communication Systems and IEEE Int. Workshop Intelligent Signal Processing & Communication Systems*, Singapore, Nov. 25-29 1996, vol. 3, pp. 1526–1530.

[66] Y. Sanada and M. Nakagawa, "A multiuser interference cancellation technique utilizing convolutional codes and orthogonal multicarrier modulation for wireless indoor communications," *IEEE J. Select. Areas Commun.*, vol. 14, no. 8, pp. 1500–1509, Oct. 1996.

[67] L. Vandendorpe and O. van de Wiel, "Performance analysis of linear joint equalization and multiple access interference cancellation for multitone CDMA," *Wireless Pers. Commun., Kluwer*, vol. 3, no. 1-2, pp. 17–36, 1996.

[68] R. E. Learned, A. S. Willsky, and D. M. Boroson, "Low complexity optimal joint detection for oversaturated multiple access communications," *IEEE Trans. Sign. Proc.*, vol. 45, no. 1, pp. 113–123, Jan. 1997.

[69] S. Beheshi and G. W. Wornell, "Interference cancellation and decoding in spread-signature CDMA systems," in *Proc. IEEE Vehic. Tech. Conf.*, Phoenix, AZ, May 4–7 1997, vol. 1, pp. 26–30.

[70] S. Verdú, *Optimum Multiuser Signal Detection*, Ph.D. thesis, Dept. of Elec. and Comp. Eng., University of Illinois, Urbana-Champaign, 1984.

[71] Z. Zvonar and D. Brady, "Multiuser detection in single-path fading channels," *IEEE Trans. Commun.*, vol. 42, no. 2/3/4, pp. 1729–1739, Feb./Mar./Apr. 1994.

[72] Z. Zvonar, "Multiuser detection in asynchronous CDMA frequency-selective fading channels," *Wireless Pers. Commun., Kluwer*, vol. 3, no. 3–4, pp. 373–392, 1996.

[73] L. Lu and W. Sun, "The minimal eigenvalues of a class of block-tridiagonal matrices," *IEEE Trans. Inform. Th.*, vol. 43, no. 2, pp. 787–791, Mar. 1997.

[74] U. Fawer and B. Aazhang, "Multiuser receivers for code-division multiple-access systems with trellis-based modulation," *IEEE J. Select. Areas Commun.*, vol. 14, no. 8, pp. 1602–1609, Oct. 1996.

[75] T. R. Giallorenzi and S. G. Wilson, "Multiuser ML sequence estimator for convolutionally coded asynchronous DS-CDMA systems," *IEEE Trans. Commun.*, vol. 44, no. 8, pp. 997–1007, Aug. 1996.

[76] S. D. Gray, M. Kocic, and D. Brady, "Multiuser detection in mismatched multiple-access channels," *IEEE Trans. Commun.*, vol. 43, no. 12, pp. 3080–3089, Dec. 1995.

[77] H. V. Poor, "On parameter estimation in DS/SSMA formats," in *Advances in Communications and Signal Processing*, W. A. Porter and S. C. Kak, Eds., vol. 129 of *Lecture Notes in Control and Information Sciences*, pp. 59–70. Springer-Verlag, Berlin Heidelberg, Germany, 1989.

[78] Z. Xie, C. K. Rushforth, R. T. Short, and T. K. Moon, "Joint signal detection and parameter estimation in multiuser communications," *IEEE Trans. Commun.*, vol. 41, no. 7, pp. 1208–1216, Aug. 1993.

[79] M. K. Varanasi and S. Vasudevan, "Multiuser detectors for synchronous CDMA communication over non-selective Rician fading channels," *IEEE Trans. Commun.*, vol. 42, no. 2/3/4, pp. 711–722, Feb./Mar./Apr. 1994.

[80] S. Vasudevan and M. K. Varanasi, "Optimum diversity combiner based multiuser detection for time-dispersive Ricean fading CDMA channels," *IEEE J. Select. Areas Commun.*, vol. 12, no. 4, pp. 580–592, May 1994.

[81] S. Vasudevan and M. K. Varanasi, "Achieving near-optimum asymptotic efficiency and fading resistance over the time-varying Rayleigh-faded CDMA channel," *IEEE Trans. Commun.*, vol. 44, no. 9, pp. 1130–1143, Sept. 1996.

[82] P.-A. Sung and K.-C. Chen, "A linear minimum mean square error multiuser receiver in Rayleigh-fading channels," *IEEE J. Select. Areas Commun.*, vol. 14, no. 8, pp. 1583–1594, Oct. 1996.

[83] H. L. van Trees, *Detection, Estimation, and Modulation Theory, Part III*, John Wiley, New York, 1971.

[84] J. Lilleberg, E. Nieminen, and M. Latva-aho, "Blind iterative multiuser delay estimator for CDMA," in *Proc. IEEE Int. Symp. Personal, Indoor and Mobile Radio Commun.*, Taipei, Taiwan, Oct. 15–18 1996, vol. 2, pp. 565–568.

[85] H. V. Poor and S. Verdú, "Single-user detectors for multiuser channels," *IEEE Trans. Commun.*, vol. 36, no. 1, pp. 50–60, Jan. 1988.

[86] A. Klein and P. W. Baier, "Linear unbiased data estimation in mobile radio systems applying CDMA," *IEEE J. Select. Areas Commun.*, vol. 11, no. 7, pp. 1058–1066, Sept. 1993.

[87] R. Lupas and S. Verdú, "Linear multiuser detectors for synchronous code-division multiple-access channels," *IEEE Trans. Inform. Th.*, vol. 34, no. 1, pp. 123–136, Jan. 1989.

[88] M. K. Varanasi and B. Aazhang, "Optimally near-far resistant multiuser detection in differentially coherent synchronous channels," *IEEE Trans. Inform. Th.*, vol. 37, no. 4, pp. 1006–1018, July 1991.

[89] M. K. Varanasi, "Noncoherent detection in asynchronous multiuser channels," *IEEE Trans. Inform. Th.*, vol. 39, no. 1, pp. 157–176, Jan. 1993.

[90] Z. Zvonar and D. Brady, "Suboptimal multiuser detector for frequency-selective Rayleigh fading synchronous CDMA channels," *IEEE Trans. Commun.*, vol. 42, no. 2/3/4, pp. 1729–1739, Feb./Mar./Apr. 1994.

[91] A Klein, *Multi-user Detection of CDMA Signals – Algorithms and Their Application to Cellular Mobile Radio*, Ph.D. thesis, University of Kaiserslautern, Düsseldorf, Germany, 1996.

[92] M. Stojanovic and Z. Zvonar, "Linear multiuser detection in time-varying multipath fading channels," in *Proc. Conf. Inform. Sciences Systems*, Princeton University, Princeton, NJ, Mar. 20–22 1996, vol. 1, pp. 349–354.

[93] M. Stojanovic and Z. Zvonar, "Performance of linear multiuser detectors in time-varying multipath fading CDMA channels," in *Proc. Commun. Th. Mini-Conf. in conj. IEEE Glob. Telecommun. Conf.*, London, U.K., Nov. 18-22 1996, pp. 163–167.

[94] T. Kawahara and T. Matsumoto, "Joint decorrelating multiuser detection and channel estimation in asynchronous CDMA mobile communications channels," *IEEE Trans. Vehic. Tech.*, vol. 44, no. 3, pp. 506–515, Aug. 1995.

[95] H. C. Huang, *Combined Multipath Processing, Array Processing, and Multiuser Detection for DS-CDMA Channels*, Ph.D. thesis, Princeton University, Princeton, NJ, Jan. 1996.

[96] S. Y. Miller, *Detection and Estimation in Multiple-Access Channels*, Ph.D. thesis, Princeton University, Princeton, NJ, 1989.

[97] S. Y. Miller and S. C. Schwartz, "Integrated spatial-temporal detectors for asynchronous Gaussian multiple-access channels," *IEEE Trans. Commun.*, vol. 43, no. 2/3/4, pp. 396–411, Feb./Mar./Apr. 1995.

[98] P. Jung, J. Blanz, M. Nasshan, and P. W. Baier, "Simulation of the uplink of JD-CDMA mobile radio systems with coherent receiver antenna diversity," *Kluwer, Wireless Personal Communications*, vol. 1, no. 2, pp. 61–89, Jan. 1994.

[99] P. Jung and J. Blanz, "Joint detection with coherent receiver antenna diversity in CDMA mobile radio systems," *IEEE Trans. Vehic. Tech.*, vol. 44, no. 1, pp. 76–88, Feb. 1995.

[100] T. Brown and M. Kaveh, "A decorrelating detector for use with antenna arrays," *Int. J. Wireless Inform. Networks*, vol. 2, no. 4, pp. 239–246, 1995.

[101] Z. Zvonar, "Combined multiuser detection and diversity reception for wireless CDMA systems," *IEEE Trans. Vehic. Tech.*, vol. 45, no. 1, pp. 205–211, Feb. 1996.

[102] S. Kandala, E. S. Sousa, and S. Pasupathy, "Multi-user multi-sensor detectors for CDMA networks," *IEEE Trans. Commun.*, vol. 43, no. 2/3/4, pp. 946–957, Feb./Mar./Apr. 1995.

[103] S. Kandala, E. S. Sousa, and S. Pasupathy, "Decorrelators for multi-sensor systems in CDMA networks," *European Trans. Telecommun.*, vol. 6, no. 1, pp. 29–40, Jan.-Feb. 1995.

[104] M. J. Juntti and J. O. Lilleberg, "Comparative analysis of conventional and multiuser detectors in multisensor receivers," in *Proc. IEEE Military Commun. Conf., submitted*, Monterey, CA, Nov. 2–5 1997.

[105] M. Saquib, R. Yates, and N. Mandayam, "Decorrelating detectors for a dual rate synchronous DS/CDMA system," in *Proc. IEEE Vehic. Tech. Conf.*, Atlanta, GA, Apr. 28 - May 1 1996, vol. 1, pp. 377–381.

[106] M. J. Juntti and J. O. Lilleberg, "Linear FIR multiuser detection for multple data rate CDMA systems," in *Proc. IEEE Vehic. Tech. Conf.*, Phoenix, AZ, May 4–7 1997, vol. 2, pp. 455–459.

[107] D. S. Chen and S. Roy, "An adaptive multiuser receiver for CDMA systems," *IEEE J. Select. Areas Commun.*, vol. 12, no. 6, pp. 808–816, May 1994.

[108] U. Mitra and H. V. Poor, "Analysis of an adaptive decorrelating detector for synchronous CDMA channels," *IEEE Trans. Commun.*, vol. 44, no. 2, pp. 257–268, Feb. 1996.

[109] U. Mitra and H. V. Poor, "Adaptive decorrelating detectors for CDMA systems," *Wireless Pers. Commun., Kluwer*, vol. 2, no. 4, pp. 415–440, 1996.

[110] T. R. Giallorenzi and S. G. Wilson, "Suboptimum multiuser receivers for convolutionally coded asynchronous DS-CDMA systems," *IEEE Trans. Commun.*, vol. 44, no. 9, pp. 1183–1196, Sept. 1996.

[111] A. Kajiwara and M. Nakagawa, "Microcellular CDMA system with a linear multiuser interference canceler," *IEEE J. Select. Areas Commun.*, vol. 12, no. 4, pp. 605–611, May 1994.

[112] F. van Heeswyk, D. D. Falconer, and A. U. H. Sheikh, "Decorrelating detectors for quasi-synchronous CDMA," *Wireless Pers. Commun., Kluwer*, vol. 3, no. 1-2, pp. 129–147, 1996.

[113] F. van Heeswyk, D. D. Falconer, and A. U. H. Sheikh, "A delay independent decorrelating detector for quasi-synchronous CDMA," *IEEE J. Select. Areas Commun.*, vol. 14, no. 8, pp. 1619–1626, Oct. 1996.

[114] R. A. Iltis and L. Mailaender, "Multiuser detection of quasisynchronous CDMA signals using linear decorrelators," *IEEE Trans. Commun.*, vol. 44, no. 11, pp. 1561–1571, Nov. 1996.

[115] R. A. Iltis, "Demodulation and code acquisition using decorrelator detectors for QS-CDMA," *IEEE Trans. Commun.*, vol. 44, no. 11, pp. 1553–1560, Nov. 1996.

[116] F.-C. Zheng and S. K. Barton, "On the performance of near-far resistant CDMA detectors in the presence of synchronization errors," *IEEE Trans. Commun.*, vol. 43, no. 12, pp. 3037–3045, Dec. 1995.

[117] S. Parkvall, E. Ström, and B. Ottersten, "The impact of timing errors on the performance of linear DS-CDMA receivers," *IEEE J. Select. Areas Commun.*, vol. 14, no. 8, pp. 1660–1668, Oct. 1996.

[118] B.-P. Paris, "Finite precision decorrelating receiver for multiuser CDMA communication systems," *IEEE Trans. Commun.*, vol. 44, no. 4, pp. 496–507, Apr. 1994.

[119] S. Bulumulla and S. S. Venkatesh, "On the quantized input decorrelating detector," in *Proc. Conf. Inform. Sciences Systems*, Princeton University, Princeton, NJ, Mar. 20–22 1996, vol. 1, pp. 595–598.

[120] A. Duel-Hallen, "A family of multiuser decision-feedback detectors for asynchronous code-division multiple-access channels," *IEEE Trans. Commun.*, vol. 43, no. 2/3/4, pp. 421–434, Feb./Mar./Apr. 1995.

[121] A. Duel-Hallen, "Decorrelating decision-feedback multiuser detector for synchronous code-division multiple-access channel," *IEEE Trans. Commun.*, vol. 41, no. 2, pp. 285–290, Feb. 1993.

[122] G. H. Golub and C. F. Van Loan, *Matrix Computations*, The Johns Hopkins University Press, Baltimore, 2nd edition, 1989.

[123] L. Wei and C. Schlegel, "Synchronous DS-SSMA system with improved decorrelating decision-feedback multiuser detection," *IEEE Trans. Vehic. Tech.*, vol. 43, no. 3, pp. 767–772, Aug. 1994.

[124] L. Wei and L. K. Rasmussen, "A near ideal noise whitening filter for an asynchronous time-varying CDMA system," *IEEE Trans. Commun.*, vol. 44, no. 10, pp. 1355–1361, Oct. 1996.

[125] C. Schlegel, S. Roy, P. D. Alexander, and Z.-J. Xiang, "Multiuser projection receivers," *IEEE J. Select. Areas Commun.*, vol. 14, no. 8, pp. 1610–1618, Oct. 1996.

[126] S. Kay, *Fundamentals of Statistical Signal Processing: Estimation Theory*, Prentice-Hall, Englewood Cliffs, NJ, 1993.

[127] Z. Xie, R. T. Short, and C. K. Rushforth, "A family of suboptimum detectors for coherent multiuser communications," *IEEE J. Select. Areas Commun.*, vol. 8, no. 4, pp. 683–690, May 1990.

[128] A. Klein, G. K. Kaleh, and P. W. Baier, "Zero forcing and minimum mean-square-error equalization for multiuser detection in code-division multiple access channels," *IEEE Trans. Vehic. Tech.*, vol. 45, no. 2, pp. 276–287, May 1996.

[129] W.-C. Wu and K.-C. Chen, "Linear multiuser detectors for synchronous CDMA communication over Rayleigh fading channels," in *Proc. IEEE Int. Symp. Personal, Indoor and Mobile Radio Commun.*, Taipei, Taiwan, Oct. 15-18 1996, pp. 578–582.

[130] X. Bernstein and A. M. Haimovich, "Space-time optimum combining for CDMA communications," *Wireless Pers. Commun., Kluwer*, vol. 3, no. 1-2, pp. 73–89, 1996.

[131] S. D. Gray, J. C. Preisig, and D. Brady, "Multiuser detection in a horizontal underwater acoustic channel using array observations," *IEEE Trans. Sign. Proc.*, vol. 45, no. 1, pp. 148–160, Jan. 1997.

[132] M. L. Honig and W. Veerakachen, "Performance variability of linear multiuser detection for DS/CDMA," in *Proc. IEEE Vehic. Tech. Conf.*, Atlanta, GA, Apr. 28 - May 1 1996, vol. 1, pp. 372–376.

[133] H. V. Poor and S. Verdú, "Probability of error in MMSE multiuser detection," *IEEE Trans. Inform. Th.*, vol. 43, no. 3, pp. 858–871, May 1997.

[134] S. L. Miller, "Training analysis of adaptive interference suppression for direct-sequence code-division multiple-access systems," *IEEE Trans. Commun.*, vol. 44, no. 4, pp. 488–495, Apr. 1996.

[135] K. B. Lee, "Orthogonalization based adaptive interference suppression for direct-sequence code-division multiple-access systems," *IEEE Trans. Commun.*, vol. 44, no. 9, pp. 1082–1085, Sept. 1996.

[136] G. Woodward, P. Rapajic, and B. S. Vucetic, "Adaptive algorithms for asynchronous DS-CDMA receivers," in *Proc. IEEE Int. Symp. Personal, Indoor and Mobile Radio Commun.*, Taipei, Taiwan, Oct. 15-18 1996, pp. 583–587.

[137] M. Latva-aho and M. Juntti, "Modified adaptive LMMSE receiver for DS-CDMA systems in fading channels," in *Proc. IEEE Int. Symp. Personal, Indoor and Mobile Radio Commun., to appear*, Helsinki, Finland, Sept. 1–4 1997.

[138] V. V. Veeravalli and B. Aazhang, "On the coding-spreading tradeoff in CDMA systems," in *Proc. Conf. Inform. Sciences Systems*, Princeton University, Princeton, NJ, Mar. 20–22 1996, vol. 2, pp. 1136–1141.

[139] I. Oppermann and B. S. Vucetic, "Capacity of a coded direct sequence spread spectrum system over fading satellite channels using an adaptive LMS-MMSE receiver," *IEICE Trans. Fundamentals Elec., Commun. and Comp. Sc.*, vol. E79-A, no. 12, pp. 2043–2049, Dec. 1996.

[140] L.-C. Chu and U. Mitra, "Improved MMSE-based multi-user detectors for mismatched delay channels," in *Proc. Conf. Inform. Sciences Systems*, Princeton University, Princeton, NJ, Mar. 20–22 1996, vol. 1, pp. 326–331.

[141] M. L. Honig, U. Madhow, and S. Verdú, "Blind adaptive multiuser detection," *IEEE Trans. Inform. Th.*, vol. 41, no. 3, pp. 944–960, July 1995.

[142] J. B. Schodorf and D. B. Williams, "A constrained optimization approach to multiuser detection," *IEEE Trans. Sign. Proc.*, vol. 45, no. 1, pp. 258–262, Jan. 1997.

[143] M. K. Tsatsanis, "Inverse filtering criteria for CDMA systems," *IEEE Trans. Sign. Proc.*, vol. 45, no. 1, pp. 102–112, Jan. 1997.

[144] X. Wang and H. V. Poor, "Multiuser diversity receivers for frequency-selective Rayleigh fading CDMA channels," in *Proc. IEEE Vehic. Tech. Conf.*, Phoenix, AZ, May 4–7 1997, vol. 1, pp. 198–202.

[145] U. Madhow, "Blind adaptive interference suppression for the near-far resistant acquisition and demodulation of direct-sequence CDMA," *IEEE Trans. Sign. Proc.*, vol. 45, no. 1, pp. 124–136, Jan. 1997.

[146] M. Rupf, F. Tarkoy, and J. L. Massey, "User-separating demodulation for code-division multiple-access systems," *IEEE J. Select. Areas Commun.*, vol. 12, no. 6, pp. 786–795, June 1994.

[147] A. M. Monk, M. Davis, L. B. Milstein, and C. W. Helstrom, "A noise-whitening approach to multiple access noise rejection–Part I: Theory and background," *IEEE J. Select. Areas Commun.*, vol. 12, no. 5, pp. 817–827, June 1994.

[148] M. Davis, A. Monk, and L. B. Milstein, "A noise whitening approach to multiple-access noise rejection–Part II: Implementation issues," *IEEE J. Select. Areas Commun.*, vol. 14, no. 8, pp. 1488–1499, Oct. 1996.

[149] Y. C. Yoon and H. Leib, "Matched filters with interference suppression capabilities for DS-CDMA," *IEEE J. Select. Areas Commun.*, vol. 14, no. 8, pp. 1510–1521, Oct. 1996.

[150] L. Mailander and R. Iltis, "Single user CDMA detection using the whitened matched filter," in *Proc. Conf. Inform. Sciences Systems*, Princeton University, Princeton, NJ, Mar. 20–22 1996, vol. 2, pp. 846–851.

[151] M. K. Tsatsanis and G. B. Giannakis, "Optimal decorrelating receivers for DS-CDMA systems: A signal processing framework," *IEEE Trans. Sign. Proc.*, vol. 44, no. 12, pp. 3044–3055, Mar. 1996.

[152] M. J. Juntti and B. Aazhang, "Finite memory-length linear multiuser detection for asynchronous CDMA communications," *IEEE Trans. Commun.*, vol. 45, no. 5, May 1997.

[153] F.-C. Zheng and S. K. Barton, "Near-far resistant detection of CDMA signals via isolation bit insertion," *IEEE Trans. Commun.*, vol. 43, no. 2/3/4, pp. 1313–1317, Feb./Mar./Apr. 1995.

[154] S. Moshavi, E. G. Kanterakis, and D. L. Schilling, "Multistage linear receivers for DS-CDMA systems," *Int. J. Wireless Inform. Networks*, vol. 3, no. 1, pp. 1–17, 1996.

[155] S. S. H. Wijayasuriya, G. H. Norton, and J. P. McGeehan, "A sliding window decorrelating receiver for multiuser DS-CDMA mobile radio networks," *IEEE Trans. Vehic. Tech.*, vol. 45, no. 3, pp. 503–521, Aug. 1996.

[156] A. Kajiwara and M. Nakagawa, "Crosscorrelation cancellation in SS/DS block demodulator," *IEICE Trans. Commun.*, vol. E 74, no. 9, pp. 2596–2602, Sept. 1991.

[157] S. S. H. Wijayasuriya, G. H. Norton, and J. P. McGeehan, "A novel algorithm for dynamic updating of decorrelator coefficients in mobile DS-CDMA," in *Proc. IEEE Int. Symp. Personal, Indoor and Mobile Radio Commun.*, Yokohama, Japan, Sept. 8-11 1993, pp. 292–296.

[158] L. L. Yang and R. A. Scholtz, "$\delta$-adjusted $m$th order multiuser detector," in *Proc. IEEE Glob. Telecommun. Conf.*, London, U.K., Nov. 18-22 1996, vol. 3, pp. 1555–1560.

[159] M. J. Juntti, B. Aazhang, and J. O. Lilleberg, "Iterative implementation of linear multiuser detection," in *Proc. Conf. Inform. Sciences Systems*, Princeton University, Princeton, NJ, Mar. 20–22 1996, vol. 1, pp. 343–348.

[160] M. J. Juntti, B. Aazhang, and J. O. Lilleberg, "Linear multiuser detection for R-CDMA," in *Proc. Commun. Th. Mini-Conf. in conj. IEEE Glob. Telecommun. Conf.*, London, U.K., Nov. 18–22 1996, pp. 127–131.

[161] M. K. Varanasi, *Multiuser Detection in Code-Division Multiple-Access Communications*, Ph.D. thesis, Department of Electrical and Computer Engineering, University of Rice, Houston, TX, 1989.

[162] M. K. Varanasi and B. Aazhang, "Near-optimum detection in synchronous code-division multiple-access systems," *IEEE Trans. Commun.*, vol. 39, no. 5, pp. 725–736, May 1991.

[163] R. S. Mowbray, R. D. Pringle, and P. M. Grant, "Increased CDMA system capacity through adaptive cochannel interference regeneration and cancellation," *IEE Proc.-I*, vol. 139, no. 5, pp. 515–524, Oct. 1992.

[164] S. Tachikawa, "Characteristics of M-ary/spread spectrum multiple access communication systems using co-channel interference cancellation techniques," *IEICE Trans. Commun.*, vol. E76-B, no. 8, pp. 941–946, June 1992.

[165] B. S. Abrams, A. E. Zeger, and T. E. Jones, "Efficiently structured CDMA receiver with near-far immunity," *IEEE Trans. Vehic. Tech.*, vol. 44, no. 1, pp. 1–13, Feb. 1995.

[166] Y. Sanada and Q. Wang, "A co-channel interference cancellation technique using orthogonal convolutional codes," *IEEE Trans. Commun.*, vol. 44, no. 5, pp. 549–556, May 1996.

[167] R. Kohno, H. Imai, M. Hatori, and S. Pasupathy, "Combination of an adaptive array antenna and a canceller of interference for direct-sequence spread-spectrum multiple-access system," *IEEE J. Select. Areas Commun.*, vol. 8, no. 4, pp. 675–682, May 1990.

[168] Y. C. Yoon, R. Kohno, and H. Imai, "Cascaded co-channel interference cancelling and diversity combining for spread-spectrum multi-access

system over multipath fading channels," *IEICE Trans. Commun.*, vol. E76-B, no. 2, pp. 163–168, Feb. 1993.

[169] Y. C. Yoon, R. Kohno, and H. Imai, "A spread-spectrum multiaccess system with cochannel interference cancellation for multipath fading channels," *IEEE J. Select. Areas Commun.*, vol. 11, no. 7, pp. 1067–1075, Sept. 1993.

[170] A. Saifuddin, R. Kohno, and H. Imai, "Integrated receiver structures of staged decoder and CCI canceller for CDMA with multilevel coded modulation," *European Trans. Telecommun.*, vol. 6, no. 1, pp. 9–19, January-February 1995.

[171] A. Saifuddin and R. Kohno, "Performance evaluation of near-far resistant receiver DS/CDMA cellular system over fading multipath channel," *IEICE Trans. Commun.*, vol. E78-B, no. 8, pp. 1136–1144, Aug. 1995.

[172] U. Fawer and B. Aazhang, "A multiuser receiver for code division multiple access communications over multipath channels," *IEEE Trans. Commun.*, vol. 43, no. 2/3/4, pp. 1556–1565, Feb./Mar./Apr. 1995.

[173] Y. Sanada and Q. Wang, "A co-channel interference cancellation technique using orthogonal convolutional codes on multipath Rayleigh fading channel," *IEEE Trans. Vehic. Tech.*, vol. 46, no. 1, pp. 114–128, Feb. 1997.

[174] A. Hottinen, H. Holma, and A. Toskala, "Performance of multistage multiuser detection in a fading multipath channel," in *Proc. IEEE Int. Symp. Personal, Indoor and Mobile Radio Commun.*, Toronto, Ont., Canada, Sept. 27–29 1995, vol. 3, pp. 960–964.

[175] H. Holma, A. Toskala, and A. Hottinen, "Performance of CDMA multiuser detection with antenna diversity and closed loop power control," in *Proc. IEEE Vehic. Tech. Conf.*, Atlanta, GA, Apr. 28 - May 1 1996, vol. 1, pp. 362–366.

[176] M. Latva-aho and J. Lilleberg, "Parallel interference cancellation in multiuser detection," in *Proc. IEEE Int. Symp. Spread Spectrum Techniques and Applications*, Mainz, Germany, Sept. 22–25 1996, vol. 3, pp. 1151–1155.

[177] A. Saifuddin and R. Kohno, "Performance evaluation of DS/CDMA scheme with diversity coding and MUI cancellation over fading multipath channel," *IEICE Trans. Fundamentals Elec., Commun. and Comp. Sc.*, vol. E79-A, no. 12, pp. 1994–2001, Dec. 1996.

[178] A. Hottinen, H. Holma, and A. Toskala, "Multiuser detection for multirate CDMA communications," in *Proc. IEEE Int. Conf. Commun.*, Dallas, TX, June 24-28 1996.

[179] M. Latva-aho and J. Lilleberg, "Parallel interference cancellation based delay tracker for CDMA receivers," in *Proc. Conf. Inform. Sciences Systems*, Princeton University, Princeton, NJ, Mar. 20-22 1996, vol. 2, pp. 852–857.

[180] M. Latva-aho and J. Lilleberg, "Delay trackers for multiuser CDMA receivers," in *Proc. IEEE Int. Conf. Universal Personal Communications*, Sept. 29 – Oct. 2 1996, pp. 326–330.

[181] R. M. Buehrer, A. Kaul, S. Striglis, and B. D. Woerner, "Analysis of DS-CDMA parallel interference cancellation with phase and timing errors," *IEEE J. Select. Areas Commun.*, vol. 14, no. 8, pp. 1522–1535, Oct. 1996.

[182] R. M. Buehrer, N. S. Correal, and B. D. Woerner, "A comparison of multiuser receivers for cellular CDMA," in *Proc. IEEE Glob. Telecommun. Conf.*, London, U.K., Nov. 18-22 1996, vol. 3, pp. 1571–1577.

[183] P. Agashe and B. Woerner, "Interference cancellation for a multicellular CDMA environment," *Wireless Pers. Commun., Kluwer*, vol. 3, no. 1-2, pp. 1–15, 1996.

[184] D. W. Chen, Z. Siveski, and Y. Bar-Ness, "Synchronous multiuser CDMA detector with soft decision adaptive canceler," in *Proc. Conf. Inform. Sciences Systems*, Princeton University, Princeton, NJ, Mar. 1994, vol. 1, pp. 139–143.

[185] Y. Bar-Ness and N. Sezgin, "Adaptive threshold setting for multiuser CDMA signal separators with soft tentative decisions," in *Proc. Conf. Inform. Sciences Systems*, The Johns Hopkins University, Baltimore, MD, Mar. 21–23 1995, pp. 174–179.

[186] V. Vanghi and B. Vojcic, "Soft interference cancellation in multiuser communications," *Wireless Pers. Commun., Kluwer*, vol. 3, no. 1-2, pp. 111–128, 1996.

[187] Y. Bar-Ness, Z. Siveski, and D. W. Chen, "Bootstrapped decorrelating algorithm for adaptive interference cancelation in synchronous CDMA communications systems," in *Proc. IEEE Int. Symp. Spread Spectrum Techniques and Applications*, Oulu, Finland, July 4–6 1994, pp. 162–166.

[188] B. Zhu, N. Ansari, and Z. Siveski, "Convergence and stability analysis of a synchronous adaptive CDMA receiver," *IEEE Trans. Commun.*, vol. 43, no. 12, pp. 3073–3079, Dec. 1995.

[189] Y. Bar-Ness and J. B. Punt, "Adaptive 'bootstrap' CDMA multi-user detector," *Wireless Pers. Commun., Kluwer*, vol. 3, no. 1-2, pp. 55–71, 1996.

[190] H. Elders-Boll, M. Herper, and A. Busboom, "Adaptive receivers for mobile DS-CDMA communication systems," in *Proc. IEEE Vehic. Tech. Conf.*, Phoenix, AZ, May 4–7 1997, vol. 3, pp. 2128–2132.

[191] D. Divsalar and M. Simon, "Improved CDMA performance using parallel interference cancellation," Tech. Rep., Jet Propulsion Laboratory, California Institute of Technology, Pasadena, CA, Oct. 1995.

[192] D. Divsalar and M. Simon, "A new approach to parallel interference cancellation for CDMA," in *Proc. IEEE Glob. Telecommun. Conf.*, London, U.K., Nov. 18-22 1996, vol. 3, pp. 1452–1457.

[193] A. Radović and B. Aazhang, "Iterative algorithms for joint data detection and delay estimation for code division multiple access communication systems," in *Proc. Annual Allerton Conf. Communication Control and Computing*, Allerton House, Monticello, IL, Sept. 29 – Oct. 1 1993.

[194] L. B. Nelson and H. V. Poor, "Iterative multiuser receivers for CDMA channels: An EM-based approach," *IEEE Trans. Commun.*, vol. 44, no. 12, pp. 1700–1710, Dec. 1996.

[195] D. Dahlhaus, H. Fleury, and A. Radović, "A sequential algorithm for joint parameter estimation and multiuser detection in DS/CDMA systems with multipath propagation," *Wireless Pers. Commun., Kluwer, to appear*, 1998.

[196] D. Dahlhaus, A. Jarosch, H. Fleury, and R. Heddergott, "Joint demodulation in DS/CDMA systems exploiting the space and time diversity of the mobile radio channel," in *Proc. IEEE Int. Symp. Personal, Indoor and Mobile Radio Commun., to appear*, Helsinki, Finland, Sept. 1–4 1997.

[197] A. J. Viterbi, "Very low rate convolutional codes for maximum theoretical performance of spread-spectrum multiple-access channels," *IEEE J. Select. Areas Commun.*, vol. 8, no. 4, pp. 641–649, May 1990.

[198] P. Patel and J. Holtzman, "Analysis of a simple successive interference cancellation scheme in a DS/CDMA system," *IEEE J. Select. Areas Commun.*, vol. 12, no. 10, pp. 796–807, June 1994.

[199] A. C. K. Soong and W. A. Krzymien, "Performance of a reference symbol assisted multistage successive interference cancelling receiver in a multi-cell CDMA wireless systems," in *Proc. IEEE Glob. Telecommun. Conf.*, Singapore, Nov. 13–17 1995, vol. 1, pp. 152–156.

[200] O. Nesper and P. Ho, "A reference symbol assisted interference cancelling hybrid receiver for an asynchronous DS/CDMA system," in *Proc. IEEE Int. Symp. Personal, Indoor and Mobile Radio Commun.*, Taipei, Taiwan, Oct. 15-18 1996, vol. 1, pp. 108–112.

[201] O. Nesper and P. Ho, "A pilot symbol assisted interference cancellation scheme for an asynchronous DS/CDMA system," in *Proc. IEEE Glob. Telecommun. Conf.*, London, U.K., Nov. 18-22 1996, vol. 3, pp. 1447–1451.

[202] A.-L. Johansson and A. Svensson, "Successive interference cancellation in multiple data rate DS/CDMA systems," in *Proc. IEEE Vehic. Tech. Conf.*, Chicago, IL, July 25–28 1995, pp. 704–708.

[203] A.-L. Johansson and A. Svensson, "Multistage interference cancellation in multirate DS/CDMA on a mobile radio channel," in *Proc. IEEE Vehic. Tech. Conf.*, Atlanta, GA, Apr. 28 - May 1 1996, vol. 2, pp. 666–670.

[204] F.-C. Cheng and J. M. Holtzman, "Effect of tracking error on DS/CDMA successive interference cancellation," Tech. Rep. WINLAB-TR-90, WINLAB, Rutgers University, Piscataway, NJ, Oct. 1994.

[205] A. C. K. Soong and W. A. Krzymien, "Robustness of the reference symbol assisted multistage successive interference cancelling receiver with imperfect parameter estimates," in *Proc. IEEE Vehic. Tech. Conf.*, Atlanta, GA, Apr. 28 - May 1 1996, vol. 2, pp. 676–680.

[206] F. van der Wijk, G. M. J. Janssen, and R. Prasad, "Groupwise successive interference cancellation in a DS/CDMA system," in *Proc. IEEE Int. Symp. Personal, Indoor and Mobile Radio Commun.*, Toronto, Ont., Canada, Sept. 27–29 1995, vol. 2, pp. 742–746.

[207] W. Haifeng, J. Lilleberg, and K. Rikkinen, "A new sub-optimal multiuser detection approach for CDMA systems in Rayleigh fading channel," in *Proc. Conf. Inform. Sciences Systems, to appear*, The Johns Hopkins University, Baltimore, MD, Mar. 19–21 1997.

[208] M. J. Juntti, "Multiuser detection in multirate CDMA systems — part II: Detector performance comparisons," in *Proc. IEEE Int. Symp. Personal,*

*Indoor and Mobile Radio Commun., submitted*, Helsinki, Finland, Sept. 1–4 1997.

[209] M. J. Juntti, "Performance of multiuser detection in multirate CDMA systems," *Wireless Pers. Commun., Kluwer, submitted*, Feb. 1997.

[210] M. K. Varanasi, "Group detection in synchronous Gaussian code-division multiple-access channels," *IEEE Trans. Inform. Th.*, vol. 41, no. 3, pp. 1083–1096, July 1995.

[211] M. K. Varanasi, "Parallel group detection for synchronous CDMA communication over frequency-selective Rayleigh fading channels," *IEEE Trans. Inform. Th.*, vol. 41, no. 3, pp. 1083–1096, July 1995.

[212] M. Abdulrahman and A. U. H. Sheikh snd D. D. Falconer, "Decision feedback equalization for CDMA in indoor wireless communications," *IEEE J. Select. Areas Commun.*, vol. 12, no. 4, pp. 698–706, May 1994.

[213] A. Hafeez and W. E. Stark, "Combined decision-feedback multiuser detection/soft-decision decoding for CDMA channels," in *Proc. IEEE Vehic. Tech. Conf.*, Atlanta, GA, Apr. 28 - May 1 1996, vol. 1, pp. 382–386.

[214] Z. Xie, C. K. Rushforth, and R.T. Short, "Multiuser signal detection using sequential decoding," *IEEE Trans. Commun.*, vol. 38, no. 5, pp. 578–583, May 1990.

[215] B. Wu and Q. Wang, "New suboptimal multiuser detectors for synchronous CDMA systems," *IEEE Trans. Commun.*, vol. 44, no. 7, pp. 782785, July 1996.

[216] M. J. Juntti, T. Schlösser, and J. O. Lilleberg, "Genetic algorithms for multiuser detection in synchronous CDMA," in *Proc. IEEE Int. Symp. Inform. Th., to appear*, Ulm, Germany, June 29 – July 4 1997.

[217] R. A. Iltis and L. Mailaender, "Multiuser code acquisition using parallel decorrelators," in *Proc. Conf. Inform. Sciences Systems*, Princeton, NJ, March 16-18 1994, vol. 1, pp. 109–114.

[218] E. G. Ström, S. Parkvall, S. L. Miller, and B. E. Ottersen, "Propagation delay estimation in asynchronous direct-sequence code-division multiple access systems," *IEEE Trans. Commun.*, vol. 44, no. 1, pp. 84–93, Jan. 1996.

[219] B. Steiner and P. Jung, "Optimum and suboptimum channel estimation for the uplink of CDMA mobile radio systems with joint detection," *European Trans. Telecommun.*, vol. 5, no. 1, pp. 39–50, Jan.-Feb. 1994.

[220] E. G. Ström, S. Parkvall, S. L. Miller, and B. E. Ottersten, "DS-CDMA synchronization in time-varying fading channels," *IEEE J. Select. Areas Commun.*, vol. 14, no. 8, pp. 1636–1642, Oct. 1996.

[221] S. E. Bensley and B. Aazhang, "Subspace-based channel estimation for code division multiple access communication systems," *IEEE Trans. Commun.*, vol. 44, no. 8, pp. 1009–1019, Aug. 1996.

[222] S. Parkvall, E. G. Ström, and L. B. Milstein, "Coded asynchronous near-far resistant DS-CDMA receivers operating without synchronization," in *Proc. Commun. Th. Mini-Conf. in conj. IEEE Glob. Telecommun. Conf.*, London, U.K., Nov. 18-22 1996, pp. 183–187.

[223] M. Torlak and G. Xu, "Blind multiuser channel estimation in asyncronous CDMA systems," *IEEE Trans. Sign. Proc.*, vol. 45, no. 1, pp. 137–147, Jan. 1997.

[224] J. Joutsensalo, J. Lilleberg, A. Hottinen, and J. Karhunen, "A hierarchic maximum likelihood method for delay estimation in CDMA," in *Proc. IEEE Vehic. Tech. Conf.*, Atlanta, GA, Apr. 1996.

[225] D. Zheng, J. Li, S. L. Miller, and E. G. Ström, "An efficient code-timing estimator for DS-CDMA signals," *IEEE Trans. Sign. Proc.*, vol. 45, no. 1, pp. 82–89, Jan. 1997.

[226] R. A. Iltis and L. Mailaender, "An adaptive multiuser detector with joint amplitude and delay estimation," *IEEE Trans. Commun.*, vol. 44, no. 11, pp. 1561–1571, Nov. 1996.

[227] T. J. Lim and L. K. Rasmussen, "Adaptive symbol and parameter estmation in asynchronous multiuser CDMA detectors," *IEEE Trans. Commun.*, vol. 45, no. 2, pp. 213–220, Feb. 1997.

[228] Y. Steinberg and H. V. Poor, "Sequential amplitude estimation in multiuser communications," *IEEE Trans. Inform. Th.*, vol. 40, no. 1, pp. 11–20, Jan. 1994.

[229] Y. Steinberg and H. V. Poor, "On sequential delay estimation in wideband digital communication systems," *IEEE Trans. Inform. Th.*, vol. 40, no. 5, pp. 1327–1333, Sept. 1994.

[230] T. K. Moon, Z. Xie, C. K. Rushforth, and R. T. Short, "Parameter estimation in a multi-user communication system," *IEEE Trans. Commun.*, vol. 42, no. 8, pp. 2553–2560, Aug. 1994.

[231] K. W. Halford and M. Brandt-Pearce, "User identification and multiuser detection of $l$ out of $k$ users in a CDMA system," in *Proc. Conf. Inform. Sciences Systems*, Princeton University, Princeton, NJ, Mar. 1994, vol. 1, pp. 115–120.

[232] K. W. Halford and M. Brandt-Pearce, "Maximum likelihood detection and estimation for new users in CDMA," in *Proc. Conf. Inform. Sciences Systems*, The Johns Hopkins University, Baltimore, MD, Mar. 21–23 1995, vol. 1, pp. 193–198.

[233] U. Mitra and H. V. Poor, "Activity detection in a multi-user environment," *Wireless Pers. Commun., Kluwer*, vol. 3, no. 1-2, pp. 149–174, 1996.

[234] J. Joutsensalo, "A subspace method for model order estimation in CDMA," in *Proc. IEEE Int. Symp. Spread Spectrum Techniques and Applications*, Mainz, Germany, Sept. 22–25 1996, vol. 2, pp. 688–692.

[235] H. Liu and G. Xu, "A subspace method for signature waveform estimation in synchronous CDMA systems," *IEEE Trans. Commun.*, vol. 44, no. 10, pp. 1346–1354, Oct. 1996.

[236] D. H. Johnson, Y. K. Lee, O. E. Kelly, and J. L. Pistole, "Type-based detection for unknown channels," in *Proc. IEEE Int. Conf. Acoustics, Speech, and Signal Processing*, 1996, pp. 2475–2478.

[237] P. M. Djuric and M. Guo, "A novel approach to multiuser detection for CDMA systems," in *Proc. IEEE Vehic. Tech. Conf.*, Phoenix, AZ, May 4–7 1997, vol. 2, pp. 563–566.

[238] B. Aazhang, B.-P. Paris, and G. Orsak, "Neural networks for multiuser detection in code-division multiple access systems," *IEEE Trans. Commun.*, vol. 40, no. 7, pp. 1212–1222, July 1992.

[239] U. Mitra and H. V. Poor, "Adaptive receiver algorithms for near-far resistant CDMA," *IEEE Trans. Commun.*, vol. 43, no. 2/3/4, pp. 1713–1724, Feb./Mar./Apr. 1995.

[240] A. Hottinen, "Self-organizing multiuser detection," in *Proc. IEEE Int. Symp. Spread Spectrum Techniques and Applications*, Oulu, Finland, July 4–6 1994, vol. 1, pp. 152–156.

[241] U. Mitra and H. V. Poor, "Neural network techniques for adaptive multiuser demodulation," *IEEE J. Select. Areas Commun.*, vol. 12, no. 9, pp. 1460–1470, Dec. 1994.

[242] T. Miyajima, T. Hasegawa, and M. Haneishi, "On the multiuser detection using a neural network in code-division multiple-access communications," *IEICE Trans. Commun.*, vol. E76-B, no. 9, pp. 961–968, Aug. 1993.

[243] T. Miyajima and T. Hasegawa, "Multiuser detection using a Hopfield network for asynchronous code-division multiple-access systems," *IEICE Trans. Fundamentals Elec., Commun. and Comp. Sc.*, vol. E79-A, no. 12, pp. 1963–1971, Dec. 1996.

[244] Telecommunication Industry Association, *Mobile Station–Base Station Compatibility Standard for Dual-Mode Wideband Spread Spectrum Cellular Systems*, July 1993.

[245] A. H. M. Ross and K. S. Gilhousen, "CDMA technology and the IS-95 north american standard," in *The Mobile Communications Handbook*, J. D. Gibson, Ed., chapter 27, pp. 430–447. CRC Press, 1996.

[246] T. Ojanperä, K. Rikkinen, H. Häkkinen, K. Pehkonen, A. Hottinen, and J. Lilleberg, "Design of a 3rd generation multirate CDMA systems with multiuser detection, MUD-CDMA," in *Proc. IEEE Int. Symp. Spread Spectrum Techniques and Applications*, Mainz, Germany, Sept. 22–25 1996, vol. 1, pp. 334–338.

[247] A. Hottinen and K. Pehkonen, "A flexible multirate CDMA concept with multiuser detection," in *Proc. IEEE Int. Symp. Spread Spectrum Techniques and Applications*, Mainz, Germany, Sept. 22–25 1996, vol. 2, pp. 556–560.

[248] T. Ojanperä, J. Castro, D. Emmer, M. Gudmundson, P. Jung, A. Klein, G. Kramer, R. Pirhonen, L. Rademacher, J. Sköld, and A. Toskala, "FRAMES – hybrid multiple access tecnology," in *Proc. IEEE Int. Symp. Spread Spectrum Techniques and Applications*, Mainz, Germany, Sept. 22–25 1996, vol. 1, pp. 320–324.

[249] K. Ohno, M. Sawahashi, and F. Adachi, "Wideband coherent DS-CDMA," in *Proc. IEEE Vehic. Tech. Conf.*, Chicago, IL, July 25–28 1995, pp. 779–783.

[250] F. Adachi, M. Sawahashi, T. Dohi, and K. Ohno, "Coherent DS-CDMA: Promising multiple access for wireless multimedia mobile communica-

tions," in *Proc. IEEE Int. Symp. Spread Spectrum Techniques and Applications*, Mainz, Germany, Sept. 22–25 1996, vol. 1, pp. 351–358.

# 4 WIRELESS COMMUNICATION TECHNOLOGY IN INTELLIGENT TRANSPORT SYSTEMS

Masao Nakagawa
Lachlan B. Michael

**Abstract**

In this paper the applications of wireless communication in the emerging intelligent transport systems (ITS) field are surveyed, applications both already in use and those that are presently at the research or experimental level. The use of ITS is expected to increase the capacities of highways and make road travel safer. In this environment the use of wireless communication is indispensable and new and different problems from cellular wireless communication must be overcome.

In this paper particular attention is paid to the uses of wireless communication in the vehicle-roadside link (where vehicles communicate in a two-way link with a roadside base station or information is received from broadcast beacons), and vehicle to vehicle wireless communication. The use of spread spectrum techniques to overcome problems of interference in ITS is examined and the problems of forming a distributed wireless network in an ever changing environment is discussed. Specific examples in this paper are taken mostly from Japanese research, though some recent research in Europe and America, and the differences in direction are examined.
wireless communication, intelligent transportation system (ITS), inter-vehicle communication, spread spectrum.

## 4.1 INTRODUCTION

The need for more capacity and better safety on our freeways can readily be understood by the average driver who encounters the worsening traffic situation in their daily commute. With the popularity of the automobile and the desire or necessity of many to live away from the city centers daily traffic jam reports have become an essential part of many peoples' mornings.

The pollution and loss of productivity caused by vehicles which are stopped in traffic is well known. Furthermore as the number of vehicles on the road has increased, traffic accidents have also increased. For example, in the Tokyo metropolitan area the number of traffic accidents increased 6.2 % from 54,994 to 58,412 from 1994 to 1995. The number of injuries increased correspondingly from 63,326 to 67,756, an increase of 7.0 % [1].

### *4.1.1 Solutions*

Many attempts to alleviate these problems have been made. One of the most common responses to capacity problems has been to build more and larger highways. However there is a limit to which this can be carried out, and it can be seen from the huge 6-lane freeways in Southern California that even this does not guarantee a city free of traffic jams!

Furthermore most cities have developed over time, rather than being planned from the beginning, and the necessity to build roads and freeways through urban and suburban areas to alleviate traffic problems has not been a well received solution. Indeed, in many cities it has become a near impossibility, either physically, economically or socially to construct more roads and freeways. In cities such as Tokyo, land can be so expensive that to buy the land alone can take tens of years.

Other attempts have been made to limit the use of the automobile, such as in Singapore, where taxes make the cost of vehicles prohibitively expensive, and special permits are needed to enter the city center. Other cheaper permits are available that allow the owner only to use the vehicle during certain hours, such as on weekends only.

To increase safety most efforts have concentrated on improving vehicles, while driver awareness campaigns are another attempt to reduce the high cost to society of traffic accidents. While all of these efforts have succeeded to some degree, the fact remains that a complete solution remains distant. It is in this scenario that ITS technologies have been proposed to overcome the preceding problems of capacity and safety.

The aim of ITS may be stated as realizing improved traffic safety, transportation efficiency and traveling comfort and also for contributing to environmental preservation through communication between roads and vehicles using the most advanced communication technologies [2].

### *4.1.2 ITS as a solution*

With this background governments have begun to seriously look for technological solutions to make the most of existing roads (that is increase capacity) to reduce traffic jams and the resulting pollution and losses, and to make same roads safer (that is to reduce accidents). Large government and industry projects such as AHS (Automated Highway System), PROMETHEUS, DRIVE and IVHS (Intelligent Vehicle Highway System) have begun to investigate the many problems associated with such a large undertaking.

Since vehicles are constantly moving the wireless communication will play an important part in any ITS network. In this paper, following a general discussion of wireless communication in ITS, the use of wireless communication in the vehicle-roadside link is examined in section 4.2. Inter-vehicle communication is examined in section 4.3. Since specific examples in this paper are taken mostly from Japanese research a short look at the direction of European and American research is presented in sections 4.4 and 4.5. The use of spread spectrum techniques to overcome problems of interference in ITS is examined in section 4.6. Future research areas of ITS wireless communication technology are presented in section 4.7.

### *4.1.3 Wireless Communication in ITS*

There are two main wireless links in ITS that have emerged. One is the vehicle to vehicle link (often termed inter-vehicle communication: IVC), and the other is the roadside-vehicle, vehicle-roadside link. Both of these links serve complementary but differing roles. That is, it is argued that the vehicle to vehicle link is necessary for speed and safety reasons, to avoid accidents, while the roadside to vehicle link is used mostly for broadcasting information useful to the driver or occupants of the vehicle.

However, depending on the system implementation, there may be a greater role for either communication beacons or the vehicle-vehicle communication link. The cost of deploying many thousands of beacons may be prohibitive, and inter-vehicle communication may be the only technology available in rural areas, since the equipment travel with the vehicle. On the other hand in beacon based systems it may be easier to implement vehicle control and therefore be more reliable and predictable than distributed IVC systems.

Both of these wireless links and the technologies that have been proposed will be examined further.

## 4.2 VEHICLE TO ROADSIDE COMMUNICATION: VRC

Vehicle to roadside communication can be divided into 2 categories:
1.broadcast (beacon based) information dissemination services
2.two-way communication between vehicle and roadside services.

By providing information on traffic conditions it is thought that drivers can reroute around trouble spots, creating a more even traffic flow and reducing the length and severity of traffic jams. At the very least drivers may know the reason for the delay (accident, road works) and have reduced stress because they can determine the time to pass through the traffic jam and thus be less likely to cause accidents.

### *4.2.1 Radio*

Broadcast information services have long been studied and indeed implemented even in ITS. The simplest type of wireless communication to disseminate information to drivers is voice based information using AM or FM radio, where periodic updates of traffic congestion give the driver information so they are able to avoid the congested areas.

However some of the major failings of this system are that the information is
1.not available when the driver wants it (time constraints)
2.not about the area in which the driver is interested (spatial constraints)

To some extent the problems of (1) have been relieved by dedicated traffic radio channels, such as the 1620 kHz traffic advisory radio available on highways in Japan. This service provides traffic information which is automatically updated every 5 minutes, 24 hours a day. The broadcast is delivered by spiral leaky coaxial cable (sLCX). By using leaky coaxial cable a uniform field strength can be obtained over the length of the cable, including areas such as tunnels where conventional radio waves have had difficulty penetrating.

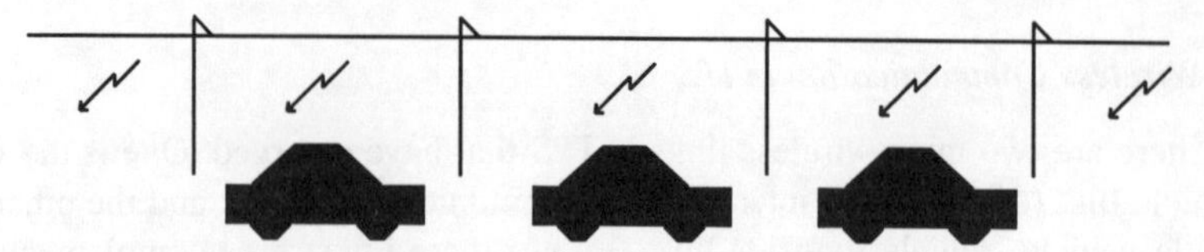

**Figure 1** Traffic information is provided to drivers via spiral leaky co-axial cable (sLCX)

Another reason that sLCX was chosen to broadcast traffic information is that to obtain a broadcasting license is very difficult in Japan, and by limiting the transmission to weak radio waves a license is not required [3].

However service areas are limited, and the information is only of a local nature.

*4.2.2 Example of broadcast VRC - VICS*

A more substantial and comprehensive driver information system which is broadcast only is the Vehicle Information and Communication System (VICS) in Japan. The VICS began public service in April 1996. It distributes road traffic information, such as traffic jams (location, length and reason) and accidents to drivers in real time.

VICS assembles and processes information received from JARTIC (JApan Road Traffic Information Center) and other sources, such as the police (who in Japan have the responsibility of controlling traffic signals and maintaining traffic flow on general roads), and disseminates this information to drivers [4].

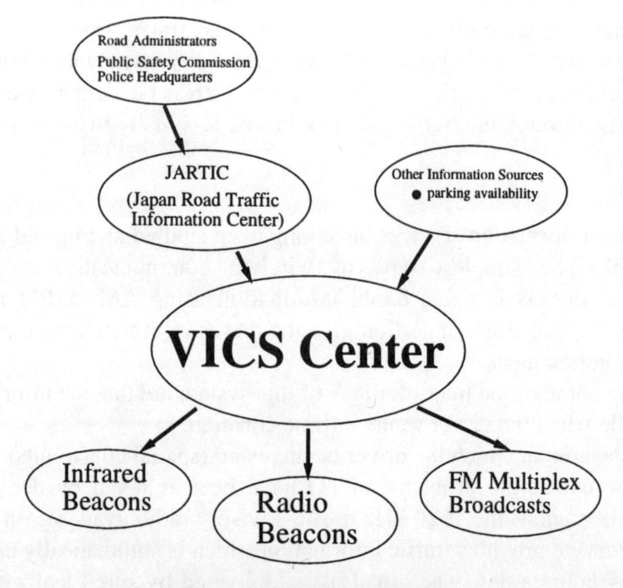


**Figure 2** Outline of VICS

There are three ways to display information:

1. superimposed over digitized maps (data stored on separate CD-ROM) and displayed on a monitor
2. a simple graphic (for example a schematic of the Tokyo Metropolitan Expressway) displayed on a monitor
3. displayed using 2 lines of simple text (15 characters/line)

With the popularity of car navigation systems in Japan, makers of these car navigation units incorporated the ability to receive VICS data in their new models. In fact it was due to the large number of car navigation units that had already been sold that allowed VICS to start with a large base of consumers who had monitors and digital maps already available in-vehicle. Sales of car navigation units rose from 130,000 in 1993 to 500,000 in 1995 [5].

In VICS, three main wireless technologies are used. The first is 64 kb/s radio wave beacons which have a transmission range of about 70 meters. They located at points 2--4 kilometers along sections of the highways, and at junctions, entrances and exits, service station and parking lots. 6000 beacons are scheduled to be in place by the end of fiscal 1997. The transmission frequency is 2.4997 GHz ± 85 kHz. Manchester coding is employed and GMSK (Gaussian filtered Minimum Shift Keying) is the modulation method.

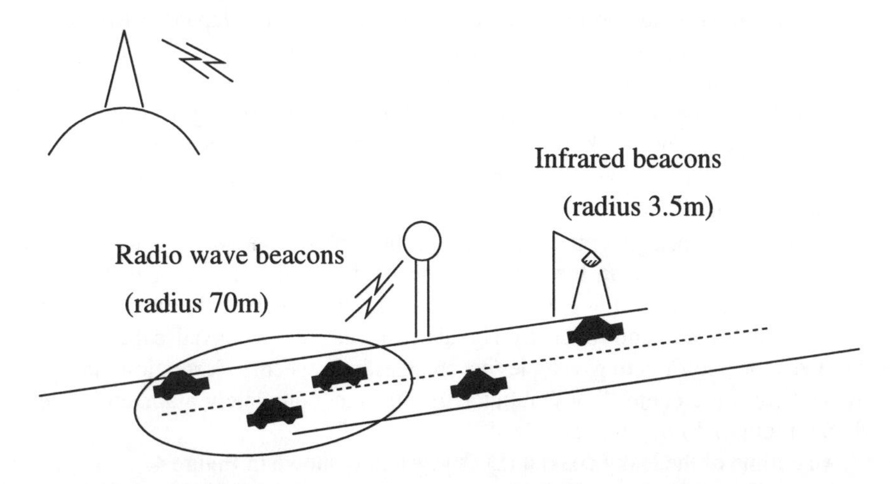


**Figure 3** Three wireless technologies used in VICS

The second wireless technology uses 1 Mb/s infrared beacons to transmit 80 k bytes of information each time a vehicle passes by. The transmission range is approximately 3.5 meters and the transmitters are installed at on poles hung above major intersections on arterial roads. There are more than 12,000 such beacons already in service. The modulation method used is PAM (Pulse Amplitude Modulation). Manchester coding is employed and error correction is done by CRC (Cyclic Redundancy Check). The BER is less than $10^{-5}$. The infrared wavelength is 850 nm ± 50 nm.

The third wireless technology uses FM multiplex at a subcarrier frequency of 76 kHz to broadcast information over a range of 20 - 50 kilometers. Approximately 50 k bytes of information are transmitted every 5 minutes. The speed of transmission is 16 kb/s and the modulation method used is L-MSK (Level controlled Minimum Shift Keying).

This use of three different media was chosen to make use of existing infrastructure as much as possible. That is NHK's (Nippon Hoso Kyokai: Japanese Broadcasting Commission) multiplex broadcast facilities were already in place, and the infrared beacons were already installed as optical vehicle detectors. The radio wave beacons were newly installed by VICS.

By providing much more detailed information drivers are better able to make correct route choices. In some advanced models the traffic data is used to predict the quickest path to the vehicles destination and show this on the monitor. Some models have computer generated voices which give directions to the destination such as "turn left after 100m".

Although the provision of information by using wireless communication is proceeding at a fast rate, more challenging problem are in the two way communication area.

*4.2.3 Example of VRC - Leaky Coax AHS*

An example of two way vehicle-road communication is the automated vehicle highway system (AVHS) project, which is supervised by the Japanese Ministry of Construction, the Highway Development Organization (HIDO) and some private companies. In a practical demonstration of their research 11 vehicles were automatically driven along a new section of the Joshinestsu highway in Nagano prefecture, Japan. The 11 kilometer round trip test between Komoro and Tobu Yunomaru interchanges was a successful application of wireless communication to automated driving. [2, 6]

In this experiment lateral control was achieved by vehicles sensing magnetic nails which were buried in the center of the road. Small left and right deviations were then easily corrected keeping the vehicle in the center of the lane. Distance between vehicles was controlled by signals from the leaky coaxial cable installed along the roadside directing vehicles to change their speed. Corrections in speed were sent from the controlling computer. In this way completely automatic (lateral and longitudinal) driving was achieved.

An outline of the leaky coaxial (LCX) system is shown in Figure 4.

The length of road was divided into sections of length 500m. For each section in the leaky coaxial cable there was a local processor and a radio transmitter/receiver. The host processor and the vehicle display equipment were installed in another building located near the test course.

There were six sections on the test course, and on the fifth section an image processing sensor was installed to detect any abnormalities, such as a vehicles which has broken down and stopped on the road. The on-road abnormality information detected by the sensor is then transmitted to the vehicles via the LCX system. Signals between the host processor and each local processor were transmitted via an optic fiber local area network (LAN).

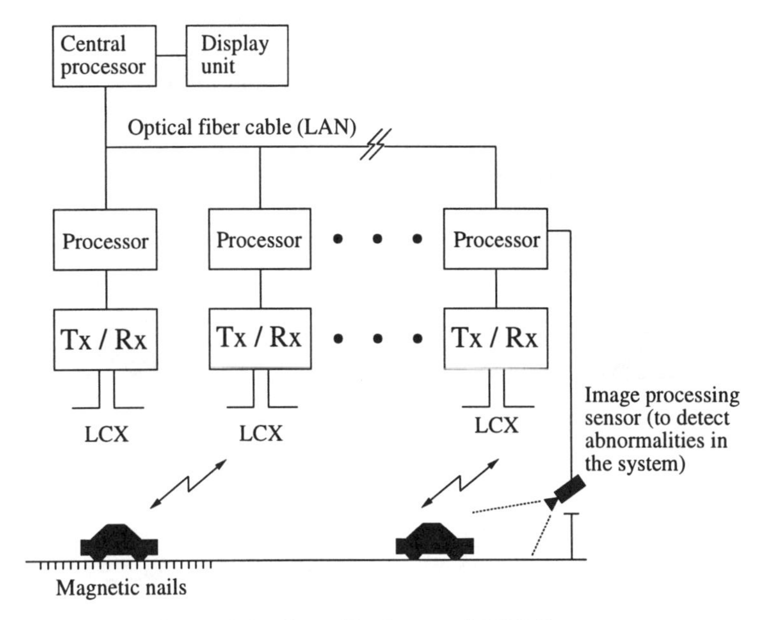


**Figure 4** Outline of leaky coaxial VRC system

The radio transmitter radiates waves from the LCX at 1W at a frequency of 2.5GHz. The downlink (roadside-vehicle) frequency is 2598 MHz and the uplink frequency is 2538 MHz. The LCX cable is designed exclusively for the 2.5 GHz band. The diameter of the cable is approximately 50 mm and the transmission loss is 0.12 dB/m. Although the experiment and the overall system was shown to be a success there are many aspects which have not been implemented yet or remain as challenges. Features such as rear-end collision prevention have not yet been implemented.

Furthermore, problems such as errors caused by interference from waves radiated from LCX laid for communication with vehicles in the opposite lanes remains. It seems that for problems such as these, and also such cases where trucks and other large vehicles block the signal to smaller vehicles, the use of inter-vehicle communication system as an additional method seems preferable. The Ministry of Construction is hoping to implement a driving aid system by the year 2000 and have an automatic driving system by 2005.

## 4.3 INTER-VEHICLE COMMUNICATION (IVC)

Apart from VRC, the other major wireless link in ITS is the cooperative driving environment often called inter-vehicle communication. In IVC vehicles share data about their progress (speed, braking situation, steering angle) and intentions (signal right/left) to warn other vehicles of possible dangerous situations. If these dangerous situations, such as side and rear collisions can be avoided and road safety increased.

### *4.3.1 Platooning*

As well as an increase in safety, a capacity benefit can be expectedfrom the use of inter-vehicle communication. When vehicles travel in a platoon the head spacing (distance) between vehicles is very small compared to the distance between conventional vehicles not equipped with the ability to communicate. The reason that vehicles can travel so close is because vehicles automatically pass data between themselves at speeds much faster than a human can react. When vehicles are able to travel at such close head spacings, then the amount of traffic on highways could be increased without the need to build new highways.

This is a very popular consequence of inter-vehicle communication in developed countries, especially where land is expensive and the construction of new highways is economically and politically not viable.

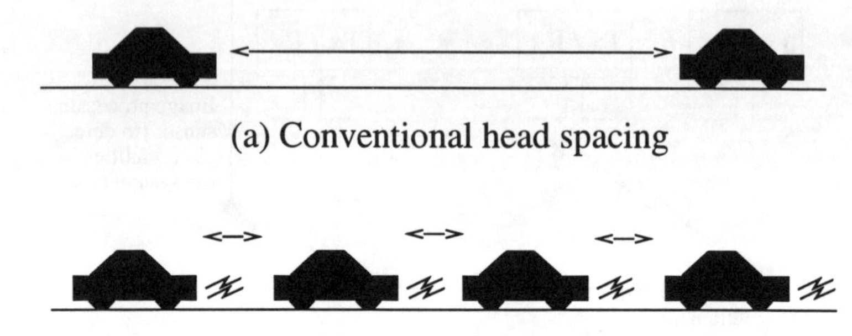


**Figure 5** Reducing distance between vehicles by forming a platoon

Platooning is also called convoying, with reference to trucks which travel together in groups called convoys.

*4.3.2 Topology of IVC network*

Inter vehicle communication is different from an ordinary wireless local area network (LAN) or cellular communication. The peculiar physical topology if the IVC network is one of the reasons that wireless IVC methods and protocols must be re-examined.

Firstly, unlike in cellular systems the movement of vehicles occurs in one direction (for one side of a highway) only. The predictable nature of movement and slow variation of speed (relative to other vehicles) offers the promise of a simplified system.

When we consider data exchange to improve safety, it is clear that vehicles traveling close by are the main target to broadcast safety information. Vehicles close by are likely to pose a safety threat than vehicles far away. Thus the area and range of communication are much more limited in space than for cellular communication.

Speed is another characteristic of IVC communication that is required. If data exachneg is slow, or the data received is old vehicles may not be able to avoid dangerous situations.

Unless vehicles travel in organised groups it is difficult for a master station to exist. Thus vehicles must form a distrubted network among themselves. Furthermore this distributed network is always changing as vehicles enter or leave.

Below are some of the issues that must be considered for any IVC network.

*4.3.3 Communication medium*

There is no firm agreement on what the communication medium for IVC should be. One reason is that there is no firm agreement on the required distance for communication between vehicles. Figures range from several meters between individual vehicles to several kilometers between platoons. Whereas in cellular communication the cell radius is relatively fixed, distances between vehicles will rapidly vary as the topology of the network changes. This change in distance affects what communication medium is chosen.

Whether the communication medium has sharp directivity or not influences the direction of the protocol. Since vehicles travel in lanes, that is in relatively straight lines, the use of directed beams can be used to reduce interference to neighboring vehicles. However in this case data cannot be transferred to vehicles in neighboring lanes without ''hopping" data via another vehicle. Such systems have been studied an shown to be effective [15, 19] but there are drawbacks. The major one is that if not all vehicles can communicate the throughput in the network falls dramatically.

Infrared laser and millimeter wave are primary candidates for the inter-vehicle communication medium.

Infrared is a popular candidate because of the lack of regulation (a license is not required) and that equipment is inexpensive. Especially for close communication of up to few tens of meters data can betransferred at a high rate. Since infrared links are basically line of sight (LOS) links, interference to other vehicles is minimized. However, several problems with infrared have been identified. One is that in conditions of rain, snow or fog the transmission distance falls dramatically. Another is that infrared systems are sensitive to transmitter/receiver alignment.

Millimeter wave communication around 60 GHz has been proposed as another one of the leading candidates for the IVC link. One of the reasons is that there is a maximum of absorption caused by the $O_2$ in the atmosphere at this frequency [7]. This means that the signal will rapidly fall away after a certain distance, allowing re-use of that particular frequency at a closer distance than would have been possible at another frequency. At millimeter frequencies there is an abundance of free bandwidth, which makes it more attractive than the crowded microwave frequencies. Also millimeter wave does not suffer from the aforementioned problems in bad weather. For these reasons future IVC systems will most likely use millimeter wave as the communication medium.

*4.3.4 Synchronization*

Synchronization is another problem in IVC systems. Since there is no external synchronizing signal, the distributed vehicles must synchronize their signals by themselves. Protocols have been proposed [8] overcome this by each vehicle transmitting a pulse periodically, and each vehicle measuring the power of other vehicles pulses. Each vehicle then calculates the average of these values and shifts its own pulse toward this average.

Other asynchronous protocols are being studied based on spread spectrum techniques.

*4.3.5 Slot reservation*

Once vehicles are synchronized, then slots must be reserved. Most protocols that have been suggested are variations of the R-ALOHA (reservation ALOHA) protocol [9, 10]. One protocol suggestion involves using the relative positions of vehicles to achieve higher throughput and packet success probability when reserving slots [11].

Other possibilities include giving control to one vehicle for a certain time based on relative positions [12].

*4.3.6 Example of IVC - Tsukuba Demonstration*

To show the the possibilities of an AHS (Automated Highway System) by using IVC, the JSK (Jidosha Soko Kyokai) recently conducted experiments on a test track in Tsukuba, Japan. Several vehicles exchanged information such as speed, acceleration, distance and intention to change lanes [13].

Six experiments were conducted:

1. testing of communication/data transmission links
2. merging of vehicles after driver signals intention
3. leaving group after driver signals intention
4. automatic speed control
5. automatic braking and stopping
6. lateral distance measurement using optical SS

In this experiment infrared was the physical transmission media used for exchange of data between vehicles. The reasons for this choice included the lack of regulation and that inexpensive devices are available [14].

The transmission system parameters are specified in Table1.

**Table 1** Transmission parameters for Tsukuba IVC demonstration

| | |
|---|---|
| Transmission range | 2.5 - 100 m |
| Directivity | ± 20° |
| Transmission rate | 1.544 Mb/s |
| Transmitter | Infrared LEDs (870nm, 20 mW) |
| Receiver | Silicon PIN photo diode |
| Optics | Tapered light concentrating tube and infra-red bandpass filter IR-80 |

The features of the IVC system include (1) there is no master station (2) the topology of the network changes frequently (3) local stations only interact with each other.

The communication method uses a distributed topology with slots reserved in a TDMA like manner. Communication takes place in the longitudinal direction.

Since the system has a narrow beam width it is necessary to use vehicles in the forward or reverse direction to pass on or "hop" data to another vehicle traveling to the side. The value of around ± 20° was shown by computer simulation to give good performance for this type of narrow beam network [15].

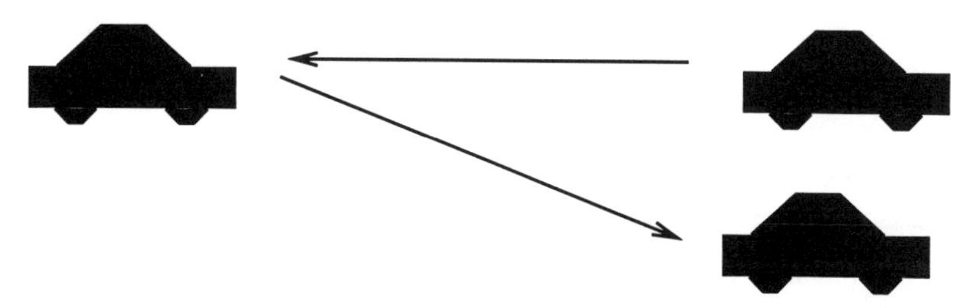

**Figure 6** Data is transferred to another vehicle.

Data is displayed on a monitor in each vehicle. As the demonstration was still on the experimental stage, data was displayed to enable the participants to learn more about the system. In the final version, data will be displayed in simple graphic form on a HUD (Head Up Display) on the inside of the windshield or the dashboard.

Each frame of data consists of 29 slots each of 20 ms in duration. Each slot or packet of data that is transmitted consists of a 40 byte preamble, 4 bytes of synchronization, 79 bytes of data, 2 bytes of CRC checking code and a postamble of 3 bytes for a total of 128 bytes (1024 bits).

The 79 bytes of data in each packet is divided up as shown in Table2.

**Table 2** Structure of each packet

| Data name | Length [bytes] | From vehicles | Contents of Data |
|---|---|---|---|
| SVD | 7 | self | ID, slot number, speed acceleration etc |
| AVD | 9 | side | SVD from vehicles to the side |
| FVD | 9 | front | SVD from vehicles forward to be relayed backward |
| BVD | 9 | behind | SVD from vehicles behind to be relayed forward |
| AXD | 9 | | other data |

A lateral wireless measurement of the distance to the guard rail was also conducted as part of the same experiment. Using optical spread spectrum (SS) technology, excellent result were obtained in the 1 - 5m range. The major components of the system are highlighted in Table3.

**Table 3** Lateral distance measurement using optical SS

| | |
|---|---|
| Transmitter | Infrared LED (870 nm 18 mW) |
| Directivity | ±5° |
| PN Code chip rate | 5.115 MHz |
| PN Code | M sequence, chip period 511 |
| Modulation method | On-off keying of laser light |
| Measuring distance | 7 m |
| Receiver | APD |
| Synchronization Acquisition | Digital matched filter |
| Synchronization Tracking | Delay locked loop |
| Data rate | 9600 b/s |

While this system demonstrated the feasibility of IVC some aspects have yet to be addressed. These include network issues, such as what happens when the network crowds and reliability of the data transmission between vehicles.

## 4.4 WIRELESS COMMUNICATION IN ITS IN EUROPE

In Europe there has been one way traffic information provided by FM radio, by the ARI (Automotive Road Information) system and more recently by the RDS-TMC (Radio Data System - Traffic Message Channel) [16].

RDS-TMC transmits at 1200 b/s using FM modulation. This had originally been intended to provide information such as name of radio station and alternative frequencies but can also be used to provide information to in-car navigation systems.

However all these systems are based on one-way data or voice transmission. Two way communication enables vehicles to function as sensors and send this information to a traffic control center. By uses existing infrastructure such as Mobitex, GSM data service and future packet oriented services and TETRA (Trans-European Trunked RAdio) traffic management can be improved without the need for expensive infrastructure.

GSM cellular service allow data rates from 2400 to 9600 b/s. Messages can be sent and received by individual vehicles, but when many vehicles wish to use such a service, there is the problem of not enough capacity in the cellular network.

In the short term, since the infrastructure for GSM is already in place it will provide a good medium to receive traffic information. However, since the user must ask for data this type of system cannot be considered very convenient or even safe (unless automatic driving were taking place!) Compared to the VICS system the applications are also limited, and the user may be unwilling to pay for the cost of mobile telephone calls as well as some possible data charge.

Mobitex is a cellular packet switched data network developed by Ericsson. The system includes point-to-point, point-to-multipoint and broadcast mode services. The data is transmitted at a maximum of 8000 b/s and the bandwidth of a channel is 12.5 kHz. Mobitex is now under construction in Germany and traffic management system using Mobitex have been built in Sweden and France.

There have been vehicle to vehicle experiments as part of the PROMETHEUS program [17]. The two wireless IVC systems (developed by MANET and THOMSON) used dynamic TDMA/CSMA at a transmission frequencies of 2.45 and 57 GHz while the bandwidth was 3 and 2 MHz respectively. The system was low power, transmitter power being 0.2 W but with an option to transmit up to 2 W. The transmission area was 500m and the reception range was > 80 dB. Synchronization was achieved by recursive convergence using a weighted average.

The overall success of the system was determined by examining the data from experiment and noting that (1) no less than 10 % of the applicative data is missing and (2) no gap in the recording data exceeds 10 seconds. The following performance was obtained:

- reception rate of correct messages is around 98%
- time for first access to channel is 4 frame cycles
- synchronization stability of slots > 99.9%

The main characteristic parameters were determined to be

- channel efficiency

- distance between vehicles sharing a slot
- number of slots changes

In Italy research has been carried out as part of the TELCO (Telecommunication Network for Cooperative Driving) program aims to define a communication system for cooperative driving application. Most research has centered on the inter-vehicle and beacon-vehicle communications at 63.5 GHz in a highway environment.

As mentioned before the main reasons for choosing this particular band are the abundance of bandwidth, the presence of $O_2$ absorption and good performance of the millimeter band in the presence of adverse weather conditions. Furthermore, although the millimeter band is quasi-optical, the possibility of multipath propagation is seen as an advantage in non line-of-sight (NLOS) conditions such as when a truck blocks communication between two vehicles. Infra-red, on the other hand, is generally considered to have only line-of-sight transmission.

Multiple hop networks are also being considered, although as yet in only a single lane environment [19].

Europe is moving toward a single integrated system, but with the difficulties involved in getting many countries to use a single standard, a solution seems still a long way off. In the meantime GSM packet data services may be able to fill the gap since infrastructure is already in place across many borders.

## 4.5 WIRELESS COMMUNICATION IN ITS IN THE U.S.

One of the most well known American ITS programs is the California PATH (Partners for Advanced Transit and Highways) program. Run by the Institute of Transportation Studies, University of California, Berkeley in collaboration with Caltrans, they have focused on an ITS model with three separate wireless links.

The first is for control, that is inter-vehicle communication. The second is for maneuver, that is communication among groups of vehicles (platoons) and advisory communication, that is vehicle-roadside communication [18].

The PATH program envisages groups of up to twenty vehicles traveling at closely spaced intervals (less than 2 meters) while each group is separated from other groups of vehicles by more than 50 meters. The main motivation for the PATH program stems from the congestion experienced on highways, particularly Southern Californian highways. With the difficulties, both politically and economically in building new freeways their aim is make the best use of the existing highway infrastructure. The platoons of vehicles would travel in special lanes separated from non-automated vehicles by barriers.

This assumption is particularly appropriate to American highways, where many lanes are available. If this is the case, one or two lanes devoted to AHS is feasible, as even now one lane is often already devoted to car-pooling. However this assumption does not necessarily hold in other countries. For example in Japan most highways are only two lane, and even in major centers like Tokyo 3-lane highways exist for only a few kilometers. In this situation is unlikely that even one lane could be set aside for platoon-only travel. The problems of several vehicles with IVC capability intertwined with vehicles without has yet to be addressed.

The three types of communication for the PATH program have been identified to have the characteristics shown in Table 4.

**Table 4** Wireless communication characteristics identified by PATH

| Type of Communication | Characteristics |
|---|---|
| Control | 1. Mostly line of sight<br>2. Channel access should be deterministically based<br>3. Should not use much bandwidth<br>4. Packet delays and loss should be minimized |
| Maneuver | 1. Should not use much bandwidth<br>2. Packet loss and delay are not critical<br>3. Channel access is random in source, destination and time |
| Advisory/Navigation | 1. Packet loss and delay are very non-critical<br>2. Channel access is random in source, destination and time<br>3. Communication distances are very long<br>4. High relative velocities |

In this structure each vehicle would communicate only with its nearest longitudinal neighbors and/or the lead vehicle of the platoon. This suggest a line-of-sight system. The PATH program chose an infrared (IR) system in the 830 nm band modulated with on/off keying. It has a useful range of about 30 meters. The data rate varies inversely with distance from 19.2 kb/s to 1.2 Mb/s, depending on the BER. The maximum channel bandwidth is 3 MHz.

There are two systems under consideration for the maneuver communication system. Both are omni-directional broadcast radio systems.

The pulse radio channel uses frequency hopping spread spectrum modulation and has a bit rate of 1 Mb/s and a range of 500 meters. The WaveLAN radio channel uses direct sequence spread spectrum modulation. It has a data rate of 2 Mb/s and the maximum packet length is 100 bytes at a relative velocity of 30 m/s.

There is still no specification for the roadside-vehicle wireless link as yet.

## 4.6 SPREAD SPECTRUM IN ITS

The use of spread spectrum technologies has already become well-known in the wireless field for its robustness against interference, particularly deep frequency selective fades. By spreading the signal out over a very wide bandwidth, and using a particular code (pseudo-noise:PN sequence) for each user better performance than other multiple access schemes have been reported.

However in those cellular based systems a base station has knowledge of each individual users PN sequence. In a distributed system this is not possible.

One of the first proposals suggested which would overcome the difficulty of not knowing each users PN sequence was the boomerang transmission system (BTS) for IVC. In this system a vehicle sends out its signal using its own pseudo-noise (PN) code. This is received by the other vehicle and transmitted back to the previous vehicle, after multiplying its own data on the original sequence. The second vehicle does not have to have knowledge of the PN sequence of the original vehicle.

Furthermore because direct sequence spread spectrum (DS-SS) is used the distance between vehicles can be determined by comparing the time delay between the received signal and the original transmission.

A broad outline of the BTS is shown in Figure 7.

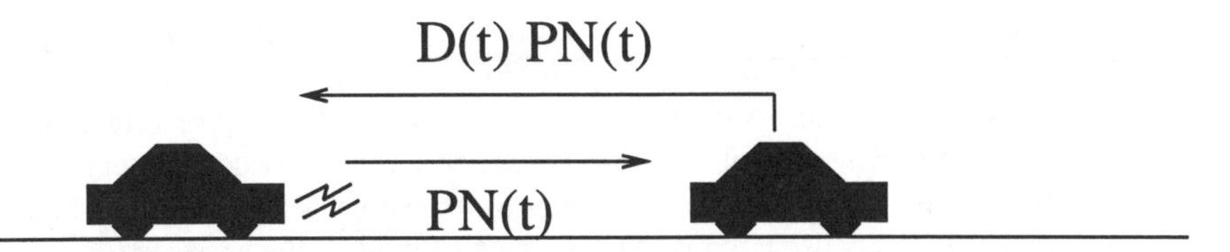


**Figure 7** Boomerang system for vehicle to vehicle communications

The boomerang transmission system has recently been refined by the useof multicarrier SS technology. By using multiple carriers the problem of large errors in range and in the data BER due to strong interference (especially when a fast PN signal is used to gain sufficient resolution) can be avoided.

In this scheme the data stream is serial to parallel converted into P streams. Each stream has S branches and the same PN code is used in each. The frequency spectrum of successive sub-carriers is allowed to overlap and each of them is orthogonal. The PN code chip duration is defined as

$$T_c = \frac{PS+1}{2} T_{c1} \tag{1}$$

where $T_{c1}$ is the PN code chip duration for the single carrier case ($P$=$S$=1).

Based on this idea simulation confirmed that this system can prevent rapid degradation under strong interference compared to the conventional boomerang operation.

Other research using SS has concentrated on improving the channel access protocol. First, CSMA (Carrier Sense Multiple Access) is used to sense if the channel is idle. A common spreading code is assigned to all vehicles for the modulation of their transmission data. By utilizing a common code each vehicle can sense whether it is receiving packets or not. Therefore a very simple communication system without knowing the spreading codes of other vehicles becomes possible. Since SS is used, the hidden terminal problem can also be solved. That is even if two vehicles sense that the channel is empty and begin to transmit packets, demodulation of the data is often successful. Even in the rare case that two vehicles transmit at exactly the same time, if they are different distances apart from the vehicle which is trying to demodulate that data, there will be a different transmission delay, and thus there will be more than one chip separation of the codes at the receiving end. Generally, the auto-correlations of the spreading codes become low when their phases are shifted more than one chip, and two peaks will appear at the matched filter of the receiving vehicle and both packets can be demodulated successfully.

Another problem that may be successfully overcome by the use of spread spectrum technology is avoidance of interference caused by vehicles in opposite lane. Previous simulation models have assumed travel in only one direction, and severe problems of interference can be expected if there not some barrier between lanes. If slots must be shared among both lanes then problems of lack of capacity can be expected to occur. Because SS systems are robust against noise such problems can be overcome more easily.

## 4.7 FUTURE RESEARCH

One of the areas which has not been addressed in ITS communication is quality of service (QoS) when a network becomes crowded. In a conventional cellular or wireless LAN system, when networks become crowded they delay or refuse service to some customers. However in a IVC network, if a vehicle is denied service or if transmission of data is delayed, the usefulness of the whole system is diminished.

Another area to be discussed is the provision of multimedia services for those in vehicles. If vehicle-roadside technology and infrastructure is developed it may rival regular cellular service in the bandwidth and types of service that can be provided. Especially if people are in vehicles that are driven automatically, some entertainment or links to database about their destination might be provided.

For IVC as well, communication between individual vehicles by way voice and television might be appropriate. One can only imagine the possibilities, though it is certain that the development of wireless communication for ITS will continue to develop at a fast rate.

## 4.8 CONCLUSION

In this paper the wireless technology in ITS has been examined. The authors have presented a review of the technologies either being used or under consideration. Some of the issues facing the development of the wireless link of ITS has been presented, and the use of spread spectrum technologies to overcome some of those issues were presented.

## References

[1] 1996 Anzen no shiori (Safety Manual), 1996. (in Japanese).

[2] M. Nakamura, H. Tsunomachi, and R. Fukui. Road vehicle communication system for vehicle control using leaky coaxial cable. *IEEE Communications Magazine,* 34(10):84-89, October 1996.

[3] T. Sakai and C. Arimatsu. Metropolitan expressway radio system. In *Second World Congress on Intelligent Transport Systems,* pages 650 655, November 1995.

[4] Shinsaku Yamada. The strategy and deployment plan for VICS. *IEEE Communications Magazine,* 34(10):94-97, October 1996.

[5] Jay Tate. The Institute. Technical Report 21(2), IEEE, February 1997. (A news supplement to IEEE Spectrum).Ministry of Construction. *ITS Handbook in Japan,* 1996.

[7] Wolfgang Schafer. Channel modeling of short-range radio links at 60 GHz for mobile intervehicle communication. In *41st IEEE International Vehicular Technology Conference,* pages 314-319, May 1991.

[8] E.A. Sourour and M. Nakagawa. Mutual decentralized synchronization for inter-vehicle communications. In *IEEE Intelligent Vehicles Symposium,* pages 272-277, September 1996.

[9] W. Kremer, D. Hubner, S. Holf, T. Benz, and W. Schafer. Computer aided design and evaluation of mobile local radio network in RTI/IVHS environments. *IEEE Journal of Selected Areas in Communications,* 11(3):407 421, April 1993.

[10] Y. Inoue and M. Nakagawa. MAC protocol for inter-vehicle communication network using spread spectrum technique. In *Vehicle Navigation and Information Systems Conference,* pages 149 52, September 1994.

[11] Y. Kim and M. Nakagawa. it-Aloha protocol for SS intervehicle communication network using head spacing information. In *IEEE Intelligent Vehicles Symposium,* pages 278-283, September 1996.

[12] M. Aoki and H. Fuji. Inter-vehicle communication: Technical issues on vehicle control application. *IEEE Communications Magazine,* 34(10):90-93, October 1996.

[13] JSK. Intervehicle data transmission system feasibility study: Explanative meeting for general experiment. Technical report, JSK, February 1997. (In Japanese).

[14] H. Fuji, O. Hayashi, and N. Nakagata. Experimental research on inter-vehicle communication using infrared rays. In *IEEE Intelligent Vehicles Symposium,* pages 266 271, September 1996.

[15] L.B. Michael and M. Nakagawa. Multi-hopping data considerations for inter-vehicle communication over multiple lanes. In *IEEE 47th International Vehicular Technology Conference,* pages 121-125, May 1997.

[16] Wolfgang Schultz. Traffic management improvement by integrating modern communication systems. *IEEE Communications Magazine,* 34(10):56-60, October 1996.

[17] Jean-Michel Valade. Vehicle to vehicle communications: Experimental results and implementation perspectives. In *Second World Congress on Intelligent Transport Systems,* pages 1606 1613, November 1995.

[18] Bret Foreman. A survey of wireless communications technologies for automated vehicle control. Technical report, Institute of Transportation Studies, University of California, 1995. (Available at http://www-path.eecs.berkeley.edu/ webed/wireless/papers/index.html).

[19] Roberto Verdone. Communications systems at millimeter waves for ITS applications. In *IEEE 47th International Vehicular Technology Conference,* pages 914-918, May 1997.

[20] K. Mizui, M. Uchida, and M. Nakagawa. Vehicle to vehicle communication and ranging system using spread spectrum technique. In *43rd IEEE Vehicular Technology Conference,* pages 335-338, May 1993.

[21] N. Morinaga, M. Nakagawa, and R. Kohno. New concepts and technologies for achieving highly reliable and high capacity multimedia wireless communications systems. *IEEE Communications Magazine,* 35(1):34 40, January 1997.

# 5 RFIC DESIGN FOR WIRELESS COMMUNICATIONS

Christian Kermarrec

**Abstract**

Three basic topics should be addressed for the design of RFICs targeting a specific system application: 1) transceiver architecture for Eoptimum system and chip partitioning as well as optimum use of sortware techniques, 2) semiconductor technology for optimum performance, cost and integration path, 3) circuit implementation. At this point in time, there is no practical universal RF IC transceiver architecture based on only one technology that can satisfy any system requirement. Every application has its own set of solutions (and some time several sets of solutions), but it is not inconceivable that "programmable" mobile radio transceivers serving multistandard applications will emerge in the next few years.

This paper addresses some of these issues and presents results that illustrate the trends and options in developing RF IC components for wireless commuications.

## 5.1 INTRODUCTION

Digital mobile communication systems are being deployed at a rate which far exceeds the deployment rate of any other consumer product ever introduced. It is anticipated that more than 200 million digital communication mobile terminals (up from 35 million in 1996) will be produced in year 2000 to serve the needs for digital cellular, PCS and cordless telephony systems worldwide. As an example, GSM has already been adopted by more than one hundred operators in about fifty countries and more 30 million handsets will be sold in 1997, up from 15 million units in 1996 (about 100 million terminals expected in 2000).

The Bill Of Material of these handsets drops at about 30% every year while demand for better performance (standby time, talk time, size) and functionality (voice/data, multi-band, multimode) is increasing drastically.

The cost and performance of digital CMOS technology is responding to these demands mainly driven by the requirement of the computer industry. However, the RFIC technologies are not driven by such a high-performance, high-volume industry and as a result is still in the infancy phase when it comes to the design and development of high-volume, low-cost ICs.

Three basic topics should be addressed for the design of RF ICs targeting a specific system application: 1) transceiver architecture for optimum system and chip partitioning as well as optimum use of software techniques, 2) semiconductor technology for optimum performance, cost and integration path, 3) circuit implementation.

At this point in time, there is no practical universal RF IC transceiver architecture based on only one technology that can satisfy any system requirement. Every application has its own set of solutions (and sometimes several sets of solutions), but it is not inconceivable that "programmable" mobile radio transceivers serving multi-standard applications will emerge in the next few years.

This paper addresses some of these issues and presents results that illustrate the trends and options in developing RF IC components for wireless communications.

## 5.2 RF TRANSCEIVER ARCHITECTURE

The transceiver architecture ultimately defines the handset performance, cost and manufacturability. The transceiver includes the following functions and interfaces: 1) the RF, IF and mixed-signal functions of the physical layer, 2) the interfaces with the equalizer and the data receiver (demodulation, carrier recovery), 3) the interface with the control, calibration and resource management software functions, and 4) the interface with the power management functions ( it can include some of the power management functions such as LDOs).

### *5.2.1 Receiver Architectures*

Figure 1 shows the most common receiver architectures. It includes dual, single, and direct down-conversion with baseband I and Q sampling, dual down-conversion with phase and RSSI sampling and dual down-conversion with IF sampling.

Most of the transceivers currently in production are still based on a double or single downconversion receiver followed by two I and Q ADCs, two FIR filters, an equalizer/ demodulator.

The dual and single down-conversion architecture is somewhat costly due to its high component count as well as for the need for several external bulky filters (more so for the dual downconversion) that unfavourably impact the cost and size of the receiver. However, the filtering and gain distribution among the RF, IF and BB sections allows for the production of high sensitivity, easy to calibrate, and highly manufacturable receivers.

Figure 2 illustrates the adjacent channel and blocking signal filtering through the RF, IF1 and IF2 chain of a typical dual down-conversion GSM receiver. The carrier to adjacent channel interferer ratio is knocked down from -49dB to -9dB, and the carrier to blocking interferer ratio is reduced from -99dB to -9dB, thus preserving most of the ADCs dynamic range. The FIR filters easily filter the remaining interferers to a level below the receiver noise floor.

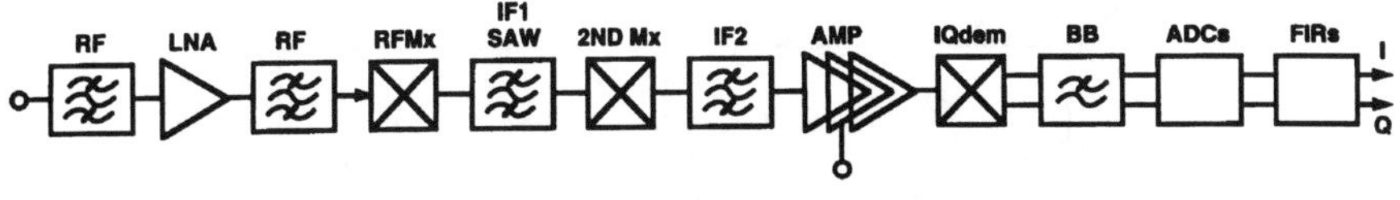


*Dual Down-conversion to IQ Baseband*

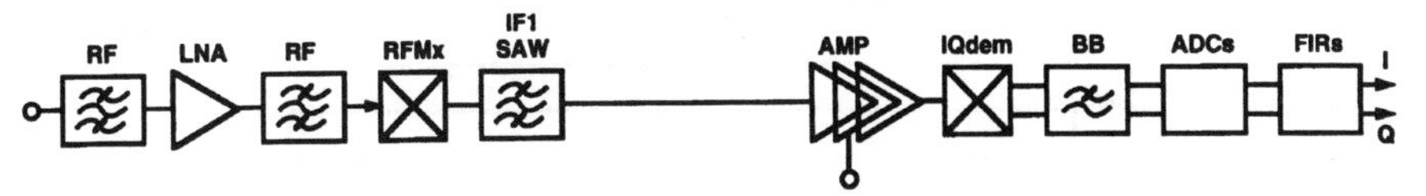


*Single Down-conversion to IQ Baseband*

*Direct Down-conversion to IQ Baseband*

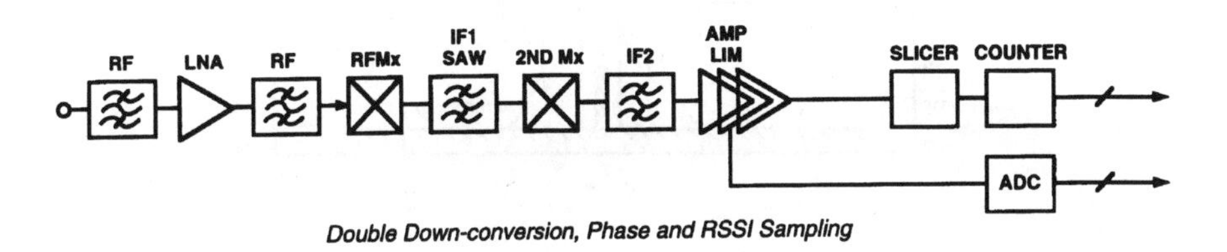


*Double Down-conversion, Phase and RSSI Sampling*

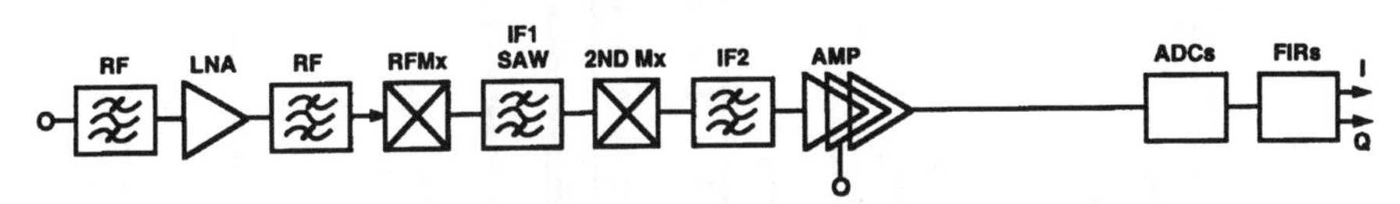


*Double Down-conversion to IQ Baseband*

*Figure 1. Receiver Architectures*

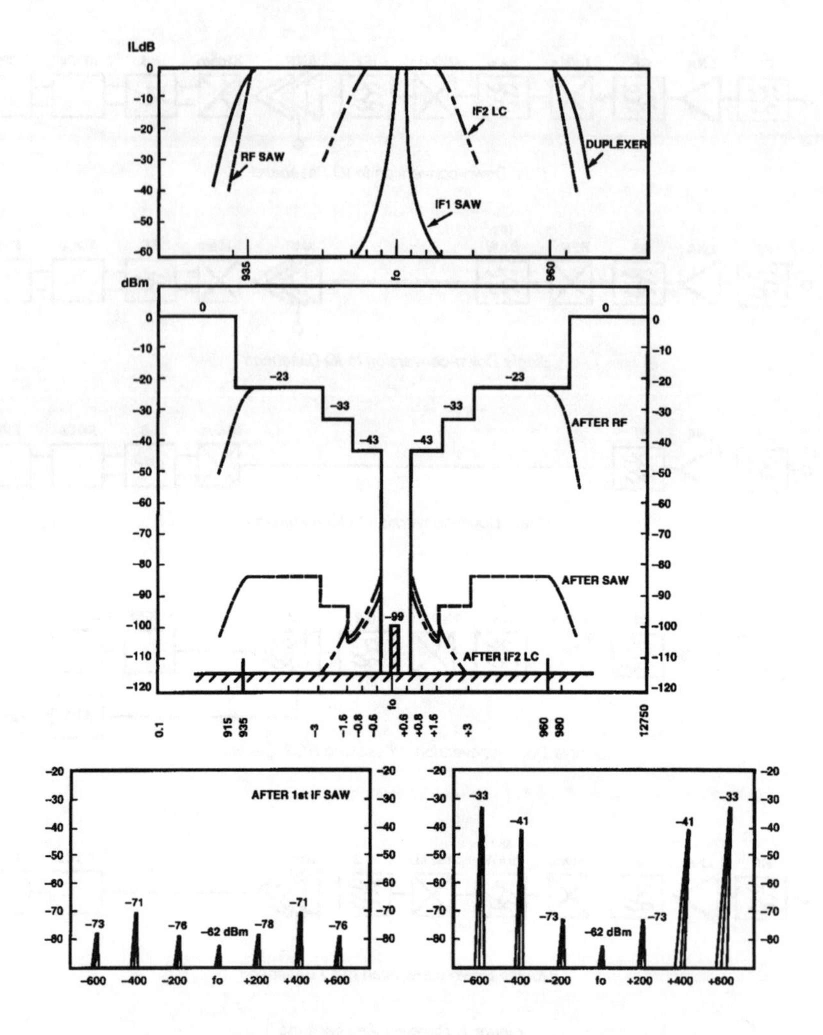


**Figure 2** Adjacent Channel and Blocking Signal Rejection (Dual Down-conversion)

As shown in Figure 1, log/polar demodulation can be substituted to IQ demodulation. The limiter fast RSSI provides the amplitude information, and the data slicer and associated counter provide the phase information. The benefit is a very simple, low-cost demodulation scheme at the expense of higher IF filter selectivity (overall adjacent channel rejection should be performed before limiting.)

Two other architectures are aggressively being pursued to reduce the cost of existing radios and, as important, to provide a path for multi-standard radios development: the direct downconversion/ zero IF architecture [1] and the single (or dual) down-conversion IF sampling architecture [2].

In zero IF solution, filtering and analog processing are performed at baseband and as such do not require expensive and cumbersome RF filters or high frequency processing functions. However, it suffers from lower sensitivity due to baseband amplifier and filter noise contribution and large DC offsets due to self converting LO and near-channel interferer to DC. It requires extremely high LO reverse isolation at the antenna port in the receive mode (LO acts as an interferer since it operates at the same frequency as the F signal), and it needs very linear mixers. Despite extensive efforts from numerous design teams, these issues have generally not yet been resolved.

Figure 3 shows the adjacent channel and blocking signal level of a hypothetical zero IF GSM receiver. The carrier to adjacent channel interferer ratio is -49dB prior to baseband filtering

(-9dB for a dual down-conversion receiver) and the carrier to blocking interferer ratio is -76dB in the same conditions (-9dB for dual conversion receiver). These figures certainly highlight the high level of linearity required for the RF and baseband circuitry.

The IF sampling architecture virtually eliminates the zero IF design issues but requires a highfrequency, high-dynamic range A to D converter [3] [4]. ADCs providing over 90dB SFDR at 200Mhz are now available in narrowband or wideband IF sampling solutions for base station applications and later on for handset/pager applications. IF and baseband processing are performed digitally (IQ demodulation, filtering) in hardware or software form. This approach drastically reduces the complexity and cost of multichannel radio receivers.

### *5.2.2 Transmitter Architecture*

Figure 4 shows the most common transceiver architectures. It includes architectures based on an IQ modulator operating at the output frequency or followed by a single upconversion stage as well as architectures based on VCO phase-locked to an IQ modulator.

Most of the transceivers currently produced are based on the IQ modulator followed by an upconverter, filter, power amplifier and duplexer (or antenna filter for TDD systems).

This architecture provides excellent phase accuracy (IQ modulation at "reasonably" low IF frequency), good immunity to VCO pulling (RF VCO frequency is not at output frequency), good transmit noise floor performance (provided the IF IQ modulator is designed for low noise floor) and can be used for any type of modulations regardless if it is a constant or non-constant envelop modulation.

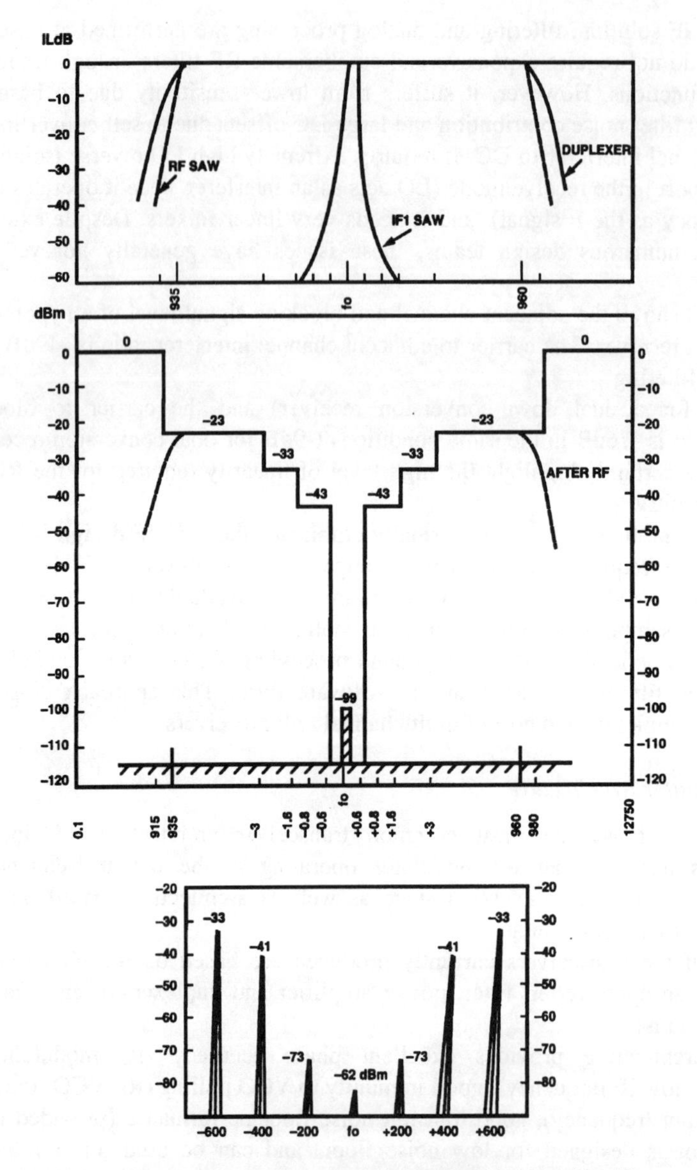


*Figure 3. Adjacent Channel and Blocking Signal Rejection (Zero IF)*

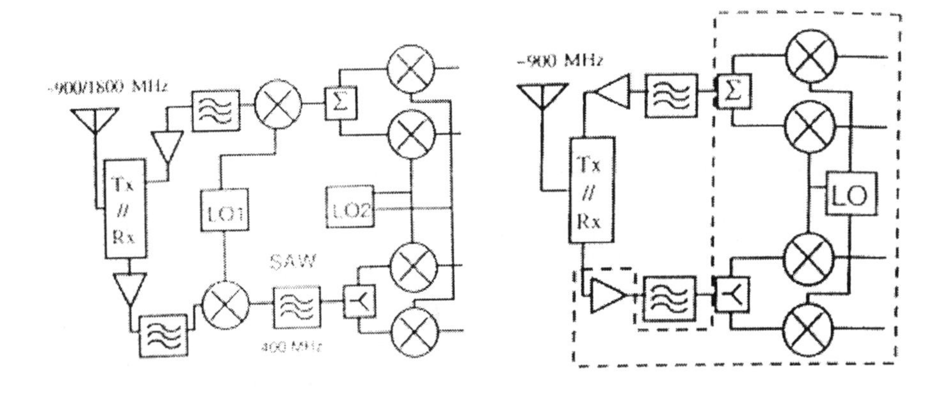


Single Up-conversion

Direct IQ Modulation

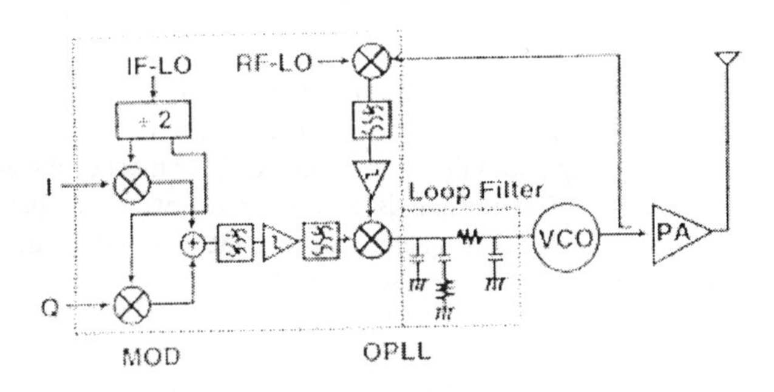


Offset Phase-Locked Loop

Figure 4. Transceiver Architectures

However, the RF VCO should be designed for its phase noise not to degrade the modulation phase accuracy and an image reject filter aimed at rejecting the image and spurious frequencies generated in the upconversion process as well as filtering the LO upconverter LO leakage should be included after the upconverter (resulting in cost and size increase).

As an example, Figure 5 shows the performance of a 280Mhz IF IQ modulator measured as a single side band modulator. LO and side band signals are respectively 47dB and 38dB below the desired signal (translating in a better than 1 degree phase accuracy).

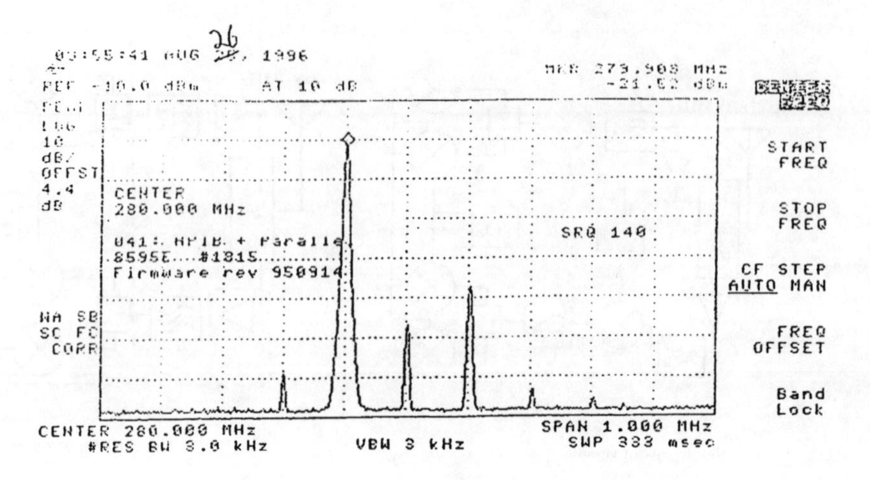


Figure 5. 280 MHz IQ Modulator Performance

The direct IQ modulation architecture does not require any IF and upconversion circuitry, but it suffers from such a poor immunity to VCO pulling (VCO frequency is at the output frequency) that it is impractical for most of the applications. One way to overcome the VCO pulling is to use an offset VCO operating at a frequency shifted from the IF VCO frequency and use the sum (or the difference) of these two frequencies to drive the direct frequency IQ modulator.

Another attractive architecture is based on a VCO operating at the output frequency and phaselocked to an IF IQ modulator and an offset VCO [5]. In principle, the phase-locked VCO does not generate any spurious frequencies and as such does not require any transmit filter. The VCO noise floor can also be made far better that the noise floor of any other architecture. Furthermore, the phase accuracy is good (IF IQ modulator) as well as the immunity to VCO pulling provided the frequency offset is large enough. However, this architecture is only applicable to constant envelop modulation (the VCO amplitude is constant regardless of its phase).

As an example, Figure 6 shows the output spectrum of a GSM transmitter using an offset phaselocked loop concept [5]. It can be seen that the spectrum is compliant with the GSM specification.

Figure 7 shows the phase error and the TX noise spectrum density as function of the loop bandwidth for the same phase-locked VCO. The phase error can be about 2 degrees and the noise floor less than -165dBc/Hz.

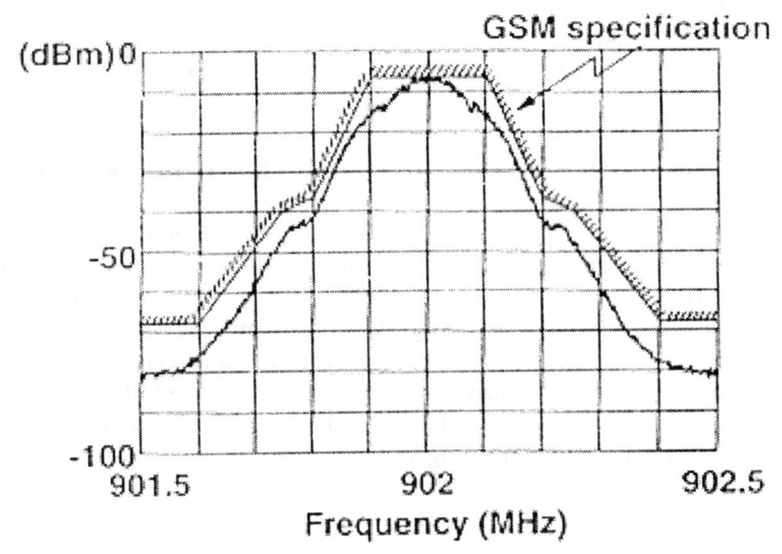


Figure 6. Offset Phase-Locked Loop GSMK Output Spectrum

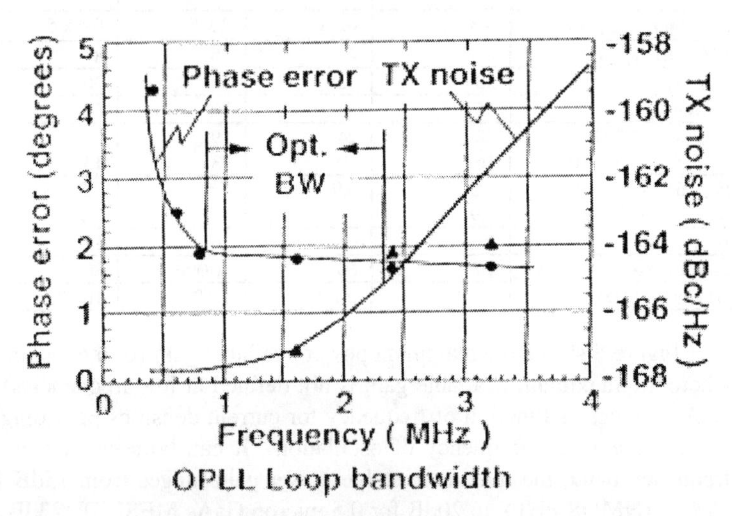


Figure 7. Offset Phase-Locked Loop
phase error and Tx noise (GSM Transmitter)

## 5.3 TECHNOLOGY

Almost all the basic semiconductor technologies silicon as well as GaAs are now competing for the fast growing RF market segment. Silicon based technologies have drastically improved their performance in the last few years and are now the technologies of choice for most of RF IC transceiver functions with the exception of T/R switches and power amplifier ICs.

New silicon bipolar processes tailored for low-voltage, low-power consumption RF and mixedsignal applications have reached the performance and cost required for mass production of RF transceivers operating in the 1 to 2ghz frequency band.

Table 1 summarizes the performance of competing technologies addressing RF IC applications. The table includes silicon homojunction BJTs, silicon heterojunction HBTs, GaAs HBTs, GaAs MESFETs, and silicon CMOS technologies. The table compares the critical design parameters for size, operation voltage, gain, noise, linearity, efficiency, and 1/f noise for these various technologies. It can be seen that silicon bipolar technology compares extremely well with GaAs in terms of performance (despite the well-known advantage of electron velocity in GaAs) with the advantage of providing an existing low-cost, high-volume production base. Silicon CMOS technology is just "awakening" to RF applications and should be seen as a major contender in the near future for high performance functions (highly linear mixers, LNAs, medium power amplifiers) as well as higher level of integration (RF, mixed-signal, logic, digital filters, some level of DSP?).

**Table 1** Technology Comparison

| | SiGe HBT | Si BJT | AlGaAs/ GaAs HBT | GaAs MESFET | CMOS |
|---|---|---|---|---|---|
| Minimum size (μm) | 0.5 x 1 | 0.5 x 1 | 2 x 5 | 0.5 x 5 | 0.5 x ? |
| $BV_{CEO}/BV_{DS}$ (V) | 4 | 4 | 15 | 8 | 8 |
| $f_t$(GHz) | 50 | 32 | 50 | 30 | 12 |
| $f_{max}$(GHz) | 55 | 35 | 70 | 60 | 18 |
| $G_{max}$(dB) @ 2 GHz<br>@ 10 GHz | 28<br>16 | 24<br>10 | 19<br>13 | 20<br>13 | 15<br>5 |
| $F_{min}$(dB) @ 2 GHz<br>@ 10 GHz | 0.5<br>0.9 | 0.9 | 1.5 | 0.3<br>0.9 | 1.5 |
| $IP_3/P_{1dB}$ (dB) | 9 | 9 | 16 | 12 | 12 |
| $P_{add}$Efficiency (%)@3V | 70 | 60 | 60@5v | 80 | >40 |
| 1/f comer (kHz) | 0.1-1 | 0.1-1 | 1-10 | 10,000 | 10-50 |

Figure 8 shows the maximum power available gain (or the maximum stable gain where the maximum available gain is not defined at low frequencies) of silicon and GaAs devices as function of frequency for current densities providing the maximum F max (maximum frequency of oscillation). It can be seen that in the 1 to 2Ghz frequency band, the maximum stable power gain ranges from 15dB for 0.5 micron CMOS (NMOS FET), to 20dB for 0.5 micron GaAs MESFET, 23dB for 0.8 micron silicon BJT, and 27dB for SiGe HBTs. Assuming CMOS technology performance will further improve in the next few years ( low threshold voltage, low drain and source contact resistance, low gate resistance, thick metallization), we should expect CMOS to play a significant role in the RF IC market.

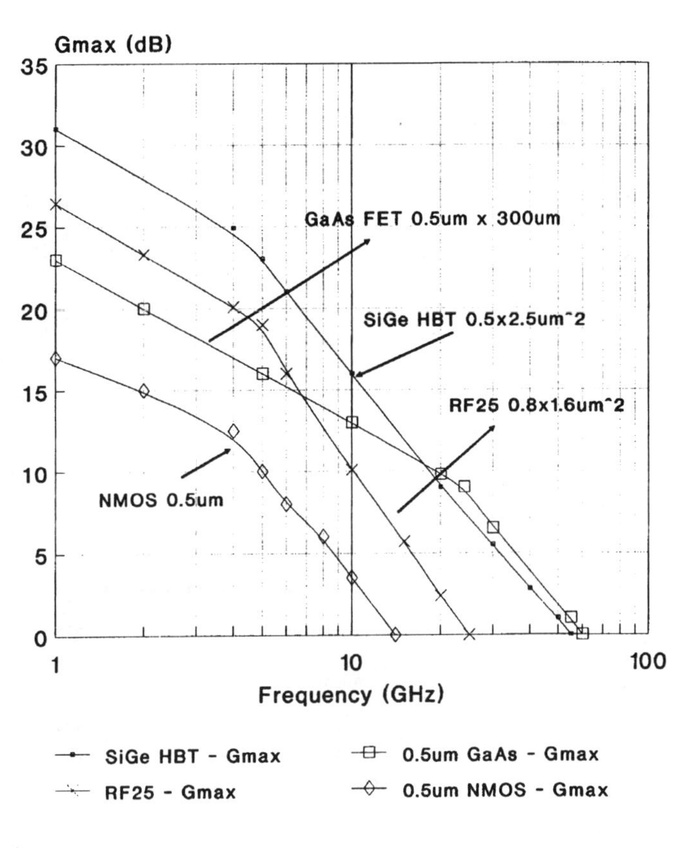


Figure 8. MSG and MAG as function of frequency for Si BJT, Si HBT, GaAs MESFET and NMOS devices

Figure 9 compares the linearity performance of three LNAs respectively based on Si BJT (20Ghz Ft), CMOS (lOGhz Ft) and GaAs MESFET (25Ghz Ft) technologies. It can be seen that, as expected, GaAs exhibits superior performance but also that CMOS provides an excellent 12dB IP3/ P-ldB ratio. Si BJT linearity performance is basically limited by the base-emitter diode exponential transfer function. (For LNAs, resistive emitter degeneration is not an option but some inductive degeneration can help improve linearity.)

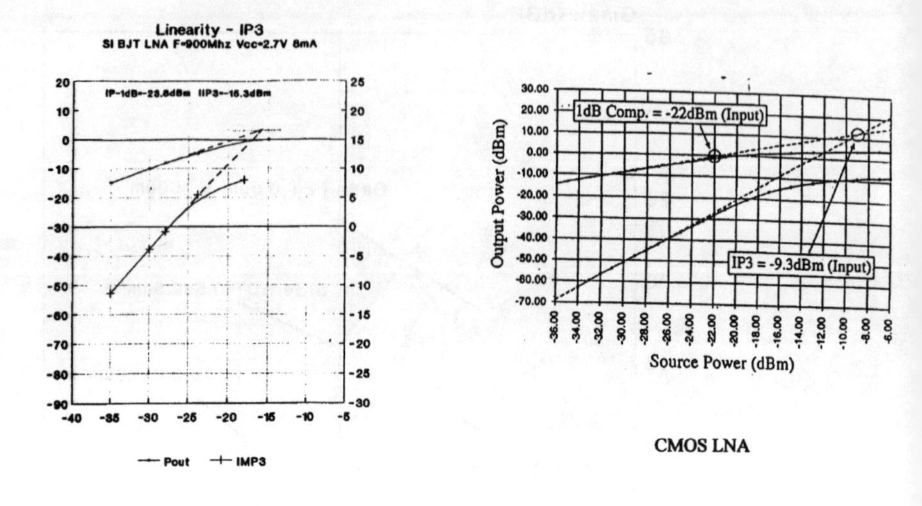


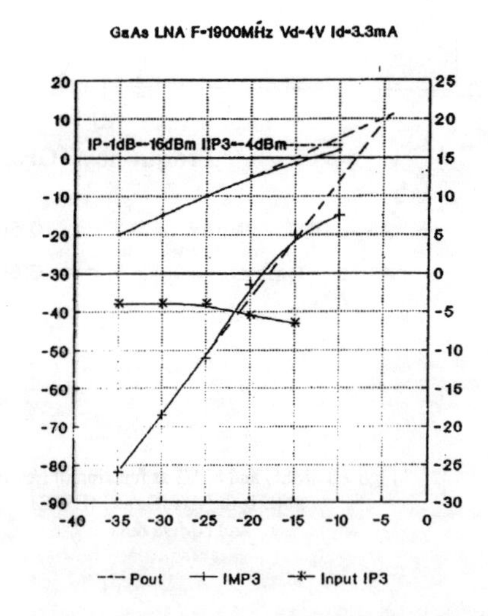


Figure 9. Linearity- IP3 for 900MHz BJT, 1500MHz CMOS, and 1900MHz GaAs LNAs

## 5.4 CIRCUIT IMPLEMENTATION/EXAMPLES

This sections presents two RF IC solutions for DECT handsets. The first solution is a complete two-St BJT chip DECT transceiver based on a dual down-conversion narrowband architecture, and the second one is a CMOS single-chip DECT receiver based on a dual down-conversion wideband approach.

*5.4.1 Narrowband Dual DownConversion Approach*

Figure 10 shows the block diagram of an RF transceiver chipset developed at Analog Devices for DECT (Digital Enhanced Cordless Telephone). DECT operates at l.9Ghz in a TDMA/TDD mode and uses 10 frequency channels spaced 1.728Mhz apart. Each frequency is divided in 24 time slots (12 receive/12 transmit) receiving and transmitting GFSK (Gaussian Frequency Shift Keying) modulated signal at a 1.152 Mbit/s data rate.

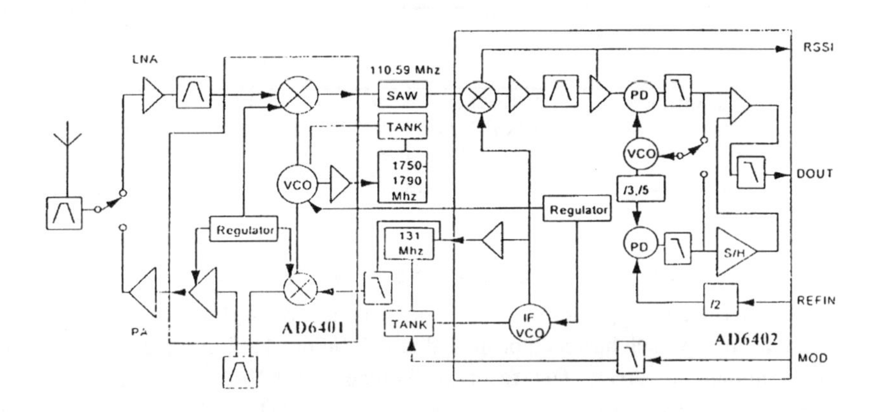


Figure 10. DECT Transceiver Block Diagram

A 25Ghz Ft junction-isolated Si BJT process is used to fabricate the RF chip and a 5Ghz SOI (Silicon On Insulator) Si complementary BJT process is used for the IF chip.

The two chips include all the RF/IF functions as well as the voltage regulators, ESD protection diodes, and the power management circuitry. They can operate from 2.7V to 5V.

The RF chip includes an Rx and Tx mixer, a VCO, a buffer, and driver amplifier. An external transmit filter is required to filter out the undesired side band of the transmit mixer. The mixer design is based on mod)fied Gilbert cells. The VCO consists of cross-coupled differential pair and an external inductor/varactor tank.

The IF chip includes a Rx mixer converting the first llOMhz IF frequency down to 20.7Mhz; a PLL demodulation loop, which outputs l.lMhz signal to a 6 bit ADC; and an IF VCO providing 131Mhz GFSK modulated signal to the RF Tx mixer.

The two RF/IF chips together with the baseband interface and DSP unit achieve a -94dBm sensitivity level for a 1E-3 BER (DECT specification is -83dBm) and provide 0 dBm bursted output power compliant with the DECT spectrum and switching transient mask. In the receive mode, the transceiver power consumption is 220mW (including on-chip VCOs) from a 2.7V battery.

As an example, Table 2 shows the RF receive chain performance.

**Table 2** RF Receive Chain Performance

| Rx Mixer | Performance | | Specification | | Unit 1A |
|---|---|---|---|---|---|
| | | | | | |
| | Conversion gain | | 8 to 12 | | 11.7 |
| | SSB Noise Figure | | 15 | | 14 |
| | RF RL | | -15 | | -15 |
| | IF RL | | -15 | | -25 |
| | i/p p1dB | | -12 | | -14 |
| | IIP3 | | -3 | | -5.2 |
| | LO Leak | at RF | -24 | | -31 |
| | | at IF | -19 | | -32.8 |
| | RF-IF ISO | | 10 | | 41 |
| | Spurious | RFx2LO | -28 | | -44 |
| | | 2RFx2LO | -50 | | -72 |
| | | 3RFx3LO | -36 | | -44 |
| | | RFx3LO | -44 | | -68 |

## 5.5 WIDEBAND DUAL DOWN-CONVERSION APPROACH

An alternative approach to the narrowband, dual down-conversion, PLL demodulation approach has been proposed in [6]. Figure 11 shows the block diagram of the complete receiver. The receiver includes an LNA, two series of image rejection mixers, two baseband filters, and two 10-bit ADCs. No external filters are required. The RF LO operates at fix frequency and converts the 1.9Ghz incoming signal (20Mhz bandwidth) down to 182 to 197MHz. The IF VCO controls channel pre-selection while the channel filtering is done at baseband using gmC anti-aliasing filters and switched-capacitor for final channel filtering. The overall power consumption is 200mW (excluding off-chip VCOs) from a 3.3V battery.

The receiver achieves -9OdBm sensitivity and proves RF CMOS technology is already viable for RF wireless applications.

## 5.6 CONCLUSION

RF ICs are critical components of a wireless communications transceiver. They process the signal coming from the antenna and deliver bits to the digital data receiver.

The RF IC technology, including architecture design, semiconductor processes, and circuit implementation design is evolving very rapidly. It has already enabled RF transceivers to meet the stringent requirements of high-performance, high-volume, low-cost wireless applications.

New transceiver architectures such as IF sampling and CMOS technologies will allow for further dramatical performance, integration, and cost improvement.

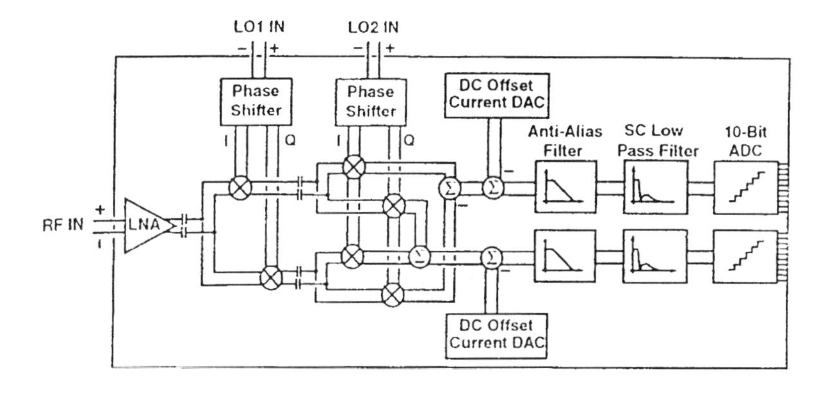


Figure 11. Wide-band IF Double Conversion Receiver Architecture

## References

[1] J. Sevenhans, et.al, "An Analog Radio Front-End Chipset for l.9ghz Mobile Radio TelephoneApplications, *International Solid-State Cicuits Conference,* San Francisco, 1994, pp.44-45

[2] Baines, R., "The DSP Bottleneck," IEEE *Communications magazine, (vol. 33) no. 5* May, 1995, pp. 46-53

[3] Murden, F.; and Gosser, R., "12b 50 MSample/s Two-Stage A/D Converter," *IEEE Solid-State Circuits conference,* Feb., 1995, pp. 278-279

[4] Jantzi, S.A.; Snelgrove, W.M.; Ferguson, P.F., Jr., "A Fourth-Order Bandpass SigmaDelta Modulator," *IEEE Journal of Solid-State Circuits,* March, 1993, Vol. 28, (No. 3) pp. 282-91

[5] K. Irie, et al, "A 2.7V GSM RF Transceiver IC," *ISSCC Digest,* pp302-303

[6] JC Rudell,et al, "A l.9Ghz Wide-Band Double Conversion CMOS IC for Cordless Telephone Applications," *ISSCC Digest,* 1997, pp304

# 6 VLSI SIGNAL PROCESSING SOLUTIONS FOR WIRELESS COMMUNICATIONS, A TECHNOLOGY OVERVIEW

Frederic Boutaud

**Abstract**

Two main ingredients for the development of wireless communication VLSI circuits are signal processing technology and Digital CMOS technology. A review of main characteristics and associated metrics is proposed. Evolution and future trends are discussed. Impact on DSP and baseband processors are presented.

## 6.1 INTRODUCTION

Digital wireless communication is one of the strongest growing area of the telecommunication market. This growth implies coordinated efforts from all development contributors including service providers, original equipment manufacturers and semiconductor suppliers. In this area, end equipment (base stations and terminals) makes a wide usage of electronic components and semiconductors such as VLSI (Very Large Scale Integrated) circuits.

In the terminal, VLSI semiconductor circuits implement the functions of a complete signal processing chain that is by nature included in a cellular phone, from antenna to microphone. In the following paragraphs we will concentrate only on one section of this signal processing chain: the digital baseband processor. We will consider two of the key technologies that are pivotal in the design and implementation of the signal processing tasks incorporated into the digital baseband processor. After a first overview of the functions included in the digital baseband processor, we will then address the digital CMOS IC technology and the Digital Signal Processing technology. We will see how they impact phone characteristics such as cost, power consumption, performance. We will also indicate trends for future technology developments and draft sketch on how this will apply on new products.

## 6.2 DIGITAL BASEBAND PROCESSOR

An overview of a generic digital cellular phone block diagram is presented Figure:1. The baseband function is divided into two parts, one is the so called mixed signal voiceband/baseband converter and the other is the digital baseband processor.

Voiceband/baseband converter section interfaces with microphone and earpiece on one side and interfaces with radio module on the other side. In most of the cases the mixed signal section includes,
in addition of the voice converters and baseband converters, a number of DAC's and ADC's performing conversion for system control and monitoring functions such as automatic gain control or battery and temperature sensing.

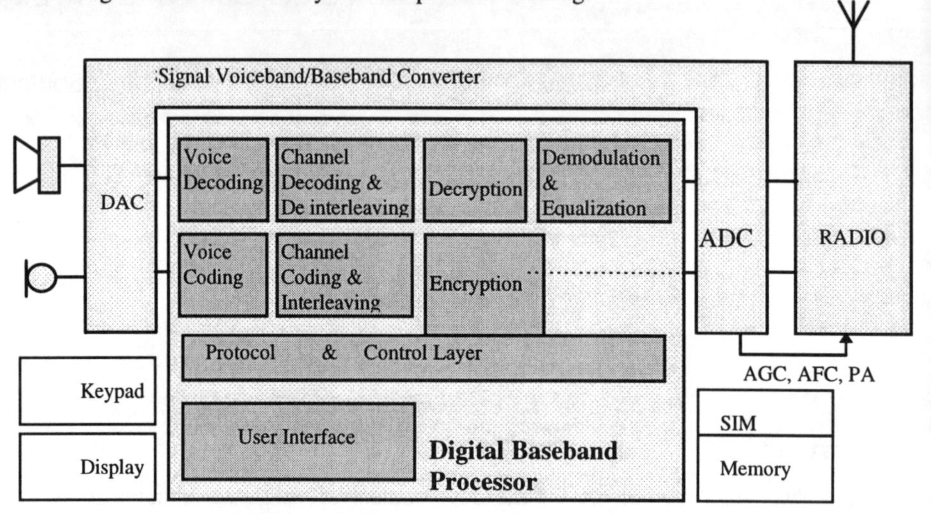


**Figure 1** Cellular Phone Block Diagram

The baseband processor regroups voice, communication channels, protocol and user interface functions. It generates control signals for the radio module and interface with SIM module, display, and keyboard. It can be integrated in a single chip such as AD6422 from Analog Devices that maps all required functions on DSP processor, Micro-controller and logic gates. Signal processing resides mainly in the voice coding and decoding functions and in the channel coding, decoding, and equalization. When developing the complete baseband processor, wireless designers have to face a challenging equation whose terms are flexibility, cost, power consumption, time to market and performance. Digital CMOS technology and Signal Processing technology are two of the main factors that drive the resolution of the equation and will allow to reach the right balance as well as providing enough reserve to enable future standards and new features integration.

## 6.3 DIGITAL CMOS IC TECHNOLOGY

The fast evolution of fabrication technology is, without any contest, one of the key enabler of the cellular phone expansion. This rapid evolution toward smaller geometry technology provides a path toward lower cost, lower energy consumption and higher performance. This evolution is represented on graph:1 . Transistor technology taken here as reference is the technology used for volume production ( of digital integrated circuits for hand-held applications ). We can observe a reduction of transistor geometry (L drawn) by almost 0.7X every two years in average.

Several metrics can be used to qualify and compare fabrication process geometry: transistor drawn length (as used on the previous graph), transistor effective length, metal pitch, ...; All may be valid when used consistently, however one can notice confusion may arise if a process is named by it's L effective value without any reference to it's L drawn value. When talking about evolution of integration capability, one key parameter to be considered is the integration density, that is, number of transistors per unit area (number of gates per unit area may be used as well).

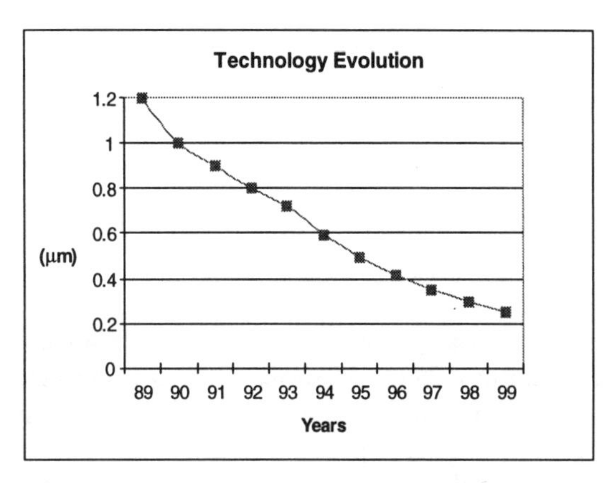


**Graph 1** Evolution in time of drawn transistor size (L in μm )

From process generation to process generation interconnect metal system is improving in two directions: one direction is the increasing number of metal levels and the other is a tighter pitch of metal lines. Combined with the evolution of transistor feature size, this allows to reduce silicon area of a given function by a factor of more than 0.5X every two years. This is illustrated in table:1 that shows evolution of number of transistor devices, in millions per square cm (transistor integration density). This number of devices is representative of the complexity of the functions that can be integrated in a square of 1cm per side; over a period of 6 years, from 1993 to 1999, this complexity will be multiplied by twenty. During the same period of time, the transistor technology shrinks by a ratio of 10 in area. That means that the metal system, and other interconnect capability such as contacts, bring and additional contribution of 2X in area reduction.

This evolution of the technology toward higher integration works in favor of the digital cellular phones' requirements: Low Cost, Medium Performance and Low Energy Consumption. We will review in the next paragraphs some of the key elements of this equation.

## 6.4 COST

In term of cost, smaller geometry allows higher system integration and therefore reduces total number of components that has direct impact on total cost. We can also consider that for a given functionality, the die area reduction obtained by migration (shrink) from one technology to another ( reduction factor x20 on a 6 years period as mentioned previously) can more than compensate higher cost of smaller geometry fabrication process.

**Table 1** impact of feature size, number of metals and metal pitch on transistor device density

| | *1993* | *1995* | *1997* | *1999* |
|---|---|---|---|---|
| Feature (μm) | 0.8 | 0.5 | 0.35 | 0.25 |
| Metal Pitch | 2.4 | 1.6 | 1.0 | 0.7 |
| Metal Levels | 2 | 3 | 4 | 5 |
| Devices/cm2 | 0.7M | 2M | 6M | 14M |

## 6.5 PERFORMANCE

Graph:2 is showing speed variations for different process geometry from 0.5μm to 0.25μm, in function of the supply voltage. The figure of merit taken here as example is based on raw transistor speed, the graph traces normalized frequency of ring oscillator for each process in function of the supply voltage. This graph gives some indications on speed improvement with smaller process. At a fixed supply voltage of 2.5V, speed goes up in more than a ratio of 2x when changing the process geometry from 0.5μm to 0.25μm. However one can understand that for deep sub-micron technology, relative contribution of interconnect capacitance on total delay increases when going to a smaller geometry. This has to be taken into account when evaluating performance of a shrunk circuit especially when speed is limited by propagation delay such as bus transfer delay.

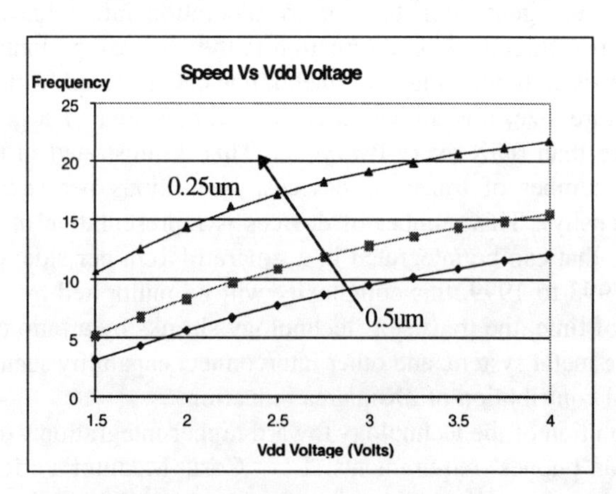


**Graph 2** Process performance variation

One can notice that the smaller the geometry, the smaller is the supply voltage necessary to achieve a given performance. This is described in table:2 generated using data extracted from graph 2. Lowering the voltage from 3.5V to 1.5V, It is still possible to maintain the same level of performance ( speed ) when going from 0.5μm to 0.25μm process.

**Table 2** Voltage power supply for a given performance

| Feature (μm) | *0.5μm* | *0.35μm* | *0.25μm* |
|---|---|---|---|
| Voltage | 3.3 | 2.5 | 1.5 |
| Speed | 10 | 10 | 10 |

## 6.6 POWER CONSUMPTION

In CMOS circuitry, the main component of power consumption relates to the switching capacitance that can be expressed by : Power = Capacitance * Voltage$^2$ * Frequency where voltage is the value of the supply voltage and frequency is the switching frequency of this capacitance. If we assume that the frequency is fixed for a given task, CMOS technology can influences energy consumption directly on the two other parameters: capacitance values and supply voltage. As described on graph:3, a smaller geometry process gives the benefit of smaller power consumption, also as mentioned previously, in same condition of speed, a smaller process allows a lower power supply voltage. The two combined together drops drastically the energy consumed to perform a given task.

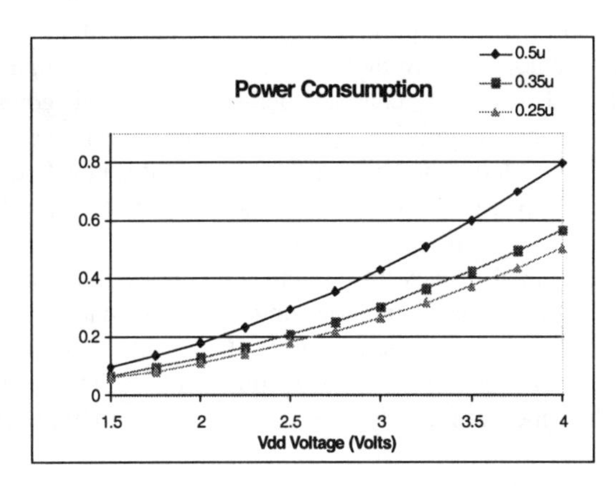


**Graph 3** Power consumption versus voltage and process

Table:3 give an example of such variation, going from 0.5μm to 0.25μm, power consumption is divided by 10; since the ratio of V$^2$ is about 5, the reduction of the capacitance term in the power equation is about 2. This is very close from the process geometry shrink factor ( 0.5μm down to 0.25μm ).

**Table 3** power consumption improvement for a fixed performance level of a given logic

| Technology | *0.5μm* | *0.25μm* | *Ratio* |
|---|---|---|---|
| Voltage (V) | 3.3 | 1.5 | 2.2x |
| Area | 100% | 15% | 6.5x |
| $V^2$ | 10.9 | 2.25 | 4.8x |
| Power consumption | 5 | 0.5 | 10x |

## 6.7 DEPENDENCIES

There is strong dependency and commonalties between energy consumption requirements and performance requirements. Reducing capacitance values by changing some process parameters such as dielectric constant or metal thickness or by increasing metal spacing will help not only to lower power consumption but also to increase maximum running frequency. This will be however at the detriment of cost.

We saw in the previous paragraphs that there is tremendous potential and motivation to apply smaller CMOS technology since it improve the three main requirements : Cost, performance and power consumption. However, for a given technology all three are linked together , for example, supply voltage links energy consumption and maximum frequency. IC architects and designers must fully understand those links in order to conceive and develop products that best utilize process technology characteristics and offer an attractive cost/energy/performance balance.

## 6.8 DIGITAL SIGNAL PROCESSING TECHNOLOGY

Digital Signal Processing plays a large role in the implementation of the digital baseband processor functions. Some of the tasks that involve signal processing are : voice encoder and decoder, channel demodulation and equalization, channel encoding and decoding. In addition, speech recognition, echo cancellation and noise suppression may be included in the baseband processor if new features, such as voice dialing or hands-free phone, are required in a search for better product differentiation and added value.

Designers of signal processing system for wireless communication are facing conflicting requirements that summarize into low cost, adequate performance, low power consumption and short time to market. Those requirements force tradeoff and drive the choice among implementation possibilities. The evolution of CMOS technology as described previously may lead to different choice from generation to generation.

Digital Signal Processing tasks can be accomplished either in dedicated hardware logic functions or on a programmable Digital Signal Processor (DSP). Today, on existing solutions, both approaches coexist.

## 6.9 HARDWARE LOGIC

Hardware logic takes various forms, from stand alone modules to tight coupled units working together with the DSP. The driving factors behind hardware solutions have been mainly either power consumption, performance or in some cases cost.

A simple example is summarized in table 4: Hardware implementation of a Viterbi function, part of the channel codec, compared to Software implementation of the same function. In this case, the hardware implementation requires about 5000 logic gates, and a small memory (64x16 bits). Silicon area is directly linked to the integration density of the Digital CMOS technology used. On the other end, the software implementation requires about 120 words of program code and some logic inside the CPU to improve number of cycles. As we can see, hardware implementation has a strong power consumption advantage at the cost of a much larger area and lower flexibility. When going to smaller geometry and lower voltage power supply, those differences are narrowing but are still significant. This is a good example of how low power consumption and cost may be conflicting requirements, thus making trade-off difficult.

**Table 4** Hardware implementation versus Software implementation

| Process | HW Viterbi | | SW Viterbi | |
|---|---|---|---|---|
| | Area | Power | Area | Power |
| *0.5μm, 3V* | 0.73mm2 | 1.25mW | 0.15mm2 | 17mW |
| *0.35μm, 2.5V* | 0.34mm2 | 0.62mW | 0.07mm2 | 7.8mW |
| *0.25μm, 1.8V* | 0.16mm2 | 0.21mW | 0.03mm2 | 2.2mW |

Hardware logic modules are rigid and not very well modifiable. Therefore, the priority being more and more on cost and flexibility, trend is to have those hardware modules targeted only at fixed well known functions, or designed in such a way that they offer more flexibility through accessibility and programmability by the DSP.

## 6.10 DSP PERFORMANCE

DSP provides instrumental capabilities: programmability and performance, that fulfill baseband processor needs. DSP's programmability resides in an instruction set that is used to implement on the same processor various tasks such as vocoders, channel coder, encryption or even speech recognition. MIPS (Million Instructions Per Second) and MHz (clock frequency) are often used to report performance level of Processors. However from processor to processor, instructions and clock cycle are not performing the same amount of work. Therefore even if this can give a good indication, it is not always accurate to use MIPS or MHz as performance metric. Instead, what really matters for a DSP is the time it takes to execute a critical algorithm or in other words, how many input samples can be processed in one second. The best approach is to be as close as possible from the real application and use most significant algorithms as reference tasks.

Task duration used as performance metric can be decomposed into the product of two factors : number of cycles needed to perform a task, and cycle execution period. Although there is dependency between architecture design and processor frequency speed, let's focus only on the number of cycles needed to perform a task (sometimes referred as algorithm) and how architecture choices and instruction set can influence this performance metric.

Number of cycles per task "Nc" is representative of how well the architecture suites the execution of this type of algorithm. This performance metric regroups two aspects of the execution: parallelism and instruction efficiency.

Parallelism means how many instructions can be executed in a single clock cycle. In most cases, one cycle means one instruction; however on the low end side some instruction may take two cycles or more. On the high end side, a double issue machine for example, two instructions may be executed in a single cycle. In this case, the parallelism rate will be slightly below 2 depending of resource availability and conflict.

Instruction efficiency means how many instructions have to be executed to complete the task; the lower this number, the more efficient the instruction set. For example, some instruction may be optimized specifically to execute wireless signal processing algorithm like Viterbi. If such instruction is made available, then this reduces the number of instructions needed to perform the algorithm. If Np is the parallelism rate and Ni is the number of instructions to be executed, then the performance metric (the total number of cycles consumed by the task) is

$$Nc=Ni/Np.$$

From this number of cycles, we can derive task execution time. If "Fc" is the DSP clock cycle frequency, then task execution time is:

$$Te = Ni/(Np*Fc)$$

We can also define the load index, that is the number of cycles per second required to execute the task real time, in the context of the application. If "Fo" is the occurrence frequency of the task, then load index is equal to :

$$Li = Nc * Fo = (Ni * Fo) /Np \quad \text{(cycles/seconde)}$$

Examples of architecture performance metric are described on Table:5 which represents different GSM vocoders implemented on AD6422, Analog Devices GSM baseband processor. This architecture is based on single cycle instruction set with no parallelism. In this case the load index is also equal to the number of MIPS needed by the task. This table illustrates as well the increasing variety and complexity of functions to be implemented on the DSP. Not only the number of cycles is multiplied by four but also the program size is multiplied by 8 when changing to a new enhanced speech coder.

**Table 5** Load Index and Program size for GSM vocoders running on AD6422 GSM baseband processor. Program size is given in number of 16bits words.

| Voice Encoder Decoder | *Average Load Index Cycles/Second* | *ROM Size (words)* |
|---|---|---|
| Full Rate GSM | 4.0 | 2500 |
| EFR GSM | 17.5 | 15063 |
| Half Rate GSM | 17.6 | 21959 |

## 6.11 LOW POWER DSP

For wireless portable applications, it is critical to obtain the lowest power consumption. Process technology is one of the strongest contributor in power consumption reduction roadmap since it allows to reduce not only capacitance values but also more important is the power supply value reduction (from 3.3V down to 1V as presented in table 6). Several metrics have been used to report power consumption characteristics : mA/MHz, mW/MHz, mA/MIPS, ... One must understand those are average values on a given sequence and they are easily meaningless when specified out of context. It is recommended to use mW/function or Joules/Function as metric for comparison. In table 6, GSM speech coder is taken as reference function.

## 6.12 TRENDS IN DSP PROCESSORS

Table 6 below presents projected DSP evolution as well as process technology evolution over a seven year period. Clock frequency is multiplied by a factor of five mainly thanks to process performance improvements while at the same time, power supply will be divided by more than three.

In term of instruction execution speed (MIPS), we predict that number of instructions per second will be multiplied by eight thanks to more general use of parallelism. However, we can expect even higher improvement at function level since new architecture will include application specific features such as dedicated instructions or dedicated computation unit. So in addition of the parallelism, we assume 20% efficiency improvement every 2 years. The three last rows of the table summarize the impact of those evolution on targeted wireless application, reduction in number of cycles per function by more than 50%, 90% reduction in task duration and reduction of energy per function by 97%.

**Table 6** DSP evolution for embedded portable applications, 7 year trends (High Volume Production)

| | '96 | '98 | '00 | '02 |
|---|---|---|---|---|
| Feature (μm) | 0.5 | 0.35 | 0.25 | 0.18 |
| Metal | 3 | 4 | 5 | 6 |
| Transistors/cm2 | 2.2M | 6M | 13M | 26M |
| Clock (MHz) | 60 | 100 | 200 | 300 |
| MIPS | 60 | 100 | 300 | 500 |
| Current (mA/MIPS) | 1 | 0.5 | 0.25 | 0.1 |
| Supply Voltage (V) | 3.3 | 2.5 | 1.8 | 1 |
| Power (mW/MIPS) | 3.3 | 1.25 | 0.45 | 0.1 |
| Cycle per Function | 100% | 80% | 54% | 44% |
| Task duration | 100% | 48% | 16% | 9% |
| Energy per function | 100% | 30% | 11% | 2.4% |

## 6.13 COST MODEL

Thanks to the integration improvements (up to 26 million of transistors per square cm), DSP processor will occupy only small silicon area, leaving space to integration of new functions such as large system memories. The relative weight of the CPU itself goes diminishing, and on chip memories become predominant. System architecture and instruction set have to play a key role in lowering memory cost.

## 6.14 SUMMARY

In the previous columns, we showed how digital CMOS technology and Digital Signal Processing all combined together have the potential to offer optimized baseband processor solutions in the form of highly integrated VLSI circuits. Process geometry, power supply voltage, performance are evolving so much that we can predict the emergence of new architecture that will offer enough performance to address needs of new demanding standard and features, and will enjoy very low energy consumption. We showed also how the cost model is changing opening the door to cannibalization of neighbor functions like memories.

## References

AD20msp415 - An Optimized GSM Baseband Chipset by Esben Randers, PIMRC'97

# INDEX